Modern Recording Techniques

Sixth Edition

Modern Recording Techniques

Sixth Edition

David Miles Huber
Robert E. Runstein

ELSEVIER

Amsterdam   Boston   Heidelberg   London   New York   Oxford   Paris
San Diego   San Francisco   Singapore   Sydney   Tokyo

Focal Press

The front cover photograph was taken by Eugenia Uhl (www.eu-photography.com) at Piety Street Studios in New Orleans, LA.

Acquisition Editor: Emma Baxter
Project Manager: Paul Gottehrer
Assistant Editor: Becky Golden-Harrell
Marketing Manager: Lucy Lomas-Walker
Cover Design: Eric DeCicco

Focal Press is an imprint of Elsevier
30 Corporate Drive, Suite 400, Burlington, MA 01803, USA
Linacre House, Jordan Hill, Oxford OX2 8DP, UK

∞ Recognizing the importance of preserving what has been written, Elsevier prints its books on acid-free paper whenever possible.

Library of Congress Cataloging-in-Publication Data
Application submitted

British Library Cataloguing-in-Publication Data
A catalogue record for this book is available from the British Library.

ISBN-13: 978-0-240-80625-9
ISBN-10: 0-240-80625-5

For information on all Focal Press publications
visit our website at www.books.elsevier.com

07 08 09   10 9 8 7 6 5 4

Printed in the United States of America

Table of Contents

◆

Contents

Acknowledgments

First off all, I'd like you to know that the subtle, yet seriously deranged guy on the cover is me . . . Now down to biz . . .

I'd like to thank my partner, Daniel Eric Butler, for putting up with the general rantin', ravin', and all 'round craziness that goes into writing a never-ending epic. Same goes for my best buddies: Steve "Stevo" L. Royea; Dan Stewart, Mische Eddins, Microsoft Studios; Phil & Vivian Williams, Voyager Records (www.voyagerrecords.com); Larry Crane (www.tapeop.com); Rick Mankowski and the folks @ Easy Street Records in West Seattle (www.easystreetonline.com) ; Michael "Hucky" Huckler and the folks @ M-Audio (www.m-audio.com) and Keith Medley @ Loud Technologies (www.mackie.com).

A very special mention goes to my production editor and good buddy, Terri Jadick, for "truly" being the best editor I've had the pleasure of working with. . . and for being an understanding friend that shows no remorse in supporting a lifelong vice—Chocoholism!

I'd also like to thank Eugenia Uhl (www.eu-photography.com) for photographing the front cover picture @ Piety Street Studios in New Orleans, LA (www.pietystreet.com). Thanks to Mark Bingham and the folks @ Piety Street. Marie Lee, Emma Baxter and the great folks @ Focal Press (UK and US) for getting this puppy off the ground and into your hands. Last but not least, I'd like to thank Paul Gottehrer for his amazing patience and understanding in laying out this book. A major-time pat on the back to y'all!

I'd also like to thank the following individuals and companies who've assisted in the preparation of this book by providing invaluable photographs and technical information:

Andy Murphy, Steven Smith, Rob Tyck, Jeff Knudsen, Ron Streicher; Pacific Audio-Visual Specialists, Richard Stoerger, www.ada-usa.com; Kathryn Kelly, www.adaptec.com; Hart Shafer, Adobe Systems, www.adobe.com; Jeffrey P. Fisher, Fisher Creative Group, www.jeffreypfisher.com; Tracey Roberts, Sunset Marquis Studios, www.thestudioweb.com; Larry Villella—ADK, www.adkmic.com; Cindy Parker—Acoustical Solutions, Inc., www.acousticalsolutions.com; Acoustical Physics Laboratories; Marcus Thompson—Acoustic Sciences Corporation, www.tubetrap.com; Erikk Lee, David Harbison—Auralex Acoustics, www.auralex.com; Wes Dooley—Audio Engineering Associates, www.wesdooley.com; Kevin Madden, Sarita M. Stewart—AKG Acoustics, Inc., www.akg-acoustics.com; Cliff Castle—Audix Corporation, www.audixusa.com; Bruce Borgerson—ATR Service Company, www.atrservice.com; Peter Chaikin—Alesis Studio Electronics, Inc., www.alesis.com; Dominic Cramp—Arboretum Systems, Inc., www.arboretum.com; Richard Giannini—Asimware Innovations, www.asimware.com; Randy Hargis—Akai Professional, www.akaipro.com; Dan Sheingold—Analog Devices, www.analog.com; Jennifer Schanhals, Chris Ellerby—Amek, www.amek.com; Marvin Ceasar—Aphex Systems, Inc., www.aphex.com; Altec Lansing Technologies, Inc., www.alteclansing.com; Russ Berger—Russ Berger Design Group, Inc., www.rbdg.com;

Peter Grueneisen—studio bau:ton, www.bauton.com; Alexis D. Kurtz—Beyerdynamic, www.beyerdynamic.com; Fujiko Kameda and Kurt Bujack—Bujack Audio; Dan Gallagher—Behringer International GMBH, www.behringer.de; Paul Bass, BSS Audio USA, www.bss.co.uk; Zac Wheatcroft—Bias Software, www.bias-inc.com; Daniel Buckley, James Tanner—Bryston Ltd, www.bryston.ca, Mary C. Bell; Jeff Palmer, Brian & Caron Smith—ClearSonic Mfg. Inc., www.clearsonic.com; Mick Whelan—Crown International, Inc., www.crownaudio.com; Carl Malone—CM Automation, www.cmautomation.com; Tom Johnson, Bonnie Anderson—Coda Music Technology, www.codamusic.com; Dan Stout—Colossal Mastering, Chicago, IL, www. colossalmastering.com; Paula Salvatore—Capitol Studios, Hollywood, CA, www.capitolstudios.com; Bob Katz—Digital Domain, www.digido.com; Reinel Adajar—Digidesign, www.digidesign.com; Jim Fiore—dissidents Software, www.dissidents.com; Luke Giles—Drawmer, www.transaudiogroup.com; Kenton Smith—Digitech, www.digitech.com; Connie Nomann, Kenton Smith—dbx Professional Products, www.dbxpro.com; Dolby Laboratories, Inc, www.Dolby.com; Wendy Clayton, Theresa Grant, Wendy Clayton—Euphonix Inc., www. euphonix.com; Bill Dooley—Extasy Recording Studios, www.extasyrecordingstudios.com; Derk Hagedorn—E-MU/Ensoniq, www.emu.com; Richard Factor—Eventide Inc., www.eventide.com; Juels Thomas—Event Electronics, www.event1.com; Tom Dambly—Focusrite, www.focusrite.com; Christos Desalernos—Furman Sound, Inc., www.furmansound.com; David Miles Huber—51bpm, www.51bpm.com; Focal Press, www.focalpress.com; Fostex Corp. of America, www.fostex.com; Scott Colburn—Gravelvoice Studios, www.gravelvoice.com; Erkki Myllynen—Genelec OY, www.genelec.com; Jack W. Gilfoy; Charlie Leib—hafler, www.hafler.com; Tomiko Jones Photography, Seattle, Washington; Chuck Thompson—JLCooper Electronics, www.jlcooper.com; Julian Colbeck—Keyfax Software, www.keyfax.com; KRK Systems, Inc., www.krksys.com; John Rotondo—Lexicon, Inc., www.lexicon.com; Keith Medley, Christopher Buttner, Steve Eborall, Andrew Gruner—Mackie Designs Inc., www.mackie.com; Focal Press, David Miles Huber—51bpm, www.51bpm.com; Jose Guillen—Marshall Electronics, inc., www.mxlmics.com; Jim Cooper—Mark of the Unicorn, Inc., www.motu.com; Jeff Wilson—Minnetonka Audio Software, www.minnetonkaaudio.com; Michael "Hucky" Huckler, Adam Castillo, Calvin Banks —M-Audio, www.m-audio.com; Ron Stein—midibrainz software, www.midibrains. com; Jim McGraw—McGraw Publishing Peripherals, www.sittingmachine.com; John La Grou—Millennia Music & Media Systems, www.mil-media.com; Aaron H. Pratt—MicroBoards Technology, Inc., www.microboards.com; John T. Mullin; Jim Van Buskirk—Nemesys Music Technology, Inc., www.nemesysmusic.com; Neumann USA, www.neumannusa.com; David Kennaugh—Neato LLC, www.neato.com; Chris Steinwand—Otari Corporation, www.otari.com; Orban Associates, Inc.; Kelly Erwin—Ocean Way Recording, www.oceanwayrecording.com; Adrienne Thompson—Primera Technology, Inc., www.primeratechnology.com; Kirt Kim—QSC Audio Products, Inc., www.qscaudio.com; Roger D'Arcy—Recording Architecture, www. aaa-design.com; John Jennings—Royer Labs, www.royerlabs.com; Sara Griggs—Roland Corp. US, www.rolandus.com; Ellen Allhands—Rane Corporation, www.rane.com;

Kathryn Kelly, Elena Bernardo—Roxio, www.roxio.com; Record Plant Recording
Studios; Sal Schandon—Sheffield Audio-Video Productions, www.sheffieldav.com;
Debra Pagen, Cathi Simpson—Solid State Logic, www.solid-state-logic.com; Walters-
Storyk—Walters-Storyk Design Group, www.wsdg.com; Sandy Schroeder—Shure
Incorporated, www.shure.com; Studer North America, www.studer.ch; Courtney
Spencer, Bob Tamburri—Sony Professional Audio, www.sony.com/proaudio; Bob
Ellison—Syntrillium Software Corp., www.syntrillium.com; Rebecca Grow—Sonic
Foundry, www.sonicfoundry.com; David Fabian—Steinberg, www.steinberg.de; Ken
Dewar—Soundscape Digital Technology LTD., www.soundscape-digital.com;
Dave—Soundtrek, www.soundtrek.com; Michael Lambie—Sound Quest Inc.,
www.squest.com; Kimberly J. Cahail—Symetrix Inc., www.symetrixaudio.com; Phil
Sanchez—Tascam, www.tascam.com; Larry Anschell—Turtle Recording, www.
turtlerecording.com; Karyn Ormiston—tc electronic, www.tcelectronic.com, TC Works,
www.tcworks.de; Chris Rice—Twelve Tone Systems, www.cakewalk.com; Clivia
Schiebel—Tannoy/TGI North America, Inc., www.Tannoy.com; Terri Aberg—Telex
Communications, Inc., www.telex.com; Bryan Ekus, Tony Denning—Tapematic USA,
Inc., www.tapematic.com; Nancy Enge —Universal Audio, www.uaudio.com; Bob Kratt,
Ken Camozzi—Versadyne International, www.versadyne.com; Marsha Vdovin,
Sheri Ragan—Waves, www.waves.com; Ken Centofante—Westlake Audio, www.
westlakeaudio.com; Bonnie Reed and Reed Ruddy—Studio X, www.studioxinc.com,
Susan Hart—Yamaha Corporation of America, www.yamaha.com/proaudio.

About the Authors

David Miles Huber is widely acclaimed in the recording industry as a digital audio consultant,
author, engineer, university professor, and guest lecturer. He received a degree in recording
techniques (I.M.P.) from Indiana University and was the first American to be admitted into
the prestigious Tonmeister program at the University of Surrey in Guildford, Surrey, England.
As well as being a member of NARAS (www.grammy.com) and the NARAS Producers and
Engineers Wing (P&E Wing), he's a regular contributing writer for numerous magazines
and websites, Dave has written such books as *The MIDI Manual* (www.focalpress.com) and
Professional Microphone Techniques (www.artistpro.com).

In addition, he's a professional musician in the ambient dance/relaxational field, having
written, produced, and engineered CDs that have sold over the million mark. You can check
out his music at www.51bpm.com.

Robert E. Runstein has been associated with all aspects of the recording industry, working as a
performer, sound mixer, electronics technician, A&R specialist, and record producer. He has
served as chief engineer and technical director of a recording studio and has taught several
courses in modern recording techniques. He is a member of the Audio Engineering Society.

Trademarks

All terms mentioned in this book that are known to be trademarks or service marks have been appropriately capitalized. Focal Press cannot attest to the accuracy of this information. Use of a term in this book should not be regarded as affecting the validity of any trademark or service mark.

ADAT and QuadraVerb are registered trademarks of Alesis Studio Electronics.

Aphex Compellor is a trademark of Aphex Systems Ltd.

Apple; Macintosh Plus, SE and II; Hypercard; Videoworks; and MacRecorder are registered trademarks of Apple Computer, Inc.

Cakewalk and Cakewalk Pro Audio are trademarks of Twelve Tone Systems.

dbx is a registered trademark of dbx, Newton, MA, USA, Division of Harmon International.

Dolby, Dolby SR, Dolby A, Dolby B, Dolby C, Dolby Surround Sound and Dolby Tone are registered trademarks of Dolby Laboratories Licensing Corporation.

Event 20/20bas is a trademark of Event Electronics.

Harmonizer and Ultra-Harmonizer are registered trademarks of Eventide, Inc.

Hum X is a registered trademark of Sound Enhancements, Inc.

PZM is a registered trademark of Crown International, Inc.

Mackie is a registered trademark of Loud Technologies Inc.

Sony is a registered trademark of Sony Corporation of America.

Sound Designer II, Sound Tools, Pro Tools, NuBus and DINR are registered trademarks of Digidesign.

Circle Surround II and TruSurround XT are registered trademarks of SRS Labs, Inc.

Tannoy is a registered trademark of Tannoy LTD. (North America Inc.).

Tube Trap is a trademark of Acoustic Sciences Corp.

U-Boats, MoPAD, T'Fusor and LENRD are registered trademarks of Auralex Acoustics, Inc.

XGEN is a trademark of Telex Communications, Inc.

Yamaha is a registered trademark of Yamaha Corporation of America.

CHAPTER 1

Introduction

The world of modern music and sound production is multifaceted. It's an exciting world of creative individuals: musicians, engineers, producers, manufacturers, and businesspeople who are experts in such fields as music, acoustics, electronics, production, broadcast media, multimedia, marketing, graphics, law, and the day-to-day workings of the business of music. The combined efforts of these talented people work together to create a single end product: marketable music. The process of turning a creative spark into a final product takes commitment, talent, a creative production team, a marketing strategy and, often, money. Over the history of recorded sound, the process of capturing music and transforming it into a marketable product has radically changed. In the past, the process of turning one's own music into a final product required the use of a commercial recording studio, which was (and still is) equipped with professional staff and was a specialized facility that was often hired for big bucks. With the introduction of the large-scale integrated circuit (LSI), mass production, and mass marketing (three of the most powerful forces in the "information age"), another option has come onto the scene: the radical idea that musicians, engineers, and/or producers can have their own project, desktop, and/or laptop studio. Along with this concept comes the realization that almost anyone can afford, construct, and learn to master a personal audio production facility . . . In short, we're living in the midst of a techno-artistic revolution that puts more power, artistic control, and knowledge directly into the hands of creative individuals from all walks of life in the music business.

On the techno side, those who are new to the world of modern multitrack recording, musical instrument digital interface (MIDI), digital audio, and

Figure 1.1. *Studio Mega of Suresnes, France, has installed a 96-channel SSL XL 9000 K Series console in Studio A's 90-square-meter control room—the premier studio in the facility's four rooms. (Courtesy of Solid State Logic, www.solid-state-logic.com.)*

Figure 1.2. *Latin award-winning producer Sergio George's private production facility. (Courtesy of Professional Audio Design, www.proaudiodesign.com.)*

their production environments should be aware that years of dedicated practice are often required to develop the skills that are needed to successfully master the art and application of these technologies. A person new to the recording or project studio environment (Figures 1.1 and 1.2) might easily be awestruck by the amount and variety of equipment that's involved in the process; however, when we become familiar with the tools, toys, and techniques of recording technology, a definite order to the studio's makeup soon begins to appear, with each piece of equipment being designed to play a role in the overall scheme of music and audio production.

The goal of this book is to serve as a guide and reference tool to help you become familiar with the recording and production process. When used in conjunction with mentors, lots of hands-on experience, further reading, and simple common sense, this book will, I hope, help you understand the equipment and day-to-day practices of sound recording and production.

Although it's taken the modern music studio about 80 years to evolve to its current level of technological sophistication, we have begun to move into the adolescent stage of an important evolutionary process in the business of music and music production: the dawning of the digital age. Truly, this is an amazing time in production history...as we've moved to a time when we can choose between an amazing array of cost-effective and powerful tools for fully realizing our creative and human potential. As always, patience and a nose-to-the-grindstone attitude are needed in order to learn how to use them effectively; however, this knowledge can help free you for the really important stuff—making music and audio productions. In my opinion, these are definitely the good ol' days!

The Recording Studio

The commercial music studio (Figures 1.3 through 1.6) is made up of one or more acoustic spaces that are specially designed and tuned for the purpose of capturing the best possible sound onto a recording medium. In addition, these facilities are often structurally isolated in order to keep outside sounds from entering the room and getting into the recording (as well as to keep inside sounds from leaking out and disturbing the surrounding neighbors). In effect, the most important characteristics of that go into the making of such a facility include:

◆ A professional staff
◆ Professional equipment
◆ Professional, yet comfortable working environment

Figure 1.3. Four Seasons Media Productions; St. Louis, MO. (Courtesy of Russ Berger Design Group, Inc., www.rbdg.com.)

Figure 1.4. CTS Studio 2, London. (Courtesy of Recording Architecture, www.aaa-design.com.)

Figure 1.5. Synchrosound Studio A, Kuala Lumpur, Malaysia. (Courtesy of Walters-Storyk Design Group—designed by Beth Walters and John Storyk, www.wsdg.com, photo credit—Robert Wolsch.)

◆ Optimized acoustic and recording environment
◆ Optimized control-room mixing environment

Recording studio spaces vary in size, shape, and acoustic design (Figures 1.6 through 1.8) and usually reflect the personal taste of the owner and/or are designed to accommodate the music styles and production needs of their clients, as shown by the following examples:

◆ A studio that records a wide variety of music (ranging from classical to rock) might have a large main room with smaller, isolated rooms off to the side for unusually loud or soft instruments, vocals, etc.

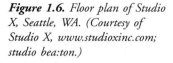

Figure 1.6. *Floor plan of Studio X, Seattle, WA. (Courtesy of Studio X, www.studioxinc.com; studio bea:ton.)*

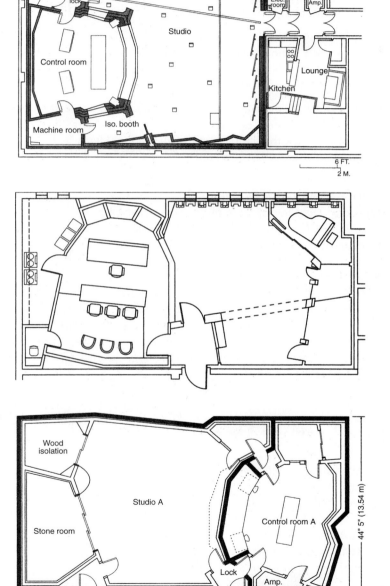

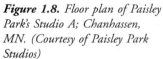

Figure 1.7. *Floor plan of Sony/Tree's Music Studio; Nashville, TN. (Courtesy of Russ Berger Design Group, Inc., www.rbdg.com.)*

Figure 1.8. *Floor plan of Paisley Park's Studio A; Chanhassen, MN. (Courtesy of Paisley Park Studios)*

◆ A studio designed for orchestral film scoring might be larger than other studio types. Such a studio will often have high ceilings to accommodate the large sound buildups that are often generated by a large the number of studio musicians.

◆ A studio used to produce audio for video, film dialog, vocals, and mixdown might only incorporate a single small recording space off the control room for overdub purposes.

In fact, there is no secret formula for determining the perfect studio design. Each studio design has its own sonic character, layout, feel, and decor that are based on the personal tastes of its owners, the designer (if any), and the going studio rates (based on the studio's investment return and the supporting market conditions).

During the 1970s, studios were generally small. Because of the advent of (and over reliance on) artificial effects devices, they tended to be acoustically "dead" in that the absorptive materials tended to suck the life right out of the room. The basic concept was to eliminate as much of the original acoustic environment as possible and replace it with artificial ambience.

Fortunately, since the mid-1980s, many commercial studios that have the physical space have begun to move back to the original design concepts of the 1930s and 1940s, when studios were larger. This increase in size (along with the addition of one or more smaller iso-booths and/or rooms) has revived the art of capturing the room's original acoustic ambience along with the actual sound pickup. In fact, through improved studio design techniques, we have learned how to achieve the benefits of both earlier and modern-day recording eras by building a room that absorbs sound in a controlled manner (thereby reducing unwanted leakage from an instrument to other mics in the room) . . . while dispersing reflections in a way that allows the room to retain a well-developed reverberant and sonic personality of its own. This effect of combining direct and natural room acoustics is often used as a tool for "livening up" an instrument (when recorded at a distance), a technique that has become popular when recording live rock drums, string sections, electric guitars, choirs, and so on.

The Control Room

A recording studio's *control room* (Figures 1.9 through 1.11) serves a number of purposes in the recording process. Ideally, the control room is acoustically isolated from the sounds that are produced in the studio, as well as from the surrounding, outer areas, and are optimized to act as a critical listening environment that uses carefully placed and balanced monitor speakers. This room also houses the majority of the studio's recording, control, and effects-related equipment. At the heart of the control room is the recording console.

The *recording console* (also referred to as the *board* or *desk*) can be thought of as an artist's palette for the recording engineer, producer, and artist. The console allows the engineer to combine, control, and distribute the input and output signals of most (if not all) of the devices found in the control room. The console's basic function is to allow for any combination of

Figure 1.9. *Paragon Studios, nestled in the rolling hills of Franklin, TN, just outside the famed Music Row in Nashville. (Courtesy of Russ Berger Design Group, Inc., www.rbdg.com.)*

Figure 1.10. *George Martin at Air Lyndhurst, London. (Courtesy of Solid State Logic, www.solid-state-logic.com.)*

mixing (variable control over relative amplitude and signal blending between channels), spatial positioning (left/right or surround-sound control over front, center, rear, and sub), routing (the ability to send any input signal from a source to a destination), and switching for the multitude of audio input/output signals that are commonly encountered in an audio production facility.

Tape machines are generally located at the rear or to the side of a control room, while digital audio workstations (DAWs) are often located at the side of the console or at the center (if the DAW serves as the room's main recording/mixing device). Because of the added noise and heat generated by recorders, computers, power supplies, amplifiers, and other devices, it is becoming more common for equipment to be housed in an isolated machine room that has an adjoining window and door for easy access and visibility. In either

Figure 1.11. Studio F at The Village Recorders; Los Angeles, CA. (Courtesy of The Village Recorders, www.villagestudios. com.)

case, remote control, autolocator devices (which are used for locating tape and media position cue points) and DAW controller surfaces (which are used for computer-based remote control and mixing functions) are often situated in the control room, near the engineer, for easy access to all recording, mixing, and transport functions. Effects devices (used to electronically alter and/or augment the character of a sound) and other signal processors are also often placed nearby for easy accessibility (often being designed into an effects island or bay that's located directly behind the console). In certain situations, a facility might not have a large recording space at all but simply a small or mid-sized iso-room for recording overdubs (this is often the case for rooms that are used in audio-for-visual post and/or music remixing).

As with recording studio designs, every control room will usually have its own unique sound, feel, comfort factor, and studio booking rate. Commercial control rooms often vary in design and amenities...from a room that's basic in form and function to one that is lavishly outfitted with the best toys, design, and fully stocked kitchens in the business. Again, the style and layout are a matter of personal choice; however, as you'll see throughout this book, there are numerous guidelines that can help make the most of a recording space. In addition to the layout, feel, and equipment, it is important to remember that the people (the staff, musicians, and you) will often play a prominent role in capturing the feel of a performance...not just the equipment.

The Project Studio

With the advent of affordable analog and digital audio recording systems, it's a foregone conclusion that the vast majority of music and audio recording/production systems have

Figure 1.12. *A home project studio. (Courtesy of Loud Technologies, Inc. www.mackie.com.)*

been built and designed for private use. The rise of the personal *project studio* (Figure 1.12) has brought about monumental changes in the business of music and professional audio, in a way that has affected and altered almost every facet of the audio production community.

As a result of this production revolution, it is now possible (if not downright common) for the artist, producer, and/or engineer to cost-effectively record, compose, and produce music in the comforts of his or her own bedroom, garage, or specially designed space . . . all being carried out under the motto "You don't have to have a million dollar studio to make good music." Literally, the modern-day project and portable studio system offer such a degree of cost-effective power and audio fidelity that they can often match the production quality of a professional recording facility . . . all it takes is knowledge, care, and patience.

The Portable Studio

One of the most amazing advances in recording technology is related to the fact that, yes, you *can* take it with you! With the advent of high-powered laptops with a USB or FireWire audio interface, instrument, and controller technology (Figure 1.13), . . . as well as all-inclusive portable recording systems that incorporate digital audio/mixing and effects technology (Figure 1.14), it's now literally possible to record and produce high-quality audio anywhere and at any time.

Knowledge Is Power!

One of the most important aspects of putting together a high-quality, cost-effective project studio is definitely *knowledge* . . . which can be gained by:

◆ Reading about the equipment choices that are available to you

Figure 1.13. *Playing with a Powerbook and Tracktion on the front lawn. (Courtesy of Loud Technologies, Inc. www.mackie.com.)*

Figure 1.14. *Yamaha AW4416 Professional Audio Workstation. (Courtesy of Yamaha Corporation of America, www.yamaha.com.)*

- Visiting and talking with others of like (and dissimilar) minds about their equipment and working styles
- Enrolling in a recording course that best fits your needs, working style, and budget
- Getting your hands on equipment before you make your final purchases (for example, checking out the tools and toys at your favorite music store).

By taking the time to familiarize yourself with the options and possibilities that are available to you, the less likely you are to be unhappy with the way you'll spent your hard-earned bucks. It is also important to point out that having the right equipment for the job isn't always enough...it's important to take the time to learn how to use your tools to their fullest potential. Whenever possible, read the manual and get your feet wet by taking the various settings, functions, and options for a test spin...before you dive into a session.

Whatever Works for You

As you begin to research the various types of recording and supporting systems that can be put to use in a project studio, you'll find that a wide variety of options are available. There are literally hundreds, if not thousands, of choices for recording media, hardware types, software systems, speakerss, effects devices...and the list goes on. This should automatically tell us that no "one" tool is right for the job. As with everything in art (even the business of an art), there are many personal choices that can combine into the making of a working environment that's right for you. Whether you:

- Work with a hard-disk or tape-based system
- Choose to use analog or digital (or both)
- Are a Mac or PC kind of person
- Use this type of software or that

...It all comes down to the bottom line of how does it sound? How does it move the audience? How can we sell it? In truth, no record executive will turn down a potential gold or platinum seller because it was not recorded on such-n-such a machine.

Making the Project Studio Pay for Itself

Beyond the obvious advantage of being able to record when, where, and how you want to in your own project studio, there are several additional benefits to working in a personal environment. A few of the following are but a sampling of how a project studio can help to subsidize itself...at any number of levels:

- Setting your own schedule and saving money while you're at it! An obvious advantage of a project studio revolves around the idea that you can create your own music on your own schedule. The expense incurred in using a professional studio requires that you be practiced and ready to roll on a specific date or range of days. A project studio can free you up to lay down practice tracks and/or record when the mood hits, without having to worry about punching the studio's time clock.

◆ For those who are in the business of music, media production, the related arts or for those who wish to use their creativity as an outlet for their personal business, it's possible to write off the acquisition of equipment, building and utility payments as a tax deductible expense (see Appendix B-Tax Tips for Musicians).

◆ A group or individual artist might consider preproducing a production at their project studio. These time and expense billings could likewise be tax-deductible expenses.

◆ The same artists might consider recording part or all of their production at their own project studio. The money saved (and deducted) could be spent on a better mixdown facility, production, music law, and/or marketing.

◆ The "signed artist/superstar approach" refers to the mega-artist who, instead of blowing the advance royalties on his or her next big hit, spends the bucks on a professional-grade project studio (Figure 1.15). After the project has been recorded, the artist will still have a tax-deductible facility that can be operated as a business enterprise. When the next project comes along, the artist will still have a personal facility in which to record.

Live/On-Location Recording: A Different Animal

Unlike the traditional multitrack recording environment where overdubs are often used to build up a song over time, *live/on-location recording* is created on the spot, in real time...often during a single in-studio or on-stage performance, with little or no studio postproduction other than mixing. A live recording might be very simple, possibly being recorded using only a few mics that are mixed directly to two tracks, or a more elaborate gig might call for a multitrack setup, requiring the use of a temporary control room or fully equipped mobile recording van or truck (Figure 1.16) that can record to tape and/or hard disk. The latter type of setup obviously requires a great deal of preparation and expertise, including a combined knowledge of sound reinforcement and live recording techniques necessary to capture instruments in a manner that can isolate the tracks, so as to yield the highest degree of control over the individual instruments during the mixdown phase. Although the equipment and system setups will be familiar to any studio engineer, live recording differs from its more controlled studio counterpart in that it exists in a world where the motto is "you only get one chance." When you're recording an event where the artist is spilling his or her guts to hundreds or even thousands of fans, it's critical that everything runs smoothly. Live recording usually requires a unique degree of preparedness, system setup, patience...and, above all, experience.

Figure 1.15. *Whitney Houston's home studio: (a) control room; (b) recording studio. (Courtesy of Russ Berger Design Group, Inc., www.rbdg.com)*

a

b

Audio for Video and Film

In recent years, audio has become a much more important part of video, film, and broadcast production. Previous to the advent of multichannel television sound (MTS), the DVD, and surround sound — broadcast audio was almost an afterthought. With the introduction of these new technologies, audio has grown from its relatively obscure position to being a highly respected part of film and video media production (Figure 1.17). With the common use of surround sound in the creation of movie sound tracks (Figure 1.18), along with the growing popularity of surround in-home entertainment systems (and an ever-growing number of playback systems for sound, visual media, and computer media), the public has come to expect

Figure 1.16.
Sheffield mobile recording truck. (Courtesy of Sheffield Audio-Video Productions, www.sheffieldav. com). a. Control room. b. Truck.

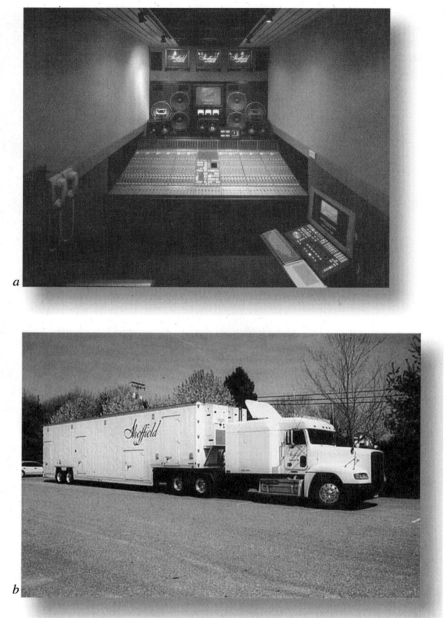

higher levels in audio production quality. In this day and age, MIDI, hard-disk recording, time code, automated mixdown, and advanced processing have become everyday components of the audio environment . . . requiring that professionals be highly specialized and skilled in order to meet the demanding schedules and production complexities.

Figure 1.17. Studio 3 at One Union Recording Studios, San Francisco, CA. (Courtesy of Euphonix, www.euphonix.com; photograph by Edward Colver.)

Figure 1.18. A small studio. (Nathan Whiehead) (Courtesy of Loud Technologies, Inc., www.mackie.com.)

Multimedia

With the integration of text, graphics, MIDI, digital audio, and digitized video into almost every facet of the personal computer environment, the field of *multimedia* audio has become a fast-growing, established industry that represents an important and lucrative source of income for both creative individuals and production facilities alike. In addition to the use of audio in

Figure 1.19. Electronic Arts, Vancouver, Canada. (Courtesy of Walters-Storyk Design Group; designed by Beth Walters and John Storyk, www.wsdg.com; photograph by Robert Wolsch.)

multimedia products for education, business, and entertainment, most of the robot-zappin', dare-devil flyin' addicts are probably aware that one of the largest and most lucrative areas of multimedia audio production is the field of scoring, designing, and producing audio for computer games (Figure 1.19). Zaaaaaaappppppppp!

The Web

It almost goes without saying that the World Wide Web has changed almost every facet of distributing and acquiring information in the modern-day world. Gigatons of sites can be found that are dedicated to even the most offbeat ideas and equipment, while search engines can give you instant access to information on how to fix the stem-bolt on a 1905 steam-driven nut cracker. It's a rare person who isn't aware of the fact that the list of resources, opinions, insights, tips, and articles relating to almost anything continues to grow at a staggering pace. It's pretty obvious to any cyber explorer that the Web is ever changing. Those of you who will be reading this book 5 years, a year, or even a few months from now could easily find that a link has been changed and that some or all of the mentioned have disappeared, having been swallowed up or made obsolete by the passage of time. This leaves me with the task of pointing out a few tips for finding sites or informational resources for equipment, selling or buying gear, magazines, software downloads, leads on independent music . . . you name it (Figure 1.20). These include:

◆ www.companyname.com—Adding ".com" onto a company's name will more often than not lead you to their site.

◆ Search engines—Any popular search engine can help you to track down any information or site that might be needed.

Figure 1.20. ModRec.com, a site that can help you with your search and reference needs in the field of recording. (Courtesy of ModRec.com, www.modrec.com.)

◆ Industry magazines—Looking through trade advertisements can be one of the best ways to find a resource. Often the classifieds ads at the back of a magazine can be a resource that's both enlightening and full of fun things.

The People Who Make It All Happen

"One of the most satisfying things about being in the professional audio [and music] industry is that sense that you are part of a community." –Frank Wells, editor, *Pro Sound News*

When you get right down to the important stuff, the recording field is built around pools of talented individuals and service industries who work together for a common goal—producing, selling, and enjoying music. As such, it's the *people* in the recording industry who make the business of music happen. Recording studios and other businesses in the industry aren't only known for the equipment that they have but are often judged by the quality, knowledge, vision, and personalities of their staff. The following sections describe but a few of the ways in which a person can be involved in this multifaceted industry.

The Artist

The strength of a recorded performance begins and ends with the artist. All the technology in the world is of little use without the existence of the central ingredients of human creativity, emotion, and technique. Just as the overall sonic quality of a recording is no better than its weakest link, it's the performer's job to see that the foundation of all music—its inner soul—is laid out for all to experience and hear. After this has been done, a carefully planned and well-produced recording can act as a gilded framework for the music's original drive, intention, and emotion.

Studio Musicians and Arrangers

A project often requires additional musicians to add extra spice and depth to the artist's recorded performance. For example:

◆ A member of a group might not be available or might not be up to the overall musical standards that are required by the project. In such situations, it is not uncommon for a professional studio musician to be called in.

◆ An entire group of studio musicians might be called on to provide the best possible musical support for a high-profile artist or vocalist.

◆ A project might require musical ensembles (such as a choir, string section, or background vocals) for a particular part or to give a piece a fuller sound.

◆ If a large ensemble is required, it might be necessary to call in a professional music contractor to coordinate all the musicians and make the financial arrangements. The project might also require a music arranger, who can notate and possibly conduct the various musical parts.

The Producer

Beyond the scheduling and budgetary aspects of coordinating a recording project, it is the job of a producer to help the artist and record company create the best possible recorded performance and final product that reflect the artist's vision. A producer can be hired onto a project to fulfill a number of possible duties and may even be given complete control over a project's artistic, financial, and programmatic content. More likely, however, a producer acts collaboratively with an artist or group to guide them through the recording process. This type of producer will often:

◆ Assist in the selection of songs

◆ Help to focus the artistic goals and performance in a way that best conveys the music to the targeted audience

◆ Help to translate that performance into a final, salable product (with the technical and artistic help of an engineer and mastering engineer)

A producer can also be chosen for his/her ability to understand the process of creating a final recorded project from several perspectives: business, musical performance, creative insight, and mastery of the recording process. Because engineers spend much of their working time with musicians and industry professionals with the intention of making their clients sound good, it's not uncommon for an engineer to take on the role of producer or co-producer (by default or by mutual agreement). Conversely, as producers become increasingly more knowledgeable about recording technology, it is also increasingly common to find them sitting behind the controls of a console.

The Engineer

The job of an engineer can best be described as an interpreter in a techno-artistic field. The engineer must be able to express the artist's music and the producer's concepts through the medium of recording technology. This job is actually best classified as an art form, as both music and recording are subjective in nature and rely on the tastes and experience of those involved. During a recording session, one or more engineers can be used on a project to:

- Place the musicians in the desired studio positions.
- Choose and place the microphones.
- Set levels and balances on the recording console or DAW mixing interface.
- Record the performance onto tape or disk.
- Overdub additional musical parts into the session that might be needed at a later time.
- Mix the project into a final master recording in any number of media and mono, stereo, and/or surround-sound formats.

In short, engineers use their talent and artful knowledge of recording media technology to convey the best possible sound for the intended media and buying public.

Assistant Engineer

Many studios often train future engineers (or build up a low-wage staff) by allowing them to work as assistants to staff and freelance engineers. The assistant engineer often does microphone and headphone setups, runs tape machines and or DAW setup, helps with session documentation, does session breakdowns, and (in certain cases) performs rough mixes and balance settings for the engineer on the console. With the proliferation of freelance engineers (those not employed by the studio but are retained by the artist, producer, or record company to work on a particular project), the role of the assistant engineer has become even more important. It is often the assistant engineer's role to guide freelance engineers through the technical aspects and quirks which are peculiar to that studio.

Maintenance Engineer

The maintenance engineer's job is to see that the equipment in the studio is maintained in top condition, regularly aligned, and repaired when necessary. Larger organizations (those with more than one studio) might employ a full-time staff maintenance engineer, while freelance maintenance engineers and technical service companies are often called in to service smaller studios and those in non-major markets.

Mastering

It's not uncommon to have to tweak a final master recording in terms of overall and relative level, equalization (EQ), and volume dynamics so as to present the final "master" recording in the best possible sonic and marketable light. This job falls to a mastering engineer, who will listen to and process the recording (often but not always in the digital domain) in a specialized, fine-tuned monitoring environment.

The DJ

Let's not forget one of the more important links for getting the musical word out to the buying public: the disc jockey (DJ). Actually, the role of disc jockey can take many modern-day forms:

◆ On the air —The DJ of the airwaves is still a very powerful voice for getting the word out about a particular musical product.

◆ On the floor—This DJ form often reinforces the messages from the airwaves or helps to promote the musical word in a counter-cultural environment.

◆ On the Web—Probaby one of the more up-n-coming voices for getting the promotional word out to a large number of specially targeted audiences (both in the mainstream and in specific counter-cultures).

Studio Management

Running a business in the field of music requires the special talents of businesspeople who are knowledgeable about the inner workings of promotion, the music studio, the music business, and—above all—people. It requires constant attention to quirky details that would probably be totally foreign to someone outside the "biz". These tasks include:

◆ Studio management—This person (who might or might not be the owner) is responsible for managerial and marketing decisions.

◆ Bookings—This staff person keeps track of most of the details relating to studio usage and billing.

◆ Competent secretarial staff—These assistants keep everyone happy and everything running as smoothly as possible.

Although any or all of these functions can easily vary from studio to studio, these and other equally important roles are required in order to successfully operate a commercial production facility on a day-to-day basis.

As we have noted, the role of the professional recording studio has begun to change as a result of upsurges in project studios, audio for video and/or film, multimedia, and Internet audio. These market forces have made it necessary for certain facilities to rethink their operational business strategies. Often, these changes have met with some degree of success, as is illustrated by the following examples:

◆ Personal production and home project studios have greatly reduced the need for an artist or producer to have constant and costly access to a professional facility. As a result, many pro studios now cater to artists and project studio owners who might have an occasional need for a larger space or better-equipped recording facility (*e.g.*, for string overdubs or an orchestral session). In addition, after an important project has been completed in a private studio, a professional facility might be needed to mix the production down into its final form. Most business-savvy studios are only too happy to capitalize on these new and constantly changing market demands.

◆ Upsurges in audio for video and film postproduction have created new markets that allow professional recording studios to provide services to the local, national, and international broadcast and visual production communities. Studios with both the equipment and creative staff necessary to enter into lasting relationships with the audio for visual and broadcast production markets are often able to thrive in the tough business of music, when music production alone might not provide enough income to keep a studio afloat.

◆ Studios are also taking advantage of Internet audio distribution techniques by offering Web development, distribution, and other services as an incentive to their clients.

These and other aggressive marketing strategies (many of which may be unique to a particular area) are being widely adopted by commercial music and recording facilities to meet the changing market demands of new and changing media. No longer can a studio afford to place all its eggs in one media basket. Tapping into changes in market forces and meeting them with new solutions is an important factor for making it in the business of music production and distribution. I'd like to say that all-important word again . . . "business." Make no mistake about it—starting, staffing, and maintaining a production facility and/or getting the clients' music heard is serious work that requires dedication, stamina, innovation, and guts.

Music Law

These lawyers serve a wide range of purposes, ranging from the primary duties of looking after their clients' interests and ensuring that they don't sign their careers away by entering into a life of indentured, nonprofit servitude . . . to introducing an artist to the best-possible music label. A good music lawyer is often the unsung hero of many a career.

Women in the Industry

Ever since its inception, males have dominated the recording industry. I remember many a session in which the only women on the scene were female artists, secretaries, or studio groupies. Fortunately, over the years, women have begun to play a more prominent role, both in front of and behind the glass . . . in every facet of studio production and the business of music as a whole (Figure 1.21). No matter who you are, where you're from, or what your race, gender, sexual, or planetary orientation is, remember this universal truth: If your heart is in it and you're willing to work hard enough, you'll make it.

Behind the Scenes

In addition to the positions listed above, there are scores of professionals who serve as a backbone for keeping the business of music alive and functioning. Without the many different facets of music business—technology, production, distribution, and law—the biz of music

Figure 1.21. *Women's Audio Mission, an organization formed to assist women in the industry. (Courtesy of the Women's Audio Mission, www.womensaudiomission. org.)*

would be very, very different. A small sampling of the additional professional fields that help make it happen includes:

- Artist management
- Artist booking agents
- A&R (artist and repertoire)
- Manufacturing
- Music and print publishing
- Distribution
- Web development
- Graphic arts and layout
- Equipment design
- Audio company marketing
- Studio management
- Live sound tour management
- Acoustics
- Audio instruction
- Club management
- Sound system installation for nightclubs, airports, homes, etc.
- . . . and a lot more!

This incomplete listing serves as a reminder that the business of making music is full of diverse possibilities and extends far beyond the notion that in order to make it in the biz you'll have to sell your soul or be someone you're not. In short, there are many paths to take in this techno-artistic business. Once you've found the one that best suits your own personal style, you can then begin the lifelong task of gaining knowledge and experience and pulling together a network of those who are currently working in the field. As the advertisement says, "Just do it!"

Career Development

It's a sure bet that those who are interested in getting into the business of audio will quickly find out that it can be a tough nut to crack. For every person that makes it, a large number don't. In short, there are a lot of people who are waiting in line to get into what is perceived by many to be a glamorous biz. So, how *do* you get to the front of the line? Well, folks, here's the key...

- ◆ A ton of self-motivation
- ◆ Good networking skills

The business of art (the techno-arts of recording and music being no exception) is one that's generally reserved for self-starters. Even if you get a degree from XYZ college or recording school, there's absolutely no guarantee that your dream studio will be knocking on the door with an offer in hand. It takes a large dose of perseverance, talent, and personality to make it. In fact, one of the best ways to get into the biz is to "knight" yourself on the shoulder with a sword (figuratively or literally) and say: "I am now a _____!" Whatever it is you want to be, become it... Shazammm! Make up a business card, start a business, begin contacting artists to work with (or make the first step toward becoming the artist you want to be). There are many ways to get to the top of your own personal mountain—for example, you could get a diploma from a school of education or the school of hard knocks (it usually ends up being from both)—but the goals and the paths are up to you. Just remember, being in the right place often means being in the wrong place several hundred times! Like a mentor of mine says: "Failure isn't a bad thing... not trying is!"

The other half of the success equation lies in your ability to network with other people. Like the venerable expression says: "It's not [only] what you know... it's who you know." Maybe you have an uncle or a friend in the business or a friend who has an uncle... you just never know where help might come from next. This idea of getting to know someone who knows someone else is what makes the business world go around. Don't be afraid to put your best face forward and start meeting people. If you want to work at XYZ Studios, hang out without being in the way. You never know—the engineer might need some help or might know someone who can help get you in the proverbial door. The longer you stick with it, the more people you'll meet... thus making a bigger and stronger network than you thought could be possible.

So, when do you start this grand adventure? When do you start building your career? The obvious answer is *now*. If you're in school, you have already started the process. If you're just hanging out with like-minded biz folks, that, too, is an equally strong start. Whatever you do, don't wait until you graduate or until a magic date in the future, as waiting until then will just put you that much further behind. Here are a few tips on how to get started:

- Choose a mentor (sometimes they fall out of the sky, sometimes you have to develop the relationship).
- Contact companies in your area that might be looking for interns.
- Use your school counselors for intern placement.
- Pick the areas you want to live in (if that is a factor).
- Pick the companies you are interested in.
- Target these areas and companies.
- Visit or send resumes to companies or studios that interest you.
- Send out *lots* of resumes.
- Follow up with a phone call.
- Visit these companies just to hang out and see what they are like.
- Follow up with another phone call.

Try not to be afraid when sending out a resume, demo tape, or CD of your work...or when asking for a job. The worst thing they can do is say "No." You might even keep in mind that "No" could actually mean "No, not right now." You might clarify to see if this is the case. If so, they might take your persistence into account before saying "No" two or three times. By picking a market and particular area, blanketing that area with resumes, and knocking on doors...you just never know what might happen. If nothing materializes, just remember the self-motivation factor. I know it's not easy, but pick yourself up (again), re-evaluate your strategies, and start pounding the streets (again). Just remember, "Failing at something isn't a bad thing...not trying is!"

The Recording Process

In this age of pro studios, project studios, digital audio workstations, groove tools, and personal choices, it's easy to understand how the "different strokes for different folks" adage applies to recording...in that differences between people and the tools they use allow the process of recording to be approached in many different ways. The cost-effective environment of the project studio has brought music and audio production to a much wider audience, thus making the process much more personal. If we momentarily set aside the monumental process of creating music in its various performance and notation forms—as well as in an electronic music environment (a process that involves MIDI composition/production, hard/software synths and samplers, digital groove editors, and a ton of other tools and toys)—then the process of capturing sound onto a recorded medium will generally occur in six distinct steps:

- Preparation
- Recording

◆ Overdubbing
◆ Mixdown
◆ Mastering
◆ Product manufacture

Preparation

One of the most important aspects of the recording process occurs *before* the artist and production team step into the studio: preparation. Questions like the following must be addressed:

◆ What is our goal?
◆ What is our budget?
◆ What are the estimated studio costs?
◆ Will we have enough time to work on vocals, mixing, and other important issues before running out of money?
◆ How much will it cost to manufacture the CDs and/or records?
◆ What are our advertising costs?
◆ How will we sell our music? And to whom?
◆ Are we practiced enough?
◆ If we won't have a producer on this project, who will speak for the group when the going gets rough?
◆ Are our instruments, our voices, and our heads ready for the task ahead?
◆ Are there any legal issues to consider?
◆ How do we get our Web site up and running?

These questions and a whole lot more will have to be addressed before it comes time to press the big red button. Further info on this topic can found in Chapter 18 (Studio Session Procedures).

Recording

The first phase in multitrack production is the recording process. In this phase, one or more sound sources are picked up by a microphone or are recorded directly (as often occurs when recording electric or electronic instruments) to one or more of the isolated tracks of a multitrack recording system. Of course, multitrack and hard-disk recording technologies have added an amazing degree of flexibility to the process by allowing multiple sound sources to be captured onto and played back from isolated tracks in a tape- and/or disk-based production environment. Because the recorded tracks are isolated from each other—with tape capabilities usually being offered in track groups of eight (*e.g.*, 8, 16, 24, 32, 48) and

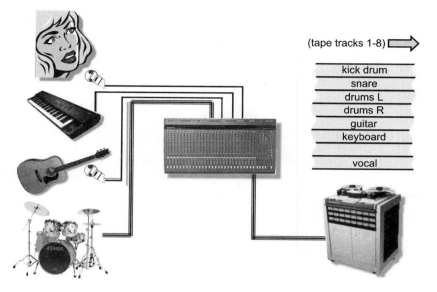

Figure 1.22. *Basic representation of how isolated sound sources can be recorded to a multitrack recorder.*

(tape tracks 1-8)

| kick drum |
| snare |
| drums L |
| drums R |
| guitar |
| keyboard |
| vocal |

with disk-based DAWs topping out at 256 tracks or higher—any number of instruments can be recorded and rerecorded without affecting other instruments. In addition, recorded tracks can be altered, added, edited, or erased at any time in order to augment or "sweeten" the original soundtrack.

Beyond the concept of capturing the best performance and sound to a recorded media, one of the key ideas within multitrack production is *isolation*. By recording a single instrument to a dedicated track (or group of tracks), it is possible to vary the level, spatial positioning (such as left/right or surround panning), EQ, signal processing, and routing without affecting the level or tonal qualities of an adjacent track or tracks (Figure 1.22). This isolation allows leakage from nearby instruments or mic pickups to be reduced to such an insignificant level that individual tracks can be rerecorded and/or processed at a later time without affecting the overall mix.

The basic tracks of a session can, of course, be built up in any number of ways. For example, the foundation of a session might be recorded in a traditional fashion, involving such acoustic instruments as drums, guitar, piano, and a scratch vocal (used as a rough guide throughout the session until the final vocals can be laid down). Alternatively, these tracks might be made up of basic electronic music loops, samples, or synth tracks that will need to be transferred to tape or imported into the studio's own digital audio workstation. Literally, the combinations of workingstyles, studio miking, isolation, and instrument arrangements are limitless and (whenever possible) are best discussed or worked out in the preparation stages. Further information on and references to isolation and the recording process can found in Chapters 3 and 18, as well as at various points throughout the book.

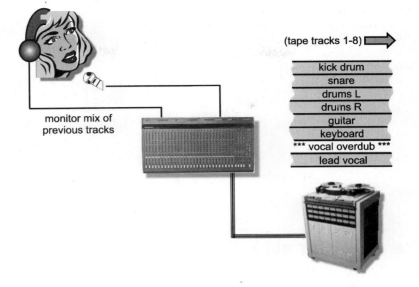

Figure 1.23. *Basic*
representation of the
overdubbing process.

(tape tracks 1-8)

kick drum
snare
drums L
drums R
guitar
keyboard
*** vocal overdub ***
lead vocal

monitor mix of
previous tracks

Overdubbing

Once the basic tracks have been laid down, the process of adding additional parts can begin in a process known as *overdubbing*. During this phase, the previously recorded tracks are monitored in the studio over headphones while one or more new musical parts are laid down onto the separate and available tracks of a recorder or DAW (Figure 1.23). During the overdub (OD) phase, individual parts are added to an existing project until the song or soundtrack is complete. If the artist makes a mistake ... no problem—simply rewind the tape or recue the workstation back to the where the instrument begins and repeat the process until you've captured the best possible take. If a take goes *almost* perfectly except for a bad line or a few flubbed notes, it's possible to go back and rerecord the offending segment onto the same or a different track in a process known as *punching in*. If the musician lays down his or her part properly and the engineer dropped in and out of record at the correct times (either manually or under automation), the listener won't even know that the part was recorded in multiple takes ... such is the magic of the recording process! More in-depth information on this topic can found in Chapter 18 (Studio Session Procedures).

Mixing Down

When all of the tracks of a project have been recorded, assembled, and edited, the time has come to individually *mix* the songs into their final media forms (Figure 1.24). The mixdown process occurs by routing the various tracks or tape- or disk-based system through a

Figure 1.24. *Basic representation of the mixdown process.*

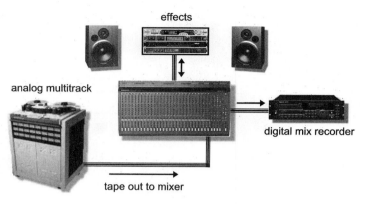

effects

analog multitrack

digital mix recorder

tape out to mixer

professional or project studio's hardware or virtual mixing console to alter the session's musical or audio program with respect to:

◆ Relative level

◆ Spatial positioning (the physical placement of a sound within a stereo or surround field)

◆ Equalization (affecting the relative frequency balance of a track)

◆ Dynamics processing (altering the dynamic range of a track, group, or output bus to optimize levels or to alter the dynamics of a track so that it "fits" within a mix)

◆ Effects processing (adding reverb-, delay-, or pitch-related effects to a mix in order to augment or alter the piece in a way that is natural, unnatural, or just plain interesting.

When all the songs in a music project have been mixed down, they can be edited into a final, sequenced order that can then be sent to a mastering engineer or manufactured into a final, commercially salable product. Additional information on this topic can found in Chapter 18 (Studio Session Procedures).

Mastering

Often, the final, edited mixdown of a project will be sent to a mastering engineer so that fine-tuning adjustments can be made to the overall recording with respect to:

◆ Relative level balancing between songs within the project

◆ Dynamic level (altering the dynamics of a song so as to maximize its level for the intended media or to tighten up the dynamic balance, overall or within certain frequency bands)

◆ Equalization

◆ Overall level

In essence, it is the job of a qualified mastering engineer to smooth over any level and spectral imbalances within a project and to present the final, recorded product in the best possible light for its intended media form. Further information on this topic can found in Chapter 16 (Mastering).

Product Manufacture

Last but never least in the production chain is the process of manufacturing the master recording into a final, salable product. Whether the final product is a compact disc or vinyl record, this process should be carefully overseen to ensure that the final product doesn't compromise all of the blood, sweat, tears, and bucks that have gone into the creation of a project. The manufacturing phases of . . .

- Creating a manufacture master
- Art layout and printing
- Product packaging

should be carefully scrutinized, checked, and rechecked. Whenever possible, ask for a proof copy of the final duplicated recording and artwork *before* it is mass duplicated. Receiving 1000 or 10,000 copies of your hard-earned project that aren't really what you wanted is worse than being handed an accordion in hell. Further info on this topic can found in Chapter 17 (Product Manufacture).

The Transducer

Before we jump into the heart of this book, I'd like to take a moment to look at a concept that is central to all music, sound, electronics, and the art of sound recording: the *transducer*. If any conceptual tool can help you to understand the technological underpinnings of the art and process of recording, this is it! Quite simply, a transducer is any device that changes one form of energy into another, corresponding form of energy. For example, a guitar is a transducer in that it takes the vibrations of picked or strummed strings (the medium), amplifies them through a body of wood, and converts these vibrations into corresponding sound-pressure waves . . . which are then perceived as sound (Figure 1.25).

A microphone is another example of a transducer. Here, sound-pressure waves (the medium) act on the mic's diaphragm and are converted into corresponding electrical voltages. The electrical signal from the microphone can then be amplified (not a process of transduction because the medium stays in its electrical form) and is then fed to a recording device. The recorder will then change these voltages into analogous magnetic flux signals on magnetic tape or into representative digital data that can be encoded onto tape, hard disk, or disc. On playback, the stored magnetic signals or digital data are converted back to their original electrical form, amplified, and then fed to a speaker system. The speakers convert the electrical signal into a mechanical motion (by way of magnetic induction), which, in turn, recreates the original air-pressure variations that were picked up by the microphone.

As can be seen from Table 1.1, transducers can be found practically everywhere in the audio environment. In general, transducers (and the media they use) are often the weakest link in the

Figure 1.25. The guitar and microphone as transducers.

Table 1.1. Media used by transducers in the studio to transfer energy.

Transducer	From	To
Ear	Sound waves in air	Nerve impulses in the brain
Microphone	Sound waves in air	Electrical signals in wires
Record head	Electrical signals in wires	Magnetic flux on tape
Playback head	Magnetic flux on tape	Electrical signals in wires
Phonograph cartridge	Grooves cut in disc surface	Electrical signals in wires
Speaker	Electrical signals in wires	Sound waves in air

audio system chain. Given our current technology, the process of changing the energy in one medium into a corresponding form of energy in another medium can't be accomplished perfectly. Noise, distortion, and (often) coloration of the sound are introduced to some degree, and unfortunately these effects can only be minimized, not eliminated. Differences in design are another major factor that can affect sound quality. Even a slight design variation between two microphones, speaker systems, digital audio converters, guitar pickups, or other transducers can cause them to sound quite different. This factor, combined with the complexity of music and acoustics, helps makes the field of recording the subjective and personal art form that it is.

It's interesting to note that fewer transducers are used in an all or largely digital recording system (Figure 1.26). In this situation, the acoustic waveforms that are picked up by a microphone are converted into electrical signals and are then converted into digital form by an analog-to-digital (A/D) converter. The A/D converter changes these continuous electrical waveforms into corresponding discrete numeric values that represent the waveform's instantaneous, analogous voltage levels. Arguably, digital information has the distinct advantage over analog in that data can be transferred between electrical, magnetic, and optical media with virtually no degradation in quality. Because the information continues to be

Figure 1.26. Example of an
all-digital production studio.
(Courtesy of Digidesign, a
division of Avid Technology,
Inc., www.digidesign.com.)

stored in its original, discrete binary form, no transduction process is involved (*i.e.*, only the
medium changes, while the data representing the actual information doesn't change). Does this
mean that digital's better? Nope! It's just another way of expressing sound through a
medium . . . which in the end, is just one of the many possible artistic and technological choices
in the making and recording of sound and music.

CHAPTER

Sound and Hearing

When a recording is made, we're actually capturing and storing sound so that an original event can be recreated at a later date. If we start with the idea that *sound* is actually a concept that describes the brain's perception and interpretation of a physical auditory stimulus, the examination of sound can be divided into four areas:

◆ The basics of sound
◆ The characteristics of the ear
◆ How the ear is stimulated by sound
◆ The psycho-acoustics of hearing

By understanding the physical nature of sound and how the ears change a physical phenomenon to a sensory one, we can discover how to best convey this science into the subjective art forms of music, sound recording, and production.

The Basics of Sound

Sound arrives at the ear in the form of periodic variations in atmospheric pressure called *sound-pressure waves*. This is the same atmospheric pressure that's measured by the weather service with a barometer; however, the changes are too small in magnitude and fluctuate too rapidly to be observed on a barometer. An analogy of how sound waves travel in air can be demonstrated by bursting a balloon in a silent room. Before we stick it with a pin, the molecular motion of the room's atmosphere is at a normal resting pressure.

Figure 2.1. *Wave movement in air as it moves away from its point of origin. (a) Intact balloon contains the high pressure. (b) When the balloon is popped, the compressed molecules exert a force upon outer neighbors in an effort to move to areas of lower pressure. (c) The outer neighbors exert a force on the next set of molecules in an effort to move to areas of lower pressure, and the process continues.*

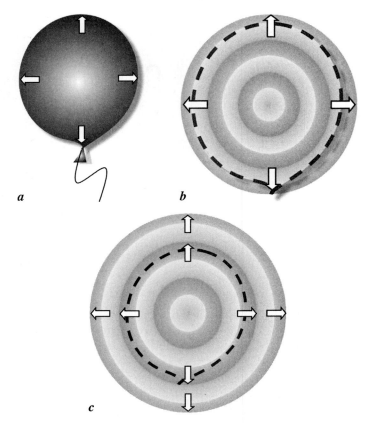

The pressure inside the balloon is much higher, though, and the molecules are compressed much more tightly together...like people packed into a crowded subway car (Figure 2.1a). When the balloon is popped..."POW!" (Figure 2.1b), the tightly compressed molecules under high pressure begin to exert an outward force upon their neighbors in an effort to move toward areas of lower pressure. When the neighboring set of molecules has been compressed, they will continue to exert an outward force upon the next set of lower-pressured neighbors (Figure 2.1c)...in an ongoing outward motion that continues until the molecules have used up their energy in the form of heat.

Likewise, as a vibrating mass (such as a guitar string, a person's vocal chords, or loudspeaker) moves outward from its normal resting state, it squeezes air molecules into a compressed area, away from the sound source. This causes the area being acted upon to have a greater than normal atmospheric pressure, a process called *compression* (Figure 2.2a). As the vibrating mass moves inward from its normal resting state, an area with a lower than normal atmospheric pressure will be created, a process called *rarefaction* (Figure 2.2b). As the vibrating body cycles through its inward and outward motions, areas of higher and lower compression states are generated. These areas of high pressure will cause the wave to move outward from the sound

Figure 2.2. *Effects of a vibrating mass on air molecules and their propagation. (a) Compression—air molecules are forced together to form a compression wave. (b) Rarefaction—as the vibrating mass moves inward, a rarefied area of lower atmospheric pressure is created.*

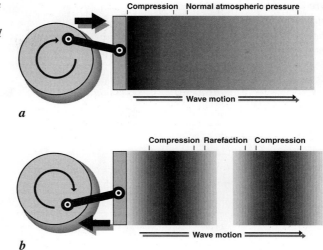

source in the same way as was caused by the balloon. It's interesting (and important) to note that the molecules themselves don't move through air at the velocity of sound—only the sound wave itself moves through the atmosphere in the form of high-pressure compression waves that continue to push against areas of lower pressure (in an outward direction). This outward pressure motion is known as *wave propagation*.

Waveform Characteristics

A *waveform* is essentially the graphic representation of a sound-pressure level or voltage level as it moves through a medium over time. In short, a waveform lets us see and explain the actual phenomenon of wave propagation in our physical environment and will generally have the following fundamental characteristics:

- Amplitude
- Frequency
- Velocity
- Wavelength
- Phase
- Harmonic content
- Envelope

These characteristics allow one waveform to be distinguished from another. The most fundamental of these are amplitude and frequency (Figure 2.3). The following sections describe each of these characteristics. Although several math formulas have been included, it is by no means important that you memorize or worry about them. It's far more important that you grasp the basic principles of acoustics rather than fret over the underlying math.

Figure 2.3.
Amplitude and frequency ranges of human hearing.

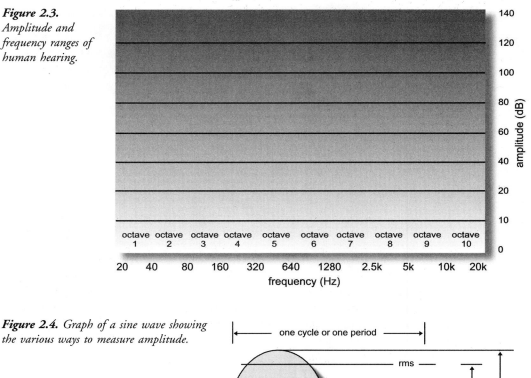

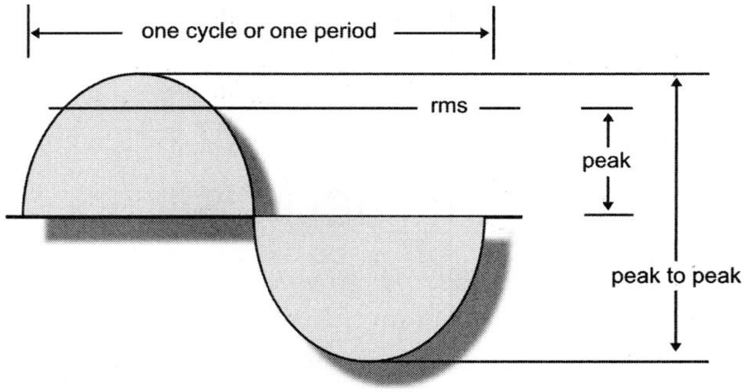

Figure 2.4. *Graph of a sine wave showing the various ways to measure amplitude.*

Amplitude

The distance above or below the centerline of a waveform (such as a pure sine wave) represents the *amplitude* level of that signal. The greater the distance or displacement from that centerline, the more intense the pressure variation, electrical signal level, or physical displacement will be within a medium. Waveform amplitudes can be measured in several ways (Figure 2.4). For example, the measurement of either the maximum positive or negative signal level of a wave is called its *peak amplitude value* (or peak value). The total measurement of the positive and negative peak signal levels is called the *peak-to-peak value*. The *root-mean-square* (*rms*) *value* was developed to determine a meaningful average level of a waveform over time (one that more closely approximates the level that's perceived by our ears and gives a better real-world measurement of overall signal amplitudes). The rms value of a sine wave can be calculated by squaring the amplitudes at points along the waveform and then taking the mathematical average of the combined results. The math isn't as important as the concept that the rms value of a perfect sine wave is equal to 0.707 times its instantaneous peak amplitude level. Because the square of a positive or negative value is always positive, the rms value will always be positive. The following simple equations show the relationship between a waveform's peak and rms values:

$$\text{rms voltage} = 0.707 \times \text{peak voltage}$$

$$\text{peak voltage} = 1.414 \times \text{rms voltage}$$

Frequency

The rate at which an acoustic generator, electrical signal, or vibrating mass repeats within a cycle of positive and negative amplitude is known as the *frequency* of that signal. As the rate of repeated vibration increases within a given time period, the frequency (and thus the perceived pitch) will likewise increase ... and *vice versa*. One completed excursion of a wave (which is plotted over the 360° axis of a circle) is known as a *cycle* (Figure 2.5). The number of cycles that occur within a second (frequency) is measured in hertz (Hz). The diagram in Figure 2.6 shows the value of a waveform as starting at zero (0°). At time $t = 0$, this value increases to a positive

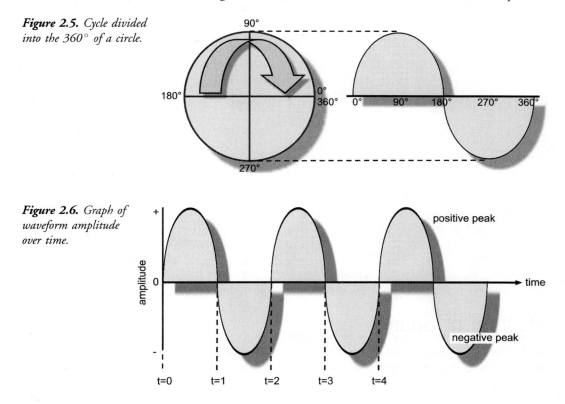

Figure 2.5. *Cycle divided into the 360° of a circle.*

Figure 2.6. *Graph of waveform amplitude over time.*

maximum value and then decreases back through zero, where the process begins all over again in a repetitive fashion. A cycle can begin at any angular degree point on the waveform; however, to be complete, it must pass through a single 360° rotation and end at the same point as its starting value. For example, the waveform that starts at $t=0$ and ends at $t=2$ constitutes a cycle, as does the waveform that begins at $t=1$ and ends at $t=3$.

Velocity

The *velocity* of a sound wave as it travels through air at 68°F (20°C) is approximately 1130 feet per second (ft/sec) or 344 meters per second (m/sec). This speed is temperature dependent and increases at a rate of 1.1 ft/sec for each Fahrenheit degree increase in temperature (2 ft/sec per Centigrade degree).

Wavelength

The *wavelength* of a waveform (frequently represented by the Greek letter lambda, λ) is the physical distance in a medium between the beginning and the end of a cycle. The physical length of a wave can be calculated using:

$$\lambda = V/f$$

where λ is the wavelength in the medium, V is the velocity in the medium, and f is the frequency (in hertz).

The time it takes to complete 1 cycle is called the *period* of the wave. To illustrate, a 30-Hz sound wave completes 30 cycles each second or 1 cycle every 1/30th of a second. The period of the wave is expressed using the symbol T:

$$T = 1/f$$

where T is the number of seconds per cycle.

Assuming that sound propagates at the rate of 1130 ft/sec, all that's needed is to divide this figure by the desired frequency. For example, the simple math for calculating the wavelength of a 30-Hz waveform would be $1130/30 = 37.6$ feet long, while a waveform having a frequency of 300 Hz would be $1130/300 = 3.76$ feet long (Figure 2.7). Likewise, a 1000-Hz waveform would work out as being $1130/1000 = 1.13$ feet long, while a 10,000-Hz waveform would be $1130/10,000 = 0.113$ feet long. From these calculations, you can see that whenever the frequency is increased, the wavelength decreases.

Reflection of Sound

Much like light waves, sound reflects off a surface boundary at an angle that is equal to (and in an opposite direction of) its initial angle of incidence. This basic property is one of the cornerstones of the complex study of acoustics. For example, Figure 2.8a shows how a sound

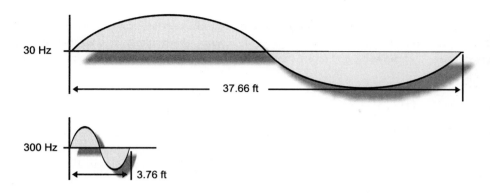

Figure 2.7. *Wavelengths decrease in length as frequency increases (and vice versa).*

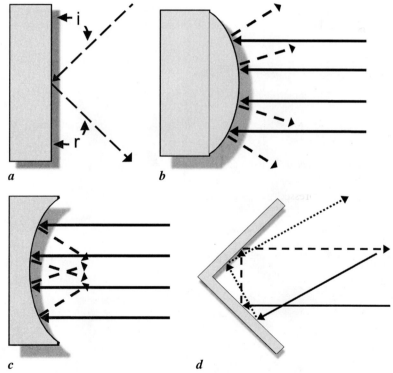

Figure 2.8. *Incident sound waves striking surfaces with varying shapes: (a) single-planed, solid, smooth surface; (b) convex surface; (c) concave surface; (d) 90° corner reflection.*

wave reflects off a solid smooth surface in a simple and straightforward manner (at an equal and opposite angle). Figure 2.8b shows how a convex surface will splay the sound outward from its surface, radiating the sound outward in a wide dispersion pattern. In Figure 2.8c, a concave surface is used to focus a sound inward toward a single point, while a 90° corner (as shown in Figure 2.8d) reflects patterns back at angles that are equal to their original incident direction. This holds true both for the 90° corners of a wall and for intersections where the wall and floor meet. These corner reflections help to provide insights into how volume levels often build up in the corners of a room (particularly at wall-to-floor corner intersections).

Diffraction of Sound

Sound has the inherent ability to diffract around or through a physical acoustic barrier. In other words, sound can bend around an object in a manner that reconstructs the signal back to its original form in both frequency and amplitude. For example, in Figure 2.9a, we can see how a small obstacle will scarcely impede a larger acoustic waveform. Figure 2.9b shows how a larger obstacle can obstruct a larger portion of the waveform; however, past the obstruction, the signal bends around the area in the barrier's wake and begins to reconstruct itself. Figure 2.9c shows how the signal is able to radiate through an opening in a large barrier. Although the signal is greatly impeded (relative to the size of the opening), it nevertheless begins to reconstruct itself in wavelength and relative amplitude and begins to

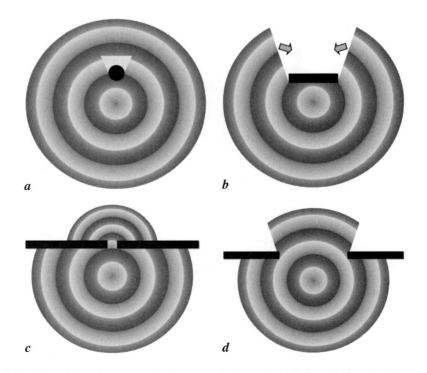

Figure 2.9. The effects of obstacles on sound radiation and diffraction. (a) A small obstacle will scarcely impede a longer wavelength signal. (b) A larger obstacle will obstruct the signal to a greater extent; the waveform will also reconstruct itself in the barrier's wake. (c) A small opening in a barrier will greatly impede a signal; the waveform will emanate from the opening and reconstruct itself as a new source point. (d) A larger opening allows sound to pass unimpeded, allowing it to quickly diffract back into its original shape.

radiate outward as though it were a new point of origin. Finally, Figure 2.9d shows how a large opening in a barrier lets much of the waveform pass through relatively unimpeded.

Frequency Response

The charted output of an audio device is known as its *frequency response curve.*(when supplied with a reference input of equal level over the 20–20,000 Hz range of human hearing). This curve is used to graphically represent how a device will respond to the audio spectrum and, thus, how it will affect a signal's overall sound. As an example, Figure 2.10 shows the frequency response of several unidentified devices. In these and all cases, the *x*-axis represents the signal's measured frequency, while the *y*-axis represents the device's measured output signal. These curves are created by feeding the input of an acoustic or electrical device with a constant-amplitude reference signal that sweeps over the entire frequency spectrum. The results are then charted on an amplitude *versus* frequency graph that can be easily read at a glance. If the measured signal is the same level at all frequencies, the curve will be drawn as a flat, straight line from left to right (known as a *flat frequency response curve*). This indicates that the device passes all frequencies equally (with no frequency being emphasized or de-emphasized). If the output lowers or increases at certain frequencies, these changes will easily show up as dips or peaks in the chart.

Phase

Because we know that a cycle can begin at any point on a waveform, it follows that whenever two or more waveforms are involved in producing a sound, their relative amplitudes can (and often will) be different at any one point in time. For simplicity's sake, let's limit our example to two pure tone waveforms (sine waves) that have equal amplitudes and frequency . . . but start their cyclic periods at different times. Such waveforms are said to be *out of phase* with respect to each other. Variations in *phase,* which is measured in degrees (°), can be described as a time delay between two or more waveforms. These delays are often said to have differences in relative phase degree angles (over the full rotation of a cycle . . . e.g., 90°, 180° or any angle between 0° and 360°). The sine wave (so named because its amplitude follows a trigonometric sine function) is usually considered to begin at 0° with an amplitude of zero; the waveform then increases to a positive maximum at 90°, decreases back to a zero amplitude at 180°, increases to a negative maximum value at 270°, and finally returns back to its original level at 360°, simply to begin all over again.

Whenever two or more waveforms arrive at a single location out of phase, their relative signal levels will be added together to create a combined amplitude level at that one point in time. Whenever two waveforms having the same frequency, shape, and peak amplitude are completely *in phase* (meaning that they have no relative time difference), the newly combined waveform will have the same frequency, phase, and shape . . . but will be double in amplitude (Figure 2.11a). If the same two waves are combined completely out of phase (having a phase difference of 180°), they will cancel each other out when added,

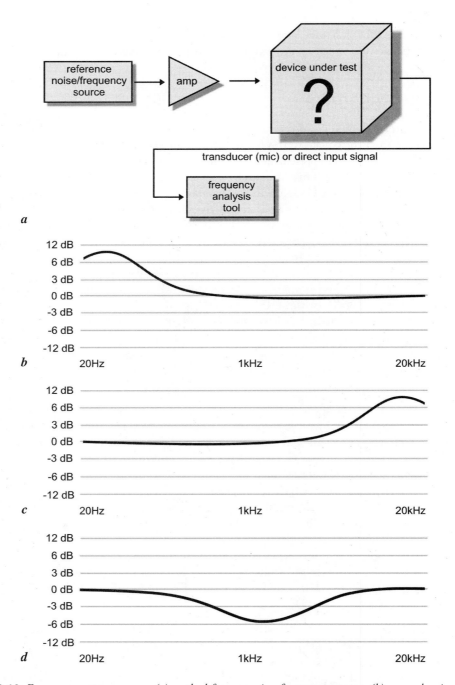

Figure 2.10. *Frequency–response curves: (a) method for measuring frequency response; (b) curve showing a bass boost; (c) curve showing a boost at the upper-end; (d) curve showing a dip in the midrange.*

which results in a straight line of zero amplitude (Figure 2.11b). If the second wave is only partially out of phase (by a degree other than 180°), the levels will be added at points where the combined amplitudes are positive and reduced in level where the combined result is negative (Figure 2.11c).

Do-It-Yourself Tutorial: Phase

1. Go to the "Tutorial" section of www.modrec.com, click on "Ch. 2—Phase Tutorial," and download the 0° and 180° sound files.
2. Load the 0° file onto track 1 of the digital audio workstation (DAW) of your choice, making sure to place the file at the beginning of the track, with the signal panned center.
3. Load the same 0° file again into track 2.
4. Load the 180° file into track 3.
5. Listen to tracks 1 and 2 (by muting track 3) and listen to the results. It should produce a summed signal that is 3 dB louder.
6. Listen to tracks 1 and 3 (by muting track 2) and listen to the results. It should cancel, producing no output.
7. Offsetting track 3 (relative to track 1) should produce varying degrees of cancellation.
8. Feel free to zoom in on the waveforms, mix them down, and view the results Cool, huh?

Figure 2.11. Combining sine waves of various phase relationships. (a) The amplitudes of in-phase waves increase in level when mixed together. (b) Waves of equal amplitude cancel completely when mixed 180° out of phase. (c) When partial phase angles are mixed together, the combined signals will add in certain places and subtract in others.

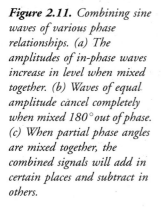
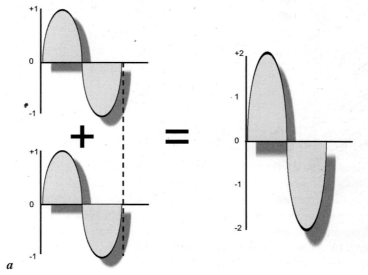

Figure 2.11. *Continued.*

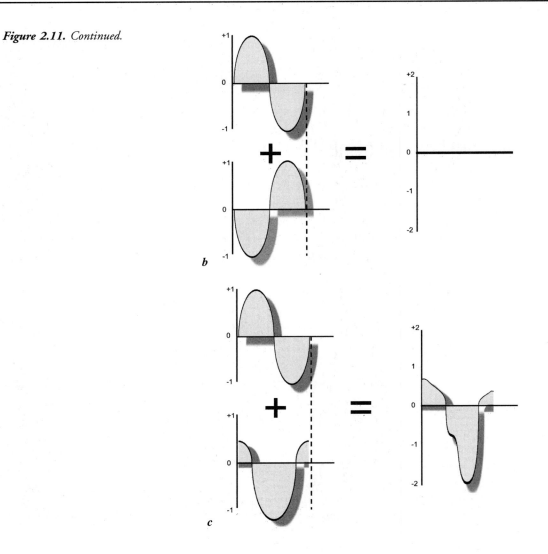

b

c

Phase Shift

Phase shift is a term that describes one waveform's lead or lag in time with respect to another. Basically, it results from a time delay between the two (or more) waveforms (with differences in acoustic distance being the most common source of this type of delay). For example, a 500-Hz wave completes one cycle every 0.002 sec. If you start with two in-phase, 500-Hz waves and delay one of them by 0.001 sec (half the wave's period), the delayed wave will lag the other by one-half a cycle . . . or 180°. Another example might include a single source that's being picked up by two microphones that have been placed at different distances (Figure 2.12), thereby creating a corresponding time delay when the mics are mixed together. Such a delay can also

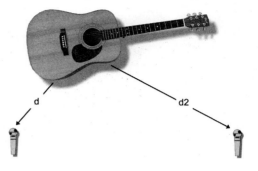

Figure 2.12. *Cancellations can occur when a single source is picked up by two microphones.*

occur when a single microphone picks up direct sounds as well as those that are reflected off a nearby boundary. These signals will be in phase at frequencies where the path-length difference is equal to the signal's wavelength, and out of phase at those frequencies where the multiples fall at or near the half-wavelength distance. In all the above situations, these boosts and cancellations combine to alter the signal's overall frequency response at the pickup. For this and other reasons, acoustic leakage between microphones and reflections from nearby boundaries should be kept to a minimum whenever possible.

Harmonic Content

Up to this point, the discussion has centered on the sine wave, which is composed of a single frequency that produces a pure sound at a specific pitch. Fortunately, musical instruments rarely produce pure sine waves. If they did, all of the instruments would basically sound the same, and music would be pretty boring. The factor that helps us differentiate between instrumental "voicings" is the presence of frequencies (called *partials*) that exist in addition to the fundamental pitch that's being played. Partials which are higher than the fundamental frequency are called *upper partials* or *overtones*. Overtone frequencies that are whole-number multiples of the fundamental frequency are called *harmonics*. For example, the frequency that corresponds to concert A is 440 Hz (Figure 2.13a). An 880-Hz wave is a harmonic of the 440-Hz fundamental because it is twice the frequency (Figure 2.13b). In this case, the 440-Hz fundamental is technically the first harmonic because it is 1 times the fundamental frequency, and the 880-Hz wave is called the second harmonic because it is 2 times the fundamental. The third harmonic would be 3 times 440 Hz, or 1320 Hz (Figure 2.13c). Some instruments, such as bells, xylophones, and other percussion instruments, will often contain overtone partials that aren't harmonically related to the fundamental at all.

The ear perceives frequencies that are whole multiples of the fundamental as being specially related (a phenomenon known as the *musical octave*). For example, as concert A is 440 Hz (A3), the ear hears 880 Hz (A4) as being the next highest frequency that sounds most like concert A. The next related note above that will be 1760 Hz (A5). Therefore, 880 Hz is said to be one octave above 440 Hz, and 1760 Hz is said to be two octaves above 440 Hz, etc. Because these frequencies are even multiples of the fundamental, they're known as *even harmonics*. Not surprisingly, frequencies that are odd multiples of the fundamental are called *odd harmonics*.

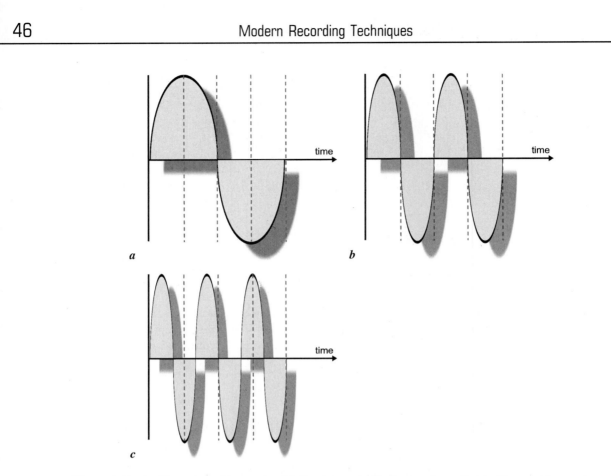

Figure 2.13. *An illustration of harmonics: (a) first harmonic (fundamental waveform); (b) second harmonic; (c) third harmonic.*

In general, even harmonics are perceived as creating a sound that is pleasing to the ear, while odd harmonics will create a dissonant, harsher tone.

Do-It-Yourself Tutorial: Harmonics

1. Go to the "Tutorial" section of www.modrec.com, click on "Ch. 2—Harmonics Tutorial," and download all of the sound files.
2. Load the 1st harmonic a440 file onto track 1 of the digital audio workstation (DAW) of your choice, making sure to place the file at the beginning of the track, with the signal panned center.
3. Load the 2nd, 3rd, 4th, and 5th harmonic files into the next set of consecutive tracks.
4. Solo the 1st harmonic track, then solo the 1st and 2nd harmonic tracks. Do they sound related in nature?
5. Solo the 1st harmonic track, then solo the 1st and 3rd harmonic tracks. Do they sound more dissonant?

6. Solo the 1st, 2nd, and 3rd harmonic tracks. Do they sound related?

7. Solo the 1st, 3rd, and 5th harmonic tracks. Do they sound more dissonant?

Because musical instruments produce sound waves that contain harmonics with various amplitude and phase relationships, the resulting waveforms bear little resemblance to the shape of the single-frequency sine wave. Therefore, musical waveforms can be divided into two categories: simple and complex. Square waves, triangle waves, and sawtooth waves are examples of *simple waves* that contain a consistent harmonic structure (Figures 2.14). They are said to be simple because they're continuous and repetitive in nature. One cycle of a square wave looks exactly like the next, and they are symmetrical about the zero line.

Complex waves, on the other hand, don't necessarily repeat and often are not symmetrical about the zero line. An example of a complex waveform (Figure 2.15) is one that's created by any naturally occuring sound (such as music or speech). Although complex waves are rarely

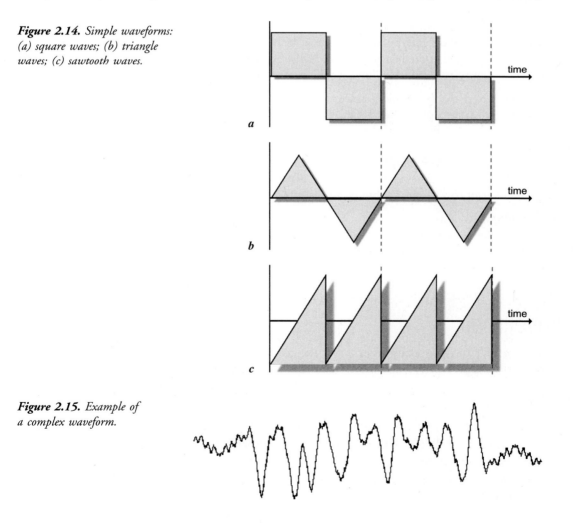

Figure 2.14. *Simple waveforms: (a) square waves; (b) triangle waves; (c) sawtooth waves.*

Figure 2.15. *Example of a complex waveform.*

repetitive in nature, all sounds can be mathematically broken down as being an ever-changing combination of individual sine wave.

Regardless of the shape or complexity of a waveform that reaches the eardrum, the inner ear is able to perceive these component waveforms and transmit the stimulus to the brain. This can be illustrated by passing a square wave through a bandpass filter that's set to pass only a narrow band of frequencies at any one time. Doing this would show that the square wave is composed of a fundamental frequency plus a number of harmonics that are made up of odd-number multiple frequencies (whose amplitudes decrease as the frequency increases). In Figure 2.16, we see how individual sine-wave harmonics can be combined to form a square wave.

If we were to analyze the harmonic content of sound waves that are produced by a violin and compare them to the content of the waves that are produced by a viola (with both playing concert A, 440 Hz), we would come up with results like those shown in Figure 2.17. Notice that the violin's harmonics differ in both degree and intensity from those of the viola. The harmonics and their relative intensities (which determine an instrument's characteristic sound) are called the *timbre* of an instrument. If we changed an instrument's harmonic balance, the sonic character of the instrument would also be changed. For example, if the violin's upper harmonics were reduced, the violin would sound a lot like the viola.

Because the relative harmonic balance is so important to an instrument's sound, the frequency response of a microphone, amplifier, speaker, and all other elements in the signal path can have an effect on the timbre (harmonic balance) of a sound. If the frequency response isn't flat, the timbre of the sound will be changed. For example, if the high frequencies are amplified less than the low and middle frequencies, then the sound will be duller than it should be. For this reason, specific devices or the use of an equalizer can be used as tools (for better or for the worse) to vary the timbre of an instrument, thereby changing its subjective sound.

In addition to the variations in harmonic balance that can exist between instruments and their families, it is common for the harmonic balance to vary with respect to direction as sound waves radiate from an instrument. Figure 2.18 shows the principal radiation patterns as they emanate from a cello (as seen from both the side and top views).

Envelope

Timbre isn't the only characteristic that lets us differentiate between instruments. Each one produces a sonic *envelope* that works in combination with timbre to determine its unique and subjective sound. The envelope of a waveform can be described as characteristic variations in level that occur in time over the duration of a played note. The envelope of an acoustic signal is composed of three sections:

1. *Attack* refers to the time taken for a sound to build up when a note is initially sounded.
2. *Sustain* describes the increases, decreases, and sustains in volume that occur after the initial attack has sounded.
3. *Decay* refers to a fade or reduction in level over time, once the note has stopped playing.

Figure 2.16. *Breaking a square wave down into its odd-harmonic components: (a) square wave with frequency f. (b) sine wave with frequency f. (c) sum of a sine wave with frequency f and a lower amplitude sine wave of frequency 3f; (d) sum of a sine wave of frequency f and lower amplitude sine waves of 3f and 5f, beginning to resemble a square wave.*

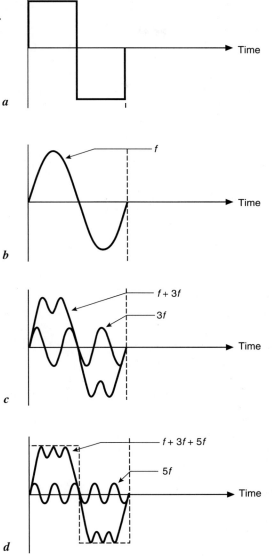

Each of these sections has three variables: *initial amplitude, time duration,* and *amplitude variation* with time. Figure 2.19a illustrates the envelope of a trombone note. The attack, decay times, and internal dynamics produce a smooth, sustaining sound. Figure 2.19b illustrates the envelope of a snare drum. Notice that the initial attack is much louder than the internal dynamics . . . while the final decay trails off very quickly, resulting in a sharp, percussive sound. A cymbal crash (Figure 2.19c) combines a high-level, fast attack with a longer sustain and decay that creates a smooth, lingering shimmer.

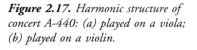

Figure 2.17. *Harmonic structure of concert A-440: (a) played on a viola; (b) played on a violin.*

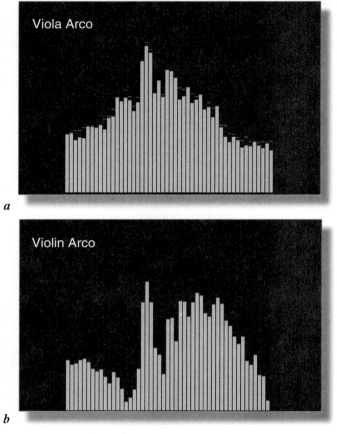

a

b

Figure 2.18. *Radiation patterns of a cello as viewed from the side (left) and top (right).*

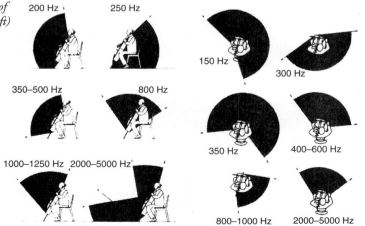

Figure 2.19. *Various musical waveform envelopes: (a) trombone, (b) snare drum, and (c) cymbal crash, where A = attack, D = decay, S = sustain, and R = release.*

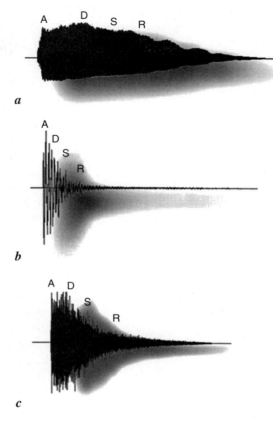

a

b

c

It's important to note that the concept of an envelope relies upon peak waveform values, while the human perception of loudness is proportional to the average wave intensity over a period of time (rms value). Therefore, high-amplitude portions of the envelope won't make an instrument sound loud unless the amplitude is maintained for a sustained period. Short high-amplitude sections tend to contribute to a sound's overall character, rather than to its loudness. By using a compressor or limiter, an instrument's character can often be modified by changing the dynamics of its envelope . . . without changing its timbre.

The waveform envelope of an electronic music instrument is similar in most respects to its acoustic counterpart and is modified with respect to its initial attack time, decay time (from the initial attack), sustain time, and final release time. This is most commonly referred to in its abbreviated form: ADSR (attack, decay, sustain, and release).

Loudness Levels: The Decibel (dB)

The ear operates over an energy range of approximately 1013:1 (10,000,000,000,000:1)—that's an extremely wide range. Because it is difficult for us humans to conceptualize number

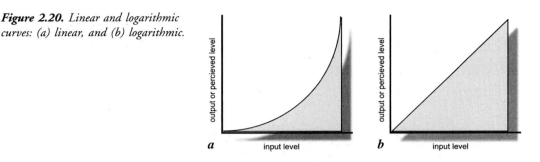

Figure 2.20. *Linear and logarithmic curves: (a) linear, and (b) logarithmic.*

ranges that are this large, a logarithmic scale has been adopted to compress the measurements into figures that are more manageable. The unit used for measuring sound-pressure level (SPL), signal level, and relative changes in signal level is the *decibel* (*dB*), a term that literally means 1/10th of a Bell . . . a telephone transmission measurement unit that was named after Alexander Graham Bell, inventor of the telephone. In order to develop an understanding of the decibel, we first need to examine logarithms and the logarithmic scale (Figure 2.20). The *logarithm* (*log*) is a mathematical function that reduces large numeric values into smaller, more manageable numbers. Because logarithmic numbers increase exponentially in a way that's similar to how we perceive loudness (*e.g.*, 1, 2, 4, 16, 128, 256, 65,536, . . .), it expresses our perceived sense of volume more precisely than a linear curve can.

Before we delve into a deeper study of this important concept and how it deals with our perceptual senses, let's take a moment to understand the basic concepts and building block ideas behind the log scale so as to get a better understanding of what examples such as "+3 dB at 10,000 Hz" really mean. Be patient with yourself! Over time, the concept of the decibel will become as much a part of your working vocabulary as ounces, gallons, and miles per hour.

Logarithmic Basics

In audio, we use logarithmic values to express the differences in intensities between two levels (often, but not always, comparing a measured level to a standard reference level). Because the differences between these two levels can be really, really big, we have to break these huge numbers down into values that are mathematical exponents of 10. To begin, finding the log of a number such as 17,386 without a calculator is not only difficult . . . in audio it's unnecessary! All that's really important to help you along are three simple guidelines:

1. The log of the number 2 is 0.3.
2. When a number is an integral power of 10 (*e.g.*, 100, 1000, 10,000), the log can be found simply by adding up the number of zeros.
3. Numbers that are greater than 1 will have a positive log value, while those less than 1 will have a negative log value.

The first one is an easy fact to remember: The log of 2 is 0.3 . . . this will make sense shortly. The second one is even easier: The logs of numbers such as 100, 1000 or 10,000,000,000,000 can be arrived at by counting up the zeros. The last guideline relates to the fact that if the measured value is less than the reference value, the log will be negative. For example:

$$\log 2 = 0.3$$
$$\log 1/2 = \log 0.5 = -0.3$$
$$\log 10,000,000,000,000 = 13$$
$$\log 1000 = 3$$
$$\log 100 = 2$$
$$\log 10 = 1$$
$$\log 1 = 0$$
$$\log 0.1 = -1$$
$$\log 0.01 = -2$$
$$\log 0.001 = -3$$

All other numbers can be arrived at by using a scientific calculator (most computers have one built in); however, it's unlikely that you will ever need to know any log values, beyond understanding the basic concepts that are listed above.

The dB

Now that we've gotten past the absolute basics, I'd like to break with tradition again and attempt an explanation of the dB in a way that's less complex and relates more to our day-to-day needs in the sound biz. First off, the dB is a logarithmic value that "expresses differences in intensities between two levels." From this, we can infer that these levels are expressed by several units of measure, the most common being sound-pressure level (SPL), voltage (V), and power (wattage, or W). Now, let's look at the basic math behind these three measurements.

Sound-Pressure Level

Sound-pressure level is the acoustic pressure that's built up within a defined atmospheric area (usually a square centimeter, or cm^2). Quite simply, the higher the SPL, the louder the sound (Figure 2.21). In this instance, our measured reference (SPL_{ref}) is the threshold of hearing, which is defined as being the softest sound that an average person can hear. Most conversations will have an SPL of about 70 dB, while average home stereos are played at volumes ranging between 80 and 90 dB SPL. Sounds that are so loud as to be painful have SPLs of about 130 to 140 dB (10,000,000,000,000 or more times louder than the 0-dB reference). We can arrive at an SPL rating by using the formula:

$$dB\, SPL = 20 \log SPL/SPL_{ref}$$

Figure 2.21. *Chart of sound-pressure levels. (Courtesy of General Radio Company.)*

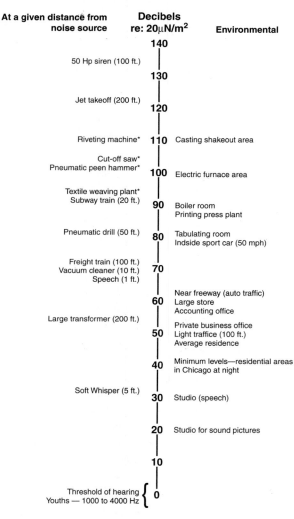

Typical A-Weighted Sound Levels

At a given distance from noise source	Decibels re: 20μN/m²	Environmental
	140	
50 Hp siren (100 ft.)		
	130	
Jet takeoff (200 ft.)		
	120	
Riveting machine*	**110**	Casting shakeout area
Cut-off saw* Pneumatic peen hammer*	**100**	Electric furnace area
Textile weaving plant* Subway train (20 ft.)	**90**	Boiler room Printing press plant
Pneumatic drill (50 ft.)	**80**	Tabulating room Indside sport car (50 mph)
Freight train (100 ft.) Vacuum cleaner (10 ft.) Speech (1 ft.)	**70**	
	60	Near freeway (auto traffic) Large store Accounting office
Large transformer (200 ft.)	**50**	Private business office Light traffice (100 ft.) Average residence
	40	Minimum levels—residential areas in Chicago at night
Soft Whisper (5 ft.)	**30**	Studio (speech)
	20	Studio for sound pictures
	10	
Threshold of hearing Youths — 1000 to 4000 Hz	**0**	

*Operation position

where *SPL* is the measured sound pressure (in dyne/cm²), and *SPL*$_{ref}$ is a reference sound pressure (0.0002 dyne/cm², the threshold of hearing).

From this, I feel that the only concept that needs to be understood is the idea that SPL levels change with the square of the distance (hence, the 20 log part of the equation). This means that whenever a source/pickup distance is doubled, the SPL level will reduce by 6 dB (20 log .5/1 = 20 × −0.3 = −6 dB SPL); as the distance is halved, it will increase by 6 dB (20 log 2/1 = 20 × 0.3 = 6dB SPL), as shown in Figure 2.22.

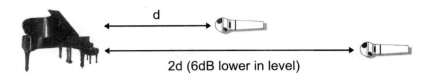

Figure 2.22. *Doubling the distance of a pickup will lower the direct signal level by 6 dB SPL.*

Voltage

Voltage can be thought of as the pressure behind electrons within a wire. As with acoustic energy, comparing one voltage level to another level (or reference level) can be expressed as dBv using the equation:

$$dBv = 20 \log V/V_{ref}$$

where V is the measured voltage, and V_{ref} is a reference voltage (0.775 volts).

Power

Power is usually a measure of wattage or current and can be thought of as the flow of electrons through a wire over time. Power is generally associated with audio signals that are carried throughout an audio production system. Unlike SPL and voltage, the equation for signal level (which is often expressed in dBm) is:

$$dBm = 10 \log P/P_{ref}$$

where P is the measured wattage, and P_{ref} is referenced to 1 milliwatt (0.001 watt = 1mW).

The Simple Heart of the Matter

I am going to stick my neck out and state that, when dealing with decibels, it's far more common for working professionals to deal with the concept of power. The dBm equation expresses the spirit of the decibel term when dealing with the markings on an audio device or the numeric values in a computer dialog box. This is due to the fact that power is the unit of measure that's most often expressed when dealing with audio equipment controls; therefore, it's my personal opinion that the average working stiff only needs to grasp the following basic concepts:

◆ A 1-dB change is barely noticeable to most ears.

◆ Turning something up by 3 dB will double the signal's level (believe it or not, doubling the signal level won't increase the perceived loudness as much as you might think).

◆ Turning something down by 3 dB will halve the signal's level (likewise, halving the signal level won't decrease the perceived loudness as much as you might think).

◆ The log of an exponent of 10 can be easily figured by simply counting the zeros (*e.g.*, the log of 1000 is 3). Given that this figure is multiplied by 10 (10 log P/P_{ref})...turning something up by 10 dB will increase the signal's level 10-fold, 20 dB will yield a 100-fold increase, 30 dB will yield a 1000-fold increase, etc.

Most pros know that turning a level fader up by 3 dB will effectively double its energy output (and vice versa). Beyond this, it's unlikely that anyone will ever ask, "Would you please turn that up a thousand times?" It just won't happen! However, when a pro asks his/her assistant to turn the gain up by 20 dB, that assistant will often instinctively know what 20 dB is...and what it sounds like. I guess I'm saying that the math really isn't nearly as important as getting an instinctive feel for the dB and how it relates to relative levels within audio production.

The Ear

A sound source produces acoustic waves by alternately compressing and rarefying the air molecules between it and the listener, causing fluctuations that fall above and below normal atmospheric pressure. The human ear is a sensitive transducer that responds to these pressure variations by way of a series of related processes that occur within the auditory organs...*our ears*. When these variations arrive at the listener, sound-pressure waves are collected in the aural canal by way of the outer ear's pinna. These are then directed to the eardrum, a stretched drum- like membrane (Figure 2.23), where the sound waves are changed into mechanical vibrations, which are transferred to the inner ear by way of three bones known as the hammer,

Figure 2.23. *Outer, middle, and inner ear.*

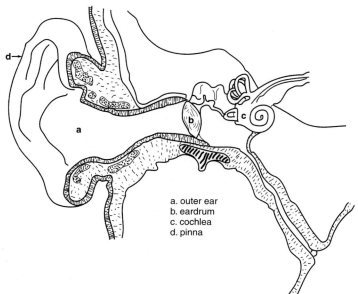

a. outer ear
b. eardrum
c. cochlea
d. pinna

anvil, and stirrup. These bones act both as an amplifier (by significantly increasing the vibrations that are transmitted from the eardrum) and as a limiting protection device (by reducing the level of loud, transient sounds such as thunder or fireworks explosions). The vibrations are then applied to the inner ear (cochlea)—a tubular, snail-like organ that contains two fluid-filled chambers. Within these chambers are tiny hair receptors that are lined in a row along the length of the cochlea. These hairs respond to certain frequencies depending on their placement along the organ which results in the neural stimulation that gives us the sensation of hearing. Permanent hearing loss generally occurs when these hair/nerve combinations are damaged or as they deteriorate with age.

Threshold of Hearing

In the case of SPL, a convenient pressure-level reference is the *threshold of hearing*, which is the minimum sound pressure that produces the phenomenon of hearing in most people and is equal to 0.0002 microbar. One microbar is equal to 1-millionth of normal atmospheric pressure, so it's apparent that the ear is an amazingly sensitive instrument. In fact, if the ear were any more sensitive, the thermal motion of molecules in the air would be audible! When referencing sound-pressure levels to 0.0002 microbar, this threshold level usually is denoted as 0 dB SPL, which is defined as the level that an average person can hear a specific frequency only 50% of the time.

Threshold of Feeling

An SPL that causes discomfort in a listener 50% of the time is called the *threshold of feeling*. It occurs at a level of about 118 dB SPL between the frequencies of 200 Hz and 10 kHz.

Threshold of Pain

The SPL that causes pain in a listener 50% of the time is called the *threshold of pain* and corresponds to an SPL of 140 dB in the frequency range between 200 Hz and 10 kHz.

Taking Care of Your Hearing

During the 1970s and early 80s, recording studio monitoring levels were often turned so high as to be truly painful. In the mid-90s, a small band of powerful producers and record executives banded together to successfully reduce these average volumes down to tolerable levels (85–95 dB)...a practice that continues into the new millennia. Live sound venues and acts often continue the practice of raising house and stage volumes to chest-thumping levels. Although these levels are exciting, long-term exposure can lead to temporary or permanent hearing loss.

Here are a few hearing conservation tips (courtesy of the House Ear Institute, www.hei.org) that can help reduce hearing loss due to long-term exposure of sounds over 115 dB:

- Avoid hazardous sound environments; otherwise, wear hearing protection devices (HPDs), such as foam earplugs, custom-molded earplugs, or in-ear monitors.
- Monitor sound-pressure levels at or around 85 dB. The general rule to follow is if you're in an environment where you must raise your voice to be heard, then you're monitoring too loudly and should limit your exposure times.
- Take 15-minute "quiet breaks" every few hours if you're being exposed to levels above 85 dB.
- Musicians and other live entertainment professionals should avoid practicing at concert-hall levels whenever possible.
- Have your hearing checked by a licensed audiologist.

Psycho-acoustics

The area of *psycho-acoustics* deals with how and why the brain interprets a particular sound stimulus in a certain way. Although a great deal of study has been devoted to this subject, the primary device in psycho-acoustics is the all-elusive brain . . . which is still largely unknown to present-day science.

Auditory Perception

The ear is a nonlinear device (what's received at your ears isn't always what you'll hear); as a result, it produces harmonic distortion that doesn't exist in the original signal whenever it is subjected to sound waves that are above a certain loudness level. For example, the ear can cause a loud 1-kHz sine wave to be perceived as being a combination of 1 kHz, 2 kHz, 3 kHz waves, and so on. Although the ear might hear the overtone structure of a violin (if the listening level is loud enough), it might also perceive additional harmonics (thus changing the timbre of the instrument). This implies that sound monitored at very loud levels could sound quite different when played back at lower levels.

The terms *linear* and *nonlinear* are used to describe the input *versus* output level characteristics of a transducer or device. A linear device or medium is one whose input and output amplitudes have the same input/output ratio at all signal levels. For example, if the input of a linear amplifier were doubled, the output signal amplitude would also double; however, if the doubling of the input signal would result in an increase in the signal's output level by a factor of more or less than two . . . the amplifier would be said to be nonlinear at those amplitudes.

In addition to being nonlinear with respect to perceived tones, the ear's frequency response (that is, its perception of timbre) also changes with the loudness of the perceived signal. The loudness compensation switch found on many hi-fi preamplifiers is an attempt to

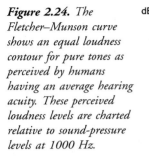

Figure 2.24. The Fletcher–Munson curve shows an equal loudness contour for pure tones as perceived by humans having an average hearing acuity. These perceived loudness levels are charted relative to sound-pressure levels at 1000 Hz.

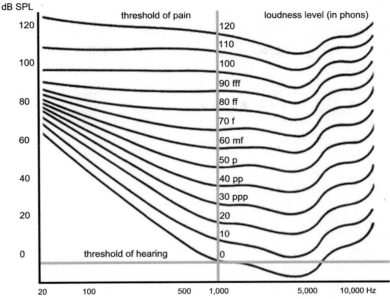

compensate for this decrease in the ear's sensitivity to low- and high-frequency sounds at low listening levels.

Fletcher–Munson equal-loudness contour curves (Figure 2.24) indicate the ear's average sensitivity to different frequencies at various levels. These indicate the sound-pressure levels that are required for our ears to hear frequencies along the curve as being equal in level to a 1000-Hz reference level (phon). Thus, to equal the loudness of a 1-kHz tone at 110 dB SPL (a level typically created by a trumpet-type car horn at a distance of 3 feet), a 40-Hz tone has to be about 6 dB louder, while a 10-kHz tone must be 4 dB greater in order to be perceived as being equally loud. At 50 dB SPL (the noise level present in the average private business office), the level of a 40-Hz tone must be 30 dB greater and a 10-kHz tone 13 dB greater than a 1-kHz tone to be perceived as having the same volume. Thus, if a piece of music is mixed to sound great at a level of 85 to 95 dB, its bass and treble balance will actually be boosted when turned up (often a good thing). If the same piece were mixed at 110 dB SPL, it would sound both bass and treble shy when played at lower levels . . . as no compensation for the ear's response was added to the mix. Over the years, it has generally been found that changes in apparent frequency balance are less apparent when monitoring at levels of 85 dB SPL.

The loudness of a tone can also affect our ear's perception of pitch. For example, if the intensity of a 100-Hz tone is increased from 40 to 100 dB SPL, the ear will hear a pitch decrease of about 10%. At 500 Hz, the pitch will change about 2% for the same increase in sound-pressure level. This is one reason why musicians find it difficult to tune their instruments when listening through loud headphones.

As a result of the nonlinearities in the ear's response, tones will often interact with each other rather than being perceived as being separate. Three types of interaction effects can occur:

- Beats
- Combination tones
- Masking

Beats

Two tones that differ only slightly in frequency and have approximately the same amplitude will produce an effect known as *beats*. This effect sounds like repetitive volume surges that are equal in frequency to the difference between these two tones. The phenomenon is often used as an aid for tuning instruments, because the beats slow down as the two notes approach the same pitch and finally stop when the pitches match. In reality, beats are a result of the ear's inability to separate closely pitched notes. This results in a third frequency that's created from the phase sum and difference values between the two notes.

Do-It-Yourself Tutorial: Beats

1. Go to the "Tutorial" section of www.modrec.com, click on "Ch. 2—Beats Tutorial," and download all of the sound files.
2. Load the 440-Hz file onto track 1 of the digital audio workstation (DAW) of your choice, making sure to place the file at the beginning of the track, with the signal panned center.
3. Load the 445-Hz and 450-Hz files into the next two consecutive tracks.
4. Solo and play the 440-Hz tone.
5. Solo both the 440- and the 445-Hz tones and listen to their combined results. Can you hear the 5-Hz beat tone? (445 Hz−440 Hz = 5 Hz)
6. Solo both the 445- and 450-Hz tones and listen to their combined results. Can you hear the 5-Hz beat tone? (450 Hz−445 Hz = 5 Hz)
7. Now, solo both the 440- and 450-Hz tones and listen to their combined results. Can you hear the 10-Hz beat tone? (450 Hz−440 Hz = 10 Hz)

Combination Tones

Combination tones result when two loud tones differ by more than 50 Hz. In this case, the ear perceives an additional set of tones that's equal to both the sum and the difference between the two original tones…as well as being equal to the sum and difference between their harmonics. The simple formulas for computing the fundamental tones are:

$$\text{Sum tone} = f_1 + f_2$$

$$\text{Difference tone} = f_1 - f_2$$

Difference tones can be easily heard when they are below the frequency of both the tone's fundamentals. For example, the combination of 2000 and 2500 Hz produces a difference tone of 00 Hz.

Masking

Masking is the phenomenon by which loud signals prevent the ear from hearing softer sounds. The greatest masking effect occurs when the frequency of the sound and the frequency of the masking noise are close to each other. For example, a 4-kHz tone will mask a softer 3.5-kHz tone but has little effect on the audibility of a quiet 1000-Hz tone. Masking can also be caused by harmonics of the masking tone (*e.g.*, a 1-kHz tone with a strong 2-kHz harmonic might mask a 1900-Hz tone). This phenomenon is one of the main reasons why stereo placement and equalization are so important to the mixdown process. An instrument that sounds fine by itself can be completely hidden or changed in character by louder instruments that have a similar timbre. Equalization, mic choice, or mic placement might have to be altered to make the instruments sound different enough to overcome any masking effect.

Do-It-Yourself Tutorial: Masking

1. Go to the "Tutorial" section of www.modrec.com, click on "Ch. 2—Masking Tutorial" and download all of the sound files.
2. Load the 1000-Hz file onto track 1 of the digital audio workstation (DAW) of your choice, making sure to place the file at the beginning of the track, with the signal panned center.
3. Load the 3800-Hz and 4000-Hz files into the next two consecutive tracks.
4. Solo and play the 1000-Hz tone.
5. Solo both the 1000- and the 4000-Hz tones and listen to their combined results. Can you hear both of the tones clearly?
6. Solo and play the 3800-Hz tone.
7. Solo both the 3800- and the 4000-Hz tones and listen to their combined results. Can you hear both of the tones clearly?

Perception of Direction

Although one ear can't discern the direction of a sound's origin, two ears can. This capability of two ears to localize a sound source within an acoustic space is called *spatial* or *binaural localization*. This effect is the result of three acoustic cues that are received by the ears:

◆ Interaural intensity differences

◆ Interaural arrival-time differences

◆ The effects of the pinnae (outer ears)

Middle to higher frequency sounds originating from the right side will reach the right ear at a higher intensity level than the left ear, causing an interaural intensity difference. This volume difference occurs because the head casts an acoustic block or shadow, allowing only reflected sounds from surrounding surfaces to reach the opposite ear (Figure 2.25). Because the reflected sound travels farther and loses energy at each reflection—in our example the intensity of sound perceived by the left ear will be greatly reduced, resulting in a signal that's perceived as originating from the right.

This effect is relatively insignificant at lower frequencies, where wavelengths are large compared to the head's diameter, allowing the wave to easily bend around its acoustic shadow. For this reason, a different method of localization (known as *interaural arrive-time differences*) is employed at lower frequencies (Figure 2.26). In both Figures 2.25 and 2.26, small time differences occur because the acoustic path length to the left ear is slightly longer than the path to the right ear. The sound pressure therefore arrives at the left ear at a later time than the right. This method of localization (in combination with interaural intensity differences) helps to give us lateral localization cues over the entire frequency spectrum.

Intensity and delay cues allow us to perceive the direction of a sound's origin but not whether the sound originates from front, behind, or below. The pinna (Figure 2.27), however, makes use of two ridges that reflect sound into the ear. These ridges introduce minute time delays between

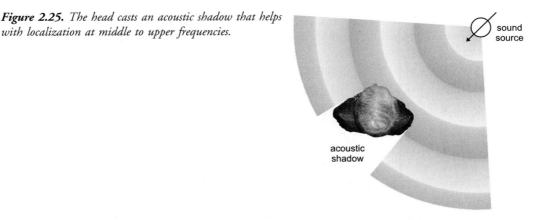

Figure 2.25. *The head casts an acoustic shadow that helps with localization at middle to upper frequencies.*

sound source

acoustic shadow

Figure 2.26. *Interaural arrival-time differences occurring at lower frequencies.*

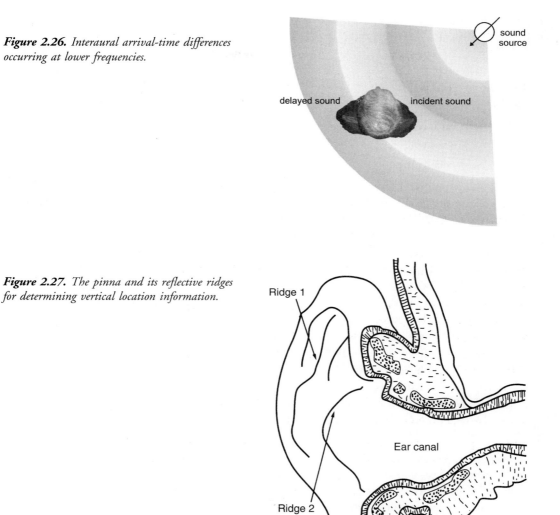

Figure 2.27. *The pinna and its reflective ridges for determining vertical location information.*

the direct sound (which reaches the entrance of the ear canal) and the sound that's reflected from the ridges (which varies according to source location). It's interesting to note that beyond 130° from the front of our face, the pinna is able to reflect and delay sounds by 0 and 80 microseconds (μ sec), making rear localization possible. Ridge number two has been reported to produce delays of between 100 and 330 μsec which help us to locate sources in the vertical plane. The delayed reflections from both ridges are then combined with the direct sound to

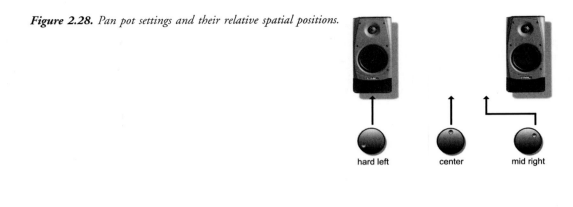

Figure 2.28. Pan pot settings and their relative spatial positions.

produce frequency–response colorations that are compared within the brain to determine source location. Small movements of the head can also provide additional position information.

If there are no differences between what the left and right ears hear, the brain assumes that the source is the same distance from each ear. This phenomenon allows us to position sound not only in the left and right loudspeakers but also monophonically between them. If the same signal is fed to both loudspeakers, the brain perceives the sound identically in both ears and deduces that the source must be originating from directly in the center. By changing the proportion that's sent to each speaker, the engineer changes the relative interaural intensity differences and thus creates the illusion of physical positioning between the speakers. This placement technique is known as *panning* (Figure 2.28).

Perception of Space

In addition to perceiving the direction of sound, the ear and brain combine to help us perceive the size and physical characteristics of the acoustic space in which a sound occurs. When a sound is generated, a percentage reaches the listener directly, without encountering any obstacles. A larger portion, however, is propagated to the many surfaces of an acoustic enclosure. If these surfaces are reflective, the sound is bounced back into the room and toward the listener. If the surfaces are absorptive, less energy will be reflected back to the listener. Three types of reflections are commonly generated within an enclosed space (Figure 2.29):

◆ Direct sound
◆ Early reflections
◆ Reverberation

Direct Sound

In air, sound travels at a constant speed of about 1130 feet per second, so a wave that travels from the source to the listener will follow the shortest path and arrive at the listener's ear first.

Figure 2.29. *The three soundfield types that are generated within an enclosed space.*

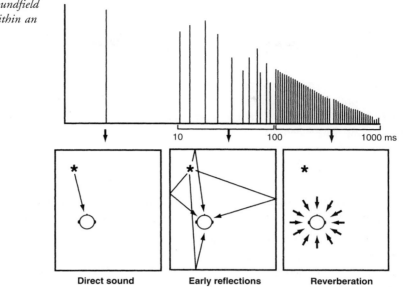

This is called the *direct sound*. Direct sounds determine our perception of a sound source's location and size and convey the true timbre of the source.

Early Reflections

Waves that bounce off of surrounding surfaces in a room must travel further to reach the listener and therefore arrive after the direct sound and from a multitude of directions. These waves form what are called *early reflections*. Early reflections give us clues as to the reflectivity, size, and general nature of an acoustic space. These sounds generally arrive at the ears less than 50 msec after the brain perceives the direct sound and are the result of reflections off of the largest, most prominent boundaries within a room. The time elapsed between hearing the direct sound and the beginning of the early reflections helps to provide information about the size of the performance room. Basically, the farther the boundaries are from the source and listener, the longer the delay before it's reflected back to the listener.

Another aspect that occurs with early reflections is called *temporal fusion*. Early reflections arriving at the listener within 30 msec of the direct sound are not only audibly suppressed, but are also fused with the direct sound. In effect, the ear can't distinguish the closely occurring reflections and considers them to be part of the direct sound. The 30-msec time limit for temporal fusion isn't absolute; rather, it depends on the sound's envelope. Fusion breaks down at 4 msec for transient clicks, whereas it can extend beyond 80 msec for slowly evolving sounds (such as a sustained organ note or legato violin passage). Despite the fact that the early reflections are suppressed and fused with the direct sound, they still modify our perception of the sound, making it both louder and fuller.

Reverberation

Whenever room reflections continue to bounce off of room boundaries, a randomly decaying set of sounds can often be heard after the source stops in the form of *reverberation*. A highly reflective surface absorbs less of the wave energy at each reflection and allows the sound to persist longer after the initial sound stops (and *vice versa*). Sounds reaching the listener 50 msec later in time are perceived as a random and continuous stream of reflections that arrive from all directions. These densely spaced reflections gradually decrease in amplitude and add a sense of warmth and body to a sound. Because it has undergone multiple reflections, the timbre of the reverberation is often quite different from the direct sound (with the most notable difference being a rolloff of high frequencies and a slight bass emphasis).

The time it takes for a reverberant sound to decrease to 60 dB below its original level is called its *decay time* or *reverb time* and is determined by the room's absorption characteristics. The brain is able to perceive the reverb time and timbre of the reverberation and uses this information to form an opinion on the hardness or softness of the surrounding surfaces. The loudness of the perceived direct sound increases rapidly as the listener moves closer to the source, while the reverberation levels will often remain the same, because the diffusion is roughly constant throughout the room. This ratio of the direct sound's loudness to the reflected sound's level helps the listener judge their distance from the sound source.

Whenever artificial reverb and delay units are used, the engineer can generate the necessary cues to convince the brain that a sound was recorded in a huge, stone-walled cathedral . . . when, in fact, it was recorded in a small, absorptive room. To do this, the engineer programs the device to mix the original unreverberated signal with the necessary early delays and random reflections. Adjusting the number and amount of delays on an effects processor gives the engineer control over all of the necessary parameters to determine the perceived room size, while decay time and frequency balance can help to determine the room's perceived surfaces. By changing the proportional mix of direct-to-processed sound, the engineer/producer can place the sound source at either the front or rear of the artificially created space.

Doubling

By repeating a signal using a short delay of 4 to 20 msec (or so), the brain can be fooled into thinking that the apparent number of instruments being played is doubled. This process is called *doubling*. Often, acoustic doubling and tripling can be physically recreated during the overdub phase by recording a track and then going back and laying down one or more passes while the musicians listen to the original track. When this isn't possible, delay devices can be cost-effectively and easily used to simulate this effect. If a longer delay is chosen (more than about 35 msec), the repeat will be heard as discrete echoes . . . causing the delay (or series of repeated delays) to create a *slap echo* or *slap back*. This and other effects can be used to double or thicken up a sound . . . anybody wanna sound like Elvis or Annette Funicello?

CHAPTER 3

Studio Acoustics and Design

The *Audio Cyclopedia* defines the term *acoustics* as "a science dealing with the production, effects, and transmission of sound waves; the transmission of sound waves through various mediums, including reflection, refraction, diffraction, absorption, and interference; the characteristics of auditoriums, theaters, and studios, as well as their design." We can see from this description that the proper acoustic design of a music recording, project, and audio-for-visual or broadcast studio is no simple matter. A wide range of complex variables and interrelationships often come into play in the making of a successful acoustic and monitoring design. When designing or redesigning an acoustic space, the following basic requirements should be considered:

◆ *Acoustic isolation*—This prevents external noises from transmitting into the studio environment through the air, ground, or building structure. It can also prevent feuds that can arise when excessive volume levels leak out into the surrounding neighborhood.

◆ *Frequency balance*—The frequency components of a room shouldn't adversely affect the acoustic balance of instruments and/or speakers. Simply stated, the acoustic environment shouldn't alter the sound quality of the original or recorded performance.

◆ *Acoustic separation*—The acoustic environment should not interfere with intelligibility and should offer the highest possible degree of acoustic separation within the room (often a requirement for ensuring that sounds from one instrument aren't unduly picked up by another instrument's microphone).

◆ *Reverberation*—The control of sonic reflections within a space is an important factor for maximizing the intelligibility of music and speech. No matter how short the early reflections and reverb times are, they will add an important psycho-acoustic sense of "space" in the sense that they can give our brain subconscious cues as to a room's size, number of reflective boundaries, distance between the source and listener, and so on.

◆ *Cost factors*—Not the least of all design and construction factors is cost. Multimillion-dollar facilities often employ studio designers and construction teams to create a plush decor that has been acoustically tuned to fit the needs of both the owners and their clients. Owners of project studios and budget-minded production facilities, however, can also take full advantage of the same basic acoustic principles and construction techniques and apply them in cost-effective ways.

This chapter discusses many of the basic acoustic principles and construction techniques that should be considered in the design of a music or sound production facility. I'd like to emphasize that any or all of these acoustical topics can be applied to any type of audio production facility and aren't only limited to professional music studio designs. For example, owners of modest project and bedroom studios should know the importance of designing a control room that is symmetrical. It doesn't cost anything to know that if one speaker is in a corner and the other is on a wall, the perceived center image will be off balance. As with many techno-artistic endeavors, studio acoustics and design are a mixture of fundamental physics (in this case, mostly dimensional mathematics) and an equally large dose of common sense and luck. More often than not, acoustics is an artistic science that melds science with the art of intuition and experience.

Studio Types

Although the acoustical fundamentals are the same for most studio design types, differences will often follow the form, function, and budgets of the required tasks at hand. Some of the more common studio types include:

◆ Professional music studios

◆ Audio-for-visual production environments

◆ Project studios

◆ Portable studios

The Professional Music Studio

The *professional recording studio* (Figures 3.1 and 3.2) is first and foremost a commercial business, so its design, decor and acoustical construction requirements are often much more demanding than those of a privately owned project studio. In some cases, an acoustical designer

Figure 3.1. BiCoastal Music, Ossining, NY: (a) control room; (b) recording studio. (Courtesy of Russ Berger Design Group, Inc., www.rbdg.com.)

a

b

and experienced construction team are placed in charge of the overall building phase of a professional facility. In others, the studio's budget precludes the hiring of such professionals, which places the studio owners and staff squarely in charge of designing and constructing the entire facility. Whether you happen to have the luxury of building a new facility from the ground up or are renovating a studio within an existing shell, you'd probably benefit from a professional studio designer's experience and skills. Such expert advice often proves to be cost effective in the long run, as errors in design judgment can lead to cost overruns, lost business due to unexpected delays, or the unfortunate state of living with mistakes that could have been easily avoided.

The Audio-for-Visual Production Environment

An audio-for-visual production facility is used for video, film, and game postproduction (often simply called "post") and includes such facets as music recording for film or other media (scoring); score mixdown; automatic dialog replacement (ADR . . . which is the replacement of

Figure 3.2. Turtle Recording Studio:
(a) control room; (b) recording studio.
(Courtesy of Turtle Recording Studio,
www.turtlerecording.com.)

a

b

on- and off-screen dialog to visual media); and Foley (the replacement and creation of on- and off-screen sound effects). As with music studios, audio-for-visual production facilities can range from high-end facilities that can accommodate the posting needs of network video or feature film productions to a simple, budget-minded project studio that's equipped with video and a digital audio workstation. As with the music studio, audio-for-visual construction and design techniques often span a wide range of styles and scope in order to fit the budget needs at hand (Figure 3.3).

The Project Studio

It goes without saying that the vast majority of audio production studios fall into the project studio category. This basic definition of such a facility is open to interpretation. It's

Figure 3.3. *Lansdowne Studios, London. (Courtesy of Recording Architecture, www.aaa-design.com.)*

Figure 3.4. *Example of a bedroom project studio. (Courtesy of Primeacoustic Studio Acoustics, www.primacoustic.com.)*

usually intended as a personal production resource for the recording music, audio-for-visual production, multimedia production, voice-overs...you name it. Project studios can range from being fully commercial in nature to smaller setups that are both personal and private (Figure 3.4). All of these possible studio types have been designed with the idea of giving artists the flexibility of making their art in a personal, off-the-clock environment that's both cost and time effective. The design and construction considerations for creating a privately owned project studio often differ from the design considerations for a professional music facility in two fundamental ways:

- Building constraints
- Cost

Generally, a project studio's room (or series of rooms) is built into an artist's home or a rented space where the construction and dimensional details are already defined. This fact (combined

Figure 3.5. *The "Rocket Ship" workstation furniture system. (Courtesy of Argosy Console, Inc., www.argosyconsole.com.)*

with inherent cost considerations) often leads the owner/artist to employ cost-effective techniques for sonically treating a room. Even if the room has little or no treatment (or if it's not deemed necessary), keep in mind that a basic knowledge of acoustical physics and room design can be a handy and cost-effective tool as your experience, production needs, and (you hope) your business grow.

Modern-day digital audio workstations (DAWs) have squarely placed the Mac and PC within the ergonomics and functionality of the project studio (Figure 3.5) . . . in many cases, the DAW *is* the project studio. With the advent of self-powered speaker monitors, cost-effective microphones, and hardware DAW controllers, it's a relatively simple matter to design a powerful production system into any existing space. With regard to setting up this or any production/monitoring environment, I'd like to draw your attention to the need for symmetry in the monitoring environment. A symmetrical acoustic environment about the central mixing axis can work wonders toward creating a balanced left/right and surround image. Most importantly, with care, this often isn't a difficult goal to achieve. An acoustical and speaker placement environment that isn't balanced between the left- and right-hand sides allows for differing reflections, absorption coefficients, and variations in frequency response that can adversely effect the imaging and balance of your final mix. Further information on this important subject can be found later in this chapter.

The Portable Studio

The laptop and "porta" recording systems have decreased in size and cost while increasing in power to the point where they've literally become a powerful studio a-go-go. Again, with the advent of self-powered speaker monitors, cost-effective microphones, and hardware DAW interface/controller devices, these small-fry systems offer up tremendous amounts of power while being light in the backpack and on the pocketbook. When a laptop is placed at your main working desk, it's wise to acoustically tune your environment as best you can, making sure to follow the above-mentioned symmetry rules. When you're on the go with self-powered monitors, simply do your best to adjust to your current monitoring conditions Those of you with your favorite headphones can create a monitoring environment which excludes any problems that might be added by the room's acoustics (although you might not be able to judge how the recording will sound later through speakers). As always, when recording and monitoring environments change, it's up to you to choose the best tools, toys, and techniques for the given situation. Of course, further discussions on portable and desktop digital audio can be found in Chapter 6.

Primary Factors Governing Studio and Control Room Acoustics

Regardless of which type of studio facility is being designed, built and used...there are a number of primary concerns that should be addressed in order to achieve the best possible acoustic results. In this section, we'll take a close look at such important and relevant aspects of acoustics as:

◆ Acoustic isolation
◆ Symmetry in control room and monitoring design
◆ Frequency balance
◆ Absorption
◆ Reflection
◆ Reverberation

In addition to these factors, the following considerations should also be taken into account:

◆ Equipment noise in the control room
◆ Power conditioning in the studio and control room
◆ Proper grounding techniques in the studio and control room

Although several mathematical formulas have been included in the following sections, it's by no means necessary that you memorize or worry about them. By far, I feel that it's more important that you grasp the basic principles of acoustics rather than fret over the underlying math. Remember—more often than not, acoustics is an artistic science that blends math with the art of intuition and experience.

Acoustic Isolation

Because most commercial and project studio environments make use of an acoustic space to record sound, it's often wise and necessary to employ effective isolation techniques into their design in order to keep external noises to a minimum. Whether that noise is transmitted through the medium of air (*e.g.*, from nearby auto, train, or jet traffic) or through solids (*e.g.*, from air-conditioner rumble, underground subways, or nearby businesses), special construction techniques will often be required to dampen down these extraneous sounds (Figure 3.6).

Figure 3.6. *Various isolation and acoustical treatments for a recording/monitoring environment. (Courtesy of Auralex Acoustics, www.auralex.com.)*

Figure 3.7. *Transmission loss refers to the reduction of a sound signal (in dB) as it passes through an acoustic barrier.*

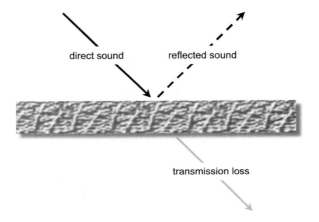

direct sound   reflected sound

transmission loss

If you happen to have the luxury of building a studio facility from the ground up, a great deal of thought should be put into selecting the studio's location. If a location has considerable neighborhood noise, you might have to resort to extensive (and expensive) construction techniques that can "float" the rooms (a process that effectively isolates and uncouples the inner rooms from the building's outer foundations). If there's absolutely no choice of studio location and the studio happens to be located next to a factory, just under the airport's main landing path, or over the subway's uptown line . . . you'll simply have to give in to destiny and build acoustical barriers to these outside interferences.

The reduction in the sound-pressure level (SPL) of a sound source as it passes through an acoustic barrier of a certain physical mass (Figure 3.7) is termed the *transmission loss* (TL) of a signal. This attenuation can be expressed (in dB) as:

$$TL = 14.5 \log M + 23$$

where *TL* is the transmission loss in decibels, and *M* is the surface density (or combined surface densities) of a barrier in pounds per square foot (lb/ft^2).

Table 3.1. Surface densities of common building materials.

Material	Thickness (inches)	Surface Density (lb/ft²)
Brick	4	40.0
	8	80.0
Concrete (lightweight)	4	33.0
	12	100.0
Concrete (dense)	4	50.0
	12	150.0
Glass	—	3.8
	—	7.5
	—	11.3
Gypsum wallboard	—	2.1
	—	2.6
Lead	1/16	3.6
Particleboard	—	1.7
Plywood	—	2.3
Sand	1	8.1
	4	32.3
Steel	—	10.0
Wood	1	2.4

Because transmission loss is frequency dependent, the following equation can be used to calculate transmission loss at various frequencies with some degree of accuracy:

$$TL = 14.5 \log Mf - 16$$

where f is the frequency (in hertz).

Both common sense and the preceding two equations tell us that heavier acoustic barriers will yield a higher transmission loss. For example, Table 3.1 tells us that a 12-inch-thick wall of dense concrete (yielding a surface density of 150 lb/ft²) offers a much greater resistance to the transmission of sound than can a 4-inch cavity filled with sand (which yields a surface density of 32.3 lb/ft²). From the second equation ($TL = 14.5 \log Mf - 16$), we can also draw the conclusion that, for a given acoustic barrier, transmission losses will increase as the frequency rises. This can be easily proven by closing the door of a car that has its sound system turned up or by shutting a single door to a music studio's control room. In both instances, the high frequencies will be greatly reduced in level, while the bass frequencies will be impeded to a much lesser extent. From this, the goal would seem to be to build a studio wall, floor, ceiling, window, or door out of the thickest and most dense material that we can get; however, expense and physical space often play roles in determining just how much of a barrier can be built to achieve the desired isolation. In most cases, a balance must be struck when using both space- and cost-effective building materials.

Walls

When building a studio wall or reinforcing an existing structure, the primary goal is to reduce leakage (increase the transmission loss) through a wall as much as possible over the audible frequency range. This is generally done by:

◆ Building a wall structure that is as massive as is practically possible (both in terms of cubic and square foot density)

◆ Eliminating open joints that can easily transmit sound through the barrier

◆ Dampening structures, so that they are well supported by reinforcement structures and are free of resonances

The following guidelines can be beneficial in the construction of framed walls that have high transmission losses:

◆ If at all possible, the inner and outer wallboards should not be directly attached to the same stud. The best way to avoid this is to alternately stagger the studs on the frame, so that the front/back facing walls aren't in physical contact with each other; see Figure 3.8.

◆ Each wall facing should have a different density to reduce the likelihood of increased transmission due to resonant frequencies that might be sympathetic to both sides. For example, one wall might be constructed of two 5/8-inch gypsum wallboards, while the other wall might be composed of soft fiberboard that's surfaced with two 1/2-inch gypsum wallboards.

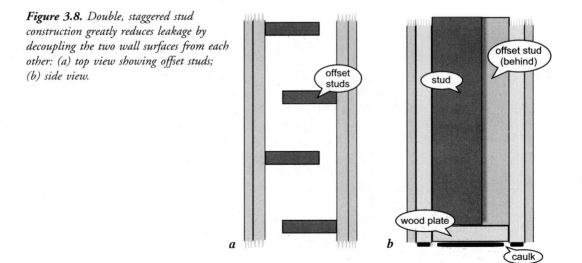

Figure 3.8. *Double, staggered stud construction greatly reduces leakage by decoupling the two wall surfaces from each other: (a) top view showing offset studs; (b) side view.*

Figure 3.9. *Single-stud wallboard construction showing the application of caulk at all joint points to reduce leakage.*

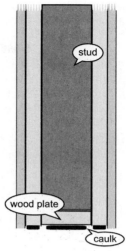

◆ If you're going to attach gypsum wallboards to a single wall face, you can increase transmission loss by mounting the additional layers (not the first layer) with adhesive caulking rather than using screws or nails.

◆ Spacing the studs 24 inches on center instead of using the traditional 16-inch spacing yields a slight increase in transmission loss.

◆ To reduce leakage that might make it through the cracks, apply a bead of nonhardening caulk sealant to the inner gypsum wallboard layer at the wall-to-floor, wall-to-ceiling, and corner junctions (Figure 3.9).

Generally, the same amount of isolation is required between the studio and the control room as is required between the studio's interior and exterior environments. The proper building of this wall is important, so that an accurate tonal balance can be heard over the control-room monitors without promoting leakage between the rooms or producing resonances within the wall that would audibly color the signal. Optionally, a *soffit* can be designed into the wall between the control room and studio (Figure 3.10). This super-structure allows the main, farfield studio monitors to be mounted directly into the control room/studio wall.

It's important that the soffit is constructed to high standards, using a multiple-wall or high-mass design that maximizes density with acoustically tight construction techniques in order to reduce leakage between the two rooms. Cutting corners by using substandard (and even standard) construction techniques in the building of a studio soffit can lead to unfortunate side effects, such as resonances, rattles, and leakage. Typical wall construction materials include:

◆ Concrete—This is the best and most solid material, but, besides the fact that it is often expensive, it's not always possible to pour cement into an existing design.

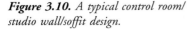

Figure 3.10. *A typical control room/ studio wall/soffit design.*

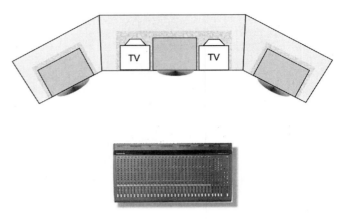

◆ Bricks (hollow-form or solid-facing)—This excellent material is often easier to place into an existing room than concrete.

◆ Gypsum plasterboard—Building multiple layers of plasterboard onto a double-walled stud frame is often the most cost- and design-efficient approach to keeping resonances and transmission losses to a minimum. It's often a good idea to reduce these resonances by filling the wall cavities with rockwool or fiberglass, while bracing the internal structure to add an extra degree of stiffness. Once built, the surfaces can be decorated with any number of materials (such as wood or paint).

Studio monitors can be designed into the soffit in a number of ways. In one expensive approach, the speakers' inner enclosures are cavities designed into walls that are made from a single concrete pour. Under these conditions, resonances are completely eliminated. Another less expensive approach has the studio monitors resting on poured concrete pedestals; in this situation, inserts can be cast into the pedestals that can accept threaded rebar rods (known as all-thread). By filing the rods to a chamfer or a sharper point, it's possible to adjust the position, slant, and height of the monitors for final positioning into the soffit's wall framing (Figure 3.11). The most common approach uses traditional wood framing in order to design a cavity into which the speaker enclosures can be designed and positioned. Extra bracing and heavy construction should be used to reduce resonances.

Floors

For many recording facilities, the isolation of floor-borne noises from room and building exteriors is an important consideration. For example, a building that's located on a busy street and whose concrete floor is tied to the building's ground foundation might experience severe low-frequency rumble from nearby traffic. Alternatively, a second-floor facility might experience undue leakage from a noisy downstairs neighbor or, more likely, might interfere with a quieter neighbor's business. In each of these situations, increasing the isolation to reduce

Figure 3.11. *Threaded rebar can be used to mount speakers and adjust their position.*

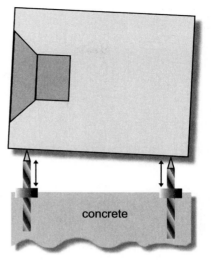

concrete

floor-borne leakage and/or transmission is essential. One of the most common ways to isolate floor-related noise is to construct a "floating" floor that is structurally decoupled from its subfloor foundation.

Common construction methods for floating a professional facility's floor uses either neoprene "hockey puck" isolation mounts (Figure 3.12a) or a continuous underlay, such as a rubberized floor mat (Figure 3.12b). In these cases, the underlay is spread over the existing floor foundation and is then covered with 1/2-inch plywood followed by a layer of plastic sheeting (for insulation and to prevent seepage). This super-structure is covered with reinforcing wire mesh and finally topped with a 4-inch layer of concrete. The isolated floor is then ready for carpeting, wood finishing, painting, or any other desired surface. A less-expensive approach for floating a floor involves placing isolation mounts on top of the framing of a standard floor structure (Figure 3.13). The flooring can then be floated over this flexible structure and covered with the desired surface material.

A more cost- and space-effective way to decouple a floor involves layering the original floor with a carpet foam pad. A 1/2- or 5/8-inch layer of tongue-and-groove plywood or oriented strand board (OSB) is then laid on top of the pad. These should not be nailed to the subfloor; instead, they can be stabilized by glue or by locking the pieces together with thin, metal braces. Another foam pad can then be laid over this structure and topped with carpeting or any other desired finishing material (Figure 3.14).

It's important that the floating super-structure be isolated from both the underflooring and the outer wall. Failing to isolate these allows floor-borne sounds to transmit through the walls to the subfloor... and *vice versa* (often defeating the whole purpose of floating the floor). This wall perimeter isolation can be made filling the isolation gaps with pliable decoupling materials such as widths of soft mineral fiberboard, neoprene, strips of rubber, or other pliable materials.

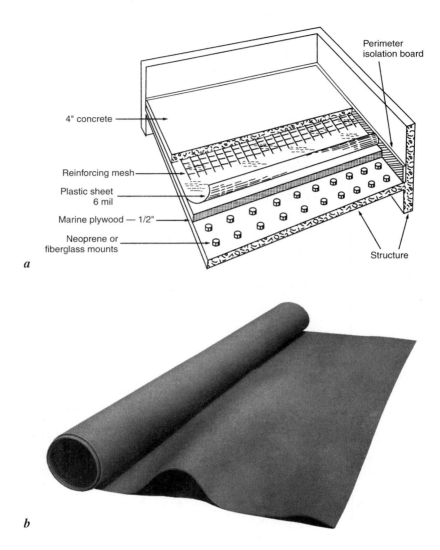

Figure 3.12. *Basic guidelines for building a floating floor: (a) using neoprene mounts; (b) roll-out matting can be substituted as an underlay. (Courtesy of Auralex Acoustics, www.auralex.com.)*

Risers

As we saw from the equation $TL = 14.5 \log Mf - 16$, low-frequency sound travels through barriers much more easily than does high-frequency sound. It stands to reason that strong, low-frequency energy is transmitted more easily than are highs between studio rooms, from the studio to the control room, or to outside locations. In general, the drum set is most likely to be the biggest leakage offender. By decoupling much of a drum set's low-frequency energy from a studio floor, many of the low-frequency leakage problems can be reduced. In

Figure 3.13. *U-Boats*™ *floor beam float channels can be placed below a standard 2 × 4-inch floor frame to increase isolation. (Courtesy of Auralex Acoustics, www.auralex.com.)*

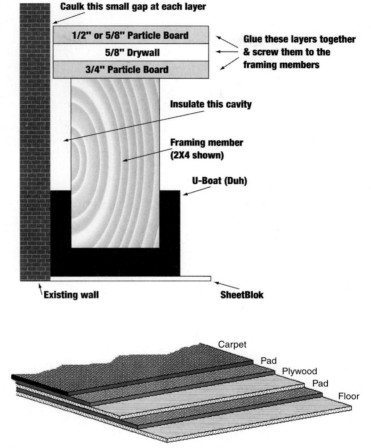

Caulk this small gap at each layer

1/2" or 5/8" Particle Board

5/8" Drywall

3/4" Particle Board

Glue these layers together & screw them to the framing members

Insulate this cavity

Framing member (2X4 shown)

U-Boat (Duh)

Existing wall

SheetBlok

Figure 3.14. *An alternative, cost-effective way to float an existing floor.*

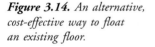

Carpet

Pad

Plywood

Pad

Floor

most cases, the problem can be fixed by using a drum riser. Drum risers are available commercially (Figure 3.15), or they can be easily constructed. In order to reduce unwanted resonances, drum risers should be constructed using 2 × 6-inch or 2 × 8-inch beams for both the frame and the supporting joists (spaced at 16 or 12 inches on center, as shown in Figure 3.16). Sturdy 1/2- or 5/8-inch tongue-and-groove plywood panels should be glued to the supporting frames with carpenter's glue (or a similar wood glue) and then nailed or screwed down (using heavy-duty, galvanizedfasteners). When the frame has dried, rubber coaster float channels or (at the very least) strips of carpeting should be attached to the bottom of the frame . . . and the riser will be ready for action.

Ceilings

Foot traffic and other noises from above a sound studio or production room are another common source of external leakage. Ceiling noise can be isolated in a number of ways. If foot

Figure 3.15. HoverDeck^{TM} 88 isolation riser. (Courtesy of Auralex Acoustics, www.auralex.com.)

Figure 3.16. General construction details for a homemade drum riser.

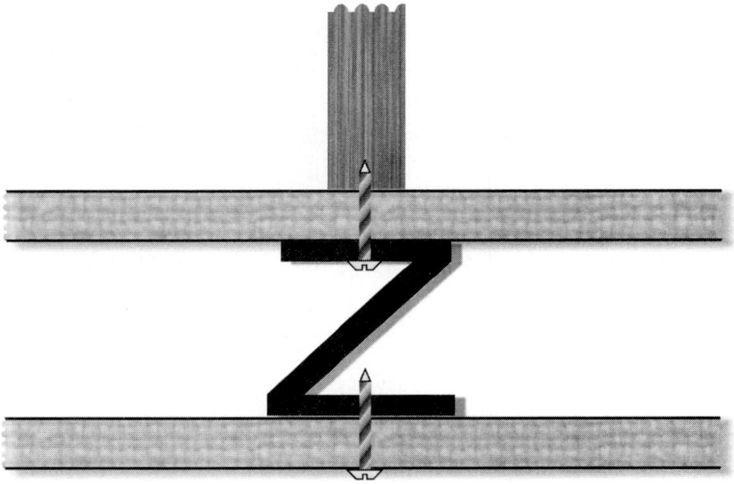

Figure 3.17. "Z" channels can be used to hang a floating ceiling from an existing overhead structure.

traffic is your problem and you're fortunate enough to own the floors above you, you can reduce this noise by simply carpeting the overhead hallway or by floating the upper floor. If you don't have this luxury, one approach to deadening ceiling-borne sounds is to hang a false structure from the existing ceiling or from the overhead joists (as is often done when a new room is being constructed). This technique can be fairly cost effective when "Z" suspension channels are used (Figure 3.17). "Z" channels are often screwed to the ceiling joists to provide a flexible yet strong support to which a hanging wallboard ceiling can be attached. If necessary, fiberglass or other sound-deadening materials can be placed into the cavities between the overhead structures. Other more expensive methods use spring support systems to hang false ceilings from an existing structure.

Windows and Doors

Access to and from a studio or production room area (in the form of windows and doors) can also be a potential source of sound leakage. For this reason, strict attention needs to be given to window and door design and construction. Visibility in a studio is extremely important within a music production environment. For example, when multiple rooms are involved, good visibility serves to promote effective communication between the producer or engineer and the studio musician (as well as among the musicians themselves). For this reason, windows have been an important factor in studio design since the beginning. The design and construction details for a window often vary with studio needs and budget requirements and can range from being deep, double-plate cavities that are built into double-wall constructions (Figure 3.18a) to more modest constructions that are built into a single wall (Figure 3.18b). Other designs include floor-to-ceiling windows that create a virtual "glass wall" . . . as well as those that offer sweeping vistas that have been designed into soffits constructed from poured concrete.

The glass panels used in window construction typically are 3/8 to 3/4 inch thick and are seated into the window frame with a rubber or similar type of elastic damping seal to prevent

Figure 3.18. Details for practical window construction between the control room and studio: (a) window suitable for a high-transmission-loss wall; (b) more modestly framed window.

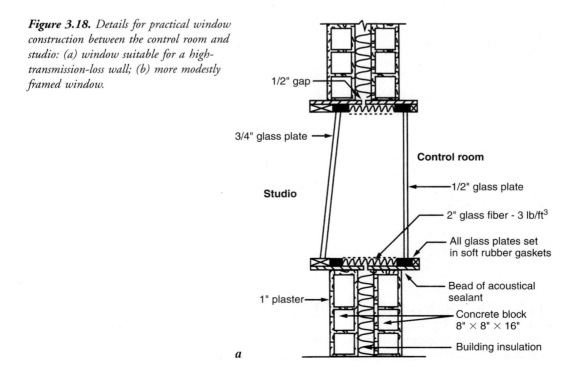

Figure 3.18. Continued.

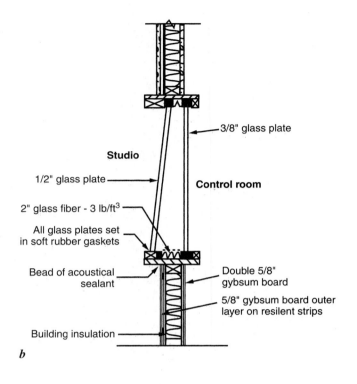

3/8" glass plate

Studio

1/2" glass plate

Control room

2" glass fiber - 3 lb/ft^3

All glass plates set
in soft rubber gaskets

Bead of acoustical
sealant

Double 5/8"
gybsum board

5/8" gybsum board outer
layer on resilent strips

Building insulation

b

structure-borne oscillations. It's important that at least one of these window panels be tilted at a 5° or greater angle with respect to the other, in order to eliminate standing waves within the sandwiched airspace. The existence of standing waves (which are discussed later in the "Frequency Balance" section) often breaks down the transmission loss at specific frequencies. As happens in wall construction, using glass panels that have varying thickness will reduce sympathetic vibrations. Other windows within the facility (such as observation and tape machine room windows) should be designed with similar isolation considerations in mind.

Access doors to and from the studio, control room, and exterior areas should be constructed of solid wood or high-quality acoustical materials (Figure 3.19), as solid doors generally offer higher TL values than their cheaper, hollow counterparts. No matter which door type is used, the appropriate seals, weatherstripping, and doorjambs should be used throughout so as to reduce leakage through the cracks. Whenever possible, double-door designs should be used to form an acoustical *sound lock* (Figure 3.20). This construction technique dramatically reduces leakage as trapped air between two solid barriers offers up high TL values.

Iso-Rooms and Iso-Booths

Isolation rooms (*iso-rooms*) are acoustically isolated or sealed areas that are built into a music studio or just off of a control room (Figures 3.21). These recording areas can be used to

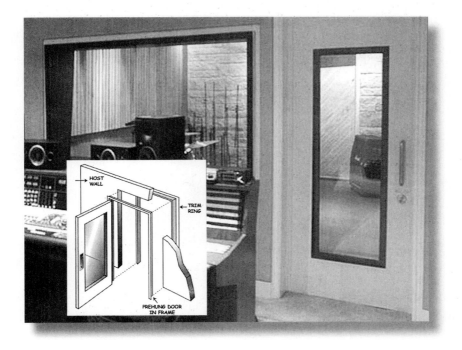

Figure 3.19. A SoundSecure™ studio door. (Courtesy of Acoustic Systems, www.acousticsystems.com.)

Figure 3.20. Example of a sound lock design.

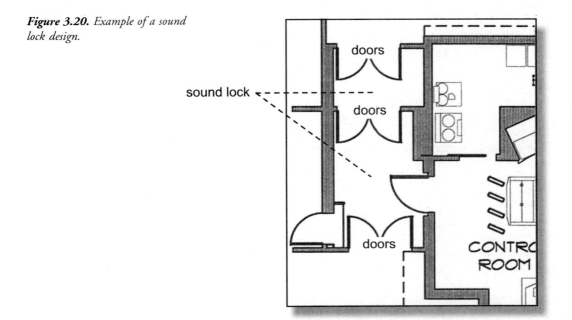

Figure 3.21. Iso-room design located to the side of the control room at Four Seasons Media Productions; St. Louis, MO. (Courtesy of Russ Berger Design Group, Inc., www.rbdg.com.)

separate louder instruments from softer ones (and *vice versa*), in order to reduce leakage and to separate instrument types by volume to maintain control over the overall ensemble balance. For example:

◆ In order to eliminate leakage when recording scratch vocals (a guide vocal track that's laid down as a session reference), a vocalist might be placed in a small room while the rhythm ensemble is placed in the larger studio area.

◆ A piano or other instrument could be isolated from the larger area that's housing a full string ensemble.

◆ A B3 organ could be blaring away in an iso-room while backing vocals are being laid down in the main room.

◆ The possibilities are endless . . . and up to you.

Isolation booths (*iso-booths*) provide the same type of isolation but are often much smaller (Figure 3.22). Often called *vocal booths*, these mini-studios are perfect for isolating vocals and single instruments from the larger studio. In fact, rooms that have been designed and built for the express purpose of mixing down a recording will only have an iso-booth . . . and nothing else. Using this space-saving option, vocals or single instruments can be easily recorded on site, and should more space be needed a larger studio can be booked to fit the bill.

The construction of an iso-room or iso-booth depends on the studio's size, design, and sonic requirements. They can vary in acoustics, thereby offering one or more environments that are more reflective (live) or absorptive (dead) than the main studio area, or they can be specifically designed to better fit a particular instrument's acoustical needs (*e.g.*, drums, piano, or vocals). They can also be designed as totally separate areas that can be accessed from the main studio or control room, or might be directly tied to the main studio by way of sliding walls or glass sliding doors. In short, their form and function will often fit the needs and personality of the studio.

Figure 3.22. *Example of an iso-booth in action. (Courtesy of www.misheeddins.com.)*

Noise Isolation within the Control Room

Isolation between rooms and the great outdoors isn't the only noise-related issue in the modern-day recording or project studio. The proliferation of multitrack tape machines, computers, digital signal processors (DSP) and cooling systems have created issues that present their own Grinch types of noise, Noise, NOISE!, NOISE!!! This usually manifests itself in the form of system fans, transport tape noises, and/or computer-related sounds from CPUs, case fans, hard drives, and the like.

When it comes to isolating tape transport and system fan sounds, should budget and size constraints permit, it is often wise to build an iso-room that's been specifically designed and ventilated for containing such equipment. An equipment room that utilizes glass paneling and easy-access doors that provides for current /future wiring needs can add a degree of peace-n-quiet and an overall professionalism that will make both you and your clients happy.

Within a smaller studio or project studio space, such a room isn't always possible; however, with care and forethought the whizzes and whirrs of the digital era can be turned into a non-issue that you'll be proud of. Here are a few examples of the most common problems and their solutions:

◆ *Replace fans with quieter ones.* By doing some careful Web searching or by talking to your favorite computer salesperson, it's often possible to install fans (especially CPU fans) that are quieter than most off-the-shelf models.

◆ *Regulate case fan speeds with variable pots.* Gamers will often regulate their computer fans in order to keep noises down. Often the controls come in a stylish, LED-lit case that fits in a PC's 5.25-inch drive space. Care needs to be taken in monitoring the CPU/case temperatures so as not to harm your system.

Figure 3.23. *Smart Drive hard drive enclosure. (Courtesy of endpcnoise.com, www.endpcnoise.com.)*

◆ *Install hard drive enclosures to reduce internal hard-drive noises.* These enclosures (Figure 3.23) are designed to acoustically encapsulate 3.5-inch hard drives into a design that fits into a 5.25-inch hard drive bay.

◆ *Place the computer in an acoustically isolated alcove.* Care needs to be taken in monitoring the CPU/case temperatures so as not to harm your system.

◆ *Connect studio computers via high-speed network to a remote server location.* By connecting a silent computer (such as a small form factor PC or Mac laptop) to a central computer via a high-speed network, not only is it possible to eliminate computer-related noises (by locating the computer and drives in another room), but it's also a relatively simple matter to connect the various production- and business-related terminals to a central server that can be backed up according to schedule. More information on networking can be found in Chapter 6.

Partitions

Movable *partitions* (also known as *flats* or *gobos*) are commonly used in studios to provide on-the-spot barriers to sound leakage. By partitioning a musician and/or instrument on one or more sides and then placing the mic inside the temporary enclosure, isolation can be greatly improved in a flexible way that can be easily changed as new situations arise. Acoustic partitions are currently available on the commercial market in various design styles and types for use in a wide range of studio applications (Figure 3.24). For those on a budget or who have particular isolation needs, it's a relatively simple matter to get out the workshop tools and make your own flats that area based around wood frames, fiberglass, or other acoustically absorptive materials with your favorite colored fabric coverings, hinges, craftsmanship . . . and, most of all, ingenuity (Figure 3.25).

If a flat can't be found in sight, acoustic partitions can be easily built using common studio and household items. For example, a simple partition can be easily made on the spot by grabbing a mic/boom stand combination and retracting the boom halfway at a 90° angle to make a "T". Simply drape a blanket or heavy coat over the T-bar and voila—you've built a quick-n-dirty flat.

With all of these partitions, it's important to be aware of the musician's need to have good visibility with other musicians, the conductor, and the producer. Musicality and human connectivity almost always takes precedence over technical issues.

Symmetry in Control Room Design

While many professional studios are often built from the ground up to standard acoustic and architectural guidelines, most budget-minded production and project studios are often limited by their own unique sets of building, space, and acoustic considerations. Even though the

Figure 3.24. Acoustic partition flat examples: (a) Guitar amp setup (courtesy of ClearSonic Mfg., Inc., www.clearsonic.com); (b) ClearSonic IsoPac A Drum Shields with AX12 Height Extenders (courtesy of ClearSonic Mfg., Inc., www.clearsonic.com); (c) piano panel setup (courtesy of Auralex Acoustics, www.auralex.com); (d) setup within the control room (courtesy of Auralex Acoustics, www.auralex.com); (e) S5-2L "Sorber" baffle system (courtesy of ClearSonic Mfg., Inc., www.clearsonic.com).

a

b

Figure 3.24.
Continued.

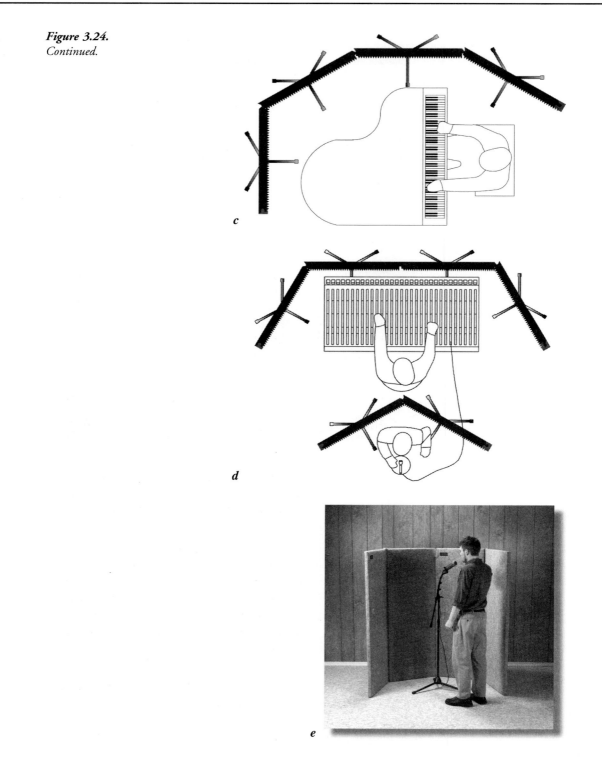

c

d

e

Figure 3.25. *Examples of a homemade flat.*

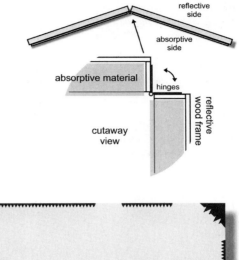

Figure 3.26. *Orienting a control room along the long dimension\can extend its low-end response.*

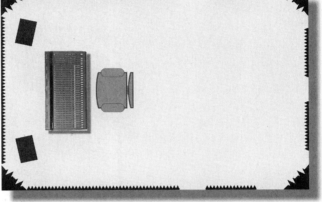

design of a budget, project, or bedroom control room might not be acoustically perfect, if the monitoring environment is to be centered around speakers (and not just headphones), certain ground rules of acoustical physics must be followed in order to create a proper working environment.

Of these ground rules, the one that's arguably more important than control over the room's frequency response is the need for symmetrical reflections on all axes within the design of a control room or single-room project studio. In short, the center and acoustic *imaging* (ability to discriminate placement and balance in a stereo or surround field) is best when the listener, speakers, walls, and other acoustical boundaries are symmetrically centered about the listener's position (often in an equilateral triangle). In a rectangular room, the best low-end response can be obtained by orienting the console and loudspeakers into the room's long dimension (Figure 3.26).

Should any primary boundaries of a control room (especially wall or ceiling boundaries near the mixing position) be asymmetrical from side to side, sounds heard by one ear will receive one combination of direct and reflected sounds, while the other ear will hear a different acoustic balance (Figure 3.27). This condition alters the sound's center image characteristics,

Figure 3.27. *Various symmetries in a monitoring environment. (a) Placing the monitoring environment off-center and in a corner will affect the audible center image, and placing one speaker in a 90° corner can cause an off-center bass buildup and adversely affect the mix's imagery. (b) Shifting the listener/monitoring position into the center will greatly improve the left/right imagery. (c) Centering the listener/ monitoring position at a 45° angle within a symmetrical corner is another example of how the left/right imagery can be improved over the first example.*

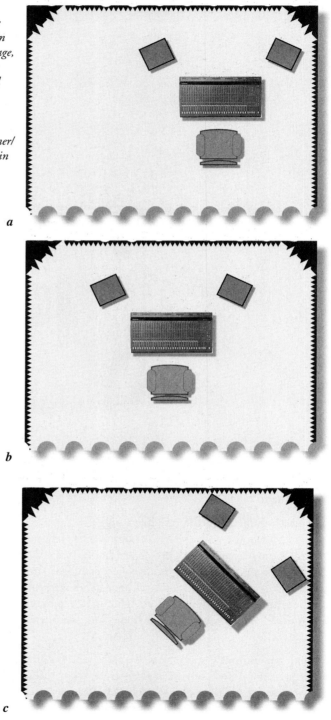

a

b

c

so when a sound is actually panned between the two monitor speakers the sound will appear to be centered; however, when the sound is heard in another studio or standard listening environment the imaging may be off center. In order to avoid this problem, care should be take to ensure that both the side and ceiling boundaries are symmetrical with respect to each other. The splayed sidewall (Figure 3.28) is one example of a symmetrical construction pattern that could reduce acoustic reflections which might cause acoustic cancellations at the listener's position. Likewise, large ceiling reflections can also interfere with the direct acoustic path (Figure 3.29).

While we're on the subject of the relationship between the room's acoustic layout and speaker placement, it's generally wise to place nearfield and all other speaker enclosures at points that are equidistant to the listener in the stereo field. Whenever possible, speaker enclosures should be placed 1 to 2 feet away from the nearest wall and/or corner. This helps avoid bass buildups that acoustically occur at boundary and corner locations.

In addition to strategic speaker placement, homemade or commercially available isolation pads (Figure 3.30) can be used to reduce resonances that often occur whenever enclosures are placed directly onto a table or flat surface.

Frequency Balance

Another important factor in room design is the need for maintaining the original *frequency balance* of an acoustic signal. In other words, the room should exhibit a relatively flat frequency

Figure 3.28. Side reflections at the mixing position. (a) Asymmetrical side reflections cause an acoustic imbalance at the listener's position. (b) Symmetrically splayed sidewalls and ceilings walls can improve the acoustic balance. (Courtesy of APL & CCE Production Technology, www.cceinc.net.)

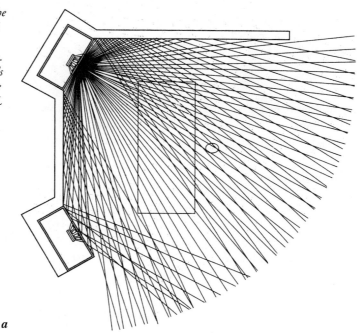

a

Figure 3.28.
Continued

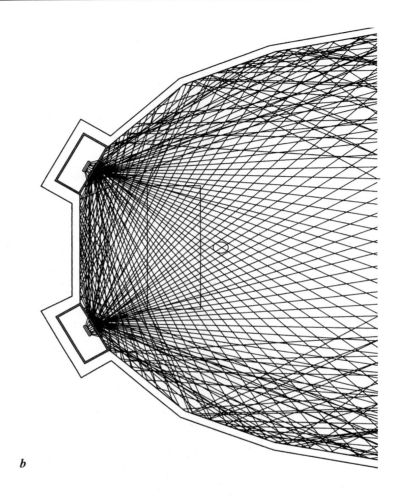

b

response over the entire audio range without adding its own particular sound coloration. The most common way to control the tonal character of a room is to use materials and design techniques that govern the acoustical reflection and absorption factors.

Reflections

One of the most important characteristic of sound as it travels through air is its ability to reflect off a boundary's surface at an angle that's equal to (and opposite of) its original angle of incidence (Figure 3.31). Just as light bounces off a mirrored surface or multiple reflections appear within a mirrored room, sound reflects throughout room surfaces in ways that are often amazingly complex. Nonetheless, sound can still be controlled in ways can add to (or detract from) a room's sonic character.

In Chapter 2 (Sound and Hearing), we learned that sonic reflections can be controlled in ways that disperse the sound outward in a wide-angled pattern (through the use of a convex

Figure 3.29. *Ceiling reflections at the mixing position. (a) Ceiling reflections can cause acoustic interference at the listener's position. (b) A splayed ceiling can reduce unwanted reflections. (Courtesy of APL & CCE Production Technology, www.cceinc.net.)*

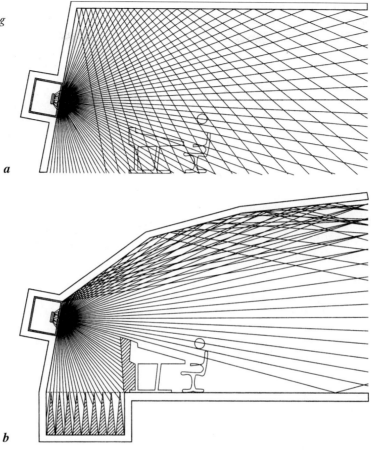

a

b

Figure 3.30. *Auralex MoPAD™ speaker isolation pads. (Courtesy of Auralex Acoustics, www.auralex.com.)*

Figure 3.31. *Sound reflects off a surface at an angle equal (and opposite) to its original angle of incidence.*

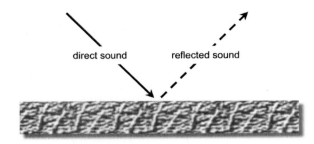

surface), or be focused to a specific point (through the use of a concave surface). Other surface shapes, on the other hand, can reflect sound back at various other angles. For example, a 90° corner will reflect sound back in the same direction as its incident source (a fact that accounts for the additive acoustic buildups at various frequencies at or near a wall-to-corner or corner-to-floor intersection).

The all-time winner of the "avoid this at all possible costs" award goes to constructions that include opposing parallel walls in its design. Such conditions give rise to a phenomenon known as *standing waves*. Standing waves (also known as room modes) occur when sound is reflected off of parallel surfaces and travels back upon its own path, thereby causing phase differences to interfere with a room's amplitude response. Room modes are expressed as integer multiples of the length, width, and depth of the room and indicate which multiple is being referred to for a particular reflection.

Walking around a room with moderate to severe mode problems produces the sensation of increasing and/or decreasing volume levels at various frequencies throughout the room. These perceived volume changes are due to amplitude (phase) cancellations and reinforcements of the combined reflected waveforms at the listener's position. The distance between parallel surfaces and the signal's wavelength determines the nodal points that can potentially cause sharp peaks or dips at various points in the response curve (up to or beyond 19 dB) at the affected fundamental frequency (or frequencies) and upper harmonic intervals (Figure 3.32). This condition exists not only for opposing parallel walls but also for all parallel surfaces (such as between the floor and ceiling or between two reflective flats).

From the above discussion, it's obvious that the most effective way to prevent standing waves is to construct walls, boundaries, and ceilings that are nonparallel. Figure 3.33 shows various standing wave modes as they occur within two rooms that have equal square footage areas. The rectangular room on the left shows how unwanted standing waves can uniformly build up throughout the entire length or width of the room (and probably the ceiling for that matter), while the modes within the nonrectangular room have been broken up and appear at irregular spaces and intervals within the room (thereby reducing the chances of high-level, destructive interference).

If the room is rectangular or if further sound-wave dispersion is desired, diffusers can be attached to the wall and/or ceiling boundaries. Diffusers (Figure 3.34) are acoustical

Figure 3.32. The reflective, parallel walls create an undue number of standing waves, which occur at various frequency intervals (f_1, f_2, f_3, f_4, and so on).

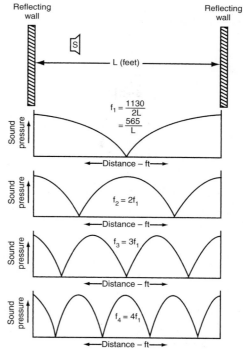

boundaries that reflect the sound wave back at various angles that are wider than the original incident angle (thereby breaking up the energy-destructive standing waves). In addition, the use of both nonparallel and diffusion wall construction can reduce a condition known as *flutter echo* and can smooth out the reverberation characteristics of a room by building further and more complex acoustical pathways.

Flutter echo (also called *slap echo*) is a condition that occurs when parallel boundaries are spaced far enough apart that the listener is able to discern a number of discrete echoes. Flutter echo often produces a "boingy," hollow sound that greatly affects a room's sound character as well as its frequency response. A larger room (which might contain delayed echo paths of 50 msec or more) can have its echoes spaced far enough apart in time that the discrete reflections produce echoes that actually interfere with the intelligibility of the direct sound, often resulting in a jumble of noise.

Absorption

Another factor that often has a marked effect upon an acoustic space involves the use of surface materials and designs that can absorb unwanted sounds (either across the entire audible band or at specific frequencies). The *absorption* of acoustic energy is, effectively, the inverse of reflection (Figure 3.35). Whenever sound strikes a material, the amount of acoustic energy that's absorbed relative to the amount that's reflected can be expressed as a

Figure 3.33. A comparison of two-dimensional soundfields (in pressure level) between rectangular and nonrectangular rooms that have the same surface area: (a) 1,0,0 mode of the rectangular room (34.3 Hz) compared to the nonrectangular room (31.6Hz); (b) 3,1,0 mode of the rectangular room (81.1 Hz) compared to the nonrectangular room (85.5 Hz); (c) 4,0,0 mode in the rectangular room (98 Hz) compared to the nonrectangular room (95.3 Hz).

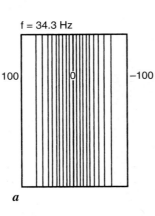
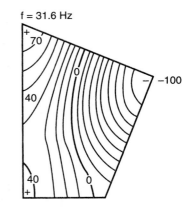

a

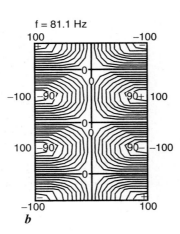
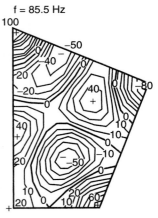

b

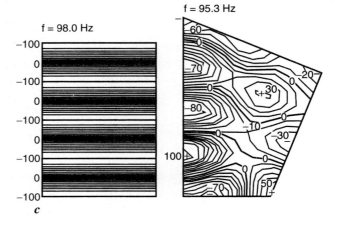

c

Figure 3.34. *Commercial diffuser examples. (a) T'Fusor*^{TM} *sound diffusors (courtesy of Auralex Acoustics, www.auralex.com); (b) Primacoustic*^{TM} *Razorblade quadratic diffuser in the control room (courtesy of Primeacoustic Studio Acoustics, www.primacoustic.com); (c) open-ended view of a Primacoustic*^{TM} *Razorblade quadratic diffuser (courtesy of Primeacoustic Studio Acoustics, www.primacoustic.com).*

a

b

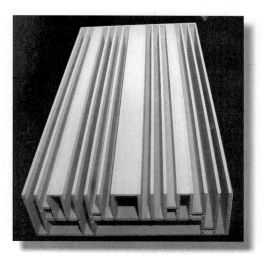

c

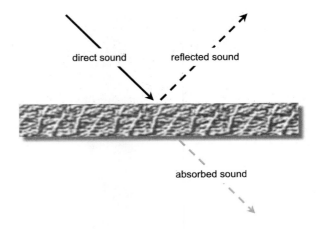

Figure 3.35. *Absorption occurs when only a portion of the incident acoustic energy is reflected back from a material's surface.*

direct sound   reflected sound

absorbed sound

simple ratio known as the material's *absorption coefficient*. For a given material, this can be represented as:

$$A = I_a / I_r$$

where I_a is the sound level (in dB) that is absorbed by the surface (often dissipated in the form of physical heat), and I_r is the sound level (in dB) that is reflected back from the surface.

The factor $(1 - a)$ is a value that represents the amount of reflected sound. This makes the coefficient a decimal percentage value between 0 and 1. If we say that a surface material has an absorption coefficient of 0.25, we're actually saying that the material absorbs 25% of the original acoustic energy and reflects 75% of the total sound energy at that frequency. A sample listing of these coefficients is provided in Table 3.2.

In order to determine the total amount of absorption that's obtained by the sum of all the absorbers within a total volume area, it's necessary to calculate the average absorption coefficient for all the surfaces together. The *average absorption coefficient* (A_{ave}) of a room or area can be expressed as:

$$A_{ave} = (s_1 a_1 + s_2 a_2 + \ldots s_n a_n)/S$$

where $s_1, s_2, \ldots, n$ are the individual surface areas; $a_1, a_2, \ldots, n$ are the individual absorption coefficients of the individual surface areas; and S is the total square surface area.

High-Frequency Absorption

The absorption of high frequencies is accomplished through the use of dense porous materials, such as cloth, fiberglass, and carpeting. These materials generally exhibit high absorption values at higher frequencies, which can be used to control room reflections in a frequency-dependent manner. Specially designed foam and acoustical treatments are also commercially available

Table 3.2. Absorption coefficients for various materials.

Coefficients (Hz)

Material	125	250	500	1000	2000	4000
Brick, unglazed	0.03	0.03	0.03	0.04	0.05	0.07
Carpet (heavy, on concrete)	0.02	0.06	0.14	0.37	0.60	0.65
Carpet (with latex backing, on 40-oz hair-felt of foam rubber)	0.03	0.04	0.11	0.17	0.24	0.35
Concrete or terrazzo	0.01	0.01	0.015	0.02	0.02	0.02
Wood	0.15	0.11	0.10	0.07	0.06	0.07
Glass, large heavy plate	0.18	0.06	0.04	0.03	0.02	0.02
Glass, ordinary window	0.35	0.25	0.18	0.12	0.07	0.04
Gypsum board nailed to 2 × 4 studs on 16-inch centers	0.013	0.015	0.02	0.03	0.04	0.05
Plywood (3/8 inch)	0.28	0.22	0.17	0.09	0.10	0.11
Air (sabins/1000 ft^3)	—	—	—	—	2.3	7.2
Audience seated in upholstered seats	0.08	0.27	0.39	0.34	0.48	0.63
Concrete block, coarse	0.36	0.44	0.31	0.29	0.39	0.25
Light velour (10 oz/yd^2 in contact with wall)	0.29	0.10	0.05	0.04	0.07	0.09
Plaster, gypsum, or lime (smooth finish on tile or brick)	0.44	0.54	0.60	0.62	0.58	0.50
Wooden pews	0.57	0.61	0.75	0.86	0.91	0.86
Chairs, metal or wooden, seats unoccupied	0.15	0.19	0.22	0.39	0.38	0.30

Note: These coefficients were obtained by measurements in the laboratories of the Acoustical Materials Association. Coefficients for other materials may be obtained from Bulletin XXII of the Association.

that can be attached easily to recording studio, production studio, or control room walls as a means of taming multiple room reflections and/or dampening high-frequency reflections (Figure 3.36).

Low-Frequency Absorption

As shown in Table 3.2, materials that are absorptive in the high frequencies often provide little resistance to the low-frequency end of the spectrum (and *vice versa*). This occurs because low frequencies are best damped by pliable materials, meaning that low-frequency energy is absorbed by the material's ability to bend and flex with the incident waveform (Figure 3.37). Rooms that haven't been built to the shape and dimensions to properly handle the low end will need to be controlled by using bass traps that are tuned to reduce the room's resonance frequencies.

Another absorber type can be used to reduce low-frequency buildup at specific frequencies (and their multiples) within a room. These attenuation devices (known as *bass traps*) are available in a number of design types, such as:

◆ Quarter-wavelength trap
◆ Pressure-zone trap
◆ Functional trap

Figure 3.36. *Commercial absorption examples: (a) various absorption and diffusion wall treatments (courtesy of Auralex Acoustics, www.auralex.com); (b) Alphasorb® wall panels and absorption materials (courtesy of Acoustical Solutions, Inc., www.acousticalsolutions.com); (c) Primacoustic™ Polyfuser, a combination diffuser and bass trap (courtesy of Primeacoustic Studio Acoustics, www.primacoustic.com).*

a

b

c

Figure 3.37. *Low-frequency absorption.*

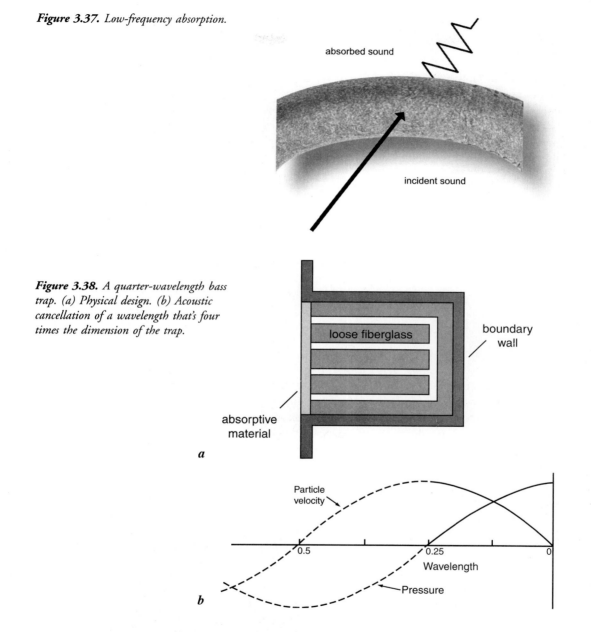

Figure 3.38. *A quarter-wavelength bass trap. (a) Physical design. (b) Acoustic cancellation of a wavelength that's four times the dimension of the trap.*

The Quarter-Wavelength Trap

The quarter-wavelength bass trap (Figure 3.38) is an enclosure with a depth that's one-fourth the wavelength of the offending frequency's fundamental frequency and is often built into the rear facing wall, ceiling, or floor structure (covered by a metal grating to allow foot traffic).

The physics behind the absorption of a calculated frequency (and many of the harmonics that fall above it) lie in the fact that the pressure component of a sound wave will be at its maximum at the rear boundary of the trap...when the wave's velocity component is at a minimum. At the mouth of the bass trap (which is at a one-fourth wavelength distance from this rear boundary), the overall acoustic pressure is at its lowest, while the velocity component (molecular movement) is at its highest potential. Because the wave's motion (force) is greatest at the trap's opening, much of the signal can be absorbed by placing an absorptive material at that point. Low-density fiberglass lining can also be placed inside the trap to increase absorption (especially at harmonic intervals of the calculated fundamental).

Pressure-Zone Trap

The pressure-zone bass trap absorber (Figure 3.39) works on the principle that sound pressure is doubled at large boundary points that are at 90° angles (such as walls and ceilings). By placing highly absorptive material at these boundary point (or points, in the case of a corner/ceiling intersection), the built-up pressure can be partially absorbed.

Functional Trap

Originally created in the 1950s by Harry F. Olson (former director of RCA Labs), the functional bass trap (Figure 3.40) uses a material generally formed into a tube or half-tube structure that is rigidly supported so as to reduce structural vibrations. By placing these devices into corners, room boundaries or in a freestanding spot, a large portion of the undesired bass buildup frequencies can be absorbed. By placing a reflective surface over the portion of the trap that faces into the room, frequencies above 400 Hz can be dispersed back into the room or focal point. Figure 3.41 shows how these traps can be used in the studio to break up reflections and reduce bass buildup.

Figure 3.39. LENRD^{TM} *bass traps. (Courtesy of Auralex Acoustics, www.auralex.com.)*

Figure 3.40. *A functional bass trap.*

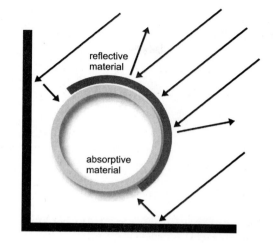

reflective
material

absorptive
material

Figure 3.41. *Quick Sound Field. (Courtesy of Acoustic Sciences Corporation, www.tubetrap.com.)*

Reverberation

Another criterion for studio design is the need for a desirable room ambience and intelligibility, which is often contradictory to the need for good acoustic separation between instruments and their pickup. Each of these factors is governed by the careful control and tuning of the reverberation constants within the studio over the frequency spectrum.

Reverberation (*reverb*) is the persistence of a signal (in the form of reflected waves within an acoustic space) that continues after the original sound has ceased. The effect of these closely spaced and random multiple echoes gives us perceptible cues as to the size, density, and nature of an acoustic space. Reverb also adds to the perceived warmth and depth of recorded sound and plays an extremely important role in the perceived enhancement of music.

Figure 3.42. *The three components of reverberation.*

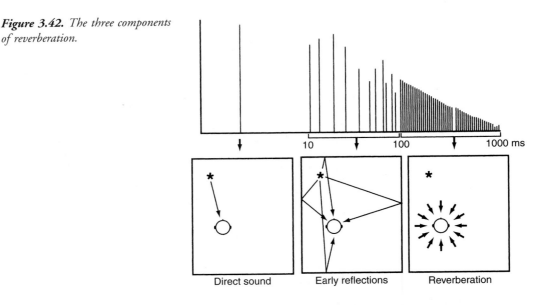

Direct sound   Early reflections   Reverberation

The reverberated signal itself (Figure 3.42) can be broken down into three components:

- Direct signal
- Early reflection
- Reverberation

The direct signal is made up of the original, incident sound that travels from the source to the listener. Early reflections consist of the first few reflections that are projected to the listener off of major boundaries within an acoustic space; these reflections generally give the listener subconscious cues as to the size of the room. The last set of signal reflections make up the actual reverberation characteristic. These signals are composed of random reflections that travel from boundary to boundary in a room and are so closely spaced that the brain can't discern the individual reflections. When combined, they are perceived as a single decaying signal.

Technically, reverb is considered to be the time that's required for a sound to die away to a millionth of its original intensity (resulting in a decrease over time of 60 dB), as shown by the following formula:

$$RT_{60} = V \times 0.049/AS$$

where RT is the reverberation time (in seconds), V is the volume of the enclosure (in ft^3), A is the average absorption coefficient of the enclosure, and S is the total surface area (in ft^2).

As you can see from this equation, reverberation time is directly proportional to two major factors: the volume of the room and the absorption coefficients of the studio surfaces. A large environment with a relatively low absorption coefficient (such as a large cathedral) will have a relatively long RT_{60} decay time, whereas a small studio (which might incorporate a heavy amount of absorption) will have a very short RT_{60}.

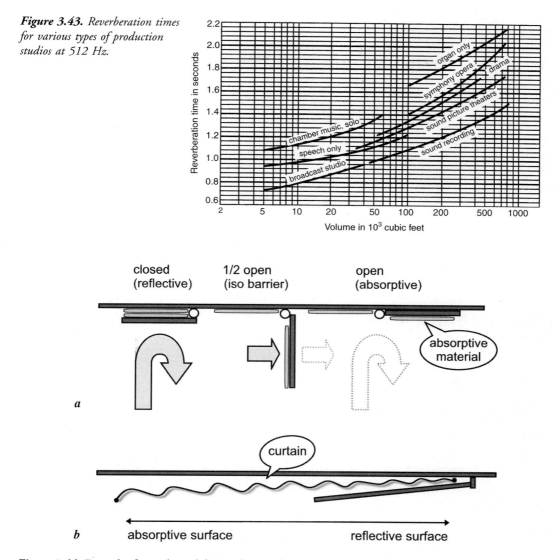

Figure 3.43. *Reverberation times for various types of production studios at 512 Hz.*

Figure 3.44. *Example of several panel designs that can be moved so as to vary the acoustics of a room: (a) hinged partitions that are absorptive on one side and reflective on the other; (b) movable curtains that can be extended or retracted to vary absorption within a room.*

The style of music and the room application will often determine the optimum RT_{60} for an acoustical environment. Figure 3.43 shows a basic guide to reverb times for different applications and musical styles. Reverberation times can range from 0.25 sec in a smaller absorptive recording studio environment to 1.6 sec or more in a larger music or scoring studio. In certain designs, the RT_{60} of a room can be altered to fit the desired application by using movable panels or louvers (Figure 3.44) or by placing carpets in a room. Other designs might separate a studio into sections that exhibit different reverb constants. One side of the studio (or separate iso-room) might be relatively nonreflective or dead, while another section or room could be much more acoustically live. The more reflective, live section is often used to bring certain instruments that rely heavily on room reflections and reverb to life . . . such as strings. The recording of any number of instruments (including drums and percussion) can also greatly benefit from an acoustically live environment.

Isolation between different instruments and their pickups is extremely important in the studio environment. If leakage isn't controlled, the room's effectiveness becomes severely limited over a range of applications. The studio designs of the 1960s and 1970s brought about the rise of the "sound sucker" era in studio design. During this time, the absorption coefficient of many rooms was raised almost to an anechoic (no reverb) condition. With the advent of the music styles of the 1980s and a return to the respectability of live studio acoustics, modern studio and control room designs have begun to increase in size and "liveness" (with a corresponding increase in the studio's RT_{60}). This has reintroduced the buying public to the thick, live-sounding music production of earlier decades . . . when studios were larger structures that were more attuned to capturing the acoustics of a recorded instrument or ensemble.

Acoustic Echo Chambers

Another physical studio design that was extensively used in the past (before the invention of artificial effects devices) for recreating room reverberation is the *acoustic echo chamber*. A traditional echo chamber is an isolated room that has highly reflective surfaces into which speakers and microphones are placed. The speakers are fed from an effects send, while the mic's reverberant pickup is fed back into the mix via an input strip of effects return. By using one or more directional mics that have been pointed away from the room speakers, the direct sound pickup can be minimized. Movable partitions also can be used to vary the room's decay time. When properly designed, acoustic echo chambers have a very natural sound quality to them. The disadvantage is that they take up space and require isolation from external sounds; thus, size and cost often make it unfeasible to build a new echo chamber, especially those that can match the caliber and quality of high-end digital reverb devices.

An echo chamber doesn't have to be an expensive, built-from-the-ground-up design. Actually, a temporary chamber can be made from a wide range of acoustic spaces to pepper your next project with a bit of "acoustic spice." For example:

- An ambient-sounding chamber can be built by placing a Blumlein (crossed figure-8) pair or spaced stereo pair of mics in the main studio space and feeding a send to the studio playback monitors (Figure 3.45).
- A speaker/mic setup could be placed in an empty garage (as could a guitar amp/mic, for that matter).
- An empty stairwell often makes an excellent chamber.
- Any vocalist could tell you what'll happen if you place a guitar speaker/mic setup in the shower.

Figure 3.45. Example of how a room or studio space can be used as a temporary echo chamber.

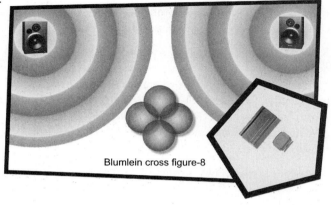

Blumlein cross figure-8

From the above, it's easy to see that ingenuity and experimentation are often the name of the makeshift chamber game. In fact, there's nothing that says that the chamber has to be a real-time effect . . . for example, you could play back a song's effects track from a laptop DAW into a church's acoustic space and record the effect back to stereo tracks on the DAW. The options and the results are totally up to you!

Power- and Ground-Related Issues

Although this falls outside the scope of acoustic design, I'd like to touch upon two final topics before we conclude this chapter: proper grounding guidelines and power conditioning.

Grounding Guidelines

Proper grounding is essential to maintaining equipment safety; however, within an audio facility, small AC voltage potentials between various devices in an audio system can leak into a system's grounding circuit. Although these potentials are small, they are sometimes large enough to induce noise in the form of hums, buzzes, or radio frequency (RF) reception that can be injected (and amplified) directly into the audio signal path. These unwanted signals generally occur whenever improper grounding allows a piece of equipment to "see" two or more different paths to ground.

Because grounding problems arise as a result of electrical interactions between any number of equipment combinations, the use of proper grounding techniques and troubleshooting within an audio production facility are by their very nature situational. As such, the following procedures are simply a set of introductory guidelines for dealing with this age-old problem. There are a great number of technical papers, books, methods, and philosophies on grounding . . . and it's recommended that you carefully research the subject further before tackling any ground-related problems. When in doubt, an experienced professional should be contacted, and care should be taken not to sacrifice safety.

Keep all studio electronics on the same AC electrical circuit—most stray hums and buzzes occur whenever parts of the sound system are plugged into outlets from different AC circuits. Plugging into a circuit that's connected to such noise-generating devices as air conditioners, refrigerators, light dimmers, neon lights, etc. will definitely invite stray noise problems. Because most project studio devices don't require a lot of current (with the possible exception of power amplifiers), it's usually safe to run all of the devices from a single, properly grounded line from the electrical circuit panel.

Keep audio wiring away from AC wiring—whenever AC and audio cables are laid side by side, portions of the 60-Hz signal might be induced into a high-gain, unbalanced circuit as hum. If this occurs, check to see if separating or shielding the lines helps reduce the noise. When all else fails:

◆ If you only hear hum coming from a particular input channel, check that device for ground-related problems. If the noise still exists when the console or mixer is turned down, check the amp circuit or any device that follows the mixer. If the problem continues to elude you, then . . .

◆ Disconnect all of the devices (both power and audio) from the console, mixer, or audio interface, then methodically plug them back in one at a time (it's often helpful to monitor through a pair of headphones).

◆ Lifting the ground lug on a power line can sometimes solve a grounding problem . . . but it often isn't safe (care to put a 100-V potential across that vocalist's mic?). Power filter devices can help solve the problem, while keeping the ground line intact (Figure 3.46).

◆ Check the cables for bad connections or improper polarity. It's also wise to keep the cables as short as possible (especially in an unbalanced circuit).

Figure 3.46. *Ebtech Hum X*^{TM} *ground loop filter. (Courtesy of Ebtech Audio, www.ebtechaudio.com.)*

◆ Another common path for ground loops is through a chassis into a 19-inch rack and then into another chassis. Test this by removing the chassis from the rack. You can isolate the offending chassis from the rack with electrical tape and then insulate the rack screws by using nylon washers.

◆ Investigate the use of a balanced power source, if traditional grounding methods don't work.

Trouble-shooting a ground-related problem can be tricky and finding the problem's source might be a needle in a haystack situation. When putting on your trouble-shooting hat, it's usually best to remain calm, be methodical, and consult with others who might know more than you do (or might simply have a fresh perspective).

Balanced Power

For those facilities that are located in areas where power lines are overtasked by heavy machinery, air conditioners, and the like, a *balanced power* source might be considered. Such a device makes use of a power transformer (Figure 3.47) that has two secondary windings, with a potential to ground on each side of 60 V. Because each side of the circuit is 180° out of phase with the other, a 120-V supply is maintained. Also, since the two 60-V legs are out of phase, any hum, noise, or RF that's present at the device's input will be canceled at the transformer's center tap (a null point that's tied to ground).

A few important points relating to a balanced power circuit include:

◆ A balanced power circuit is able to reduce line noise *if* all of the system's gear is plugged into it. As a result, the device must be able to deliver adequate power.

◆ Balanced power will not eliminate noise from gear that's already sensitive to hums and buzzes.

◆ When to use balanced power is often open to interpretation, depending on who you talk to. For example, some feel that a balanced power conditioner should be used only after all other options to eliminate noise have been explored, while others believe it is a starting point from which to build a noise-free environment.

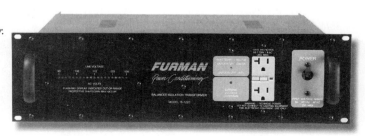

Figure 3.47. Furman IT-1220 balanced-output power conditioner. (Courtesy of Furman Sound, Inc., www.furmansound.com.)

Power Conditioning

Whether your facility is a project studio or full-sized professional facility, it's often a wise idea to regulate and/or isolate the voltage supply that's feeding one of your studio's most precious investments (besides you and your staff) . . . the equipment! This discussion of *power conditioning* can basically be broken down into three topics:

- ◆ Voltage regulation
- ◆ Eliminating power interruptions
- ◆ Keeping the lines quiet

In an ideal world, the power that's being fed to your studio outlets should be very close to the standard reference voltage of the country you are working in (*e.g.*, 120 V, 220 V, 240 V). The real fact of the matter is that these line voltages regularly fluctuate from this standard level, resulting in voltage sags (a condition that can seriously underpower your equipment), surges (rises in voltage that can harm or reduce the working life of your equipment), transient spikes (sharp, high-level energy surges from lightning and other sources that can do serious damage), and brown-outs (long-term sags in the voltage lines). Through the use of a voltage regulator (Figure 3.48), high-level, short-term spikes and surge conditions can be clamped, thereby reducing or eliminating the chance that the mains voltage will rise above a standard, predetermined level.

Certain devices that are equipped with voltage regulation circuitry are able to deal with power sags, long-term surges, and brown-outs by electronically switching between the multiple voltage level taps of a transformer so as to match the output voltage to the ideal mains level (or as close to it as is possible). One of the best approaches for regulating voltage fluctuations both above and below its nominal power levels is to use an adequately powered *uninterruptible power supply* (*UPS*). In short, a quality UPS works by using a regulated power supply to constantly charge a rechargeable battery or bank of batteries. This battery supply is again regulated and used to feed sensitive studio equipment (such as a computer, bank of effects devices, etc.) with a clean and constant voltage supply.

Just as the "uninterruptible" part of the name implies, should the power be momentarily interrupted or give out altogether, a good UPS can draw upon its battery supply to see you through a momentary power loss or to give you enough time to safely shut your system down without losing data during a total power failure. The biggest concern here is to make sure that you

Figure 3.48. A voltage regulator can be used to reduce or eliminate variations in the power line. (Courtesy of Furman Sound, Inc., www.furmansound.com.)

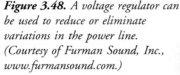

buy a UPS that has enough power reserves to adequately and continuously power the equipment during normal operation, and will give you enough time to save your session and/or file data and then safely shut the system down during an outage. In short—if you're looking to buy a UPS, make sure that it has a continuous power rating that's high enough for your supply needs.

A Few Helpful Hints

Producers, musicians, audio professionals, and engineers spend a great deal of time in the control room and studio. It only makes sense that this environment should be laid out in a manner that's esthetically, functionally, and acoustically pleasing...from a *feng shui* point of view. Creating a quality working environment that's conducive to making good music is the goal of every professional and project studio owner. Beyond the basics of creating a well-designed facility from an acoustic and electronic standpoint, a number of basic concepts should be kept in mind when building or designing a recording facility...no matter how grand or humble. Here are a few helpful hints:

◆ Given the fact that an engineer spends a huge amount of time sitting on his or her bum, it's always a good idea to make an investment in both your and your clients' posture and creature comforts by having comfortable, high-quality chairs around for both the production team and the musicians (Figure 3.49).

◆ Velcro® or tie-straps can be used to organize studio wiring bundles into groups that can be laid our in a way that can reduce clutter, increase organization (color-coded straps can be used), and make the studio look more professional.

◆ Portable label printers can be used to identify cable runs within the studio, identify patch points, I/O strip instrumentation...you name it!

Figure 3.49. *The venerable Herman Miller chair. (Courtesy of McGraw Publishing Peripherals, www.sittingmachine.com.)*

◆ Most of us are guilty of cluttering up our workspace with unused gear, papers . . . you know, junk! I know it's hard, but a clean, uncluttered working environment tells your clients a lot about you, your facility, and your work habits.

◆ Unused cables, adapters, and miscellaneous stuff can be sorted into plastic storage boxes with snap-on lids and stacked or easy storage.

◆ Important tools and items that are used every day (such as screwdrivers, masking tape, or markers) can be stored in a rack-mounted drawer that can be easily accessed without cluttering up your space.

CHAPTER

◆

Microphones: Design and Application

The Microphone: An Introduction

A *microphone* (which also goes by the name of *mic*) is often the first device in a recording chain. Essentially, a mic is a transducer that changes one form of energy (sound waves) into another corresponding form of energy (electrical signals). The quality of its pickup will often depend upon external variables (such as placement, distance, instrument, and the acoustic environment), as well as upon design variables (such as the microphone's operating type, design characteristics, and quality). These interrelated elements tend to work together to affect its overall sound quality.

In order to deal with the wide range of musical, acoustic, and situational circumstances that might come your way (not to mention your own personal taste), a large number of mic types, styles, and designs are available for use as "sonic tools." Because the particular sonic characteristics of a mic might be best suited to a specific range of applications, engineers can use their artistic talents to get the best possible sound from an acoustic source by carefully choosing the mic or mics that fit the specific pickup application at hand.

When considering the best microphone choice and placement, the road to understanding mic techniques is best approached by considering a few simple rules:

◆ Rule 1: *There are no rules, only guidelines.* Although guidelines can help you achieve a good pickup, don't hesitate to experiment in order to get a sound that best suits your needs or personal taste.

◆ Rule 2: *The overall sound of an audio signal is no better than the weakest link in the signal path.* If a mic or its placement doesn't sound as good as it could, make the changes to improve it before you commit it to tape, disc, or whatever. More often than not, the concept of "fixing it later in the mix" will often put you in the unfortunate position of having to correct a situation after the fact, rather than recording the best sound and/or performance at the initial session.

◆ Rule 3: *Whenever possible, use the "Good Rule"*: Good musician + good instrument + good performance + good acoustics + good mike + good placement = good sound. This rule refers to the fact that a music track will only be as good as the performer, the instrument, the mike placement, and the mic. If any of these elements falls short of its potential, the track will suffer accordingly; however, if all of these links are the best that they can be, the recording will be something that you can be proud of!

The miking of vocals and instruments (both in the studio and onstage) is definitely an art form. It's often a balancing act to get the most out of the Good Rule. Sometimes you'll have the best of all of the elements, but at others you'll have to work hard to make lemonade out of a situational lemon. The best rule of all is to use common sense and to trust your instincts.

Beyond giving you a basic foundation into how microphones (and their characteristics) work, the goal of this chapter is to give you a set of insights and placement guidelines that can help you get the most out of your recordings.

Microphone Design

A microphone is a device that converts acoustic energy into corresponding electrical voltages that can be amplified and recorded. In audio production, three transducer types are used:

◆ Dynamic mics
◆ Ribbon mics
◆ Condenser mics

The Dynamic Microphone

In principle, the *dynamic mic* (Figure 4.1) operates by using electromagnetic induction to generate an output signal. The *theory of electromagnetic induction* states that, whenever an electrically conductive metal cuts across the flux lines of a magnetic field, a current of a specific

Figure 4.1. *The Shure Beta 58A dynamic mic. (Courtesy of Shure Incorporated, www.shure.com.)*

Figure 4.2. *Working properties of a dynamic microphone's diaphragm.*

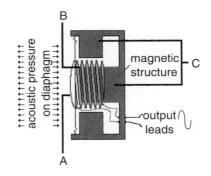

magnitude and direction will be generated within that metal. Dynamic mic designs (Figure 4.2) generally consist of a stiff Mylar diaphragm of roughly 0.35-mil thickness. Attached to the diaphragm is a finely wrapped core of wire (called a *voice coil*) that's precisely suspended within a high-level magnetic field. Whenever an acoustic pressure wave hits the diaphragm's face (A), the attached voice coil (B) is displaced in proportion to the amplitude and frequency of the wave, causing the coil to cut across the lines of magnetic flux that's supplied by a permanent magnet (C). In doing so, an analogous electrical signal (of a specific magnitude and direction) is induced into the coil and across the output leads.

The Ribbon Microphone

Like the dynamic microphone, the *ribbon mic* (Figure 4.3) also works on the principle of electromagnetic induction. Older ribbon design types, however, use a diaphragm of extremely thin aluminum ribbon (2 microns). Often, this diaphragm is corrugated along its width and is suspended within a strong field of magnetic flux (Figure 4.4). Sound-pressure variations between the front and the back of the diaphragm cause it to move and cut across these flux lines, inducing a current into the ribbon that is proportional to the amplitude and frequency of the acoustic waveform. Since the ribbon generates a small output signal (when compared to the larger output that's generated by the multiple wire turns of a moving coil), its signal is too low to drive a microphone input stage directly . . . thus, a step-up transformer must be used to boost the output signal and impedance to an acceptable range.

Figure 4.3. The AEA R44 ribbon mic. (Courtesy of Audio Engineering Associates, www.wesdooley.com.)

Figure 4.4. Cutaway detail of a ribbon microphone. (Courtesy of Audio Engineering Associates, www.wesdooley.com.)

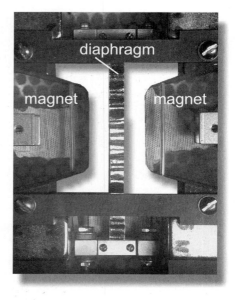

Recent Developments in Ribbon Technology

Over the past several decades, certain microphone manufacturers have made strides toward miniaturizing and improving the operating characteristics of ribbon mics. The popular M160 (Figure 4.5) and M260 ribbon mics from Beyerdynamic use a rare-earth magnet to produce a capsule that's small enough to fit into a 2-inch grill ball (much smaller than a traditional ribbon-style mic). The ribbon (which is corrugated along its length to give it added strength and at each end to give it flexibility) is 3 microns thick, about 0.08 inch wide, and 0.85 inch long and weighs only 0.000011 ounce. A plastic throat is fitted above the ribbon, which houses a pop-blast filter. Two additional filters and the grill greatly reduce the ribbon's potential for blast and wind damage, a feature that has made these designs suitable for outdoor and handheld use. Another relatively recent advance in ribbon technology has been the development of the printed ribbon mic. In principle, the printed ribbon operates in precisely

Figure 4.5. *The Beyerdynamic M160 ribbon mic. (Courtesy of Beyerdynamic, www.beyerdynamic.com.)*

the same manner as the conventional ribbon pickup; however, the rugged diaphragm is made from a polyester film that has a spiral aluminum ribbon printed onto it. Ring magnets are then placed at the diaphragm's front and back, thereby creating a wash of magnetic flux that makes the electromagnetic induction process possible.

The Condenser Microphone

Condenser mics (like the capsules that are shown in Figures 4.6 and 4.7) operate on an *electrostatic principle* rather than the electromagnetic principle used by the dynamic and ribbon mics. The capsule of a basic condenser mic consists of two very thin plates—one movable diaphragm and one fixed backplate. These two plates form a capacitor (or condenser, as it's still called in the U.K. and in many parts of the world). A capacitor is an electrical device capable of

Figure 4.6. *Exposed example of a condenser diaphragm. (Courtesy of ADK, www.adkmic.com; photograph by K. Bujack.)*

Figure 4.7. *Inner detail of an AKG C3000 B condenser microphone. (Courtesy of AKG Acoustics, Inc., www.akg.com.)*

storing an electrical charge. The amount of charge that a capacitor can store is determined by its capacitance value and the voltage that's applied to it, according to the formula:

$$Q = CV$$

where Q is the charge (in coulombs), C is the capacitance (in farads), and V is the voltage (in volts).

The capacitance of a capsule is determined by the composition and surface area of these plates (which are fixed in value), the dielectric or substance between the plates (which is air and is also fixed), and the distance between the plates (which proportionately varies with sound pressure). From this, it's fairly easy to see that plates of a condenser mic capsule form a sound-pressure-sensitive capacitor (Figure 4.8).

Commonly, the plates of a condenser capsule are connected to opposite sides of a stable DC power supply, which provides a polarizing voltage to charge the capacitor (Figure 4.9). Electrons are drawn from the plate connected to the positive side of the power supply and are forced through a high-value resistor onto the negatively charged plate. Once powered, the capsule charges to a state whereby the charge (the difference between the number of electrons on the positive and negative plates) is equal to the capsule capacitance times the polarizing voltage. Once this equilibrium is quickly reached, no further appreciable current flows through the resistor. It's at this point that the device can start acting like a microphone. Once a sound-pressure wave falls upon the diaphragm, its capacitance changes. When the distance between the plates decreases, the capacitance will increase; conversely, when the distance increases, the capacitance will decrease. According to the previous equation, Q, C, and V are interrelated . . . that is, since the charge (Q) is known to be constant and the diaphragm's capacitance (C) changes with differences in sound pressure, the voltage (V) must change in inverse proportion. Given that the capsule's voltage changes in proportion to sound waves that act upon it, *voila* . . . we now have a condenser mic!

Figure 4.8. *Output and potential relationships as a result of changing capacitance.*

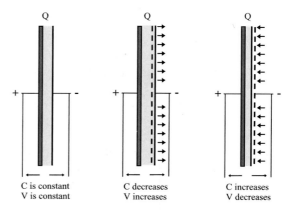

Directional Response

The *directional response* of a mic refers to its sensitivity (output level) at various angles of incidence with respect to the front (on-axis) of the microphone (Figure 4.11). This angular response can be graphically charted in a way that shows a microphone's sensitivity with respect to direction and frequency over 360°. Such a chart is commonly referred to as the mic's *polar pattern*. Microphone directionality can be classified into two categories:

◆ Omnidirectional polar response

◆ Directional polar response

The *omnidirectional mic* (Figure 4.12) is a pressure-operated device that's responsive to sounds that emanate from all directions. In other words, the diaphragm will react equally to all

Figure 4.11. *Directional axis of a microphone.*

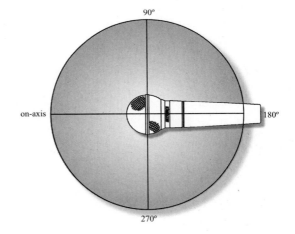

Figure 4.12. *Graphic representation of a typical omnidirectional pickup pattern.*

sound-pressure fluctuations at its surface, regardless of the source's location. Pickups that display *directional* properties are pressure-gradient devices. This means that the pickup is responsive to relative differences in pressure between the front, back, and sides of a diaphragm. For example, a purely pressure-gradient mic will exhibit a *bidirectional* polar pattern (commonly called a *figure 8* pattern), as shown in Figure 4.13. Many of the older ribbon mics exhibit a bidirectional pattern. Since the ribbon's diaphragm is often exposed to sound waves from both the front and rear axes, it's equally sensitive to sounds that emanate from either direction. Sounds from the rear will produce a signal that's 180° out of phase with an equivalent on-axis signal (Figure 4.14a). Sound waves arriving 90° off-axis produce equal but opposite pressures at both the front and rear of the ribbon (Figure 4.14b), resulting in a cancellation at the diaphragm and no output signal.

Figure 4.15 graphically illustrates how the acoustical combination (as well as electrical and mathematical combination, for that matter) of a bidirectional (pressure-gradient) and omni-directional (pressure) pickup can be combined to obtain other directional pattern types. Actually, an infinite number of directional patterns can be obtained from this mixture, with the

Figure 4.13. *Graphic representation of a typical bidirectional pickup pattern.*

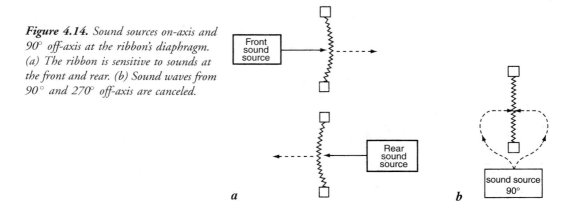

Figure 4.14. *Sound sources on-axis and 90° off-axis at the ribbon's diaphragm. (a) The ribbon is sensitive to sounds at the front and rear. (b) Sound waves from 90° and 270° off-axis are canceled.*

Figure 4.15. *Directional combinations of various bidirectional and nondirectional pickup patterns.*

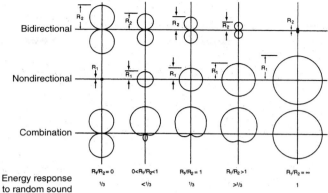

most widely known patterns being the *cardioid, supercardioid,* and *hypercardioid* polar patterns (Figure 4.16).

Often, dynamic mics achieve a cardioid response (named after its heart-shaped polar chart, as shown in Figure 4.17) by incorporating a rear port into their design. This port serves as an acoustic labyrinth that creates an acoustic resistance (delay). In Figure 4.18a, a dynamic pickup having a cardioid polar response is shown receiving an on-axis (0°) sound signal. In effect, the diaphragm receives two signals: the incident signal, which arrives from the front, and an acoustically delayed rear signal. In this instance, the on-axis signal exerts a positive pressure on the diaphragm and begins its travels 90° to a port located on the side of the pickup. At this point, the signal is delayed by another 90° (using an internal, acoustically resistive material or labyrinth). In the time that it takes for the delayed signal to reach the rear of the diaphragm (180°), the on-axis signal has moved on to the negative portion of its acoustic cycle and then begins to exert a negative pressure on the diaphragm (pulling it outward). Since the delayed rear signal is 180° out of phase at this point in time, it will also begin to push the diaphragm outward . . . resulting in an output signal.

Conversely, whenever a sound arrives at the rear of the mic, it begins its trek around to the mic's front. As the sound travels 90° to the side of the pickup, it is again delayed by another 90° before reaching the rear of the diaphragm. During this delay period, the sound continues its journey around to the front of the mic—a delay shift that's also equal to 90°. Since the acoustic pressure at the diaphragm's front and rear sides are equal and opposite, it is being simultaneously pushed inward and outward with equal force, resulting in little or no move-ment . . . and therefore will have little or no output signal (Figure 4.18b). The attenuation of such an off-axis signal, with respect to an equal on-axis signal, is known as its *front-to-back discrimination* and is rated in decibels.

Certain condenser mics can be electrically switched from one pattern to another by using a second capsule that's mounted on both sides of a central backplate. Configuring these dual-capsule systems electrically in phase will create an omnidirectional pattern, while configuring them out of phase results in a bidirectional pattern. A number of intermediate patterns (such as cardioid and hypercardioid) can be created by electrically varying between these two polar states (in either continuous or stepped degrees), as can be seen in Figure 4.15.

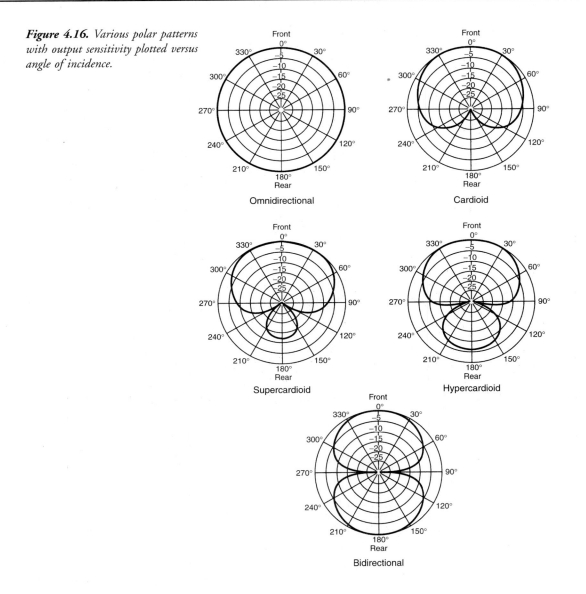

Figure 4.16. *Various polar patterns with output sensitivity plotted versus angle of incidence.*

Frequency Response

The on-axis *frequency-response curve* of a microphone is the measurement of its output over the audible frequency range when driven by a constant, on-axis input signal. This response curve (which is generally plotted in output level [dB] over the 20- to 20,000-Hz frequency range) will often yield valuable information and can give clues as to how a microphone will react at specific frequencies. It should be noted that there are a number of other variables

Figure 4.17. *Graphic representation of a typical cardioid pickup pattern.*

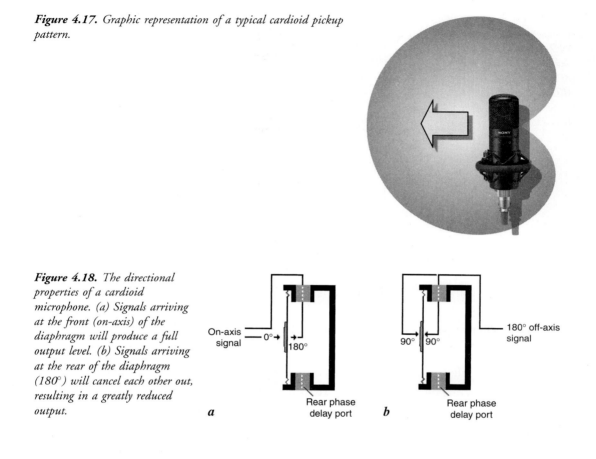

Figure 4.18. *The directional properties of a cardioid microphone. (a) Signals arriving at the front (on-axis) of the diaphragm will produce a full output level. (b) Signals arriving at the rear of the diaphragm (180°) will cancel each other out, resulting in a greatly reduced output.*

On-axis signal — 0° 180°

Rear phase delay port

a

90° 90°

180° off-axis signal

Rear phase delay port

b

that determine how a mic will sound, some of which have no measurement standards . . . . The final determination should always be your own ears.

A mic that is designed to respond equally to all frequencies is said to exhibit a flat frequency response (shown in Figure 4.19a). Others can be made to emphasize or de-emphasize the high-, middle-, or low-end response of the audio spectrum (Figure 4.19b) so as to give it a particular sonic character. The solid frequency–response curve lines (shown in Figure 4.19) were measured on-axis and exhibit totally acceptable responses. However, certain designs may have a "peaky" or erratic curve when measured off-axis. These signal colorations could affect a mic's sound when operating in an area where off-axis sounds (in the form of leakage) arrive at the pickup, often resulting in a change in tone quality when the leaked signal is mixed in with other properly miked signals. This off-axis frequency response is often charted along with the on-axis curve (shown as dotted 180° curves in Figure 4.19).

At low frequencies, *rumble* (high-level vibrations that occur in the 3- to 25-Hz region) can be easily introduced into the surface of a large unsupported floor space, studio, or hall from any number of sources (such as passing trucks, air conditioners, subways, or fans). They can be

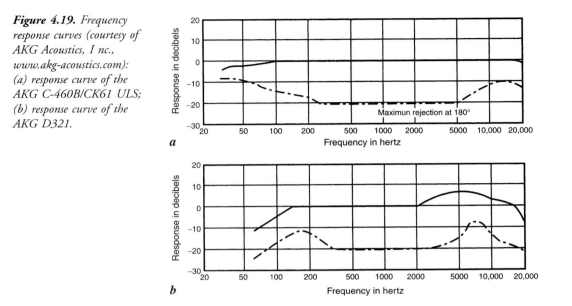

Figure 4.19. *Frequency response curves (courtesy of AKG Acoustics, I nc., www.akg-acoustics.com): (a) response curve of the AKG C-460B/CK61 ULS; (b) response curve of the AKG D321.*

reduced or eliminated in a number of ways, such as:

◆ Use a shock mount to isolate the mic from the vibrating surface and floor stand.

◆ Choose a mic that displays a restricted low-frequency response.

◆ Restrict the response of a wide-range mic by using a low-frequency rolloff filter.

Another low-frequency phenomenon that occurs in most directional mics is known as *proximity effect*. This effect causes an increase in bass response whenever a directional mic is brought within 1 foot of the sound source. This bass boost (which is often most noticeable on vocals) proportionately increases as the distance decreases. To compensate for this effect (which is somewhat greater for a bidirectional mics than for cardioids), a low-frequency roll-off filter switch (which is often located on the microphone body) can be used. If none exists, an external roll-off or equalizer can be used to reduce the low end. Any of these can be used to help restore the bass response to a flat and natural-sounding balance. Another way to reduce or eliminate proximity effect and its associated "popping" of the letters "p" and "b" is to replace the directional microphone with an omnidirectional mic when working at close distances. On a more positive note, this increase in bass response has long been appreciated by vocalists and DJs for their ability to give a full, "larger-than-life" quality to voices that are otherwise thin. In many cases, directional mics have become an important part of an engineer's, producer's, and vocalist's toolbox.

Transient Response

A significant piece of data (which currently has no accepted standard of measure) is the *transient response* of a microphone (Figure 4.20). Transient response is the measure of how quickly

Figure 4.20. *Transient response characteristics of the various microphone types.*

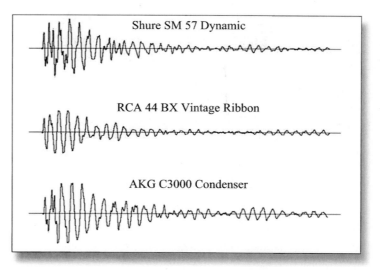

a mic's diaphragm will react when it is hit by an acoustic wavefront. This figure varies wildly among microphones and is a major reason for the difference in sound quality among the three pickup types. For example, the diaphragm of a dynamic mic can be quite large (up to 2-1/2 inches). With the additional weight of the coil of wire and its core, this combination can be a very large mass when compared to the power of the sound wave that drives it. Because of this, a dynamic mic can be very slow in reacting to a waveform—often giving it a rugged, gutsy, and less accurate sound.

By comparison, the diaphragm of a ribbon mic is much lighter, so its diaphragm can react more quickly to a sound waveform, resulting in a clearer sound. The condenser pickup has an extremely light diaphragm, which varies in diameter from 2-1/2 inches to less than 1/4 inch and has a thickness of about 0.0015 inch. This means that the diaphragm offers very little mechanical resistance to a sound-pressure wave, allowing it to accurately track the wave over the entire frequency range.

Output Characteristics

A microphone's *output characteristics* refer to its measured sensitivity, equivalent noise, overload characteristics, impedance, and other output responses.

Sensitivity Rating

A mic's *sensitivity rating* is the output level (in volts) that a microphone will produce, given a specific and standardized acoustic signal at its input (rated in dB SPL). This figure will specify the amount of amplification that's required to raise the mic's signal to line level (−10 dBV or + 4 dBm) and allows us to judge the relative output levels between any two mics.

A microphone with a higher sensitivity rating will produce a stronger output-signal voltage than one with a lower sensitivity.

Equivalent Noise Rating

The *equivalent noise rating* of a microphone can be viewed as the device's electrical self-noise. It's expressed in dB SPL or dBA (a weighted curve) as a signal that would be equivalent to the mic's self-noise voltage. As a general rule, the mic itself doesn't contribute much noise to a system when compared to the mixer's amplification stages, the recording system, or media (whether analog or digital). However, with recent advances in mic preamp/mixer technologies and overall reductions in noise levels produced by digital system, these noise ratings have become increasingly important. The internal noise of a dynamic or ribbon pickup is actually generated by the electrons that move within the coil or ribbon itself. Most of the noise that's produced by a condenser mic is generated by its built-in preamplifier (preamp). It almost goes without saying that certain microphone designs will have a higher degree of self-noise than will others; thus, care must be taken in your microphone choices for critical applications (such as with distant classical recording techniques).

Overload Characteristics

Just as a microphone is limited at low levels by its inherent self-noise, it's also limited at high sound-pressure levels (SPLs) by *overload distortion*. In terms of distortion, the dynamic microphone is an extremely rugged pickup, often capable of an overall dynamic range of 140 dB. Typically, a condenser microphone won't distort, except under the most severe sound-pressure levels; however, the condenser system differs from the dynamic in that at high acoustic levels the capsule's output might be high enough to overload the mic's preamplifier. To prevent this, most condenser mics offer a switchable attenuation pad that immediately follows the capsule output and serves to reduce the signal level at the pre-amp's input, thereby reducing or eliminating overload distortion. When inserting such an attenuation pad into a circuit, keep in mind that the mic's signal-to-noise ratio will be degraded by the amount of attenuation; therefore, it's always wise to remove the inserted pad when using the microphone under normal conditions.

Microphone Impedance

Microphones are available with different *output impedances*. Output impedance is a rating that's used to match the signal-providing capability of one device to the signal-drawing (input impedance) requirements of another. Impedance is measured in ohms (with its symbol being Ω or Z). The most commonly used microphone output impedances are 50 Ω, 150 to 250 Ω (low), and 20 to 50 kΩ (high). Each impedance range has its advantages. In the past, high-impedance mics were less expensive to use because the input impedances of most tube-type amplifiers were high. In order to use a low-impedance mic, tube-type amplifiers required an expensive input transformer. By their very nature, dynamic mics are low-impedance devices and use a built-in, step-up transformer to achieve a high-impedance output. A major disadvantage to using high-impedance mics is the likelihood that their cables will pick up electrostatic noise (like those caused by motors and fluorescent lights). In order to reduce such interference, a shielded cable is necessary, although this begins to act as a capacitor at lengths greater than 20 to 25 feet, which serves to short out much of the high-frequency information that's picked up by the mic. For these reasons, high-impedance microphones are rarely used in the professional recording process.

The lines of very-low-impedance mics (50 Ω) have the advantage of being fairly insensitive to electrostatic pickup. They are, however, sensitive to induced hum pickup from electromagnetic fields (such as those that are generated by AC power lines). This extraneous pickup can be greatly reduced through the use of a twisted-pair cable, whereby the interference that's magnetically induced into the cable will flow in opposite directions along the cable's length and be canceled out at the console or mixer's balanced microphone input stage. Mic lines of 150 to 250 Ω are less susceptible to signal losses and can be used with cable lengths of up to several thousand feet. They are also less susceptible to electromagnetic pickup than 50 Ω lines but are more susceptible to electrostatic pickup. As a result, most professional mics operate with an impedance of 200 Ω, use a shielded twisted-pair cable, and attain the lowest noise through the use of a balanced signal line.

Balanced/Unbalanced Lines

A *balanced line* uses three wires to carry the audio signal. Two of the wires are used to carry the signal voltage, while a third lead or shield is used as a neutral ground wire. Since neither of the two signal conductors is directly connected to the signal ground, balanced lines operate on the principle that the alternating current of an audio signal will travel along the two wires, with an opposite + and − polarity (as occurs in any AC audio circuit). Whenever an electrostatic or electromagnetic signal is induced across the audio leads, it will be induced into both leads at equally polarities and levels (Figure 4.21). Since the input transformer or balancing amplifier of the receiving device will only respond to the difference in voltage between the two leads, the unwanted signal (which is equal in polarity) will be canceled out, while the audio signal (which is opposite in polarity) will be allowed to pass through to the input of the next device.

The standard that has been widely adopted for the proper polarity of two-conductor, balanced, XLR connector cables specifies pin 2 as being positive (+) or "hot" and pin 3 as being negative (−) or "neutral," with the cable shield being connected to pin 1. Under certain conditions, connecting the cable grounds to pin 1 isn't enough to provide an adequate ground. Should excessive radio frequency interference (RFI) leak into the signal path, a possible remedy would be to solder pin 1 to the XLR shell's ground lug at the male end. Sometimes, a condenser mic will draw more current from the phantom supply than the small circuit trace that connects pin 1 to the preamp can handle, and grounding the shell of the mike cable to pin 1 and the male cable shield will reduce noise and RFI. At other times, the cable shields must be attached to

Figure 4.21. Wiring detail of a balanced microphone cable (courtesy of Loud Technologies Inc., www.mackie.com): (a) diagram for wiring a balanced microphone (or line source) to a balanced XLR connector; (b) physical drawings; (c) diagram for wiring a balanced 1/4-inch phone connector; (d) equivalent circuit, where the induced signals travel down the wires in equal polarities that cancel at the transformer whereby the AC audio signals are of opposing polarities that generate an output signal.

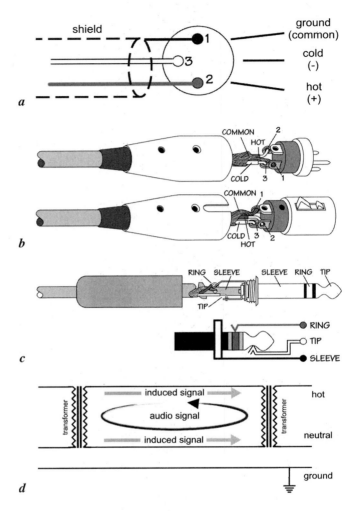

the shell at both ends of the cable instead of just on the male side . . . . It's not always an exact science.

If the hot and neutral pins of balanced mic cables are haphazardly pinned in a music or production studio, it's possible that any number of mics (and other equipment, for that matter) could be wired in opposite polarities. For example, if a single instrument is picked up by two mics that uses improperly phased cables, the instrument might end up being either totally or partially canceled when mixed to mono. For this reason, it's always wise to use a phase tester (Figure 4.22) or volt–ohm meter to check the system's cable wiring.

High-impedance mics and most line-level instrument lines use *unbalanced lines* (Figure 4.22) to transmit signals from one device to another. In an unbalanced circuit, a single signal lead carries a positive current potential to a device, while a second, grounded shield (which is tied to the chassis ground) is used to complete the circuit's return path. When working at low signal

Figure 4.22. Unbalanced microphone circuit (courtesy of Loud Technologies Inc., www.mackie.com): (a) diagram for wiring an unbalanced microphone (or line source) to a balanced XLR connector; (b) diagram for wiring an unbalanced 1/4-inch phone connector; (c) physical drawings; (d) equivalent circuit.

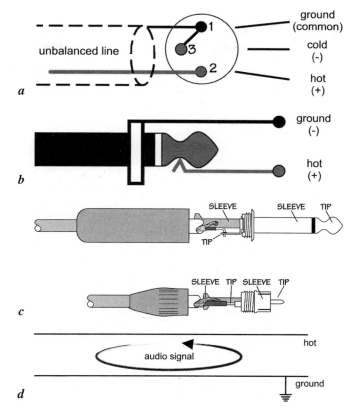

levels (especially at mic levels), any noises, hums, buzzes, or other induced interferences will enter into the signal path, where it will be amplified along with the input signal.

Microphone Preamps

Since the output signals of most microphones are at levels far too low to drive the line level input stage of most recording systems, a mic preamplifier must be used to boost its signal to acceptable levels (often by 30–60 dB). With the advent of improved technologies in analog and digital console design, hard disk recorders, digital audio workstations, signal processors, and the like, low noise and distortion figures have become more important than ever. To many professionals, the stock mic pres (pronounced "preeze") that are designed into many console types don't have that special "sound," aren't high enough in quality to be used in critical applications, or simply don't fit the task at hand. As a result, outboard mic preamps are chosen instead (Figures 4.23 through 4.26). Often offering low-noise, low-distortion specs and/or a unique sound, these devices make use of tube, FET, and/or integrated circuit technology and offer basic features such as variable input gain, phantom power, and high-pass filtering. As with most recording tools, the sound, color scheme, retro-style, tube or transistor type, and budget level are up to the individual, the producer, and the artist . . . it's totally a matter of personal

Figure 4.23. *The Audio Buddy phantom powered dual microphone preamp and direct box. (Courtesy of M-Audio, www.m-audio.com.)*

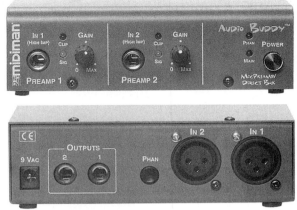

Figure 4.24. *PreSonus TubePre. (Courtesy of PreSonus Audio Electronics, www.presonus.com.)*

Figure 4.25. *Focusrite Platinum TwinTrack Pro dual mono/stereo tracking device. (Courtesy of Digidesign, a division of Avid Technology, Inc., www.digidesign.com.)*

Figure 4.26. *Xlogic SuperAnalogue Mic Amp ($4345) (www.solid-stste-logic.com/xlogic).*

style and taste. Mic pres have also tapped into the growing market of those systems that are based around a digital audio workstation, which doesn't need a console or mixer but does require a quality pre (or set of pres) for plugging mic signals directly into the soundcard.

Phantom Power

Most modern professional condenser mics don't require internal batteries, external battery packs, or individual AC power supplies in order to operate. Instead, they are designed to be powered directly from the console through the use of a *phantom power* supply. Phantom power works by supplying a positive DC supply voltage of +48V through both conductors (pins 2 and 3) of a balanced mic line to the condenser capsule and impedance preamp. This voltage is equally distributed through identical value resistors, so that no differential exists between the two leads. The −48V side of the circuit is supplied to the capsule and preamp through the audio cable's grounding wire (pin 1). Since the audio is only affected by potential differences between pins 2 and 3, the carefully matched +48V potential at these leads is therefore not electrically "visible" to the input stage of a balanced mic preamp. Instead, only the alternating audio signal that's being simultaneously carried along the two audio leads is detected (Figure 4.27).

The resistors used for distributing power to the signal leads should be 1/4-W resistors with a ±1% tolerance and have the following values for the following supply voltages (as some mics can also be designed to work at lower voltages than 48 V): 6.8 kΩ for 48 V, 1.2 kΩ for 24 V, and 680 Ω for a 12-V supply. In addition to precisely matching the supply voltages, these resistors also provide a degree of power isolation between other mic inputs on a console. If a signal lead were accidentally shorted to ground (as could happen if defective cables or unbalanced XLR cables are used), the power supply should still be able to deliver power to other mics in the system. If two or more inputs were accidentally shorted, however, the phantom voltage could drop to levels that would be too low to be usable.

Microphone Techniques

Each microphone has a distinctive sound character that's based on its specific type and design. A large number of types and models can be used for a variety of applications, and it's up to

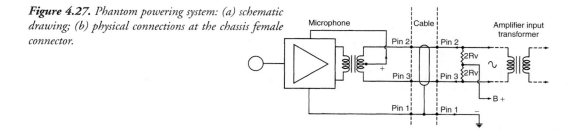

Figure 4.27. *Phantom powering system: (a) schematic drawing; (b) physical connections at the chassis female connector.*

the engineer to choose the right one for the job. At this point, I'd like to point out two particular styles for choosing the types and models of microphones that you might have in your own studio or production toolbox. These can basically be placed into the categories of:

◆ Choosing a limited range of mics that are well suited for a wide range of applications
◆ Amassing a collection of mics that are commonly perceived as being individually suited for a particular instrument or situation

The first approach not only is ideal for the project studio and those who are just starting out and are on a limited budget but is also common practice among seasoned professionals who swear by a limited collection of their favorite mics (often chosen in stereo pairs). These dynamic and/or condenser mics can be used both in the project studio and in the professional studio (it never hurts to have your own favorites around when traveling to different environments). The second approach (I often refer to it as the "Alan Sides Approach") is better suited to the professional studio (and to personal collectors) who actually have a need to amass their own "dream collection" and offer it to their clients. In the end, both approaches have their merits . . . . Indeed, it's usually wise to keep an open mind and choose the range and types of mics that best fit your needs, budget, and personal style.

Choosing the appropriate mic, however, is only half the story. The placement of a microphone can play just as important a role, and is one of an engineer's most valued tools. Because mic placement is an art form, there is no right or wrong. Placement techniques that are currently considered "bad" might easily be the accepted standard five years from now. As new musical styles develop, new recording techniques also tend to evolve. This helps to breathe new life into music and production. The craft of recording should always be open to change and experimentation, two of the strongest factors that keep the music and the biz of music alive and fresh.

Pickup Characteristics as a Function of Working Distance

In studio and sound-stage recording, four fundamental styles of microphone placement are directly related to the working distance of a microphone from its sound source. These important pickup styles (which are described in the following sections) include:

◆ Distant miking
◆ Close miking
◆ Accent miking
◆ Ambient miking

Distant Microphone Placement

With distant microphone placement (Figure 4.28), one or more mics are positioned at a distance of 3 feet or more from the intended signal source. This technique (whose distance

Figure 4.28. *Example of an overall distant pickup.*

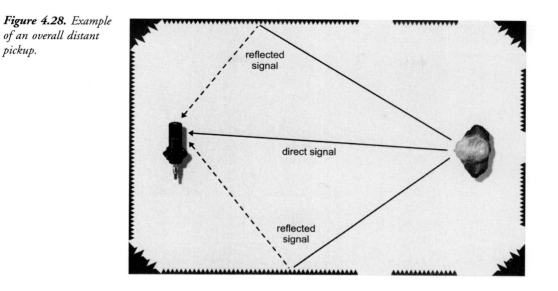

may vary with the size of the instrument) will often yield the following results:

- It can pick up a large portion of a musical instrument or ensemble, thereby preserving the overall tonal balance of that source. Often, a natural tone balance can be achieved by placing the mic at a distance that's roughly equal to the size of the instrument or sound source.

- It allows the room's acoustic environment to be picked up (and naturally mixed in) with the direct sound signal.

Distant miking is often used to pick up large instrumental ensembles (such as a symphony orchestra or choral ensemble). In this application, the pickup will largely rely upon the acoustic environment to help achieve a natural, ambient sound. The mic is placed at a distance so as to strike an overall balance between the ensemble's direct sound and the room's acoustics. This balance is determined by a number of factors, including the size of the sound source, its overall volume level, and mic distance and placement, as well as the reverberant characteristics of the room.

Distant miking techniques tend to add a live, open feeling to a recorded sound; however, this technique could put you at a disadvantage if the acoustics of a hall, church, or studio aren't particularly good. Improper or bad room reflections can create a muddy or poorly defined recording. To avoid this, the engineer can take one of the following actions:

- Temporarily correct for bad or excessive room reflections by using absorptive and/or offset reflective panels.

- Place the mic closer to its source and add artificial ambience.

If a distant mic is used to pick up a portion of the room sound, placing it at a random height can result in a hollow sound due to phase cancellations that occur between the direct sound and delayed sounds that are reflected off the floor and other nearby surfaces (Figure 4.29). If these

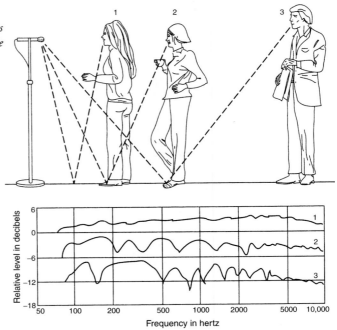

Figure 4.29. *Resulting frequency response from a microphone that receives a direct and delayed sound from a single source.*

delayed reflections arrive at the mic at a time that's equal to one-half a wavelength (or an odd multiple thereof), the reflected signal will be 180° out of phase with the direct sound. This could produce dips in the signal's pickup response that would color the signal. Since the reflected sound is at a lower level than the direct sound (as a result of traveling farther and losing energy as it bounces off a surface), the cancellation is usually only partially complete.

Although raising the mic will have the effect of reducing reflections (due to the increased distances that the reflected sound must travel), moving the mic close to the floor will conversely reduce the path length and raise the range in which the frequency cancellation occurs. In practice, a height of 1/8 to 1/16 inch will raise the cancellation above 10 kHz. One such microphone design type, known as a *boundary microphone* (Figures 4.30 and 4.31), places an electret-condenser or condenser diaphragm well within these low height restrictions. For this reason, this mic type can be a good choice for use as an overall distant pickup, when the mics need to be out of sight (*i.e.*, placed on a floor, wall, or large boundary).

Close Microphone Placement

When close microphone placement is used, the mic is often positioned about 1 inch to 3 feet from a sound source. This commonly used technique generally yields two results:

◆ It creates a tight, present sound quality.
◆ It effectively excludes the acoustic environment.

Figure 4.30. *The boundary microphone system.*

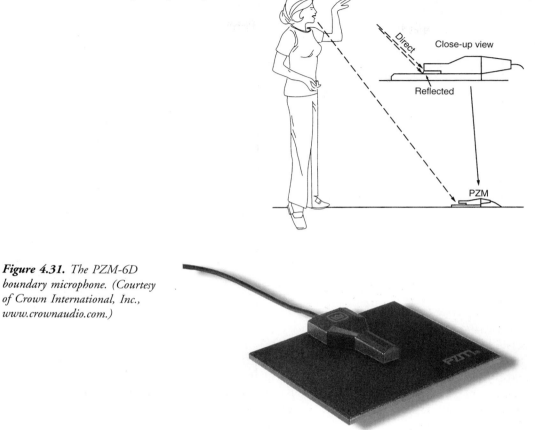

Figure 4.31. *The PZM-6D boundary microphone. (Courtesy of Crown International, Inc., www.crownaudio.com.)*

Because sound diminishes with the square of its distance from the sound source, a sound that originates 3 inches from the pickup will be much higher in level than one that originates 6 feet from the mic (Figure 4.32). Therefore, whenever *close miking* is used, only the desired on-axis sound will be recorded—while extraneous, distant sounds (for all practical purposes) won't be picked up. In effect, the distant pickup will be masked by the closer sounds and/or will be reduced to a relative level that's well below the main pickup.

Whenever an instrument's mic also picks up the sound of a nearby instrument, a condition known as *leakage* occurs (Figure 4.33). Since this "leaked" signal will be picked up by both its intended mic and a nearby mic (or mics), it's easy to see how the signals could be combined together within the mixdown process. When this happens, level and phase cancellations often make it more difficult to have control over individual tracks in mixdown without affecting the level and sound character of other tracks. As a result, try to avoid this condition during the recording process, whenever possible.

Figure 4.32. *Close miking reduces the effects of the acoustic environment.*

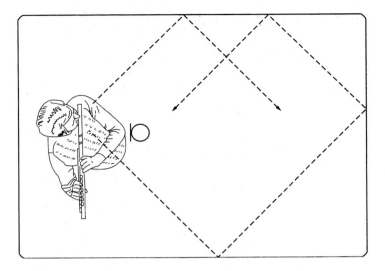

Figure 4.33. *Leakage due to indirect signal pickup: (a) distant leakage; (b) close-miked leakage.*

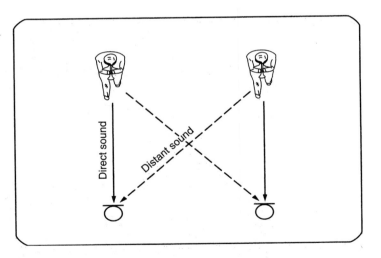

To avoid the problems of leakage, any or all of the following methods can be tried:

◆ Place the mics closer to their respective instruments (Figure 4.34a).

◆ Place an acoustic barrier (known as a flat, gobo, or divider) between the two instruments (Figure 4.34b).

◆ Use directional mics.

◆ Spread the instruments farther apart.

Whenever individual instruments are being miked close (or semi-close), it's generally wise to follow the *3:1 distance rule*. This principle states that, in order to reduce leakage and maintain

Figure 4.34. *Two methods for reducing reduce leakage. (a) Place the microphones closer to their sources. (b) Use an acoustic barrier to reduce leakage.*

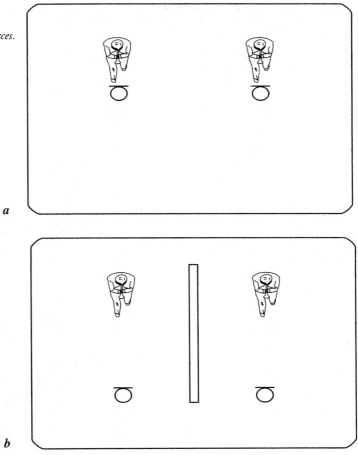

a

b

phase integrity, for every unit of distance between a mic and its source a nearby mic (or mics) should be separated by at least three times that distance (Figure 4.35). It should be noted that some err on the side of caution and avoid leakage even further by following a 5:1 distance rule. As always, experience will be your best teacher. Although the close miking of a sound source offers several advantages, a mic should be placed only as close to the source as is necessary, not as close as possible. Miking too close can color the recorded tone quality of a source.

Because such techniques commonly involve distances of 1 to 6 inches, the tonal balance (timbre) of an entire sound source often can't be picked up; rather, the mic may be so close to the source that only a small portion of the surface is actually picked up, giving it a tonal balance that's very area specific (much like hearing parts of the instrument through an acoustic microscope). At these close distances, moving a mic by only a few inches can easily change the pickup tonal balance. If this occurs, try using one of the following remedies:

1. Move the microphone along the surface of the sound source until the desired balance is achieved.

Figure 4.35. Example of the 3:1 microphone distance rule: "For every unit of distance between a mic and its source, a nearby mic (or mics) should be separated by at least three times that distance."

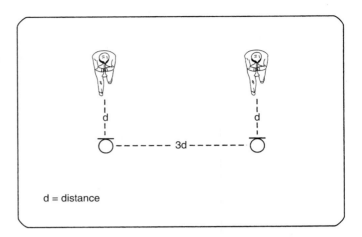

d = distance

2. Place the mic farther back from the sound source to allow for a wider angle (thereby picking up more of the instrument's overall sound).

3. Change the mic.

4. Equalize the signal until the desired balance is achieved.

Do-It-Yourself Tutorial: Close Miking

1. Mic an acoustic instrument (such as a guitar, piano or violin) at a distance of 3 inches.

2. Move (or have someone move) the mic over the instrument's body while listening to variations in the sound. Does the sound change? What are your favorite and least favorite positions?

Accent Microphone Placement

Often, the tonal and ambient qualities sound very different between a distant- and close-miked pickup. Under certain circumstances, it's difficult to obtain a naturally recorded balance when mixing the two together. For example, if a solo instrument within an orchestra needs an extra mic for more volume and presence, placing the mic too close would result in a pickup that sounds overly present, unnatural, and out of context with the distant, overall orchestral pickup. To avoid this pitfall, a compromise in distance should be struck. A microphone that has been placed within a reasonably close range to an instrument or section within a larger ensemble (but not so close as to have an unnatural sound) is known as an accent microphone (Figure 4.36). Whenever *accent miking* is used, care should be exercised in placement and pickup type. The amount of accent signal that's introduced into the mix should sound natural relative to the overall pickup. A good accent mic should only add presence to a solo passage and not stick out as separate, identifiable pickup.

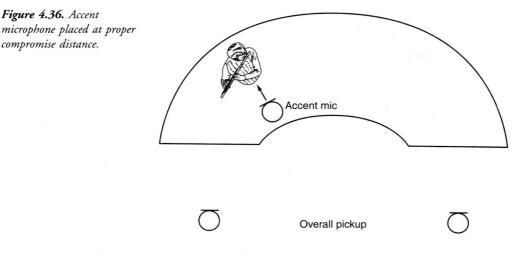

Figure 4.36. *Accent microphone placed at proper compromise distance.*

Ambient Microphone Placement

Ambient miking places the pickup at such a distance that the reverberant or room sound is more prominent than the direct signal. The ambient pickup is often a cardioid stereo pair or crossed figure 8 (Blumlein) pair that can be mixed into a stereo or surround-sound production to provide a natural reverb and/or ambience. To enhance the recording, you can use ambient mic pickups in the following ways:

◆ In a live concert recording, ambient mics can be placed in a hall to restore the natural reverberation that is often lost with close miking techniques.

◆ In a live concert recording, ambient microphones can be placed over the audience to pick up their reaction and applause.

◆ In a studio recording, ambient microphones can be used in the studio to add a sense of space or natural acoustics back into the sound.

Do-It-Yourself Tutorial: Ambient Miking

1. Mic an instrument or its amp (such as an acoustic or electric guitar) at a distance of 6 inches to 1 foot.
2. Place a stereo mic pair (in an X/Y and/or spaced configuration) in the room, away from the instrument.
3. Mix the two pickup types together. Does it "open" the sound up and give it more space? Does it muddy the sound up or breathe new life into it?

Stereo Miking Techniques

For the purpose of this discussion, the term *stereo miking techniques* refers to the use of two microphones in order to obtain a coherent stereo image. These techniques can be used

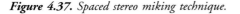

Figure 4.37. *Spaced stereo miking technique.*

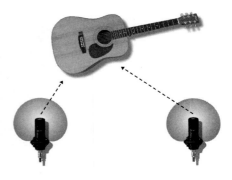

in either close or distant miking of single instruments, vocals, large or small ensembles, within on-location or studio applications...in fact, the only limitation is your imagination. The four fundamental stereo miking techniques are

◆ Spaced pair
◆ X / Y
◆ M/S
◆ Decca tree

Spaced Pair

Spaced microphones (Figure 4.37) can be placed in front of an instrument or ensemble (in a left/right fashion) to obtain an overall stereo image. This technique places the two mics (of the same type, manufacturer, and model) anywhere from only a few feet to more than 30 feet apart (depending on the size of the instrument or ensemble) and uses time and amplitude cues in order to create a stereo image. The primary drawback to this technique is the strong potential for phase discrepancies between the two channels due to differences in a sound's arrival time at one mic relative to the other. When mixed to mono, these phase discrepancies could result in variations in frequency response and even the partial cancellation of instruments and/or sound components in the pickup field.

X / Y

X / Y stereo miking is an intensity-dependent system that uses only the cue of amplitude to discriminate direction. With the X / Y coincident-pair technique, two directional microphones of the same type, manufacture, and model are placed with their grills as close together as possible (without touching) and facing at angles to each other (generally between 90 and 135°). The midpoint between the two mics is pointed toward the source, and the mic outputs are equally panned left and right. Even though the two mics are placed together, the stereo imaging is excellent—often better than that of a spaced pair. In addition, due to their close proximity, no appreciable phase problems arise. Most commonly, X / Y pickups use mics

Figure 4.38. *X/Y stereo miking technique: (a) X/Y crossed cardioid pair; (b) sideways shot of the Royer/ Speiden SF-24 phantom-powered stereo coincident ribbon microphone in the Blumlein cross-figure-8 setting. (Courtesy of Royer Labs, www.royerlabs.com.)*

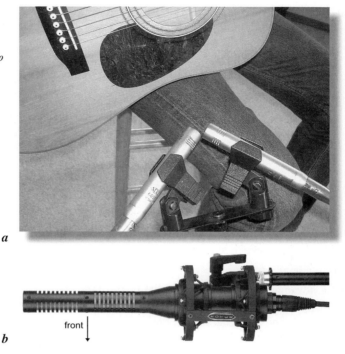

a

b

that have a cardioid polar pattern (Figure 4.38a), although the Blumlein technique is being increasingly used. This technique (which is named after the unheralded inventor, Alan Dower Blumlein) uses two crossed bidirectional mics, which are offset by 90° to each other, and often yields excellent ambient results (Figure 4.38b). Stereo microphones that contain two diaphragms in the same case housing are also available on the used and new market. These mics will either be fixed (generally in a 90° or switchable X/Y pattern) or designed so that the top diaphragm can be rotated by 180° (allowing for the adjustment of various coincident X/Y angles.

M/S

Another coincident-pair system, known as the M/S (or mid-side) technique (Figure 4.39), is similar to X/Y in that it uses two closely spaced, matched pickups. The mid-side method differs from X/Y in that it requires the use of an external transformer, active matrix, or software plug-in in order to work. In the classic M/S stereo miking configuration, one of the microphone capsules is designated to be the *M* (mid) position pickup and is generally selected as having a cardioid pickup pattern that faces forward towards the sound source. The *S* (side) capsule is generally chosen as a figure-8 pattern that's oriented sideways (90° and 270°) to the on-axis pickup (*i.e.*, with the null side facing the cardioid's main axis). In this way, the mid capsule picks up the direct sound, while the side figure-8 capsule picks up ambient and reverberant sound. These outputs are then combined through a sum-and-difference decoder matrix either electrically (through a transformer matrix) or mathematically (through a digital

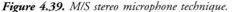

Figure 4.39. M/S stereo microphone technique.

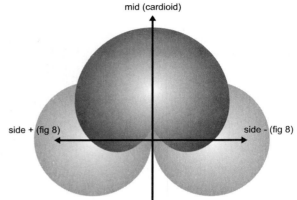

M/S plug-in), which then resolves them into a conventional X/Y stereo signal: $(M+S=X)$ and $(M-S=Y)$.

One advantage of this technique is its absolute monaural compatibility. When the left and right signals are combined, the sum of the output will be $(M+S)+(M-S)=2M$. That's to say, the side (ambient) signal will be canceled out, while the mid (direct) signal will be accentuated. Since it is widely accepted that a mono signal loses its intelligibility with added reverb, this tends to work to our advantage. Another amazing side benefit of using M/S is the fact that it lets us continuously vary the mix of mid (direct) to side (ambient) sound that's being picked up either during the recording (from the console location) . . . or even during mixdown, after it's been recorded! These are both possible by simply mixing the ratio of mid to side that's being sent to the decoder matrix (Figure 4.40). In a mixdown scenario, all that's needed is to record the mid on one track and the side on another (it's often best to use a digital recorder, as phase delays associated with analog recording can interfere with the decoding process). Upon mixdown, routing the M/S tracks to the decoder matrix allows you to make important decisions regarding stereo width and depth at a later, more controlled date.

Decca Tree

Although not as commonly used as the above techniques, the *Decca tree* is a time-tested, classical miking technique that uses both time and amplitude cues in order to create a coherent stereo image. Attributed originally to Decca engineers Roy Wallace and Arthur Haddy in 1954, the Decca tree (Figure 4.41) consists of three omnidirectional mics (originally, Neumann M50 mics were used). In this arrangement, a left and right mic pair are placed 3 feet apart, while a third mic is placed 1.5 feet out in front and panned in the center of the stereo field. Still favored by many in orchestral situations as a main pickup pair, the Decca tree is most commonly placed on a tall boom, above and behind the conductor. According to lore, when Haddy first saw the array, he remarked: "It looks like a bloody Christmas tree!" The name stuck.

Figure 4.40. *M/S decoder matrix. (a) AEA MS-38 Mark II Dual-Mode Stereo Width Controller and Matrix MS processor (courtesy of Audio Engineering Associates, www.wesdooley.com). (b) Waves S1 Stereo Imager plug-in includes True Blumlein shuffling and MS/LR processing (courtesy of Waves, www.waves.com).*

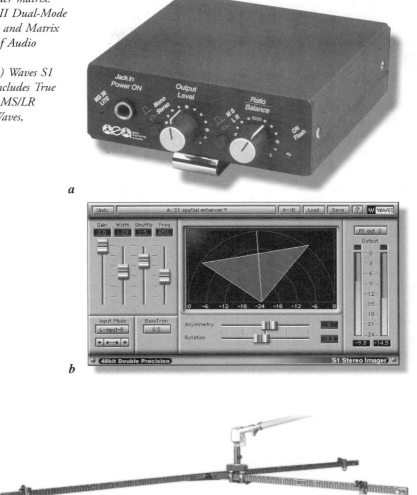

a

b

Figure 4.41. *Decca tree Microphone array. (Courtesy of Audio Engineering Associates, www.wesdooley.com.)*

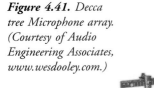

Surround Miking Techniques

With the advent of 5.1 surround sound production, it's certainly possible to make use of a surround console or DAW to place sources that have been recorded in either mono or stereo into a surround image field. Under certain situations, it's also possible to consider using multiple-pickup surround miking techniques in order to capture the actual acoustic environment and then translate that into a surround mix. Just as the number of techniques and personal styles increases when miking in stereo compared to mono...the number of placement and technique choices will likewise increase when miking a source in surround.

Although guidelines have been and will continue to be set, both placement and mixing styles are definitely an art and not a science.

Ambient Surround Mics

A relatively simple, yet effective way to capture the surround ambience of a live or studio session is to simply place a spaced or coincident mic pair out in the studio at a distance from the sound source. These can be facing toward or away from the sound source, and placement is totally up to experimentation. During a surround mixdown, these ambient mics can work wonders to add a sense of space to a group or individual overdub.

Surround Decca Tree

One of the most logical techniques for capturing an ensemble or instrument uses five mics and makes use of several variations on the Decca tree. For example, Ron Streicher has developed an ingenious and simple system of adding two rear-facing mics to the existing three-mic Decca tree system (although he has modified the system by placing a coincident M/S mic in the forward-center position), as shown in Figure 4.42a. Another simpler approach is to place five cardioid mics in a circle, such that the center channel faces toward the source, thereby creating a simple setup that can be routed L–C–R–RL–RR (Figure 4.42b).

One last approach (which doesn't actually fall under the Decca tree category) involves the use of four cardioid mics that are spaced at 90° angles, representing L–R–RL–RR, with the on-axis point being placed 45° between the L and R mics (Figure 4.43). This configuration can be easily made by mounting the mics on two stereo bars that are offset by 90°.

Recording Direct

As an alternative, the signal of an electric or electronic instrument (guitar, keyboard, and so on) can be directly "injected" into a console, recorder or DAW without the use of a microphone. This option can produce a cleaner, more present sound by bypassing the distorted components of a head/amp combination. It also reduces leakage into other mics by eliminating room sounds. In the project or recording studio, the *direct injection* (*DI*) box (Figure 4.44) serves to interface an instrument with an analog output signal to a console or recorder in the following ways:

◆ It reduces an instrument's line-level output to mic level for direct insertion into the console's mic input jack.

◆ It changes an instrument's unbalanced, high-source impedance line to a balanced, low-source impedance signal that's needed by the console's input stage.

◆ It electrically isolates audio signal paths between the instrument and mic/line preamp stages (thereby reducing the potential for ground-loop hum and buzzes).

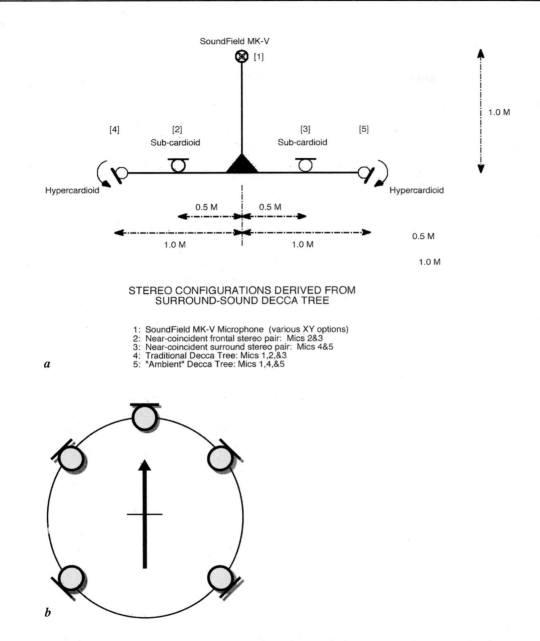

Figure 4.42. *Surround Sound Decca examples: (a) Ron Streicher's Surround Sound Decca tree (courtesy of Ron Streicher, author of* The New Stereo Soundbook *and proprietor of Pacific Audio-Visual Enterprises, Pasadena, CA); (b) placing five cardioid microphones in a circular pattern (with the center microphone facing toward the source) to create a modified, mini-surround Decca tree.*

Figure 4.43. *A four-microphone (quad) approach to surround miking.*

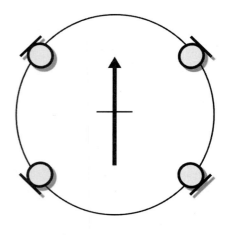

Figure 4.44. *Radial JDI Passive Direct Box. (Courtesy of Radial Engineering, www.radialeng.com.)*

Most commonly, the instrument's output is plugged directly into the DI box (where it's stepped down in level and impedance), and the box's output is then fed into the mic pre of a console or DAW. If a "dirtier" sound is desired, certain boxes will allow high-level input signals to be taken directly from the amp's speaker output jack. It's also not uncommon for an engineer, producer, and/or artist to combine the punchy, full sound of a mic with the present crispness of a direct sound. These signals can then be combined onto a single tape track or recorded to separate tracks (thereby giving more flexibility in the mixdown stage). The ambient image can be "opened up" even further by mixing a semi-distant or distant mic (or stereo pair) with the direct (and even with the close miked amp) signal. This ambient pickup can be either mixed into a stereo field or at the rear of a surround field to fill out the sound.

Microphone Placement Techniques

The following sections are meant to be used as a general guide to mic placement for various acoustic and popular instruments. It's important to keep in mind that these are only guidelines. Several general application and characteristic notes are detailed in Table 4.1, and descriptions of several popular mics are outlined in the "Microphone Selection" section to help give insights into mic placements and techniques that might work best in a particular application.

As a general rule, choosing the best mic for an instrument or vocal will ultimately depend upon the basic character of the sound that you're searching for. For example, a dynamic mic will often yield a "rugged" or "punchy" character (which is often further accentuated by a close proximity bass boost that's associated with most directional mics). A ribbon mic will often yield a mellow sound that ranges from being open and clear to slightly "croony" . . . depending on the type and distances involved. Condenser mics are often characterized as having a clear, present, and full-range sound that varies with mic design, grill options, and capsule size. Before jumping into this section, I'd like to again take time to point out the "Good Rule" to anyone who wants to be a better engineer, producer, and/or musician:

Good musician + good acoustics + good mike + good placement = good sound

Table 4.1. Microphone selection guidelines.

Needed Application	Required Microphone Choice and/or Characteristic
Natural, smooth tone quality	Flat frequency response
Bright, present tone quality	Rising frequency response
Extended lows	Dynamic or condenser with extended low-frequency response
Extended highs (detailed sound)	Condenser
Increased "edge" or mid-range detail	Dynamic
Extra ruggedness	Dynamic or modern ribbon/condenser
Boosted bass at close working distances	Directional microphone
Flat bass response up close	Omnidirectional microphone
Reduced leakage, feedback, and room acoustics	Directional microphone, or omnidirectional microphone at close working distances
Enhanced pickup of room acoustics	Place microphone or stereo pair at greater working distances
Reduced handling noise	Omnidirectional, vocal microphone, or directional microphone with shock mount
Reduced breath popping	Omnidirectional or directional microphone with pop filter
Distortion-free pickup of very loud sounds	Dynamic or condenser with high maximum SPL rating
Noise-free pickup of quiet sounds	Condenser with low self-noise and high sensitivity

As a rule, starting with an experienced, rehearsed, and ready musician who has a quality instrument that's well tuned is the best insurance toward getting the best possible sound. Let's think about this for a moment. Say that we have a live rhythm session that involves drums, piano, bass guitar, and scratch vocals. All of the players are the best around, except for the drummer, who is new to the studio process . . . . Unfortunately, you've just signed on to teach the drummer the ropes of proper drum tuning. The session will go far less smoothly than it otherwise would, as you'll have to take the extra time to work with the player to tune the drums and set up the mics in order to get the best possible sound. Once you're rolling, it'll then be up to you or the producer to pull a professional performance out of someone who's new to the field.

Don't get me wrong, musicians have to start somewhere . . . but an experienced studio musician who comes into the studio with a great instrument that's tuned and ready to go (and who might even clue you in on some sure-fire mic and placement techniques for the instrument) is simply a joy from a sound, performance, and time- and budget-saving standpoint. Simply put, if you and/or the project's producer have prepared enough to get all your "goods" lined up, the track will have a much better chance of being something that everyone can be proud of. Just as with the art of playing an instrument, the art of mic choice, placement, and style is also subjective and is often one of the calling cards of a good engineer. Experience simply comes with time and the willingness to experiment.

Brass Instruments

The following sections describe many of the sound characteristics and miking techniques that are encountered in the brass family of instruments.

Trumpet

The fundamental frequency of a trumpet ranges from E3 to D6 (165–1175 Hz) and contains overtones that stretch upward to 15 kHz. Below 500 Hz, the sounds emanating from the trumpet project uniformly in all directions; above 1500 Hz, the projected sounds become much more directional; and above 5 kHz, the dispersion emanates at a tight 30° angle from in front of the bell. The formants of a trumpet (the relative harmonic and resonance frequencies that give an instrument its specific character) lie at around 1 to 1.5 kHz and at 2 to 3 kHz. Its tone can be radically changed by using a mute (a cup-shaped dome that fits directly over the bell), which serves to dampen frequencies above 2.5 kHz. A conical mute (a metal mute that fits inside the bell) tends to cut back on frequencies below 1.5 kHz while encouraging frequencies above 4 kHz. Because of the high sound-pressure levels that can be produced by a trumpet (up to 130 dB SPL), it's best to place a mic slightly off the bell's center at a distance of 1 foot or more (Figure 4.45). When closer placements are needed, a −10- to −20-dB pad can help prevent input overload at the mic or console preamp input. Under such close working conditions, a windscreen can help protect the diaphragm from windblasts.

Figure 4.45. *Typical microphone placement for a single trumpet.*

Trombone

Trombones come in a number of sizes; however, the most commonly used "bone" is the tenor that has a fundamental note range spanning from E2 to C5 (82–520 Hz) and produces a series of complex overtones that range from 5 kHz (when played medium loud) to 10 kHz (when overblown). The trombone's polar pattern is nearly as tight as the trumpet's: Frequencies below 400 Hz are distributed evenly, whereas its dispersion angle increases to 45° from the bell at 2 kHz and above. The trombone most often appears in jazz and classical music. The *Mass in C Minor* by Mozart, for example, has parts for soprano, alto, tenor, and bass trombones. This style obviously lends itself to the spacious blending that can be achieved by distant pickups within a large hall or studio. On the other hand, jazz music often calls for closer miking distances. At 2 to 12 inches, for example, the trombonist should play slightly to the side of the mic to reduce the chance of overload and wind blasts. In the miking of a trombone section, a single mic might be placed between two players, acoustically combining them onto a single channel and/or track.

Tuba

The bass and double-bass tubas are the lowest pitched of the brass/wind instruments. Although the bass tuba's range is actually a fifth higher than the double bass, it's still possible to obtain a low fundamental of B (29 Hz). A tuba's overtone structure is limited—with a top response ranging from 1.5 to 2 kHz. The lower frequencies (around 75 Hz) are evenly dispersed; however, as frequencies rise, their distribution angles reduce. Under normal conditions, this class of instruments isn't miked at close distances. A working range of 2 feet or more, slightly off-axis to the bell, will generally yield the best results.

French Horn

The fundamental tones of the French horn range from B1 to B5 (65–700 Hz). Its "oo" formant gives it a round, broad quality that can be found at about 340 Hz, with other frequencies falling between 750 Hz and 3.5 kHz. French horn players often place their hands inside the bell to mute the sound and promote a formant at about 3 kHz. A French horn player or section is traditionally placed at the rear of an ensemble, just in front of a rear, reflective stage wall. This wall serves to reflect the sound back toward the listener's position (which tends to create a fuller, more defined sound). An effective pickup of this instrument can be achieved by placing an omni- or bidirectional pickup between the rear, reflecting wall and the instrument bells, thereby receiving both the direct and reflected sound. Alternatively, the pickups can be placed in front of the players, thereby receiving only the sound that's being reflected from the rear wall.

Guitar

The following sections describe the various sound characteristics and techniques that are encountered when miking the guitar.

Acoustic Guitar

The popular steel-strung, acoustic guitar has a bright, rich set of overtones (especially when played with a pick). Mic placement and distance will often vary from instrument to instrument and may require experimentation to pick up the best tonal balance. A balanced pickup can often be achieved by placing the mic (or an X/Y stereo pair) at a point slightly off-axis and above or below the sound hole at a distance of between 6 inches and 1 foot (Figure 4.46). Condenser mics are often preferred for their smooth, extended frequency response and excellent transient response. The smaller-bodied classical guitar is normally strung with nylon or gut and is played with the fingertips, giving it a warmer, mellower sound than its steel-strung counterpart. To make sure that the instrument's full range is picked up, place the mic closer to the center of the bridge, at a distance of between 6 inches and 1 foot.

Miking Near the Sound Hole

The sound hole (located at the front face of a guitar) serves as a bass port, which resonates at lower frequencies (around 80–100 Hz). Placing a mic too close to the front of this port

Figure 4.46.
*Typical
microphone
placement for
the guitar.*

might result in a boomy and unnatural sound; however, miking close to the sound hole is often popular on stage or around high acoustic levels because the guitar's output is highest at this position. To achieve a more natural pickup under these conditions, the microphone's output can be rolled off at the lower frequencies (5–10 dB at 100 Hz).

Surround Guitar Miking

An effective way to translate an acoustic guitar to the wide stage of surround (if a big, full sound is what you're after) is to record the guitar using X/Y or spaced techniques stereo (panned front L and R) ... and pan the guitar's electric pickup (or contact pickup, if it doesn't have one) to the rear center of the surround field. Extra ambient surround mics can be used in an all-acoustic session.

The Electric Guitar

The fundamentals of the average 22-fret guitar extend from E2 to D6 (82–1174 Hz), with overtones that extend much higher. All of these frequencies might not be amplified, as the guitar chord tends to attenuate frequencies above 5 kHz (unless the guitar has a built-in low impedance converter or low-impedance pickups). The frequency limitations of the average guitar loudspeaker often add to this effect, as their upper limit is generally restricted to below 5 or 6 kHz.

Miking the Guitar Amp

The most popular guitar amplifier used for recording is a small practice-type amp/speaker system. This amp type often helps the guitar's suffering high end by incorporating a sharp rise in the response range at 4 to 5 kHz, thus helping to give it a clean, open sound. High-volume, wall-of-sound speaker stacks are rarely used in a session, as they're harder to control in the studio and in a mix. By far the most popular mic type for picking up an electric guitar amp is the cardioid dynamic. A dynamic tends to give the sound a full-bodied character without picking up extraneous amplifier noises. Often guitar mics will have a pronounced presence peak in the upper frequency range, giving the pickup an added clarity. For increased separation, a microphone can be placed at a working distance of 2 inches to 1 foot. When miking at a distance of less than 4 inches, mic/speaker placement becomes slightly more critical (Figure 4.47). For a brighter sound, the mic should face directly into the center of the speaker's cone. Placing it off the cone's center tends to produce a more mellow sound while reducing amplifier noise.

Recording Direct

A DI box is often used to feed the output signal of an electric guitar directly into the mic input stage of a recording console or mixer. By routing the direct output signal to a track, a cleaner, more present sound can be recorded (Figure 4.48a). This technique also reduces the leakage that results from having a guitar amp in the studio and even makes it possible for the guitar to be played in the control room or project studio. A combination of direct and miked signals often results in a sound that adds the characteristic fullness of a miked amp to the extra "bite" that a DI tends to give. These may be combined onto a single track or, whenever possible, can be assigned to separate tracks . . . allowing for greater control during mixdown (Figure 4.48b). During an overdub, the ambient image can be "opened up" even further by mixing a semi-distant or distant mic (or stereo pair) with the direct (and even with the close miked amp

Figure 4.47. Miking an electric guitar cabinet directly in front of and off-center to the cone.

on center   off center

Figure 4.48. *Direct recording of an electric guitar: (a) direct recording. (b) combined direct and miked signal.*

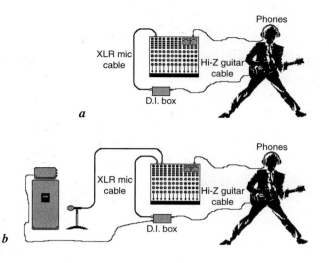

signal). This ambient pickup can be either mixed into a stereo field or at the rear of a surround field to fill out the sound.

The Electric Bass Guitar

The fundamentals of an electric bass guitar range from about E1 to F4 (41.2–343.2 Hz). If it's played loudly or with a pick, the added harmonics can range upwards to 4 kHz. Playing in the "slap" style or with a pick gives a brighter, harder attack, while a "fingered" style will produce a mellower tone. In modern music production, the bass guitar is often recorded direct for the cleanest possible sound. As with the electric guitar, the electric bass can be either miked at the amplifier or picked up through a DI box. If the amp is miked, dynamic mics usually are chosen for their deep, rugged tones. The large-diaphragm dynamic designs tend to subdue the high-frequency transients. When combined with a boosted response at around 100 Hz, these large diaphragm dynamics give a warm, mellow tone that adds power to the lower register. Equalizing a bass can sometimes increase its clarity, with the fundamental being affected from 125 to 400 Hz and the harmonic punch being from 1.5 to 2 kHz. A compressor is commonly used on electric and acoustic bass. It's a basic fact that the signal output from the instrument's notes often varies in level, causing some notes to stand out while others dip in volume. A compressor having a smooth input/output ratio of roughly 4:1, a fast attack (8–20 milliseconds), and a slower release time (1/4–1/2 second) can often smooth out these levels, giving the instrument a strong, present, and smooth bass line.

Keyboard Instruments

The following sections describe the various sound characteristics and techniques that are encountered when miking keyboard instruments.

Grand Piano

The grand piano is an acoustically complex instrument that can be miked in a variety of ways (depending on the style and preferences of the artist, producer, and/or engineer). The overall sound emanates from the instrument's strings, soundboard, and mechanical hammer system. Because of its large surface area, a minimum miking distance of 4 to 6 feet is needed for the tonal balance to fully develop and be picked up; however, leakage from other instruments often means that these distances aren't practical or possible. As a result, pianos are often miked at distances that favor such instrument parts as:

- *Strings and soundboard*, often yielding a bright and relatively natural tone
- *Hammers*, generally yielding a sharp, percussive tone
- *Soundboard holes alone*, often yielding a sharp, full-bodied sound

In modern music production, two basic grand piano styles can be found in the recording studio: concert grands, which traditionally have a rich and full-bodied tone (often used for classical music and ranging in size up to 9 feet in length), and studio grands, which are more suited for modern music production and are designed to have a sharper, more percussive edge to their tone (often being about 7 feet in length).

Figure 4.50 shows a number of possible microphone positions that are acceptable for recording a grand piano. Although several mic positions are illustrated, it's important to keep in mind that these are only guidelines from which to begin. Your own personal sound can be achieved through mic choice and experimentation with mic placement. The following list explains the numbered miking positions shown in Figure 4.49:

- *Position 1.* The mic is attached to the partially or entirely open lid of the piano. The most appropriate choice for this pickup is the boundary mic, which can be permanently attached or temporarily taped to the lid. This method uses the lid as a collective reflector and provides excellent pickup under restrictive conditions (such as on stage and during a live video shoot).
- *Position 2.* Two mics are placed in a spaced stereo configuration at a working distance of 6 inches to 1 inch. One mic is positioned over the low strings and one is placed over the high strings.
- *Position 3.* A single mic or coincident stereo pair is placed just inside the piano between the soundboard and its fully or partially open lid.
- *Position 4.* A single mic or stereo coincident pair is placed outside the piano, facing into the open lid (this is most appropriate for solo or accent miking).
- *Position 5.* A spaced stereo pair is placed outside the lid, facing into the instrument.
- *Position 6.* A single mic or stereo coincident pair is placed just over the piano hammers at a working distance of 4 to 8 inches to give a driving pop or rock sound.

A condenser or extended-range dynamic mic is most often the preferred choice when miking an acoustic grand piano, as they tend to accurately represent the transient and complex nature

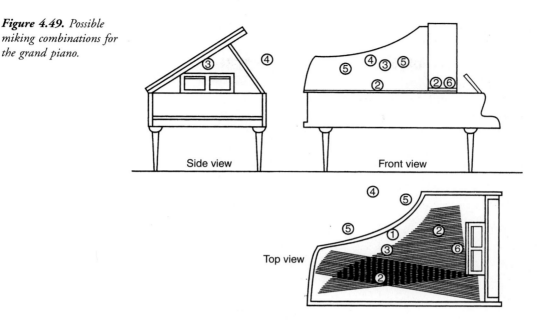

Figure 4.49. *Possible miking combinations for the grand piano.*

Side view   Front view

Top view

of the instrument. Should excessive leakage be a problem, a close-miked cardioid (or cardioid variation) can be used; however, if leakage isn't a problem, backing away to a compromise distance (3–6 feet) can help capture the instrument's overall tonal balance.

Separation

Separation is often a problem associated with the grand piano whenever it is placed next to noisy neighbors. Separation, when miking a piano, can be achieved in the following ways:

◆ Place the piano inside a separate isolation room.

◆ Place a flat (acoustic separator) between the piano and its louder neighbor.

◆ Place the mics inside the piano and lower the lid onto its short stick. A heavy moving or other type of blanket can be placed over the lid to further reduce leakage.

◆ Overdub the instrument at a later time. In this situation, the lid can be removed or propped up by the long stick, allowing the mics to be placed at a more natural-sounding, distance.

Upright Piano

You would expect the techniques for this seemingly harmless piano type to be similar to those for its bigger brother. This is partially true. However, since this instrument was designed for home enjoyment and not performance, the mic techniques are often very different. Since it's

Figure 4.50. *One possible miking combination (over the top) for an upright piano.*

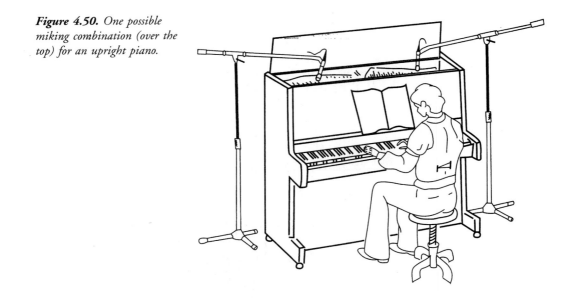

often more difficult to achieve a respectable tone quality when using an upright, you might want to try the following methods (Figure 4.50):

- *Miking over the top.* Place two mics in a spaced fashion just over and in front of the piano's open top—with one over the bass strings and one over the high strings. If isolation isn't a factor, remove or open the front face that covers the strings in order to reduce reflections and, therefore, the instrument's characteristic "boxy" quality. Also, to reduce resonances you might want to angle the piano out and away from any walls.
- *Miking the kickboard area.* For a more natural sound, remove the kickboard at the lower front part of the piano to expose the strings. Place a stereo spaced pair over the strings (one each at a working distance of about 8 inches over the bass and high strings). If only one mic is used, place it over the high-end strings. Be aware, though, that this placement can pick up excessive foot-pedal noise.
- *Miking the upper soundboard area.* In order to reduce excessive hammer attack, place a microphone pair at about 8 inches from the soundboard, above both the bass and high strings. In order to reduce muddiness, the soundboard should be facing into the room or be moved away from nearby walls.

Electronic Keyboard Instruments

Signals from most electronic instruments (such as synthesizers, samplers, and drum machines) are often taken directly from the device's line level output(s) and inserted into a console, either through a DI box or directly into a channel's line-level input. Alternatively, the keyboard's output can be plugged directly into the recorder or interface line-level inputs. The approach to

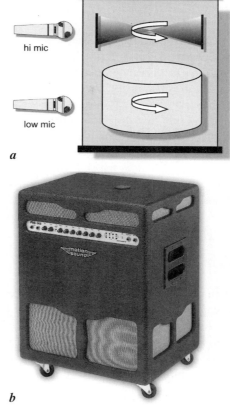

Figure 4.51. *A Leslie speaker cabinet creates a unique vibrato effect by using a set of rotating speaker baffles that spin on a horizontal axis: (a) miking the rotating speakers of a Leslie cabinet; (b) modern portable rotary amp with built-in microphones and three XLR outputs. (Courtesy of Motion Sound, www.motion-sound.com.)*

miking an electronic organ can be quite different from the techniques just mentioned. A good Hammond or other older organ can sound wonderfully "dirty" through miked loudspeakers. Such organs are often played through a Leslie cabinet (Figure 4.51), which adds a unique, Doppler-based vibrato. Inside the cabinet is a set of rotating speaker baffles that spin on a horizontal axis and, in turn, produce a pitch-based vibrato as the speakers accelerate toward and away from the mics. The upper high-frequency speakers can be miked by either one or two mics (each panned left and right), with the low-frequency driver being picked up by one mic. Motor and baffle noises can produce quite a bit of wind, possibly creating the need for a windscreen and/or experimentation with placement.

Percussion

The following sections describe the various sound characteristics and techniques that are encountered when miking drums and other percussion instruments.

Drum Set

The standard drum kit (Figure 4.52) is often at the foundation of modern music, as it provides the "heartbeat" of a basic rhythm track; consequently, a proper drum sound is extremely important to the outcome of most music projects. Generally, the drum kit is composed of the kick drum, snare drum, high-toms, low-tom (one or more), hi-hat, and a variety of cymbals. Since a full kit is a series of interrelated and closely spaced percussion instruments, it often takes real skill to translate the proper spatial and tonal balance into a project. The larger-than-life driving sound of the acoustic rock drum set that we've all become familiar with is the result of an expert balance among playing techniques, proper tuning, and mic placement. Should any of these variables fall short, the search for that "perfect sound" could prove to be a long and hard one. As a general rule, a poorly tuned drum will sound just as out-of-tune through a good mic as it will through a bad one; therefore, it's important to be sure that the individual drum components sound good to the ears, before attempting to place the mics.

Miking the Drum Set

After the drum set has been optimized for the best sound, the mics can be placed into their pickup positions (Figure 4.53). Because each part of the drum set is so different in sound and function, it's often best to treat each grouping as an individual instrument. In its most basic form, the best place to start when miking a drum set is to start with the fundamental "groups." These include placing a mic on the kick (1) and on the snare drum (4). At an absolute minimum, the entire drum set can be adequately picked up using only four mics by adding two overhead pickups... either spaced (3) or coincident (4). In fact, this "bare bones"

Figure 4.52. Peter Erskine's studio drum kit. (Courtesy of Beyerdynamic, www.beyerdynamic.com.)

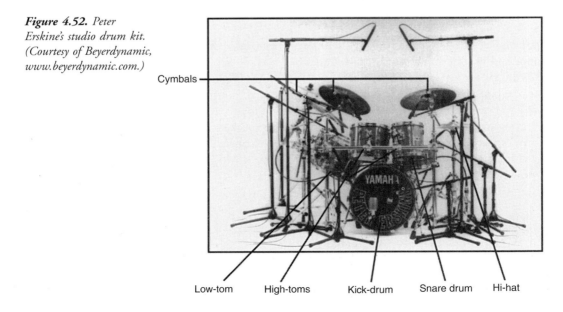

Cymbals

Low-tom   High-toms   Kick-drum   Snare drum   Hi-hat

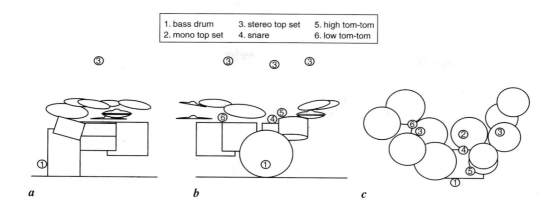

| 1. bass drum | 3. stereo top set | 5. high tom-tom |
| 2. mono top set | 4. snare | 6. low tom-tom |

Figure 4.53. *Typical microphone placements for a drum set: (a) top view; (b) front view; (c) side view.*

placement was (and continues to be) commonly used on many classic jazz recordings. If more tracks are available (or required), additional mics can be placed on the various toms, hi-hat, and even individual cymbals.

A mic's frequency response, polar response, proximity effect, and transient response should be taken into effect when matching it to the various drum groups. Dynamic range is another important consideration when miking drums. Since a drum set is capable of generating extremes of volume and power (as well as softer, more subtle sounds), the chosen mics must be able to withstand strong peaks without distorting...yet still be able to capture the more delicate nuances of a sound.

Since the drum set usually is one of the loudest sound sources in a studio setting, it's often wise to place it on a solidly supported riser. This reduces the amount of low-end "thud" that can otherwise leak through the floor into other parts of the studio. Depending on the studio layout, the following drum scenarios may occur:

◆ The drums could be placed in their own room, isolated from other instruments.

◆ To achieve a bigger sound, the drums could be placed in the large studio room while the other instruments are placed in smaller iso-rooms or are recorded direct.

◆ To reduce leakage, the drums could be placed in the studio, while being enclosed by 4-foot (or higher) divider flats (Figure 4.54).

Kick Drum

The kick drum adds a low-energy drive or "punch" to a rhythm groove. This drum has the capability to produce low frequencies at high sound-pressure levels, so it's necessary to use a mic that can both handle and faithfully reproduce these signals. Often the best choice for the job is a large-diaphragm dynamic mic. Since proximity effect (bass boost) occurs when using

Figure 4.54. ClearSonic's IsoPac B Isolation Package. (Courtesy of ClearSonic Mfg., Inc., www.clearsonic.com.)

Figure 4.55. Placing the microphone at a distance just outside the kick drumhead to bring out the low end and natural fullness.

a directional mic at close working distances and because the drum's harmonics vary over its large surface area, even a minor change in placement can have a profound effect on the pickup's overall sound. Moving the mic closer to the head (Figure 4.55) can add a degree of warmth and fullness, while moving it farther back often emphasizes the high-frequency "click." Placing the mic closer to the beater emphasizes the hard "thud" sound, whereas an off-center mic captures more of the drum's characteristic skin tone. A dull and loose kick sound can be tightened to produce a sharper, more defined transient sound, by placing a blanket or other damping material inside the drum shell firmly against the beater head. Cutting back on the kick's equalization at 300 to 600 Hz can help reduce the dull "cardboard" sound, whereas boosting from 2.5 to 5 kHz adds a sharper attack, "click," or "snap." It's also often a good idea to have a can of WD-40® or other light oil handy in case squeaks from some of the moving parts (most often the kick pedal) gets picked up by the mics.

Snare Drum

Commonly, a snare mic is aimed just inside the top rim of the snare drum at a distance of about 1 inch (Figure 4.56). The mic should be angled for the best possible separation from other drums and cymbals. Its rejection angle should be aimed at either the hi-hat or rack-toms (depending on leakage difficulties). Usually, the mic's polar response is cardioid, although bidirectional and super-cardioid responses may offer a tighter pickup angle. With certain musical styles (such as jazz), you might want a crisp or "bright" snare sound. This can be achieved by placing an additional mic on the snare drum's bottom head and then combining the two mics onto a single track. As the bottom snare head is 180° out of phase with the top, it's generally a wise idea to reverse the bottom mic's phase polarity. When playing in styles where the snare springs are turned off, it's also a good idea to keep your ears open for snare rattles and buzzes that can easily leak into the snare mic (as well as other mics). The continued ringing of an "open" snare note (or any other drum type, for that matter) can be dampened in several ways. Dampening rings, which can be purchased at music stores, are used to reduce the ring and to deepen the instrument's tone. If there are no dampening rings around, the tone can be dampened by taping a billfold or similar-sized folded paper towel to the top of a drumhead, a few inches off its edge.

Overheads

Overhead mics are generally used to pick up the high-frequency transients of cymbals with crisp, accurate detail while also providing an overall blend of the entire drum kit. Because of the cymbals' transient nature, a condenser mic is often chosen for its accurate high-end response. Overhead mic placement can be very subjective and personal. One type of placement is the spaced pair, whereby two mics are suspended above the left and right sides of the kit. These mics are equally distributed about the L–R cymbal clusters so as to pick up their respective instrument components in a balanced fashion (Figure 4.57a). Another placement method is to suspend the mics closely together in a coincident fashion (Figure 4.57b). This

Figure 4.56. *Typical microphone positioning for the snare drum.*

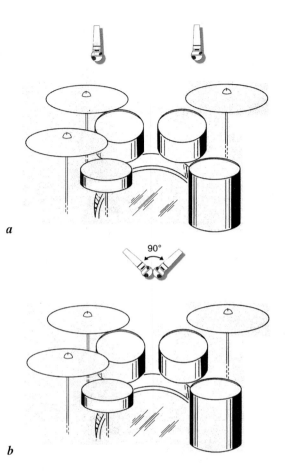

Figure 4.57. Typical stereo overhead pickup positions: (a) spaced pair technique; (b) X/Y coincident technique.

a

b

often yields an excellent stereo overhead image with a minimum of the phase cancellations that might otherwise result when using spaced mics. Again, it's important to remember that there are no rules for getting a good sound. If only one overhead mic is available, place it at a central point over the drums. If you're using a number of pickups to close mic individual components of a kit, there might be times when you won't need overheads at all (the leakage spillover just might be enough to do the trick).

Rack-Toms

The upper rack-toms can be miked either individually (Figure 4.58) or by placing a single mic between the two at a short distance (Figure 4.59). When miked individually, a "dead" sound can be achieved by placing the mic close to the drum's top head (about 1 inch above and 1 to 2 inches in from the outer rim). A sound that's more "live" can be achieved by increasing the height above the head to about 3 to 6 inches. If isolation or feedback is a consideration, a hypercardioid pickup pattern can be chosen. Another way to reduce leakage and to get a deep,

Figure 4.58. *Individual miking of a rack-tom.*

Figure 4.59. *Single microphone placement for picking up two toms.*

driving tone (with less attack) is to remove the tom's bottom head and place the mic inside, 1 to 6 inches away from the top head.

Floor-Tom

Floor-toms can be miked similarly to the rack-toms (Figure 4.60). The mic can be placed 2 to 3 inches above the top and to the side of the head, or it can be placed inside 1 to 6 iches from the head. Again, a single mic can be placed above and between the two floor-toms, or each can have its own mic pickup (which often yields a greater degree of control over panning and tonal color).

Hi-Hat

The "hat" usually produces a strong, sibilant energy in the high-frequency range, whereas the snare's frequencies often are more concentrated in the mid-range. Although moving the

Figure 4.60. *Typical microphone placement for the floor-tom.*

hat's mic won't change the overall sound as much as it would on a snare, you should still keep the following three points in mind:

◆ Placing the mic above the top cymbal will help pick up the nuances of sharp stick attacks.

◆ The open and closing motion of the hi-hat will often produce rushes of air; consequently, when miking the hat's edge, angle the mic slightly above or below the point where the cymbals meet.

◆ If only one mic is available (or desired), both the snare and hi-hat can be picked up simultaneously by carefully placing the mic between the two, facing away from the rack-toms as much as possible. Alternatively, a figure-8 mic can be placed between the two with the null axis facing toward the cymbals and the kick.

Tuned Percussion Instruments

The following sections describe the various sound characteristics and techniques that are encountered when miking tuned percussion instruments.

Congas and Hand Drums

Congas, tumbas, and bongos are single-headed, low-pitched drums that can be individually miked at very close distances of 1 to 3 inches above the head and 2 inches in from the rim . . . or the mics can be pulled back to a distance of 1 foot for a fuller, "live" tone. Alternatively, a single mic or X / Y stereo pair can be placed at a point about 1 foot above and between the drums (which are often played in pairs). Another class of single-headed, low-pitched drums (known as hand drums) isn't necessarily played in pairs but is often held in the lap or strapped across the player's front. Although these drums can be as percussive as congas, they're often deeper in tone and often require that the mic(s) be backed off in order to allow the sound to develop and/or fully interact with the room. In general, a good pickup can be achieved by placing a mic at a distance of 1 to 3 feet in front of the hand drum's head. Since a large part of the drum's sound (especially its low-end power) comes from its back hole, another mic can be placed at the lower port at a distance of 6 inches to 2 feet. Since the rear sound will be 180° out of phase from the front pickup, the mic's phase should be reversed whenever the two signals are combined.

Xylophone, Vibraphone, and Marimba

The most common way to mic a tuned percussion instrument is to place two high-quality condenser or extended-range dynamic pickups above the playing bars at a spaced distance that's appropriate to the instrument size (following the 3:1 general rule). A coincident stereo pair can help eliminate possible phase errors; however, a spaced pair will often yield a wider stereo image.

Stringed Instruments

Of all the instrumental families, stringed instruments are perhaps the most diverse. Ethnic music often uses instruments that range from being single stringed to those that use highly complex and developed systems to produce rich and subtle tones. Western listeners have grown accustomed to hearing the violin, viola, cello, and double bass (both as solo instruments and in an ensemble setting). Whatever the type, stringed instruments vary in their design type and in construction to enhance or cut back on certain harmonic frequencies. These variations are what give a particular stringed instrument its own characteristic sound.

Violin and Viola

The frequency range of the violin runs from 200 Hz to 10 kHz. For this reason, a good mic that displays a relatively flat frequency response should be used. The violin's fundamental range is from G3 to E6 (200–1300 Hz), and it is particularly important to use a mic that's flat around the formant frequencies of 300 Hz, 1 kHz, and 1200 Hz. The fundamental range of the viola is tuned a fifth lower and contains fewer harmonic overtones. In most situations, the violin or viola's mic should be placed within 45° of the instrument's front face. The distance will depend on the particular style of music and the room's acoustic condition. Miking at a greater distance will generally yield a mellow, well-rounded tone, whereas a closer position might yield a scratchy, more nasal quality...the choice will depend on the instrument's tone quality. The recommended miking distance for a solo instrument is between 3 and 8 feet, over and slightly in front of the player (Figure 4.61). Under studio conditions, a closer mic distance of between 2 and 3 feet is recommended. For a fiddle or jazz/rock playing style, the mic can be placed at a close working distance of 6 inches or less, as the increased overtones help the instrument to cut through an ensemble. Under PA (public address) applications, distant working conditions are likely to produce feedback (since less amplification is needed).

Figure 4.61. *Example of a typical microphone placement for the violin.*

In this situation, an electric pickup, contact, or clip-type microphone can be attached to the instrument's body or tailpiece.

Cello

The fundamental range of the cello is from C2 to CS (56–520 Hz), with overtones up to 8 kHz. If the player's line of sight is taken to be 0°, then the main direction of sound radiation lies between 10° and 45° to the right. A quality mic can be placed level with the instrument and directed toward the sound holes. The chosen microphone should have a flat response and be placed at a working distance of between 6 inches and 3 feet.

Double Bass

The double bass is one of the orchestra's lowest-pitched instruments. The fundamentals of the four-string type reach down to E1 (41 Hz) and up to around middle C (260 Hz). The overtone spectrum generally reaches upward to 7 kHz, with an overall angle of high-frequency dispersion being ±15° from the player's line of sight. Once again, a mic can be aimed at the "f" holes at a distance of between 6 inches and 1-1/2 feet.

Voice

From a shout to a whisper, the human voice is a talented and versatile sound source that displays a dynamic and timbral range that's matched by few other instruments. The male bass voice can ideally extend from E2 to D4 (82–293 Hz) with sibilant harmonics extending to 12 kHz. The upper soprano voice can range upward to 1050 Hz with harmonics that also climb to 12 kHz.

When choosing a mic and its proper placement, it's important to step back for a moment and remember that the most important "device" in the signal chain is the vocalist. Let's assume that the engineer/producer hasn't made the classic mistake of waiting until the last minute (when the project goes over budget and/or into overtime) to record the vocals . . . . Good, now the vocalist can relax and concentrate on a memorable performance. Next step is to concentrate on the vocalist's "creature comforts": How are the lighting and temperature settings? Is the vocalist thirsty? Once done, you can go about the task of choosing your mic and its placement to best capture the performance.

The engineer/producer should be aware of the following traps that are often encountered when recording the human voice:

◆ *Excessive dynamic range.* This can be solved either by mic technique (physically moving away from the mic during louder passages) or by inserting a compressor into the signal path. Some vocalists have dynamics that range from whispers to normal volumes to practically screaming . . . all in a single passage. If you optimize your recording levels during a moderate-volume passage and the singer begins to belt out the lines, then the levels will become too "hot" and will distort. Conversely, if you set your recording levels for the loudest passage, the moderate volumes will be buried in the music. The solution to this dilemma is to place a compressor in the mic's signal path. The compressor automatically "rides" the signal's gain and reduces excessively loud passages to a level that the system can effectively handle. (See Chapter 12 for more information about compression and devices that alter dynamic range.)

◆ *Sibilance.* This occurs when sounds such as "f," "s," and "sh" are overly accentuated. This often is a result of tape saturation and distortion at high levels or slow tape speeds. Sibilance can be reduced by inserting a frequency-selective compressor (known as a de-esser) into the chain or through the use of moderate equalization.

◆ *Popping.* Explosive popping "p" and "b" sounds result when turbulent air blasts from the mouth strike the mic diaphragm. This problem can be avoided or reduced by placing a pop filter over the mic, by placing a mesh windscreen between the mic and the vocalist, or by using an omnidirectional mic (which is less sensitive to popping).

◆ *Excessive bass boost due to proximity effect.* This bass buildup often occurs when a directional mic is used at close working ranges. It can be reduced or compensated for by increasing the working distance between the source and the mic, by using an omnidirectional mic (which doesn't display a proximity bass build up), or through the use of equalization.

Woodwind Instruments

The flute, clarinet, oboe, saxophone, and bassoon combine to make up the woodwind class of instruments. Not all modern woodwinds are made of wood nor do they produce sound in the same way. For example, a flute's sound is generated by blowing across a hole in a tube, while other woodwinds produce sound by causing a reed to vibrate the air within a tube.

Opening or covering finger holes along the sides of the instrument controls the pitch of a woodwind by changing the length of the tube and, therefore, the length of the vibrating air column. It's a common misunderstanding that the natural sound of a woodwind instrument radiates entirely from its bell or mouthpiece. In reality, a large part of its sound often emanates from the fingerholes that span the instrument's entire length.

Clarinet

The clarinet comes in two pitches: the B clarinet, with a lower limit of D3 (147 Hz), and the A clarinet, with a lower limit of C3 (139 Hz). The highest fundamental is around G6 (1570 Hz), whereas notes an octave above middle C contain frequencies of up to 1500 Hz when played softly. This spectrum can range upward to 12 kHz when played loudly. The sound of this reeded woodwind radiates almost exclusively from the finger holes at frequencies between 800 Hz and 3 kHz; however, as the pitch rises, more of the sound emanates from the bell. Often, the best mic placement occurs when the pickup is aimed toward the lower finger holes at a distance of 6 inches to 1 foot (Figure 4.62).

Flute

The flute's fundamental range extends from B3 to about C7 (247–2100 Hz). For medium loud tones, the upper overtone limit ranges between 3 and 6 kHz. Commonly, the instrument's sound radiates along the player's line of sight for frequencies up to 3 kHz. Above this frequency, however, the radiated direction often moves outward 90° to the player's right. When miking a flute, placement depends on the type of music being played and the room's overall acoustics. When recording classical flute, the mic can be placed on-axis and slightly above the player at a distance of between 3 and 8 feet. When dealing with modern musical styles,

Figure 4.62. *Typical microphone position for the clarinet.*

Figure 4.63. *Typical microphone position for the flute.*

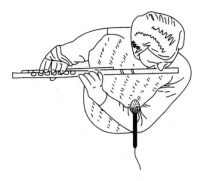

the distance often ranges from 6 inches to 2 feet. In both circumstances, the microphone should be positioned at a point 1/3 to 1/2 the distance from the instrument's mouthpiece to its footpiece. In this way, the instrument's overall sound and tone quality can be picked up with equal intensity (Figure 4.63). Placing the mic directly in front of the mouthpiece will increase the level (thereby reducing feedback and leakage); however, the full overall body sound won't be picked up and breath noise will be accentuated. If mobility is important, an integrated contact pickup can be used or a clip mic can be secured near the instrument's mouthpiece.

Saxophone

Saxophones vary greatly in size and shape. The most popular sax for rock and jazz is the S-curved B-flat tenor sax, whose fundamentals span from B2 to F5 (117–725 Hz), and the E-flat alto, which spans from C3 to G5 (140–784 Hz). Also within this family are the straight-tubed soprano and sopranino, as well as the S-shaped baritone and bass saxophones. The harmonic content of these instruments ranges up to 8 kHz and can be extended by breath noises up to 13 kHz. As with other woodwinds, the mic should be placed roughly in the middle of the instrument at the desired distance and pointed slightly toward the bell (Figure 4.64). Keypad noises are considered to be a part of the instrument's sound; however, even these can be reduced or eliminated by aiming the microphone closer to the bell's outer rim.

Harmonica

Harmonicas come in all shapes, sizes, and keys and are divided into two basic types: the diatonic and the chromatic. Their pitch is determined purely by the length, width, and thickness of the various vibrating metal reeds. The "harp" player's habit of forming his or her hands around the instrument is a way to mold the tone by forming a resonant cavity. The tone can be deepened and a special "wahing" effect can be produced by opening and closing a cavity that's formed by the palms; consequently, many harmonica players carry their preferred microphones with them (Figure 4.65) rather than being stuck in front of an unfamiliar mic and stand.

Figure 4.64. *Typical microphone positions for the saxophone: (a) standard placement; (b) typical "clip-on" placement.*

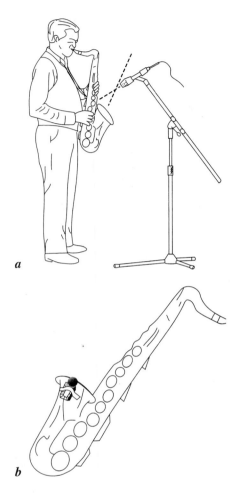

a

b

Figure 4.65. *The Shure 520DX "Green Bullet" microphone, a preferred harmonica pickup for many musicians. (Courtesy of Shure Brothers, Inc., www.shure.com.)*

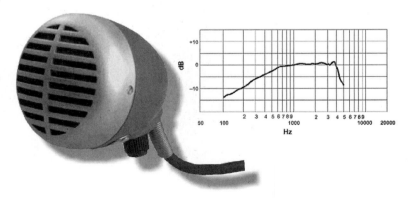

Microphone Selection

The following information is meant to provide insights into a limited number of professional mics that are used for music recording and professional sound applications. This list is by no means complete, as literally hundreds of mics are available, each with its own particular design, sonic character, and application.

Shure SM-57

The SM-57 (Figure 4.66) is widely used by engineers, artists, touring sound companies, etc., for instrumental and remote recording applications. The SM-57's mid-range presence peak and good low-frequency response make it useful for use with vocals, snare drums, toms, kick drums, electric guitars, and keyboards.

Specifications

- *Transducer type:* moving-coil dynamic
- *Polar response:* cardioid
- *Frequency response:* 40–15,000 Hz
- *Equivalent noise rating:* −7.75 dB (0 dB = 1 V/microbar)

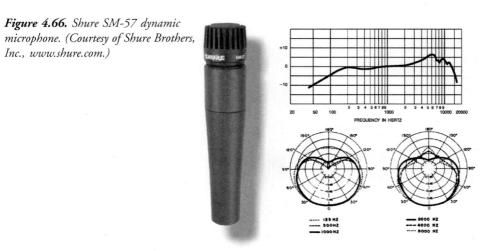

Figure 4.66. *Shure SM-57 dynamic microphone. (Courtesy of Shure Brothers, Inc., www.shure.com.)*

Audix D2

This dynamic hypercardioid drum and instrument microphone (Figure 4.67) has a warm, contoured response for added bottom and punch on drums, instruments, and brass. It features a VLM noise rejection capsule and compact design for easy placement and is milled from a solid block of aluminum.

Specifications

◆ *Transducer type:* moving-coil dynamic
◆ *Polar response:* hypercardioid
◆ *Frequency response:* 44 to 18,000 Hz
◆ *Maximum SPL rating:* 144 dB

The Ball

So named for its unique shape and inimitable Blue styling, the Ball (Figure 4.68) is a phantom-powered dynamic. Although dynamic mics don't require an external supply to operate, the Ball incorporates a phantom-powered proprietary active balancing circuit in its output stage to maintain a constant, pure-resistive, 50-ohm load across the useable frequency spectrum, which has a dramatic effects on the transducer's acoustic balance, phase coherence, noise specification, and overall sound.

Figure 4.67. *Audix D2 dynamic microphone. (Courtesy of Audix Corporation, www.audixusa.com.)*

Figure 4.68. *The Ball. (Courtesy of Baltic Latvian Universal Electronics, www.bluemic.com.)*

Specifications

- *Operating principal:* dynamic transducer with active "class A" phantom-powered solid-state circuitry
- *Polar pattern:* cardioid
- *Frequency response:* 35Hz to 16kHz
- *Sensitivity:* 3.5 mV/Pa at 1kHz (1 Pa = 94 dB SPL)
- *Output impedance:* 50 Ω
- *Recommended load impedance:* 2 kΩ
- *Maximum SPL:* 162 dB SPL (2 kΩ load at 1% THD)
- *Output noise:* 17 dB "A" weighted
- *Power requirement:* 48 V DC phantom power
- *Current draw:* 2.5 mA

AKG D112

Large-diaphragm cardioid dynamic mics, such as the AKG D112 (Figure 4.69), are often used for picking up kick drums, bass guitar cabinets, and other low-frequency, high-output sources.

Specifications

- *Transducer type:* moving-coil dynamic
- *Polar response:* cardioid

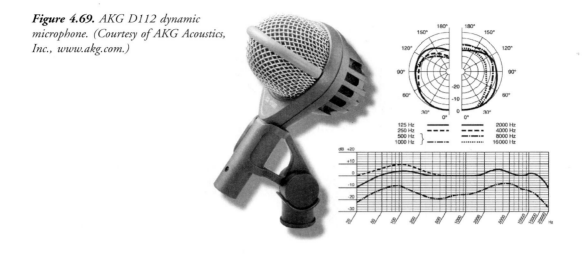

Figure 4.69. AKG D112 dynamic microphone. (*Courtesy of AKG Acoustics, Inc., www.akg.com.*)

◆ *Frequency response:* 30 to 17,000 Hz
◆ *Sensitivity:* −54 dB ±3 dB re. 1 V/microbar

Beyerdynamic M-160

The Beyer M-160 ribbon microphone (Figure 4.70) is capable of handling high sound-pressure levels without sustaining damage while providing the transparency that often is inherent in ribbon mics. Its hypercardioid response yields a wide-frequency response/low-feedback characteristic for both studio and stage.

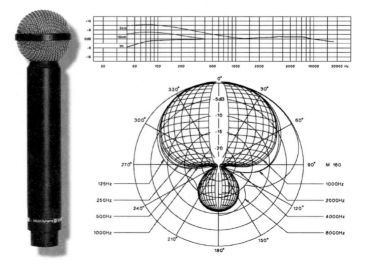

Figure 4.70. Beyerdynamic M-160 ribbon microphone. (*Courtesy of Beyerdynamic, www.beyerdynamic.com.*)

Specifications

- *Transducer type:* Ribbon dynamic
- *Polar response:* Hypercardioid
- *Frequency response:* 40–18,000 Hz
- *Sensitivity:* 52 dB (0 dB = 1 mW/Pa)
- *Equivalent noise rating:* −145 dB
- *Output impedance:* 200 Ω

Royer Labs R-121

The R-121 is a ribbon mic with a figure-8 pattern (Figure 4.71). Its sensitivity is roughly equal to that of a good dynamic mic, and it exhibits a warm, realistic tone and flat frequency response. Made using advanced materials and cutting-edge construction techniques, its response is flat and well balanced; low end is deep and full without getting boomy, mids are well defined and realistic, and the high end response is sweet and natural sounding.

Specifications

- *Acoustic operating principle:* electrodynamic pressure gradient
- *Polar pattern:* figure-8
- *Generating element:* 2.5-micron aluminum ribbon
- *Frequency response:* 30 to 15,000 Hz ±3 dB

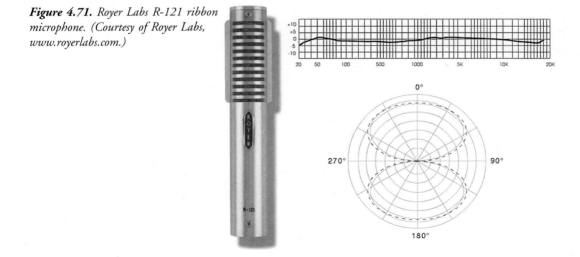

Figure 4.71. *Royer Labs R-121 ribbon microphone. (Courtesy of Royer Labs, www.royerlabs.com.)*

- *Sensitivity:* −54 dBV re. 1 V/Pa ±1 dB
- *Output impedance:* 300 Ω at 1 K (nominal); 200 Ω optional
- *Maximum SPL:* >135 dB
- *Output connector:* male XLR three-pin (pin 2 hot)
- *Finish:* Burnished satin nickel; matte black chrome optional

AEA R84

The AEA R84 ribbon mic (Figure 4.72) is a cost-effective, general-purpose large geometry ribbon (LGR) mic that's ideally suited for solo and accent work. The R84 comes standard with a bidirectional polar response and a black//bright chrome "radio" finish.

Neumann KM 180 Series

The 180 Series consists of three compact miniature microphones (Figure 4.73): the KM 183 omnidirectional and KM 185 hypercardioid microphones and the successful KM 184 cardioid microphone. All 180 Series microphones are available with either a matte black or nickel finish and come in a folding box with a windshield and two stand mounts that permit connection to the microphone body or the XLR-connector.

Specifications

- *Transducer type:* condenser
- *Polar response:* cardioid (183), cardioid (184), and hypercardioid (185)

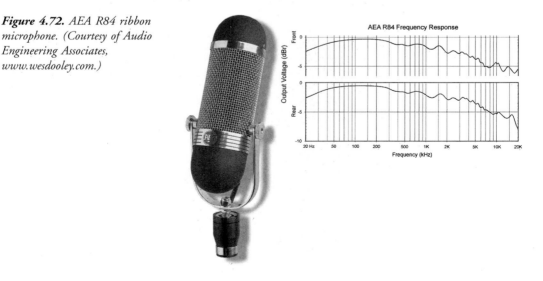

Figure 4.72. AEA R84 ribbon microphone. (Courtesy of Audio Engineering Associates, www.wesdooley.com.)

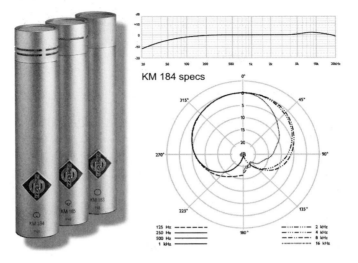

Figure 4.73. *Neumann KM 180 Series condenser microphones. (Courtesy of Georg Neumann GMBH, www.neumann.com.)*

KM 184 specs

◆ *Frequency response:* 20 Hz to 20 kHz

◆ *Sensitivity:* 12/15/10 mV/Pa

◆ *Output impedance:* 50 W

◆ *Equivalent noise level:* 16/16/18 dB −A

AKG C3000B

The AKG C3000B (Figure 4.74) is a low-cost, large-diaphragm condenser mic. Its design incorporates a bass rolloff switch, a−10-dB pad, and a highly effective internal windscreen.

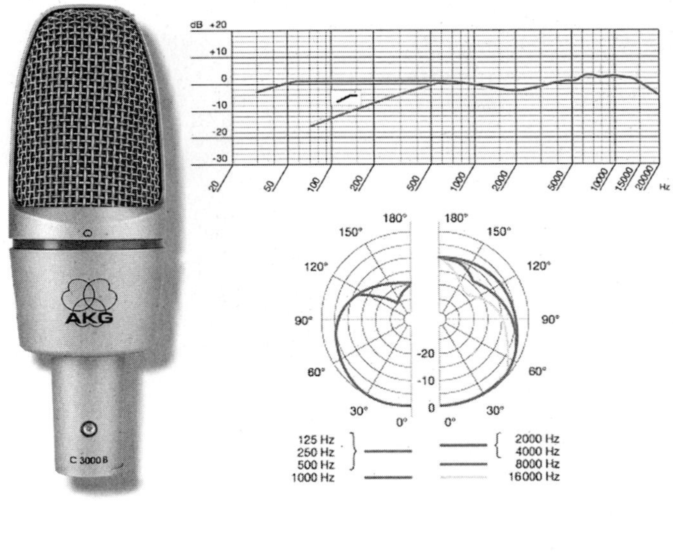

Figure 4.74. *The AKG C3000 B condenser microphone. (Courtesy of AKG Acoustics, Inc., www.akg.com.)*

The mic's dual-diaphragm capsule design is floated in an elastic suspension for improved rejection of mechanical noise.

Specifications

- *Transducer type:* condenser
- *Polar response:* cardioid
- *Frequency response:* 20 to 20,000 Hz
- *Sensitivity:* 25 mV/Pa (−32 dBV)

Marshal MXL 2001

This low-cost, large-capsule, gold-diaphragm condenser (Figure 4.75) has all the fullness and warmth characteristic of the classic large-capsule mics. It has a realistic, natural tone that makes it a good, all-around mic for both vocal and instrumental recording.

Specifications

- *Transducer type:* condenser
- *Polar response:* cardioid
- *Frequency response:* 30 to 20 kHz
- *Sensitivity:* 15 mV/Pa
- *Output impedance:* 200 W
- *Equivalent noise level:* −20 dB−A

Figure 4.75. *The Marshal MXL 2001 condenser microphone. (Courtesy of Marshall Electronics, Inc., www.mxlmics.com.)*

Neumann Solution-D Digital Mic

The system of the Solution-D digital microphone (Figure 4.76) consists of three components: the digital microphone, D-01; the digital microphone interface (DMI)-2; and the remote control software (RCS), which permits operation and thus remote control of the microphone. Signal and data transmission of the microphone conform to the AES 42-2001 Standard, which identifies the transmission of output signals, power supply for the microphones, and remote control of all typical microphone functions and parameters.

◆ *Acoustical transducer:* new double-diaphragm capsule K 07 large diaphragm, diameter 30 mm with protected internal electrodes

◆ *Sensitivity:* 12 mV/Pa at 1 kHz, 0 dB gain

◆ *Equivalent SPL:* CCIR 468-3, 18 dB; DIN/IEC 651, 7 dB-A

◆ *S/N ratio:* CCIR 468-3, 76 dB; DIN/IEC 651, 87 dB

◆ *Weight:* 700 g (approx.)

◆ *Diameter:* 63.5 mm

◆ *Length:* 185 mm

◆ *Interface:* AES 42-2001

◆ *Dynamic range:* 130 dB (complete system including capsule); 140 dB (ADC input shorted)

◆ *Internal resolution:* 28 bit

◆ *Sampling rates:* 48 kHz/96 kHz; alternatively, 44.1 kHz/88.2 kHz

◆ *Remote controllable functions:* Polar pattern (15 patterns), omni...cardioid...figure-8
Low cut, flat; 40 Hz; 80 Hz; 160 Hz
Pre-attenuation, 0 dB, −6 dB, −12 dB, 18 dB
Gain, 0–63 dB in 1-dB steps, clickless

Figure 4.76. *Neumann Solution-D digital microphone. (Courtesy of Georg Neumann GMBH, www.neumann.com.)*

- *Switch functions:* Soft muting, phase reverse, signal light
- *Synchronization:* AES 42, mode 2 (default); AES 42, mode 1 (see DMI)
- *Signals:* Blue and red LED (switchable via control software and Auxiliary User Port)
- *Output:* XLR 3 M, AES 42-2001
- *Data format:* transmission of audio and status data from the microphone, phantom powering and remote control data to the microphone

Studio Projects LSD2 Stereo Mic

The Studio Projects LSD2 stereo mic (Figure 4.77) is comprised of two separate dual-membrane solid-state microphones contained within a single housing. Its capsules are mounted in close proximity on a vertical axis—the upper capsule assembly having the ability to rotate 270° horizontally, relative to the lower capsule. Two three-way switches control the polar response, high-pass filtering, and −10 dB pad for each capsule. The switches on the front of the body correspond to the lower fixed capsule, while the switches 180° opposite on the back of the mic control the rotating upper capsule. It is the combination of capsule articulation and independent pattern switching that allows a user of the LSD2 to achieve all manner of coincident pair stereophonic recording techniques. Due to the close proximity of the capsules, there is no phase cancellation resulting from the time delay between the two signals.

Specifications

- *Type:* Stereo condenser microphone with vertically coincident 1.06-inch (27-millimeter) dual diaphragms
- *Polar pattern:* Cardioid, omnidirectional, figure 8
- *Frequency response:* 30~20,000 Hz
- *Sensitivity:* 12 mV/Pa = −38 dB (0 dB = 1 V/Pa)
- *Output impedance:* <200 Ω
- *Load impedance:* >1200 Ω
- *Maximum SPL:* 139/146 dB SPL for 1% THD @ 1000 Hz (0 dB/ −10 dB pad, 0 dB SPL = 0.00002 Pa)

Figure 4.77. *Studio Projects LSD2 stereo microphone. (Courtesy of Studio Projects USA, www.studioprojectsusa.com.)*

- ◆ *Noise:* Line, 28 dB; A weighted, 18 dB-A
- ◆ *Signal-to-noise ratio:* 76 dB
- ◆ *Power requirement:* 24 to 52.5 V phantom power
- ◆ *Current consumption:* 2.5 mA
- ◆ *Circuit:* transformerless circuit featuring extremely low self-noise and large dynamic range
- ◆ *Connector:* Gold-plated 7-pin XLR
- ◆ *Size:* 2.1-inch (53.34-millimeter) diameter
- ◆ *Low Cut:* 6 dB/octave at 150 Hz
- ◆ *Pad:* −10dB

CHAPTER 5

The Analog Tape Recorder

From its German inception in the late '20s to its American introduction by Jack Mullin in 1945 (Figure 5.1), the *analog tape recorder* (or *ATR*) had increased in quality and universal acceptance to the point that professional and personal studios had totally relied upon magnetic media for the storage of analog sound onto reels of tape. With the dawning of the project studio and the computer-based "Digital Age," the use of two-channel and multitrack ATRs has steadily dwindled to the point where only a few new analog tape machine models are currently being manufactured.... In short, it has become a high-cost, future–retro, "specialty" tool. This being said, the analog recording process is still highly regarded and even sought after by many studios as an additional sonic tool...and by others as a raised fist against the onslaught of the "evil digital empire." Without delving into the ongoing debate of the merits of analog *versus* digital, I think it's fair to say that each has its own distinct type of sound and application in audio and music production. Although professional analog recorders are usually much more expensive than their digital counterparts, as a general rule, a properly aligned, professional analog deck will have its own particular sound, often described as being "full," "punchy," "gutsy," and "raw" (when used on drums, vocals, entire mixes, or almost anything that you want to throw at it). In fact, the limitations of tape and head performance (in the form of saturation) are often used as a form of "artistic expression." From this, it's easy to see and hear why the analog tape recorder isn't dead yet...and probably won't be for some time.

Figure 5.1. John T. Mullin (on the left) proudly displaying his two WWII vintage German Magnetophones, which were the first two tape-based recorders in the United States. (Courtesy of John T. Mullin.)

To 2-Inch or Not To 2-Inch?

Before we delve into the inner workings of the analog tape recorder, let's take a moment to discuss ways that the analog tape sound can be taken advantage of in the digital studio and project studio environment. Beyond the concept of owning your own deck, other cost-effective options might include:

◆ Making use of plug-ins that can emulate (or approximate) the overdriven sound of an analog tape track.

◆ Renting out a studio that has an analog multitrack for a day or two. You could then record new tracks and/or transfer existing live or musical instrument digital interface (MIDI) tracks to tape and then dump them to hard drive, where they can be integrated into your digital session. For the cost of studio time and a reel of tape, you might inject your project with an entirely new type of sound (you might consider buying a single reel of multitrack tape, as it can be erased and reused . . . once the takes have been transferred to disk).

◆ Renting an analog machine from a local rental service. For the rental fee and basic cartage charges, you could have the benefits of having an analog ATR for the duration of a project, without any long-term financial and maintenance overhead.

A few guidelines should also be kept in mind when recording and/or transferring tracks to or from a multitrack recorder:

◆ Obviously, high recording levels add to that sought-after "overdriven" analog sound; however, driving a track too hard (hot) can actually kill a track's definition or "air." The trick is often to find a center balance between the right amount of saturation and downright distortion.

◆ Noise reduction can be a good thing, but it can also diminish what is thought of as the "classic analog sound." Newer, wide tape width recorders (such as ATR Services ATR-102 1-inch, two-track and 108C 2-inch, eight-track recorders), as well as older 2-inch, 16-track recorders, can provide improved definition at lower recording levels (and speeds) without any noise reduction.

Magnetic Recording and Its Media

Like its digital counterpart, the analog audio tape recorder can be thought of as a sound recording device that has the capacity to store audio information onto a magnetizable tape medium and then play this information back at a later time. By definition, analog refers to something that's "analogous," similar to or comparable to something else. An ATR is referred to as an analog machine due to its ability to transform an electrical input signal into a corresponding magnetic energy that can be stored onto tape in the form of magnetic remnants. Upon playback, this magnetic energy is reconverted back into a corresponding electrical signal that can be amplified, mixed, processed, and heard.

In present-day production practices, audio and other forms of information can be recorded onto a wide range of magnetic medium types (including linear tape, credit cards, and hard disks). When speaking of analog audio recording, the tape itself is composed of several layers of material, each serving a specific function (Figure 5.2). The base material that makes up most of a tape's thickness is often composed of polyester or polyvinyl chloride (PVC), which is a durable polymer that's physically strong and can withstand a great deal of abuse before being damaged. Bonded to the PVC base is the all-important layer of magnetic oxide. The molecules of this oxide combine to create some of the smallest known permanent magnets, which are called *domains* (Figure 5.3a). On an unmagnetized tape, the polarities of these domains are randomly oriented over the entire surface of the tape. The net result of this random magnetization is a general cancellation of the north and south magnetic poles of each domain at the reproduce head, resulting in no signal at the recorder's output.

Figure 5.2. *Structural layers of magnetic tape.*

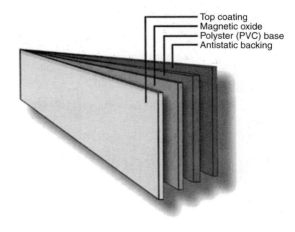

Top coating
Magnetic oxide
Polyster (PVC) base
Antistatic backing

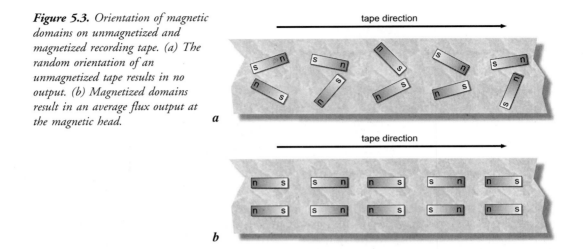

Figure 5.3. *Orientation of magnetic domains on unmagnetized and magnetized recording tape. (a) The random orientation of an unmagnetized tape results in no output. (b) Magnetized domains result in an average flux output at the magnetic head.*

When a signal is recorded, the magnetization from the record head polarizes the individual domains (at varying degrees in positive and negative angular directions) in such a way that their average magnetism produces a much larger combined magnetic flux (Figure 5.3b). When played back at the same, constant speed at which it was recorded, this alternating magnetic output can be converted into an alternating signal that can then be amplified and further processed for reproduction.

The Professional Analog ATR

Professional analog ATRs can be found in 2-, 4-, 8-, 16-, and 24-track formats. Each configuration is generally best suited to a specific production and postproduction task. For example, 2-track ATRs are generally used to record the final stereo mix of a project, whereas 8-, 16-, and 24-track machines are usually used for multitrack recording. Although only a few professional analog machines are currently being manufactured (most notably, the Otari MTR-90 2-inch, 24-track recorder), quite a few analog decks can be found on the used market . . . in varying degrees of working condition. Other decks (such as the ATR-108C 2-inch, multitrack/mastering recorder) allow you to quickly switch between tape widths and track formats. Versatile recorders such as this can be easily converted to handle multitrack, mixdown, and mastering tasks. Indeed, such a machine could be the only analog recorder needed in a facility. Several examples of both new and used machines can be found in Figures 5.4 through 5.7.

The Tape Transport

The process of recording audio onto magnetic tape depends on the transport's capability to pass the tape across a head path at a constant speed and with uniform tension. In other

Figure 5.4. *Otari MX-5050 B3 two-channel recorder. (Courtesy of Otari Corporation, www.otari.com.)*

Figure 5.5. *ATR-102 1-inch stereo mastering recorder. (Courtesy of ATR Service Company, www.atrservice.com.)*

words, recorders must uniformly pass a precise length of tape over the record head within a specific time period (Figure 5.8). During playback, this relationship is maintained by again moving the tape across the heads at the same speed, thereby preserving the program's original pitch, rhythm, and duration.

This constant speed and tension movement of the tape across a head's path is initiated by simply pressing the Play button. The drive can be disengaged at any time by pressing the Stop button, which applies a simultaneous breaking force to both the left and right reels. The Fast Forward and Rewind buttons causes the tape to rapidly shuttle in the respective directions in order to locate a specific point in the tape. Initiating either of these modes engages the tape lifters, which raise the tape away from the heads (definitely an ear-saving device). Once the play mode has been engaged, pressing the Record button allows audio to be recorded onto any selected track or tracks.

Figure 5.6. *ATR-108C 2-inch multitrack/mastering recorder. (Courtesy of ATR Service Company, www.atrservice.com.)*

Figure 5.7. *Otari MTR-90 MkIII 24-track master recorder. (Courtesy of Otari Corporation, www.otari.com.)*

Figure 5.8. *Relationship of time to the physical length of recording tape.*

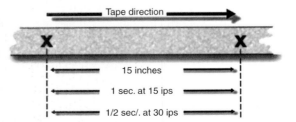

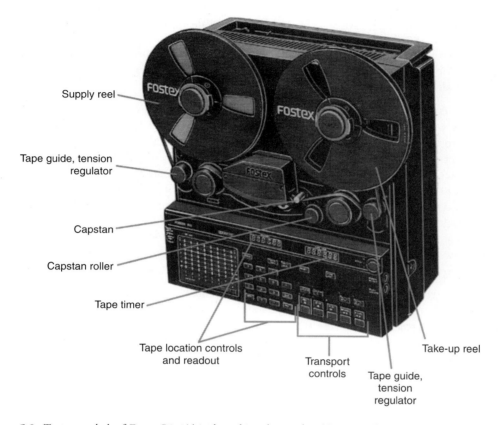

Supply reel

Tape guide, tension regulator

Capstan

Capstan roller

Tape timer

Tape location controls and readout

Transport controls

Tape guide, tension regulator

Take-up reel

Figure 5.9. Transport deck of Fostex R8 1/4-inch multitrack recorder. (Courtesy of Fostex Corporation of America, www.fostex.com.)

Beyond these basic controls, you might expect to run into several differences between transports (often depending on the machine's age). For example, older recorders may require that both the Record and Play buttons be simultaneously pressed in order to go into record; while others might begin recording when the Record button is pressed while already in the Play mode. Figure 5.9 shows the elements of the transport deck of an ATR recorder.

On certain older professional transports (particularly those wonderful Ampex decks from the 1950s and '60s), stopping a fast-moving tape by simply pressing the Stop button can stretch or destroy a master tape, because the inertia is simply too much for the mechanical brake design to deal with. In such a situation, a procedure known as "rocking" the tape is used to prevent tape damage. The deck can be rocked to its stop position by engaging the fast-wind mode in the direction opposite the current travel direction until the tape slows down to a reasonable speed . . . at which point it's safe to press the Stop button.

In recent decades, tape transport designs have incorporated total transport logic (TTL), which places monitoring and overall transport function under microprocessor control. This has a number of distinct advantages in that it lets you push the Play or Stop buttons while the tape is in fast-wind mode without fear of tape damage. With TTL, the recorder can sense the tape speed and direction and then automatically rock the transport until the tape can safely be stopped or can slow the tape to a point where the deck can automatically slip into play or record mode.

Most modern ATRs are equipped with a shuttle control that allows the tape to be shuttled at various wind speeds in either direction. This allows a specific cue point to be located by listening to the tape at varying play speeds or to gently and evenly wind the tape onto its reel at a slower speed for long-term storage. The Edit button (which can be found on certain professional machines) often has two operating modes: stop-edit and dump-edit. If the Edit button is pressed while the transport is in the stop mode, the left and right tape reel brakes are released and the tape sensor is bypassed. This makes it possible for the tape to be manually rocked back and forth until the edit point is found. Often, if the Edit button is pressed while in the play mode, the take-up turntable is disengaged and the tape sensor is bypassed. This allows unwanted sections of tape to be spooled off the machine (and into the trash can) while listening to the material that's being dumped.

A safety switch, which is incorporated into all professional transports, initiates the stop mode when it senses the absence of tape along its guide path; thus, the recorder stops automatically at the end of a reel or should the tape accidentally break. This switch can be built into the tape-tension sensor, or it might exist in the form of a light beam that's interrupted when tape is present.

Most professional ATRs are equipped with automatic tape counters that accurately read out time in hours, minutes, seconds, and sometimes frames (00:00:00:00). Many of these recorders have digital readout displays that double as tape-speed indicators when in the "varispeed" mode. This function incorporates a control that lets you vary the tape speed from fixed industry standards. On many tape transports, this control can be continuously varied over a $\pm 20\%$ range from the 7-1/2, 15, or 30 ips (inches per second) standard.

The Magnetic Tape Head

Most professional analog recorders use three magnetic tape heads, each of which performs a specialized task:

- ◆ Record
- ◆ Reproduce
- ◆ Erase

The function of a *record head* (Figure 5.10) is to electromagnetically transform the analog input signal into corresponding magnetic fields that can be permanently stored onto magnetic tape. In short, the input current flows through coils of wire that are wrapped around the head's magnetic pole pieces. Since the theory of magnetic induction states that "whenever a current is injected into metal, a magnetic field is created within that metal"... a magnetic force is

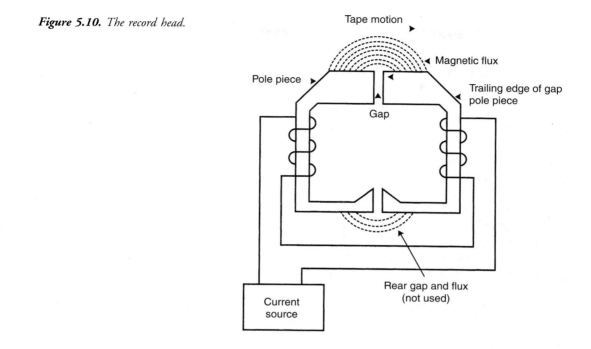

Figure 5.10. *The record head.*

Tape motion

Magnetic flux

Pole piece

Trailing edge of gap
pole piece

Gap

Rear gap and flux
(not used)

Current
source

caused to flow through the pole pieces and across the head gap. Like electricity, magnetism flows more easily through some media than through others. The head gap between poles creates a break in the magnetic field, thereby creating a physical resistance to the magnetic "circuit." Since the gap is in physical contact with the moving magnetic tape, the tape's magnetic oxide offers a lower resistance path to the field than does the nonmagnetic gap. Thus, the flux path travels from one pole piece, into the tape, and to the other pole. Since the magnetic domains retain the same polarity and magnetic intensity that they had on leaving the head gap, the tape now has an analogous magnetic "memory" of the recorded event.

The reproduce or *playback head* (Figure 5.11) operates in a way that's opposite to the record head. When a recorded tape track passes across the reproduce head gap, a magnetic flux is induced into the pole pieces. Since the theory of magnetic induction also states "whenever a magnetic field is induced into metal, a current will be set up within that metal" . . . an alternating current is caused to flow through the pickup coil windings, which can then be amplified and processed into a larger output signal.

The reproduce head's output is nonlinear because this signal is proportional to both the tape's average flux magnitude and the rate of change of this magnetic field. This means that the rate of change increases as a direct function of the recorded signal's frequency. Thus, the output level of a playback head effectively doubles for each doubling in frequency . . . resulting in a 6-dB increase in output voltage for each increased octave (Figure 5.12).

Figure 5.11. The playback head.

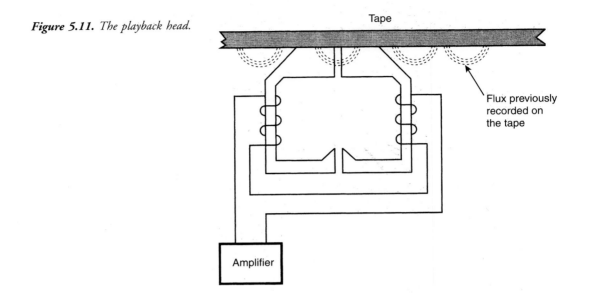

Figure 5.12. Output of a playback head, which increases proportionately with frequency.

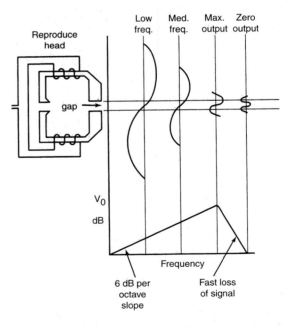

The tape speed and head gap width work together to determine the reproduce head's upper-frequency limit, which in turn determines the system's overall bandwidth. The wavelength of a signal that's recorded onto tape is equal to the speed at which tape travels past the reproduce head, divided by the frequency of the signal; therefore, the faster the tape speed, the higher the upper-frequency limit. Similarly, the smaller the head gap, the higher the upper-frequency limit. However, as the playback signal's frequency increases, more and more of the complete

Figure 5.13. *Scanning loss, which occurs when the recorded wavelengths approach or exceed the head gap's width.*

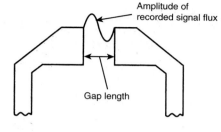

cycle will fall inside the head gap at any one point in time until the signal wavelength is equal to the gap width itself (Figure 5.13). At that point, the average output level will be zero. This reduced output (known as *scanning loss*) is also a factor in determining the system's upper-frequency limit.

The function of the *erase head* is to effectively reduce the average magnetization level of a recorded tape track to zero, thereby allowing the tape to be rerecorded. After a track is placed into the record mode, a high-frequency and high-intensity sine wave signal is fed into the erase head (resulting in a tape that's being saturated in both the positive- and negative-polarity directions). This alternating saturation occurs at such a high speed that it serves to destroy any magnetic pattern that existed on the tape. As the tape moves away from the erase head, the intensity of the magnetic field decreases, leaving the domains in a random orientation, with a resulting average magnetization or output level that's as close to zero as tape noise will allow.

Equalization

Equalization (EQ) is the term used to denote an intentional change in relative amplitudes at different frequencies. Because the analog recording process isn't linear, equalization is needed to achieve a flat frequency–response curve when using magnetic tape. The 6-dB per-octave boost that's inherent in the playback head's response curve requires that a complementary equalization cut of 6 dB per octave be applied within the playback circuit (see Figure 5.14).

Bias Current

In addition to the nonlinear changes that occur in playback output level relative to frequency, another discrepancy in the recording process exists between the amount of magnetic energy that's applied to the record head and the amount of magnetism that's retained by tape after the initial recording field has been removed. As Figure 5.15a shows, a tape's magnetization curve is only linear between points A and B ... and between points C and D. Signals greater than A (in the negative direction) and D (in the positive direction) have reached the saturation

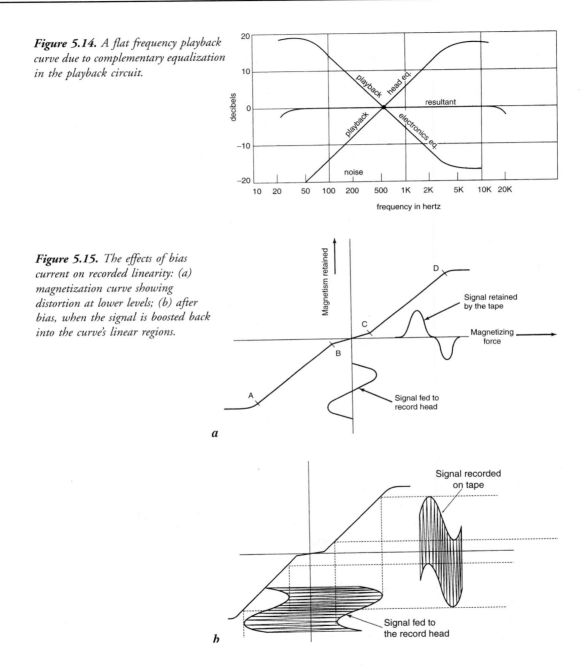

Figure 5.14. *A flat frequency playback curve due to complementary equalization in the playback circuit.*

Figure 5.15. *The effects of bias current on recorded linearity: (a) magnetization curve showing distortion at lower levels; (b) after bias, when the signal is boosted back into the curve's linear regions.*

level and are subject to clipping distortion. Signals falling within the B to C range are too low in flux level to adequately magnetize the domains during the recording process. For this reason, it's important that low-level signals be boosted so that they fall into the linear A–B and C–D ranges. This boost is applied by mixing a *bias current* or *AC bias* in with the audio signal.

Bias current is applied by mixing the incoming audio signal with an ultrasonic signal (often between 75 and 150 kHz). In effect, this causes the high-frequency bias signal to be linearly added with the lower-frequency audible waveform. Since the combined energy (bias + audio signal) is much higher than the audio signal alone, the overall magnetic flux levels are given an extra "oomph," which effectively boosts the signal above the nonlinear zero-crossover range and into the linear portion of the curve. Upon playback, the head is capable of reproducing the high-frequency bias signal at extremely low levels (due to the above-mentioned scanning loss); however, a bias trap circuit is used to eliminate this signal, allowing only the audio signal to be reproduced (Figure 5.15b).

Recording Channels and Monitoring Modes

No matter what the machine's track configuration, each recording channel of a modern ATR is designed to be electrically identical to the others; the channel circuitry is simply duplicated by the number of available channels. Just as the magnetic tape head performs three functions, the electronics of an ATR is specialized to perform the same record, reproduce, and erase processes. On older machines, each recording channel consists of a module that contains three cards, with each one performing its own function. Newer ATRs often use input/output (I/O) modules that incorporate all the adjustment controls and electronics of a recording channel onto a single, printed circuit board.

The output signal of a professional ATR channel can be switched between three working modes:

◆ Input
◆ Reproduce
◆ Sync

In the *input* (source) *mode*, the signal at the selected channel output is derived from its input signal. Thus, with the ATR transport in any mode (including stop), it's possible to meter and monitor a signal that's present at the input of the selected channel. In the reproduce mode, the output and metering signal is derived from the playback head. This mode can be useful in two ways: It allows previously recorded tapes to be played back, and it enables the monitoring of material off of the tape while in the record mode. The latter provides an immediate quality check of the ATR's entire record and reproduce process. The *sync mode* (originally known as selective synchronization, or sel-sync) is a required feature in multitrack ATRs because of the need to record new material on one or more tracks while simultaneously monitoring tracks that have been previously recorded (a process known as *overdubbing*). Here's the deal . . . recording one or more tracks while listening to previously recorded tracks through the reproduce head would actually cause the newly recorded track(s) to be out of sync with the others on final playback (Figure 5.16a). To prevent such a time lag, the tracks to be monitored must be placed in the sync mode. This function actually plays the selected tracks from their respective record head tracks. Since the record head both recording and playing back their respective tracks, there is no time lag and, thus, no signal delay (Figure 5.16b).

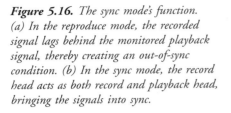

Figure 5.16. *The sync mode's function. (a) In the reproduce mode, the recorded signal lags behind the monitored playback signal, thereby creating an out-of-sync condition. (b) In the sync mode, the record head acts as both record and playback head, bringing the signals into sync.*

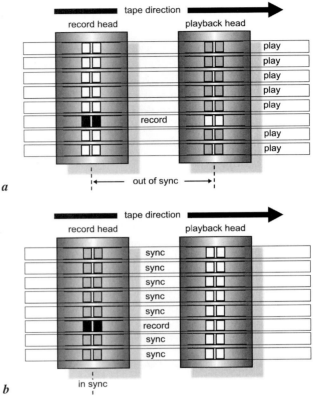

In addition to control over input, reproduce, and sync, the professional ATR has a selectable track function that's known as the *record-enable switch*. Activating this switch (which is usually labeled Safe or Ready) prevents the accidental erasure of a recorded track.

The Remote Control Unit and Autolocator

The *remote control unit* has evolved from having control over simple deck transport functions to a present-day unit that can handle all a recorder's transport, monitoring, and track-status functions. These roll-around units are usually located near the audio production console, which places the operating controls at the hand of the engineer. One feature that's built into many ATR remote control and computer control systems is the *autolocator* (Figure 5.17). This microprocessor-based system lets you enter important cue point locations into memory for recall at any time. When an engineer enters a cue point into a keypad or calls it from memory, the autolocator will go about the task of shuttling the tape to the desired position. At that point, the transport can be programmed to stop, place itself into the play mode, automatically loop (play–relocate–play), or punch-in (record–relocate–record) between two cue points. Time-code-based synchronizers (which are often used in high-level audio and video production) incorporate remote control and autolocator capabilities. Such a synchronization

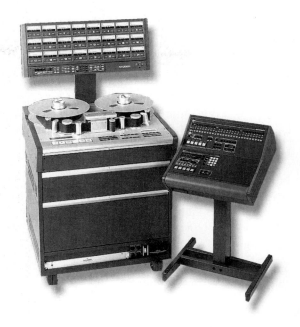

Figure 5.17. Studer A827 analog
multitrack recorder with autolocator.
(Courtesy of Studer
North America, www.studer.ch.)

system (which is explained fully in Chapter 9) uses time code to exercise full, simultaneous control over the remote location functions of one or more transports within the production studio. Whenever multiple ATR, video tape recorder (VTR), and computer-based systems are synchronously interlocked, a single master is designated, and all other units will chase or follow the master tape position until the specified time code address has been reached.

To Punch or Not To Punch

You've all heard the age-old adage... "$%&* happens." Well, it happens in the studio, a lot! Whenever a mistake or bad line occurs during a multitrack session, it's often (but not always) possible to *punch-in* on a specific track or set of tracks. Instead of going back and rerecording an entire song or overdub, performing a punch involves going back and rerecording over a specific section in order to fix a bad note, musical line... you name it. This process is done by cueing the tape at a logical point before the bad section and then pressing Play. Just before the section to be fixed, pressing the Record button will place the track into record mode. At the section's end, pressing the Play button again causes the track to fall back out of record, thereby preserving the section that follows the punch. From a monitor standpoint, the recorders begins playback in the sync mode; once placed into record, the track is switched to monitor the input source and is switched again to monitor the sync playback. This lets the performers hear themselves during the punch while listening to playback both before and after the take.

When performing a punch, it's often far better to "fix" the track immediately after the take, before the levels, mic positions, and performance vibe have had a chance to change. This also makes it easier to go back and rerecord the entire song or a larger section should the punch not work for any reason. If you aren't able to perform a punch at that time, however, it's generally a good idea to take detailed notes about mic selection, placement, preamps, and so on to recreate the session's setup without having to recreate or guess the details from memory.

As every experienced engineer/producer knows, performing a punch can be tricky (especially with an analog recorder). In certain situations, it's a complete no-brainer. These are the times when a stretch of silence the size of a Mac truck exists both before and after the bad section, giving you plenty of room to punch in and out. At other times, a punch can be very tight or problematic (*e.g.*, if there's very little time to punch in or out, when trying to keep vocal lines fluid and in-context, when it's hard to feel the beat of a song or if it has a fast rhythm). In short, punching-in shouldn't be taken too lightly... nor so seriously that you're afraid of the process. Talk it over with the producer and/or musicians.Is this an easy punch? Does the section really need fixing? Do we have the time right now?... Or, is it better just to redo the song?

The following comments and techniques can be helpful when dealing with the possibilities and solutions that are encountered when fixing a recorded section.... The process is usually situational and requires attention, skill, experience, and sometimes a great deal of luck.

- ◆ Before committing the punch to tape, it's often a wise idea to rehearse the punch, without actually committing the fix to tape. This has the advantage of giving both you and the performer a chance to practice beforehand.
- ◆ Some analog decks (and most DAWs) will let you enter the punch-in and punch-out times into their locate functions, thereby allowing the punch to be surgically performed under complete automation.
- ◆ If you're recording onto the same track, a fudged punch may leave you with few options other than rerecording the entire song or section of a song. An alternative to this dilemma would be to record the fix into a separate track and then switch between tracks in mixdown.
- ◆ The track(s) could be transferred to a DAW, where the edits could be performed in the digital domain.

MIDI Machine Control

The MIDI specification includes a protocol known as *MIDI machine control* (*MMC*) that can remotely control the transport functions of a suitably equipped analog, digital, and/or computer-controlled device from a central location (often a DAW, console, or sequencer). Using MMC, a wide range of transport commands can be cost-effectively communicated over standard MIDI lines from a central controller to a particular device or number of devices within a connected system, either manually or under automated computer control. As an example, the music production system shown in Figure 5.18 includes a computer-based MIDI sequencer that's been configured to be the master transport controller for an MMC-equipped analog tape recorder. Since MMC is MIDI based, the real beauty lies in the fact that it can

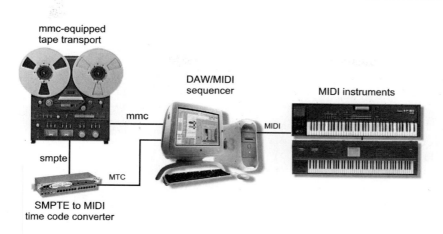

Figure 5.18. *Example of a music production system using MIDI machine control.*

be designed into software packages, controllers, audio and video recorders, mixing consoles, and even musical instruments for little or no added cost to the manufacturer. A more detailed description of MIDI and the MMC protocol can be found in Chapter 7.

Tape, Tape Speed, and Head Configurations

Professional analog ATRs are currently available in a wide range of track- and tape-width configurations. The most common analog configurations are 2-track mastering machines that use tape widths of 1/4-inch, 1/2-inch, and even 1-inch, as well as 16- and 24-track machines that use 2-inch tape. Figure 5.19 details many of the tape formats that can be currently found. Optimal tape-to-head performance characteristics for an analog ATR are determined by several parameters: track width, head-gap width, and tape speed. In general, track widths are on the order of 0.080 inch for a 1/4-inch 2-track ATR; 0.070 inch for 1/2-inch 4-track, 1-inch 8-track, and 2-inch 16-track formats; and 0.037 inch for the 2-inch 24-track format. These widths are large when compared to the 0.021-inch widths that are found in cassette recorder head design. As you might expect, with greater recorded track widths, an increased amount of magnetism can be retained by the magnetic tape...resulting in a higher output signal and an improved signal-to-noise ratio. The use of wider track widths also makes the recorded track less susceptible to signal-level dropouts. The guardband (an unrecorded area that exists between adjacent tracks) is used to reduce unwanted crosstalk between channels.

Since tape speed is directly related to the recorded signal's level and wavelength, it has a direct bearing on the performance characteristics of an ATR. At high tape speeds, the number of magnetic domains that pass over the tape head gap in a given period of time is greater than at slower speeds; thus, the average magnetization that's received by the playback head is greater, produces a stronger output signal, and requires less amplification. As a result, the signal will have a wider dynamic range and less tape noise. Because higher tape speeds

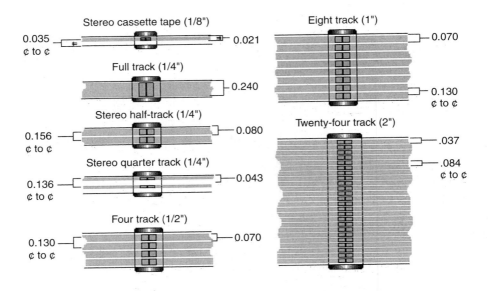

Figure 5.19. *Analog track configurations for various tape widths.*

increase the recorded signal's wavelength, less high-frequency boosting will be required by the record electronics and higher levels can be achieved before the tape will saturate. Due to the reproduce head's response characteristics, the overall bandwidth will also increase at faster tape speeds.

The most common tape speeds used in audio production are 15 ips (38 cm/sec) and 30 ips (76 cm/sec). Although 15 ips will eat up less tape, 30 ips has gained wide acceptance in recent years for having its own characteristic sound (often having a tighter bottom end), as well as a higher output and lower noise figures (which in certain cases eliminates the need for noise reduction). On the other hand, 15 ips has a reputation for having a "gutsy," rugged sound.

Print-Through

A form of deterioration in a recording's quality, known as *print-through*, can occur after a recording has been made. This effect is the result of the transfer of a recorded signal from one layer of tape to an adjacent track layer by means of magnetic induction, which gives rise to an audible false signal or pre-echo on playback. The effects of print-through are greatest when recording levels are very high, and the effect decreases by about 2 dB for every 1-dB reduction in signal level. The extent of this condition also depends on such factors as length of storage, storage temperature, and tape thickness (tapes with a thicker base material are less likely to have severe print-through problems).

Because of the directional properties of magnetic lines of flux, print-through has a greater effect on the outer-facing layers of tape (where magnetic induction is in phase) than on the adjacent inner layers (where the magnetic induction is out of phase), as shown in Figure 5.20. Therefore, if a recorded tape is stored heads-out (with the tape being rewound onto the supply reel), a "ghost" signal will transfer to the outer layer and be heard as a pre-echo, before the original signal. For this reason, the standard method of professionally storing a recorded analog tape is in the *tails-out* position. Remember,

◆ The tape should *always* be stored tails-out (on the right-hand take up reel).

◆ Upon playback, the tape should be wound onto the left-most "supply reel."

◆ During playback, feed the tape back onto the right-hand take-up reel, after which time it can again be removed for storage.

◆ If the tape has been continuously wound and rewound during the session, it's often wise to rewind the tape and then smoothly play or slow-wind the tape onto the take-up reel, after which time, it can be removed for storage.

When a tape is stored tails-out (Figure 5.21), the print-through will bleed to the outer layers, a condition that causes the echo to follow the original signal in a way that's similar to the sound's natural decay and is subconsciously perceived by the listener as reverb.

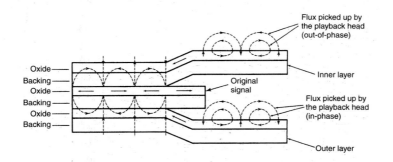

Figure 5.20. Print-through caused by magnetic induction onto a tape's outer layer.

Figure 5.21. Recorded analog tapes should always be stored in the tails-out position.

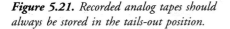

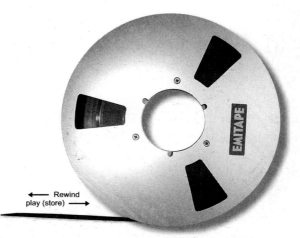

Cleanliness

It's very important that the magnetic recording heads and moving parts of an ATR transport deck be kept free from dirt and oxide shed. Oxide shed occurs when friction causes small particles of magnetic oxide to flake off and accumulate on surface contacts. This accumulation is most critical at the surface of the magnetic recording heads, since even a minute separation between the magnetic tape and heads can cause high-frequency *separation loss*. For example, a signal that's recorded at 15 ips and has an oxide shed buildup of 1 mil (0.001 inch) on the playback head will be 55 dB below its standard level at 15 kHz. Denatured (isopropyl) alcohol or an appropriate cleaning solution should be used to clean transport tape heads and guides (with the exception of the machine's pinch roller and other rubber-like surfaces) at regular intervals.

Degaussing

Magnetic tape heads are made from a magnetically soft metal alloy, which means that the alloy is easily magnetized...but once the coil's current is removed, the core won't retain any of its magnetism. Small amounts of residual magnetism, however, will build up over time, which can actually partially erase high-frequency signals from a master tape. For this reason, all of the tape heads should be demagnetized after 10 hours of operation with a head demagnetizer. This hand-held device works much like an erase head in that it saturates the magnetic head with a high-level alternating signal that randomizes residual magnetic flux. Once a head has been demagnetized (after 5 to 10 seconds), it's important to move the activated tool away from the tape heads at a speed of less than 2 inches per second so as to avoid inducing a larger magnetic flux back into the head. Before an ATR is aligned, the magnetic tape heads should always be cleaned and demagnetized in order to obtain accurate readings and to protect expensive alignment tapes.

Head Alignment

An important factor in an analog recorder's performance is the physical *alignment* of each magnetic tape head. The erase, record, and playback heads will often have five adjustments:

- ◆ Height
- ◆ Azimuth
- ◆ Zenith
- ◆ Wrap
- ◆ Rack

The *height* determines the track's vertical positioning in relation to the tape path (Figure 5.22). If the track is recorded and reproduced on heads that have different height settings, not all

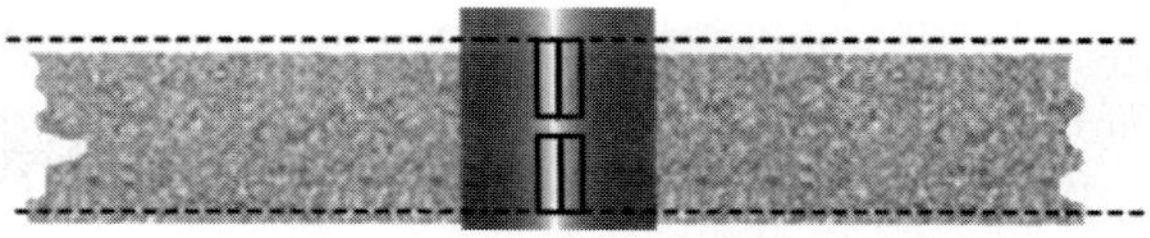

Figure 5.22. *The head gap's height must be centered on the track location.*

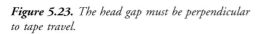

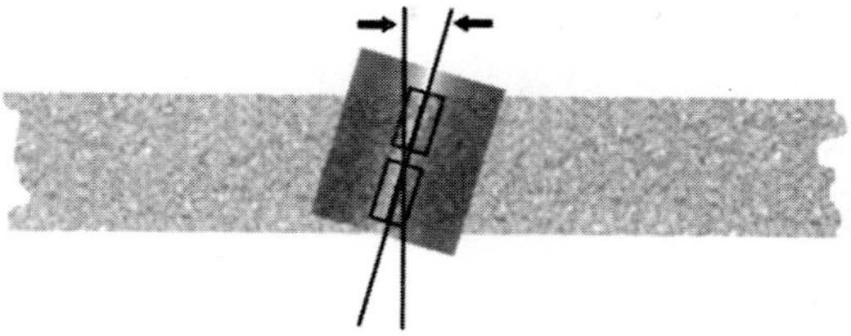

Figure 5.23. *The head gap must be perpendicular to tape travel.*

of the recorded signal will be reproduced, resulting in a compromised signal-to-noise ratio and increased crosstalk between multitrack channels.

Azimuth refers to the head's tilt in the plane parallel to the tape (Figure 5.23). The head gap should be perpendicular (90°) to the tape, so that all track gaps are electrically in phase with each other.

Zenith refers to the head's tilt toward or away from the tape. Zenith must be adjusted so that the tape contacts the top and bottom of the head with the same degree of force—otherwise, the tape will tend to skew off its path. Skewing is bad news and occurs when tape rides up, down, and even out of its guide path.

Wrap refers to the angle at which the tape bends around the head. It also determines the degree of tape-to-head contact and thus controls the head's sensitivity to dropouts.

Rack determines the pressure of the tape against the head. The pressure will increase as the head is moved forward.

Although it's highly recommended that a qualified technician make these adjustments, height can be adjusted visually or by using a test tape in order to yield a maximum output at 1 or 3 kHz. Zenith can be tested by covering the pole pieces with a white grease pencil and playing a piece of scrap tape in order to observe the pattern that's formed as the grease coating wears off. The edges of the wear pattern should be parallel. If they're not, use the screws on the headblock to adjust the zenith. The wrap angle can be checked at the same time as zenith by making sure that the wear pattern is centered on the gap. A rack adjustment may be needed if the wear pattern is wider on the record head than on the playback head, or vice versa.

Azimuth can be tested by deliberately skewing the tape across the heads (by pushing up and down on the edges of the tape near the front of the head), while playing back the 15-kHz tone off an older standard alignment tape. If the output increases, the azimuth will need to be adjusted in either of two ways (both of which use a full-track standard alignment tape). The first step is to play the 15-kHz tone and adjust the azimuth for the highest output on all channels. This setting will be a compromise because various channels on a multitrack machine will rise as others fall on either side of the proper setting. The peak at the correct setting will be fairly sharp while smaller, broader peaks occur on either side. To ensure that the proper peak is found, the head gap should be visually checked to make sure that it's perpendicular to the tape path before further adjustments are attempted. The second method uses the phase angle of a 12- or 15-kHz test tone in order to find the correct setting. After the peak is found (using the first method), the output of the top channel is fed into the vertical input of an oscilloscope, while the bottom channel is fed into the horizontal input of the scope. The resulting pattern on the face of the scope represents the relative phase between the two channels. A straight line sloping up 45° to the right indicates that the two channels are in phase, a circle indicates they are 90° out of phase, while a straight line sloping up 45° to the left indicates that the two channels are 180° out of phase. The azimuth should be adjusted so that the two channels are in phase.

It should be noted that the azimuth adjustment of the heads applies to the record head, as well as the reproduce head, and that record head azimuth adjustment requires that the playback head be adjusted first with the calibration tape . . . after which time the record head azimuth can be adjusted while recording a signal to tape and monitoring back from the playback head.

On a multitrack head, it's not possible to get all the gaps in phase at one time because of gap scatter (variations from a vertical gap line) that occurs in manufacturing. Once the proper phase adjustment has been made for the outer tracks, the inner track phase will usually be within 60° of this figure. The record-head azimuth can be set in the same way as that of the playback head: by playing the test tape while in the sync mode. The erase head's azimuth isn't nearly as critical, since its gap width is much wider. It will be correct as long as it's placed at the proper height and at right angles to the tape path. Although head adjustments don't have to be done frequently, the need for prompt care and attention is generally indicated by deterioration of the ATR's performance, and a qualified maintenance engineer should be called in.

Electronic Calibration

Since sensitivity, output level, bias requirements, and frequency response vary considerably between recorders and from one tape formulation to the next, professional ATRs require that a number of adjustments be set for record/ playback level, equalization, bias current, and so on. It's extremely important that a standard be adhered to in the adjustment of these level settings so that a recording made on one ATR will be compatible with another. The procedure used to set these controls to a standard level is called the *calibration* or electronic alignment process.

Proper ATR calibration depends on the session's tape type and formulation, as well as head alignment settings . . . so it's a good practice to recalibrate all production ATRs at regular or semi-regular intervals. In major music recording studios, analog alignments are routinely made prior to each session in order to ensure the standardization and performance reliability of each master tape. Fortunately, this process isn't necessary when using a modern digital recording system.

Assuming that the record and reproduce heads are in proper alignment, an electronic calibration of the ATR can be performed. This procedure is carried out with reference to a standard set of equalization curves for each tape speed. Such curves have been established by the National

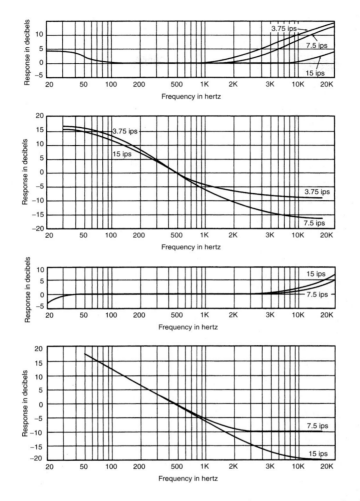

Figure 5.24. *Pre- and post-equalization curves for the NAB (top) and DIN (lower) standards.*

Association of Broadcasters (NAB) in the United States, Japan, and much of the world; throughout Europe, the Deutsche Institute Norme (DIN) is used; and the Audio Engineering Society (AES) standard is used at 30 ips. These levels and EQ curves (Figure 5.24) are established by using a standardized reproduce alignment tape, which is available in various tape speeds and track-width configurations . . . and generally contains the following set of recorded tones:

◆ *Standard reference level.* A 1-kHz signal recorded at a standard reference flux level, 185 nWb/m for standard operating level, 250 nWb/m for elevated level, or 320 nWb/m for DIN (European operating level).

◆ *Azimuth adjustment tone.* 10- to 16-kHz reference tone of 30-second duration.

◆ *Frequency response.* Tones at 16, 32, 63, 125, 250, 500, 1000, 2000, 4000, 8000, 10,000, 12,500, 16,000, and 20,000 Hz.

To anyone new to the alignment process, calibration might seem complicated, but with repetition it quickly becomes second nature and usually takes less than 10 minutes to perform. Electronic calibration is carried out in two stages: playback alignment and record alignment.

Playback Alignment

To begin, place the alignment tape on the recorder in the tails-out position and rewind it onto the supply reel. In the playback mode, you can now follow the verbal instructions on the tape and set the initial reference tone and subsequent frequencies to read at 0 VU on all tracks. Most alignment tapes are recorded in a full-track configuration; that is, the recorded signal is laid down across the entire width of the tape. Because of an effect called "fringing," these test-tape frequencies exhibit a bass boost, which ranges upward to $+3$ dB at 30 Hz. To avoid the inaccuracies of fringing, the low-frequency EQ settings may be postponed until after the record alignment has been made. With each output of the ATR being switched into the sync mode, the preceding process can now be duplicated on each track. By adjusting the appropriate sync trim pots, all level and EQ controls can be likewise aligned. After all the reproduce/sync alignment adjustments have been made, rewind the alignment test tape again and play or slow-wind it without interruption onto the take-up reel for storage in the tails-out position.

Record Alignment

Setup for the record alignment stage consists of placing a fresh reel of unrecorded tape on the machine. Ideally, this tape will be the same reel that's to be used in the actual recording session; if not, it should at least be from the same manufacturer and of the same formulation. The first setting to be made in the routine adjustment of an ATR's record electronics is the *bias adjustment control*. This important adjustment determines the amount of AC-bias signal that's to be mixed with the audio signal at the record head and has a direct relationship to the machine's noise-to-distortion ratio. Too little bias signal will result in an increased noise floor, increased distortion, and an excessive rise in high frequencies. Too much bias results in a reduced output at high frequencies; thus, bias provides for a compromise between noise, distortion, and frequency response.

To set the bias control, thread the tape onto the machine and feed a 1-kHz signal to all inputs of the ATR at operating level (0 VU on the console). With the track output selectors in the reproduce mode, switch the transport to record and use the appropriate adjustment tool to locate the individual track controls marked Bias or Bias Adjust. The ATR's output level can then be monitored from its VU meter bridge. Turn the bias trim counterclockwise until the test signal drops to its lowest level. Now, slowly turn the trim pot clockwise until the signal rises to a peak reading (after which the reading will begin to fall). Continue to increase the trim clockwise, until the VU drops 1 dB in level. The bias level will now be optimized for the specific tape head and formulation that's being used. Continue to adjust bias controls for all other channels. If the recorded signal begins to peak off the VU scale, reduce the record input-level trim pot . . . not the reproduce trim. This is not to be adjusted after it has been set to its standard operating level.

After the bias-adjust controls have been set, adjust the record-level and equalization controls. With the tape rewound to the beginning of the reel, route a 1-kHz signal at 0 VU from the console to all of the ATR's inputs. With all channel electronics switched to reproduce, place the transport in the record mode and adjust the "record level" for all channels until each meter reads 0 VU. Place the ATR in the stop mode and, with all channel electronics switched to read the input signal, adjust each "input level" control to read 0 VU. The meter and output levels between the reproduce and input modes should now be matched. The final step is to adjust the record- equalization controls. In this adjustment, place the ATR back into the record mode. While monitoring the electronics, send tone signals of 100 Hz, 1 kHz, and 10 kHz to all ATR inputs at 0 VU. The individual high-equalization (10-kHz) and low-equalization (100-Hz) trims are adjusted to 0 VU in order to attain the flattest possible frequency response over the audio spectrum. The step-by-step procedures for various speeds and tapes can be carried out using the following instructions:

Playback

1. Thread the playback alignment tape that has been especially produced for either 15 ips or 30 ips onto the machine. Always place the test tape on the take-up turntable and rewind the tape onto the supply reel (it's generally wise to store this tape tails-out, with an even wind tension).

2. Set the repro level for 0 VU at 1000 Hz.

3. Set the high-frequency playback EQ for 0 VU at 10 kHz.

Record

1. Thread the tape that's to be used during the session onto the machine. Set all output selectors to the bias position and set the machine into the record mode on all tracks.

2. Adjust the erase peak control for a maximum meter reading.

3. Feed a 1000-Hz tone into the machine inputs.

4. Set the output selector to playback (tape) and, beginning with a low bias setting, increase the amount of bias until the meter reading rises to a maximum. Continue increasing the bias level until the meter reading drops by 1 dB.

5. Feed a 1000-Hz tone from the console into all ATR inputs at a 0-VU (+ 4-dBm) level.

6. Set all output selectors on the recorder to the input position and adjust the record calibrate controls for a 0-VU meter reading.

7. Set the output selectors to Playback (tape) and feed the console's 0-VU/1-kHz oscillator to all machine inputs. Adjust the playback levels to read 0 VU.

8. Set the console's oscillator to output a 0-VU tone at 10 kHz. Adjust the playback levels to read 0 VU (at the peak of the meter's swing, if the needle is unsteady).

9. Feed a 100-Hz tone into the machine inputs at 0-VU level. Adjust the low-frequency playback EQ for a 0-VU meter reading.

Backup and Archive Strategies

In this day of hard drives, CDs and digital data... we've all come to know the importance of backing up our data. With important music and media projects, it's equally important to back up a copy in case of an unforeseen catastrophe or as added insurance that future generations can enjoy the fruits of your work.

Backing Up Your Project

The one basic truth that can be said about analog recordings that have been stored onto magnetic tape... is that this medium has withstood the test of time. With care and reconditioning, tapes that have been recorded in the 1940s have been fully restored, allowing us to preserve and enjoy some of the best music of the day. On the other hand, digital data has two points that aren't exactly its favor:

◆ Data that resides on hard drives isn't the most robust of media over time. Even CDRs (which are rated to last over 100 years) haven't really been proven to last.

◆ Even if the data remains intact, with the ever-increasing advances in computer technology... who's to say that the media, drives, programs, session formats, and file formats will be around in 5 years, let alone 50!

These warnings aren't slams against digital, just precautions against the march of technology *versus* the need for media preservation. For the above reasons, media preservation is a top priority for such groups as the Recording Academy's Producers & Engineers Wing (P&E Wing), as well as for many major record labels—so much so that many stipulate in their contracts that multitrack sessions (no matter what the original medium) are to be transferred and archived to 2-inch multitrack analog tape.

When transferring digital tracks to an analog machine, it's always wise to make sure that the recorder has been properly calibrated and that tones (1 kHz, 10 kHz, 16 kHz, and 100 Hz) have been recorded at the beginning of the tape for future reference. When copying from analog to analog, both machines should be properly calibrated, but the source for the newly recorded tones should be that of the master tape. If a SMPTE track is required, be sure to stripe the copy with a clean, jam-synched code. The backing up of analog tapes and/or digital data usually isn't a big problem... unless you've lost your original masters. In this situation, it's the difference between panic and peace.

Archive Strategies

Just as it's important to back up your media, it's also important that the original and backed-up media be treated and stored properly. Here are a few guidelines:

◆ Store your tapes tails-out.

◆ Wind the tapes onto the take-up storage reel at slow-wind or play speeds.

◆ Store the boxes vertically. If they're stored horizontally, the outer edges could get bent and become damaged.

◆ Media storage facilities exist that can store your masters or backups for a fee. If this isn't an option, store them is an area that's cool and dry (*e.g.*, no temperature extremes, in low humidity, no attics or basements).

Store your masters and backups in separate locations. In case of a fire or other disaster . . . one would be lost, not both (always a good idea with digital data, as well).

CHAPTER 6

◆

Digital Audio Technology

In recent decades, digital audio (and its related industries) has evolved from being an infant technology that was available only to a select few to its present-day position as a primary driving force in audio production, entertainment, and communication. In fact, digital audio and media production has such an impact upon our lives that it's often an integral part of both the medium and the message within modern-day communication.

Although digital audio is a varied and complex field of study, the basic theory of how the process works isn't really that difficult to understand. At its most elementary level, it is simply a process by which numeric representations of analog signals (in the form of voltage levels) are encoded, processed, stored, and reproduced over time through the use of a binary number system.

Just as English-speaking humans communicate by combining any of 26 letters together into groupings known as "words" and manipulate numbers using the decimal (base 10) system . . . the system of choice for a digital device is the binary (base 2) system. This numeric system provides a fast and efficient means for manipulating and storing digital data. By translating the alphabet, base 10 numbers, or other form of information into a binary form, a digital device (such as a computer or processor) can perform calculations and tasks that might otherwise be cumbersome, less cost effective, and/or downright impossible to perform in the analog domain. This binary data can be encoded in such media forms as:

- ◆ Logical 1 or 0
- ◆ On or off
- ◆ Voltage or no voltage

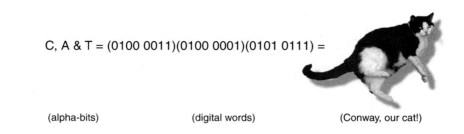

C, A & T = (0100 0011)(0100 0001)(0101 0111) =

(alpha-bits)   (digital words)   (Conway, our cat!)

Figure 6.1. Digital and analog equivalents for a strange four-legged animal (Conway, one of our cats).

- ◆ Magnetic flux or no flux
- ◆ Optical reflection off of a surface or no reflection

After the information has been recorded, stored, and/or processed, the resulting data can be converted back into an analogous form that we humans can readily understand.

Before we delve onto the basic aspects of recording, processing, and reproducing audio in the digital domain, let's take a look at how one form of information can be converted into another equivalent (analogous) form of information. For example, if we type the letters C, A, and T into a word processor, the computer quickly goes about the task of translating these keystrokes into a series of 8-bit digital words that would be represented as [0100 0011], [0100 0001], and [0101 0100]. These "alpha-bits" don't have much meaning when examined individually; however, when placed together into a group, this data represents a four-legged animal that's either seldom around or always underfoot (Figure 6.1). From this, we can deduct that whenever binary words are grouped together as a string of data that has an analogous and recognizable pattern, a meaningful message can be conveyed.

In a similar manner, a digital audio system works by sampling (measuring) the instantaneous voltage level of an analog signal at a single point in time . . . and then converting these samples into an encoded word that digitally represents that voltage level. By successively measuring changes in an analog signal's voltage level (over time) . . . this stream of representative words can be stored in a form that represents the original analog signal. Once stored, the data can then be processed and reproduced ways that have changed the face of audio production forever.

The Basics of Digital Audio

In Chapter 2, we learned about the two most basic characteristics of sound:

- ◆ Frequency (the component of time)
- ◆ Amplitude (the signal-level component)

Digital audio can be likewise broken down into two analogous components:

- ◆ Sampling (which represents the component of time)
- ◆ Quantization (which represents the signal-level component)

Sampling

In the world of analog audio, signals are passed, recorded, stored, and reproduced as changes in voltage levels that continuously change over time (Figure 6.2). The digital recording process, on the other hand, doesn't operate in a continuous manner; rather, digital recording takes periodic *samples* of a changing audio waveform (Figure 6.3) and transforms these sampled signal levels into a representative stream of binary words that can be manipulated or stored for later processing and/or reproduction.

Within a digital audio system, the *sampling rate* is defined as the number of measurements (samples) that are taken of an analog signal in one second. Its reciprocal (sampling time) is the elapsed time that occurs between each sampling period. For example, a sample rate of 48 kHz corresponds to a sample time of 1/48,000th of a second. Because sampling is tied directly to the component of time, the sampling rate of a system determines its overall bandwidth (Figure 6.4), meaning that a system with higher sample rates is capable of storing more frequencies at its upper limit.

Figure 6.2. An analog signal is continuous in nature.

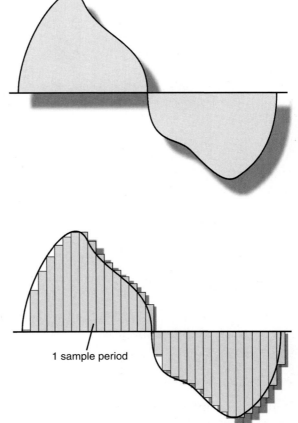

Figure 6.3. A digital signal makes use of periodic sampling to encode information.

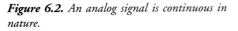

1 sample period

Figure 6.4. Discrete time sampling. (a) Whenever the sample rate is set too low, important data between sample periods will be lost. (b) As the rate is increased, more frequency-related data can be encoded. (c) Increasing the sampling frequency further can encode the recorded signal with an even higher bandwidth range.

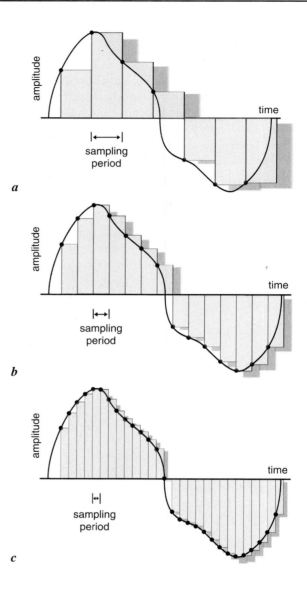

As you might expect, the sampling process can be likened to a photographer who takes a series of shots of an action sequence. As the number of pictures taken in a second increases, the accuracy of the captured event will likewise increase...until the resolution is so great, that you can't tell that the successive pictures have turned into a (hopefully) compelling movie.

During the sampling process (Figure 6.5), an incoming analog signal is sampled at discrete and precisely timed intervals (as determined by the sample rate). At each interval, this analog

Figure 6.5. *The sampling process. (a) The analog signal is momentarily "held" (frozen in time), while the converter goes about the process of determining the voltage level at that point in time and then converting that level into a binary-encoded word that's numerically equivalent to the sampled level. (b) After the converter has stored the representative word into a memory medium, the sample is released and the next sample is held, as the system again goes about the task of determining the level of the next sampled voltage . . . and so forth, and so forth, and so forth over time.*

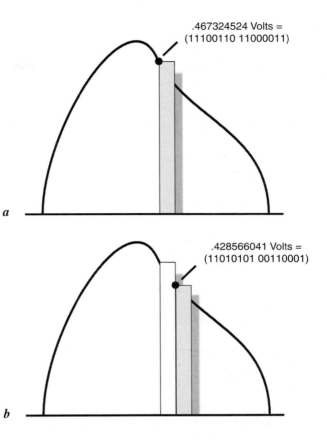

signal is momentarily "held" (frozen in time), while the converter goes about the process of determining what the voltage level actually is, with a degree of accuracy that's defined by the converter's circuitry (Figures 6.6 and 6.7) and the chosen bit rate. The converter then generates a binary-encoded word that's numerically equivalent to the analog level currently being sampled. Once done, the converter can store the representative word into a memory medium (tape, disk, disc, etc.), release its hold, and then go about the task of determining the level of the next sampled voltage. The process is then continuously repeated throughout the recording process.

The Nyquist Theorem

According to the *Nyquist theorem*, in order for the desired frequency bandwidth to be faithfully encoded in the digital domain, *the selected sample rate must be at least twice as high as the highest frequency to be recorded* (sample rate ≥2 × highest frequency). Thus, an audio signal with a bandwidth of 20 kHz would require that the sampling rate be at least 40 kHz samples/second.

Figure 6.6. Using the "comparator" circuit, during each sample and hold period, the converter compares the input signal level with a given set of reference voltages (which are successively reduced in scale by one-half for each "bit") until an equivalent digital word has been determined. (a) If the signal is greater than the first reference voltage, the first bit gets a "1," and if it's lower than the first reference voltage, the first bit gets a "0." (b) If the signal is greater than the second reference voltage, the second bit gets a "1," and if it's lower than the second reference voltage, the second bit gets a "0." (c) If the signal is greater than the third reference voltage, the second bit gets a "1," and if it's lower than the second reference voltage, the second bit gets a "0." (d) These comparisons continue to zero in on the precise voltage level in smaller and smaller steps until all of the bits in the digital word (in this case, an 8-bit word) have been defined . . . and the process can begin again to determine the next sample level.

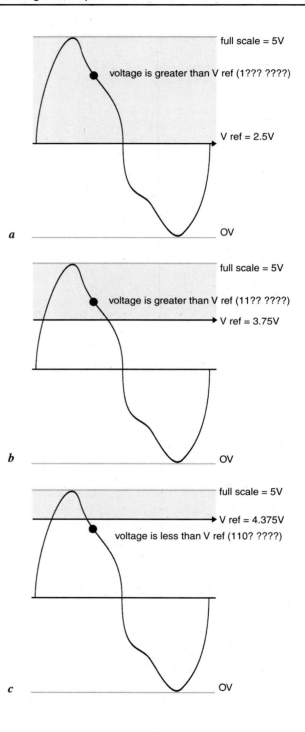

Figure 6.6. *Continued.*

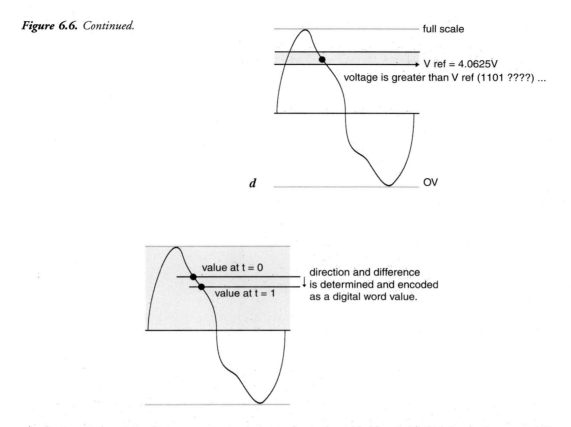

Figure 6.7. *Using a Delta Sigma-type circuit, during each sample and hold period (which has been oversampled to a higher rate; i.e., 192 kHz.), the converter compares the current input signal level against the previous sample level reference. Thus, the converter needs only to determine the level's direction (and amount of change in a multi-bit circuit) . . . compared to the previously known level. In short, it doesn't have to guess the overall voltage level for each sample period, just the direction and difference from one sample to the next.*

In addition, it's equally important that no audio signal greater than half the sampling frequency enter into the digital conversion process. If frequencies greater than one-half the sample rate are allowed to enter into the path, erroneous frequencies—known as *alias frequencies* (Figure 6.8)—could enter into the audible signal as false frequencies, which might be heard as harmonic distortion.

In order to eliminate the effects of *aliasing*, a low-pass filter is placed before the analog-to-digital (A/D) conversion stage. In theory, an ideal filter would pass all frequencies up to the Nyquist cutoff frequency and have an infinite attenuation above this point (Figure 6.9a); however, in the real world, such a "brickwall" filter doesn't exist. For this reason, a slightly higher sample rate must be chosen in order to account for an attenuation slope that's required for the filter to be effective (Figure 6.9b). For example, a sample rate of 44.1 kHz is chosen in order to accurately encode an effective bandwidth up to 20 kHz.

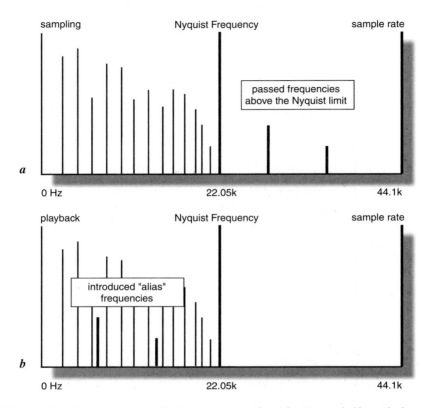

Figure 6.8. *Frequencies that enter into the digitization process above the Nyquist half-sample frequency limit could introduce harmonic distortion: (a) introduced frequencies above the limit; (b) alias frequencies that are introduced into the audio band.*

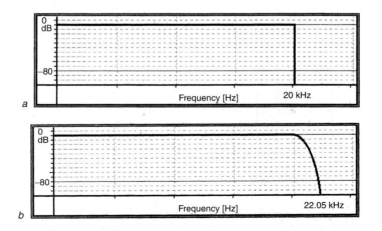

Figure 6.9. *Anti-alias filtering. (a) An ideal filter would have an infinite attenuation at the 20,000-Hz Nyquist cutoff frequency. (b) Real-world filters require an additional frequency "guardband" in order to fully attenuate unwanted frequencies that fall above the half-bandwidth Nyquist limit.*

Oversampling

Oversampling is a process that's commonly used in professional and consumer digital audio systems to improve a Nyquist filter's anti-aliasing characteristics. This process has the effect of further reducing intermodulation and other forms of audible distortion. Whenever oversampling is used, the effective sampling rate within a converter's filtering block is multiplied by a specific factor (often ranging between 12 and 128 times the original rate). This significant increase in the sample rate is accomplished by interpolating sampled level points between the original sample times. In effect, this technique makes educated guesses as to what the sample levels would be and where they'd fall at the newly generated sample time and then creates an equivalent digital word at that point. This increased sample rate likewise results in a much wider frequency bandwidth (so much so that a simple, less-expensive filter can be used to cut off the frequencies above the Nyquist limit). By down-sampling the rate back to its original value after the filter block, the Nyquist filter's bandwidth will be narrowed to such a degree that it approximates a much more complex and expensive cutoff filter.

Quantization

Quantization represents the amplitude component of the digital sampling process. It is used to translate the voltage levels of a continuous analog signal (at discrete sample points over time) into *bi*nary digi*ts* (bits) for the purpose of manipulating and/or storing audio data in the digital domain. By sampling the amplitude of an analog signal at precise intervals over time, it becomes the job of the converter to determine the exact voltage level of the signal (during the sample interval, when the voltage level is momentarily held) . . . and then output an analogous set of binary numbers (as a grouped word of n-bits length) that represents the originally sampled voltage level (Figure 6.10). The resulting word is used to encodes the original voltage level with as high a degree of accuracy as can be permitted by the word's bit length and the system's overall design.

Figure 6.10. *The instantaneous amplitude of the incoming analog signal is broken down into a series of discrete voltage steps, which are then converted into an equivalent binary-encoded word.*

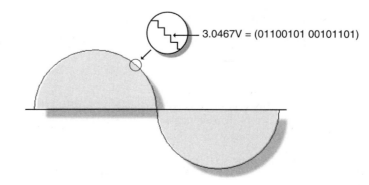

3.0467V = (01100101 00101101)

Currently, the most common binary word length for audio is 16-bit (for example, [0110010100101101]); however, professional systems having 20- and 24-bit resolution are also in common use. In addition, computers and signal-processing devices are capable of performing calculations internally at the 32- and 64-bit resolution level. This added internal headroom at the bit level, helps reduce errors in level and performance at low-level resolutions, whenever multiple audio datastreams are mixed and/or processed within a digital signal processing (DSP) system. This greater internal bit resolution is used as errors are most likely to accumulate within the least-significant bits (LSBs, the final and smallest numeric value within a digital word). As multiple signals are mixed together and multiplied (a regular occurrence in gain change and processing functions), lower-bit-resolution numbers become more and more of a component of the final result. . . . Since the internal bit depth is higher, these resolutions can be preserved (instead of being dropped within the hard- or software processing function) . . . with a final result being an *n*-bit datastream that's relatively free of errors.

This leads us to the conclusion that greater word lengths will often directly translate into an increased resolution (and thus higher quality) due to the added number of finite steps into which a signal can be digitally encoded. The following details the number of encoding steps that are encountered for the most commonly used bit lengths:

8-bit word = (nnnnnnnn) = 256 steps
16-bit word = (nnnnnnnn nnnnnnnn) = 65,536 steps
20-bit word = (nnnnnnnn nnnnnnnn nnnn) = 1,048,576 steps
24-bit word = (nnnnnnnn nnnnnnnn nnnnnnnn) = 16,777,216 steps
32-bit word = (nnnnnnnn nnnnnnnn nnnnnnnn nnnnnnnn) = 4,294,967,296 steps

where *n* = binary 0 or 1.

Signal-to-Error Ratio

Although analog signals are continuous in nature, as we've read, the process of quantizing a signal into an equivalent digital word isn't. Since the number of discrete steps that can be encoded within a digital word limits the accuracy of the quantization process, the representative digital word can only be an approximation (albeit an extremely close one) of the original analog signal level. A digital system's *signal-to-error ratio* is closely akin (although not identical) to the analog concept of signal-to-noise (S/N) ratio. Whereas a S/N ratio is used to indicate the overall dynamic range of an analog system, the signal-to-error ratio of a digital audio device indicates the degree of accuracy that's used with regard to a signal level's accuracy and its step-related effects of quantization. Given a properly designed system, the signal-to-error ratio for a signal coded with *n* bits is:

$$\text{Signal-to-error ratio} = 6n + 1.8 (\text{dB})$$

Therefore, the theoretical signal-to-error ratios for the most common bit rates will yield a dynamic range of:

8-bit word = 49.8 dB

16-bit word = 97.8 dB

20-bit word = 121.8 dB

24-bit word = 145.8 dB

32-bit word = 193.8 dB

Dither

When you come right down to it, the fundamental difference between digital audio and analog audio is one of resolution. Theoretically, the resolution of an analog signal is infinite in both its time and level components, whereas digital audio represents both time (sampling) and level (quantization) as discrete and quantifiable steps. Although these steps are quite small, they add a "squared off" component to a waveform that adds distortions and sonic properties that are similar to that of a square wave. This characteristic (which occurs at the digitized signal's least significant bit level) has the audible effect of adding low levels of harmonic distortion to the encoded signal; however, by adding a small amount of noise (whose frequency spectrum has been shaped so as to be as unobtrusive and unnoticeable as possible), it's possible to statistically improve the resolution of the conversion process below LSB level. You heard that right . . . by adding a small amount of random noise into the A/D path, we can actually:

◆ Improve the resolution of the conversion process below the least significant bit level.

◆ Reduce harmonic distortion in a way that greatly improves the signal's performance.

In order to look between the LSB steps, let's use an analogy to understand how the process works in its basic form. In Figure 6.11, the least significant bit in the encoding process can be

Figure 6.11. Values falling below the least significant bit level cannot be encoded without the use of dither.

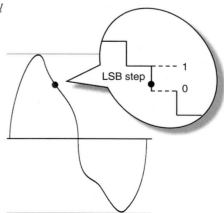

seen as falling between either the "0" or "1" value. Without dither, if the actual voltage level to be sampled falls at any point between these LSB values, there will be a 50% chance that it will end up being indiscriminately encoded as either a "0" or a "1." Whenever dither is added, the random element of noise will create a probability curve that allows the A/D circuit to detect whether the lower-level signal is closer to the least significant "0" or "1." Without going much further into the process, suffice it to say that dither is mathematically determined to provide the most accurate results at the LSB level from a statistical perspective.

In addition to the above, it is also a basic truth that *whenever a high-resolution signal is reduced in resolution a quantization error will be introduced into that signal.* In short, whenever a signal is converted to a lower-resolution signal, dither should be employed in order to reduce or avoid artifacts that would otherwise be introduced whenever bits are *truncated* (the simple dropping of lower value bits when converting to a lower-resolution bit rate). This hold true when digitally converting or mixing between any bit-rate structure. For example:

♦ When internally mixing at one rate and outputting at another lower-resolution rate (hopefully, dithering is provided within the DSP mix processing function)

♦ When converting 24 to 20 bits, 24 to 16 bits, 20 to 16 bits, etc. (in these cases, dithering can be added via a hardware or plug-in processing module)

♦ Transferring an analog tape to a 24 or 16 bit rate (in this case, dithering is naturally added via the tape noise as well as the converter's internal thermal noise)

In the following example, a system could simply drop (truncate) the least significant 8 bits within the conversion process. Alternatively, dither could be added to improve the educational guess that can be made at the new LSB level . . . thereby improving the signal resolution:

	Upper 16 bits	Lower 8 bits
Original 24-bit word	nnnnnnnn nnnnnnnn	Nnnnnnnn
Truncated bits		t t t t t t t t
Outputed 16-bit word	nnnnnnnn nnnnnnnn	
Original 24-bit word	nnnnnnnn nnnnnnnn	Nnnnnnnn
Random dither		Dddddddd
Outputed 16-bit word	nnnnnnnn nnnnnnnn	(improved resolution is passed to the 16-bit LSB value)

where n = binary 0 or 1; t = truncated bits; d = dithered bits.

The Digital Recording/Reproduction Process

The following sections provide a basic overview of the various stages that are encountered within the process of encoding analog signals into equivalent digital data (Figure 6.12a) and then converting this data back into its original analog form (Figure 6.12b).

Figure 6.12. *The digital audio chain: (a) recording; (b) reproduction.*

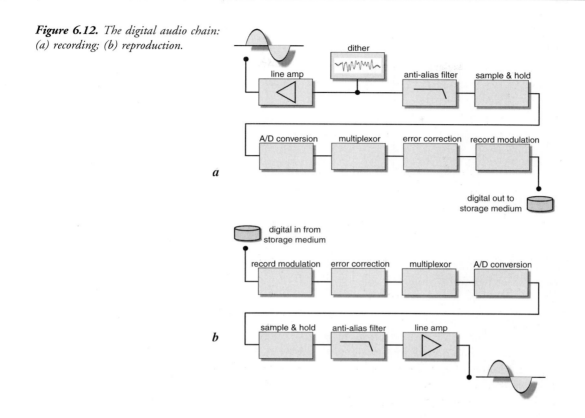

The Recording Process

In its most basic form, the *digital recording chain* includes a low-pass filter, a sample-and-hold circuit, an analog-to-digital converter, and the circuitry for signal coding and error correction. At the input of a digital sampling system, the analog signal must be band-limited with a low-pass filter so as not to allow frequencies that are greater than half the sample rate frequency to pass into the A/D conversion circuitry. Such a stop-band (anti-aliasing) filter generally makes use of a roll-off slope at its high-frequency cutoff point (as an immediate or "brickwall" slope would introduce severe signal distortion and phase shifts). In order for the full bandwidth to be accurately encoded, the Nyquist theorem requires that a sampling rate be chosen that's higher than twice the highest frequency to be recorded. For example, a system with a bandwidth that reaches into the 20-kHz range is often sampled at a rate of at least 44.1K samples per second.

Following the low-pass filter, a *sample-and-hold* (S/H) circuit holds and measures the analog voltage level for the duration of a single sample period, which is determined by the sample rate (i.e., less than 1/44,100th of a second). At this point, computations must be performed in order to translate the sampled level into an equivalent binary word. This step in the A/D conversion is a critical component of the digitization process, as the sampled DC voltage

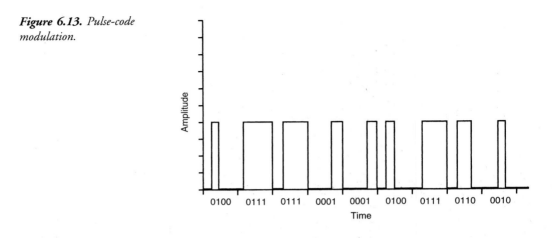

Figure 6.13. *Pulse-code modulation.*

level must be accurately quantized into an equivalent digital word (to the nearest step level) in a very short period of time.

Once the sampled signal has been converted into an equivalent digital form, the data must be conditioned for further data processing and storage. This conditioning includes data coding, data modulation, and error correction. In general, the binary digits of a digital bitstream aren't directly stored onto a recording medium as raw data; rather, data coding is used to translate the data (along with synchronization and address information) into a form that allows the data to be most efficiently and accurately stored. The most common form of digital audio data coding is *pulse-code modulation*, or PCM (Figure 6.13).

The density of stored information within a PCM recording and playback system is extremely high . . . so much so that any imperfections (such as dust, fingerprints, or scratches that might adhere to the surface of any magnetic or optical recording medium) will cause severe or irretrievable errors to be generated. In order to keep these errors within acceptable limits, several forms of *error correction* will be used (depending upon the media type). One method uses redundant data in the form of parity bits and check codes, in order to retrieve and/or reconstruct lost data; a second uses error correction that involves interleaving techniques (whereby data is deliberately scattered across the digital bitstream, according to a complex mathematical pattern). The latter has the effect of spreading the data over a larger surface of the recording media . . . thereby making the recording media less susceptible to the effects of dropouts (Figure 6.14). In fact, it's a simple truth that without error correction, the quality of

Figure 6.14. *An example of interleaved error correction.*

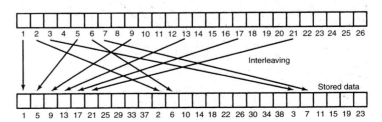

most digital audio media would be greatly reduced or (in the case of the CD) rendered almost useless.

The Reproduction Process

In many respects, the digital reproduction chain works in a manner that is complementary to the digital encoding process. Since most digital media encodes data onto media in the form of highly saturated magnetic transition states or optical reflections, the recorded data must first be reconditioned in a way that restores the digital bitstream back into its original, modulated binary state. Once this is done, the data can then be de-interleaved (reassembled) back into its original form, where it can be converted back into PCM data.

Once the signal has been reconstructed back into its original PCM form, the process of digital-to-analog (D/A) conversion can take place. Often, a stepped resistance network (sometimes called an R/2R network) is used to convert the representative words back into their analogous voltage levels within the playback phase. During a complementary sample-and-hold period, each bit within the word being converted is assigned to a leg in the network (moving from the most-significant to the least-significant bit). Each leg is designed to pass one-half the reference voltage level that can be passed by the previous step (Figure 6.15). The presence or absence of a logical "1" in each step is then used to turn on each successive voltage leg. As you might expect, summing together these voltages will yield a precise level that can be passed on to the converter's analog output.

Following the conversion process, a final, complementary low-pass filter is inserted into the signal path. This filter is used to smooth out any nonlinear steps that are introduced by the sampling process . . . resulting in a waveform that will faithfully represent the originally recorded analog waveform (assuming that the circuit has been properly designed).

Digital Audio Transmission

In this digital age, it's become increasingly common for audio data to be distributed from one device to another or throughout a connected production system in the digital domain. In this way, digital audio can be transmitted in its original numeric form and (in theory) without any

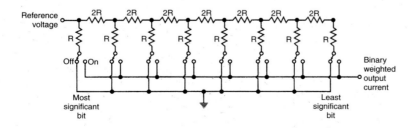

Figure 6.15. *A stepped resistance network is a common device for accomplishing D/A conversion by assigning each word bit to a series of resistors that are scaled by factors of 2.*

degradation throughout a connected path or system. When looking at the differences between the distribution of digital and analog audio, it should be kept in mind that, unlike its counterpart, the transmitted bandwidth of digital audio data occurs in the megahertz range; therefore, the transmission of digital audio actually has more in common with video signals than the lower bandwidth range that's encountered with analog audio. This means that care must be exercised to ensure that impedance is more closely matched and that quick-fix solutions don't occur (for example, using a Y-cord to split a digital signal between two recorders is a major no-no). Failure to follow these precautions could seriously degrade or deform the digital signal.

Due to these tight restrictions, several digital transmission standards have been adopted that allow digital audio data to be quickly and reliably transmitted between compliant devices. These include such protocols as:

◆ AES/EBU
◆ S/PDIF
◆ MADI
◆ ADAT lightpipe
◆ TDIF
◆ mLAN

AES/EBU

The AES/EBU (Audio Engineering Society and the European Broadcast Union) protocol has been adopted for the purpose of transmitting digital audio between professional digital audio devices. This standard (which is most often referred to as simply an "AES" digital connection) is used to convey two channels of interleaved digital audio through a single, three-pin XLR microphone cable in a single direction. This balanced configuration connects pin 1 to the signal ground, while pins 2 and 3 are used to carry signal data. AES/EBU transmission data is low impedance in nature (typically 110 Ω) and has digital burst amplitudes that range between 3 and 10 V. These combined factors allow for a maximum cable length of up to 328 feet (100 meters) at sample rates of less than 50 kHz, without encountering undue signal degradation.

Digital audio channel data and subcode information are transmitted in blocks of 192 bits that are organized into 24 words (with each being 8 bits long). Within the confines of these data blocks, two subframes are transmitted during each sample period that convey information and digital synchronization codes for both channels in an L–R–L–R... fashion. Because the data is transmitted as a self-clocking biphase code (Figure 6.16), wire polarity can be ignored, and whenever two devices are directly connected the receiving device will usually derive its reference timing clock from the digital source device.

In the late 1990s, the AES protocol was amended to include the "stereo 96k dual AES signal" protocol. This was created to address signal degradations that can occur when running

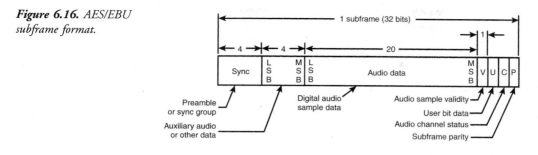

Figure 6.16. *AES/EBU subframe format.*

longer cable runs at sample rates above 50 kHz. In order to address the problem, the dual AES standard allows stereo rates above 50 kHz (such as 96/24) to be transmitted over two, synchronized AES cables (with one cable carrying the L information and the other carrying the R).

S/PDIF

The S/PDIF (Sony/Phillips Digital Interface) protocol has been widely adopted for transmitting digital audio between consumer digital audio devices and their professional counterparts (using a similar data structure). Instead of using a balanced 3-pin XLR cable, the popular S/PDIF standard has adopted the single-conductor, unbalanced phono (RCA) connector (Figure 6.17a), which conducts a nominal peak-to-peak voltage levels of 0.5 V between connected devices, with an impedance of 75 Ω. In addition to using RCA wire cable connections, S/PDIF can also be transmitted between devices using Toslink optical connection lines (Figure 6.17b), which are commonly referred to as "lightpipe" connectors.

As with the AES/EBU protocol, S/PDIF channel data and subcode information are transmitted in blocks of 192 bits consisting of 12 words that are 16 bits long. A portion of this information is reserved as a category code that provides the necessary setup information (sample rate, copy protection status, and so on) to the copy device. Another portion is set aside for transmitting audio data used to relay track indexing information (such as start ID and program ID numbers), allowing this relevant information to be digitally transferred from the master to the copy. It should be noted that the professional AES/EBU protocol isn't capable of digitally transmitting these codes during a copy transfer.

In addition to transmitting two channels in an interleaved L–R–L–R . . . fashion, S/PDIF is able to communicate multichannel data between devices. Most commonly, this shows up as

Figure 6.17. *S/PDIF connectors: (a) RCA wire connection; (b) Toslink optical connection.*

a

b

a direct digital surround-sound link between a DVD player and audio receiver/amplifier playback system (via either an RCA coax or optical connection.)

SCMS

Initially, certain digital recording devices (such as a DAT recorder) were intended to provide consumers with a way to make high-quality recordings for their own personal use. Soon after its inception, however, for better or for worse, the recording industry began to see this new medium as a potential source of lost royalties due to home copying and piracy practices. As a result, the RIAA (Recording Industry Association of America) and the former CBS Technology Center set out to create a "copy inhibitor." After certain failures and long industry deliberations, the result of these efforts was a process that has come to be known as the *Serial Copy Management System*, or SCMS. SCMS (pronounced "scums") has been incorporated into many consumer digital devices in order to prohibit the unauthorized copying of digital audio at 44.1 kHz (the standard CD sample rate). This copy inhibitor does not apply to the making of analog copies, to digital copies that are made using the AES/EBU protocol, or to sample rates other than 44.1 kHz.

So, what is SCMS? Technically, it's a digital protection flag that is encoded in byte 0 (bits 6 and 7) of the S/PDIF subcode area. This flag can have only one of three possible states:

◆ Status 00: No copy protection, allowing unlimited copying and subsequent dubbing
◆ Status 10: No more digital copies allowed
◆ Status 11: Allows a single copy to be made of this product but that copy cannot be copied

Suppose that we have two DAT machines that are equipped with SCMS (with one being used for playback and the other for recording). If we try to digitally copy a DAT that has a 10 SCMS status, we would simply be out of luck; however, suppose that we found a DAT that has an 11 status flag. By definition, the bitstream data would inform the copy machine that it's OK to record the digital signal; however, the status flag on the subsequent copy tape would then be changed to a 10 flag. If at a later time we wanted to clone this DAT copy, the machine doing the second-generation copy couldn't be placed into Record. At that point, we have two possible choices: We could record the signal using the analog ports, or a digital format converter could be used that (among other things) allows us to strip the SCMS copy protection flags from the bitstream and continue to make multigenerational copies (it should be noted that some digital audio editors are also able to strip out or reset the protection flag bits).

MADI

The MADI (Multichannel Audio Digital Interface) standard was jointly proposed as an AES standard by representatives of Neve, Sony, and SSL as a straightforward, clutter-free digital connection interface between multitrack devices (such as a digital tape recorder or mixing console). The format allows up to 56 channels of linearly encoded digital audio to be connected via a single 75-Ω, video-grade coaxial cable at distances of up to 120 feet (50 meters) or at greater distances whenever a fiberoptic interface/cable is used. MADI makes use of a serial data transmission format that's compatible with the AES/ EBU twin-channel protocol (whereby the data, Status, User, and parity bit structure is preserved) . . . and sequentially cycles through each channel (starting with Ch. 0 and ending with Ch. 55). The transmission rate of 100 Mbit/second provides for an overall bandwidth that's capable of handling audio data and numerous sync codes at various sample rate speeds (including allowances for changing pitch either up or down by 12.5% at rates between 32 and 48 kHz).

ADAT Lightpipe

A wide range of modular digital multitrack recorders, soundcards, and hardware devices use the Alesis *lightpipe* system for transmitting multichannel audio via a standardized optical cable link. Lightpipe connections make use of standard Toslink connectors and cables to transmit up to eight digital audio channels over a sequential, optical bitstream. Although these connectors are identical to those that are used to optically transmit S/PDIF stereo digital audio, the datastreams are incompatible with each other. Lightpipe data is not bidirectional in that it can only travel from a single source to a destination in one direction; therefore, two cables are needed to distribute data both to and from a device. Only digital audio data is transmitted over the serial bitstream, meaning that sync data will not be passed. Should sync be needed to lock one or more devices together (as in a multiple ADAT system or ADAT/DAW setup), separate sync cabling (often 9-pin serial connectors) will be required.

TDIF

The TDIF (Tascam Digital InterFace) is a proprietary format that uses a 25-pin D-sub cable to transmit and/or receive up to eight channels of digital audio between compatible devices. Unlike the lightpipe connection, TDIF is a bidirectional connection, meaning that only one cable is required to connect the 8 ins and outs of one device to another. Although systems that support TDIF-1 cannot send and receive sync information (a separate wordclock connection is required for that), the newer TDIF-2 protocol is capable of receiving and transmitting sync through the existing connection, without any additional cabling.

mLAN

One of the major problems that faces the professional and project studio is the vast number of interconnections that must be made between devices within a connected setup. Given that audio, MIDI, and sync are often routed separately within a production system, the mess can often result in a complex pile of spaghetti that's not only cumbersome and complex . . . but can also be unreliable and extremely difficult to reconfigure. Although several proprietary network interconnections are beginning to solve this problem in the digital domain, an innovative system from Yamaha (known as *mLAN*) seems to be leading the pack by allowing multichannel digital audio and MIDI music data to be transferred via a single IEEE

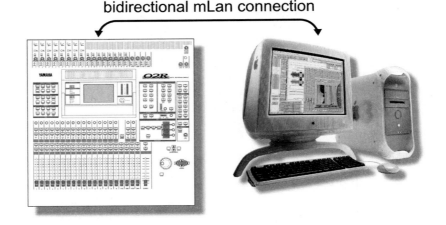

bidirectional mLan connection

Figure 6.18. Example of a simple, bidirectional mLAN multichannel connection.

1394/FireWire® or iLINK® cable (Figure 6.18). Theoretically, mLAN can transfer approximately 100 channels of digital audio data and up to 256 ports of MIDI data (16 channels × 256 connections) between digital audio workstations (DAWs), digital mixers, synths, and effects processors over a single cable in a bidirectional fashion . . . at speeds of 100, 200, or 400 Mbps!

All of this means that a mLAN-based music system could be quickly and easily configured, saved, and recalled with none of the frustration and down time that occur when reconfiguring a conventional system. Using the patch bay application that's provided with all mLAN products, devices can be routed and reconfigured within software. Since it's "hot pluggable," devices (called "nodes") can be plugged and unplugged without having to power-down or reset the system. In addition, mLAN is able to transmit and resolve world clock issues . . . even allowing devices to run at differing sample rates on the same network. Finally, this format can run with or without a computer. Should multiple mLAN hardware devices be connected, the system will establish a network link for easy connectivity.

Although Yamaha developed mLAN, it is available to other manufacturers on a royalty-free (no-cost) basis. As of this writing, over 40 manufacturers have signed on as mLAN licensees, and eight manufacturers have developed first-generation products.

Signal Distribution

If copies are to be made from a single, digital audio source or if data is to be distributed throughout a connected network using AES/EBU, S/PDIF, and MADI digital transmission cables . . . it is possible to distribute the data from one device to the next in a straightforward, daisy-chain fashion (Figure 6.19). This method works well only if a few devices are to be

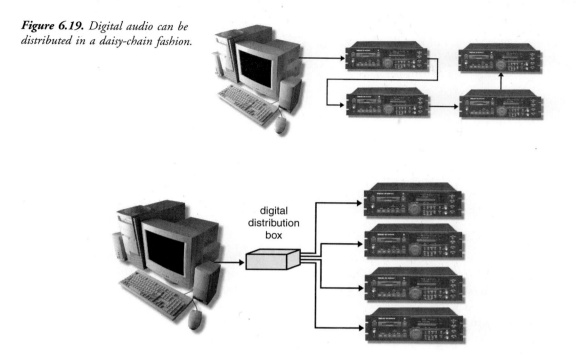

Figure 6.19. *Digital audio can be distributed in a daisy-chain fashion.*

Figure 6.20. *A distribution system can be used to route digital audio data to individual devices.*

chained together. If several devices are to be connected together in the system, time-base errors (known as *jitter*) might be introduced into the path, with the possible side effects being added noise and a slightly "blurred" signal image. One way to reduce such potential time-base errors is to use a digital audio distribution device that can route the data from a single digital audio source to a number of individual device destinations (Figure 6.20).

What Is Jitter?

Jitter is a controversial and widely misunderstood phenomenon. To my knowledge, it's been explained best by Bob Katz of Digital Domain (www.digido.com; Orlando, Florida). The following is a brief excerpt of his article "Everything You Always Wanted To Know About Jitter But Were Afraid To Ask." Further reading on digital audio and mastering techniques can be found in Bob's excellent book, *Mastering Audio: The Art and the Science*, from Focal Press (www.focalpress.com):

> Jitter is time-base error. . . . It is caused by varying time delays in the circuit paths from component to component in the signal path. The two most common causes of jitter are poorly designed Phase Locked Loops (PLLs) and waveform distortion due to mismatched impedances and/or reflections in the signal path.

Figure 6.21. Example of time-base errors: (a) a theoretically perfect digital signal source; (b) the same signal with jitter errors.

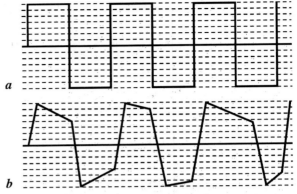

Here is how waveform distortion can cause time-base distortion: The top waveform (Figure 6.21a) represents a theoretically perfect digital signal. Its value is 101010, occurring at equal slices of time, represented by the equally spaced dashed vertical lines. When the first waveform passes through long cables of incorrect impedance, or when a source impedance is incorrectly matched at the load, the square wave can become rounded, fast rise times become slow, also reflections in the cable can cause misinterpretation of the actual zero crossing point of the waveform. The second waveform (Figure 6.21b) shows some of the ways the first might change; depending on the severity of the mismatch you might see a triangle wave, a squarewave with ringing, or simply rounded edges. Note that the new transitions (measured at the Zero Line) in the second waveform occur at unequal slices of time. Even so, the numeric interpretation of the second waveform is still 101010! There would have to be very severe waveform distortion for the value of the new waveform to be misinterpreted, which usually shows up as audible errors—clicks or tics in the sound. If you hear tics, then you really have something to worry about.

If the numeric value of the waveform is unchanged, why should we be concerned? Let's rephrase the question: "When (not why) should we become concerned?" The answer is "hardly ever." The only effect of time-base distortion is in the listening; as far as it can be proved, it has no effect on the dubbing of tapes or any digital-to-digital transfer (as long as the jitter is low enough to permit the data to be read; high jitter may result in clicks or glitches as the circuit cuts in and out). A typical D to A converter derives its system clock (the clock that controls the sample and hold circuit) from the incoming digital signal. If that clock is not stable, then the conversions from digital to analog will not occur at the correct moments in time. The audible effect of this jitter is a possible loss of low-level resolution caused by added noise, spurious (phantom) tones, or distortion added to the signal.

A properly dithered 16-bit recording can have over 120 dB of dynamic range; a D to A converter with a jittery clock can deteriorate the audible dynamic range to 100 dB or less, depending on the severity of the jitter. I have performed listening experiments on purist, audiophile-quality musical source material recorded with a 20-bit accurate A/D converter (dithered to 16 bits within the A/D). The sonic results of passing this signal through processors that truncate the signal at −110, −105, or −96 dB are increased "grain" in the image; instruments losing their sharp edges and focus; reduced sound-stage width; apparent loss of level causing the listener to want to turn up the monitor level, even though high-level signals are reproduced at unity gain. Contrary to intuition, you can hear these effects without having to turn up the listening volume beyond normal (illustrating that low-level ambience cues are very important to the quality of reproduction). Similar degradation has been observed when jitter is present. Nevertheless, the loss due to jitter is subtle and primarily audible with the highest-grade audiophile D/A converters.

Wordclock

One aspect of digital audio recording that never seems to get enough attention is the need for synchronization at the sample level within a series of interconnected digital audio devices. In order to reduce such gremlins as clicks, pops, and jitter (oh my!), it's often necessary to lock the overall sample rate timing to a single master clock signal (so that the conversion sample and hold states for all digital audio channels and devices will occur at exactly the same point in time) . . . through the use of a single timing reference known as *wordclock*.

As an example, let's assume that we're in a room that has four or five clocks, and none of them reads the same time! In places like this, you never quite know what the time really is . . . the clocks could be running at different speeds or at the same speed but are ticking at different times. Basically, trying to accurately keep track of the time while simultaneously looking at all of the clocks would end up being a jumbled nightmare. On the other hand, if all of these clocks were locked to a single, master clock (remember those self-correcting clocks that have been installed in most schools?) . . . keeping track of the time (even when moving from room to room) would be much simpler.

In effect, wordclock works in a similar fashion. If the sample clock (the timing reference that determines the sample rate and DSP traffic control) for each device was set to operate in a freewheeling, internal fashion . . . the timing references of each device within the connected digital audio chain wouldn't accurately match up. Even though the devices are all running at the same sample rate, these resulting mismatches in time will often result in clicks, ticks, excessive jitter, and other unwanted grunge. In order to correct for this, the internal clocks of all the digital devices within a connected chain must be referenced to a single "master" wordclock timing element (Figure 6.22).

Similar to the distribution of time code, there can only be one master wordclock reference within a connected digital distribution network. This reference source can be derived from a digital mixer, soundcard . . . or any desired source that can transmit wordclock. Often, this reference pulse is chained between the involved devices through the use of BNC and/or RCA connectors, using low-capacitance cables (often 75-Ω, video-grade coax cable is used, although this cable grade isn't always necessary on shorter cable runs).

It's interesting to note that wordclock isn't generally needed when making a digital copy from one device to another (via such protocols as AES, S/PDIF, MADI, or TDIF2), as the

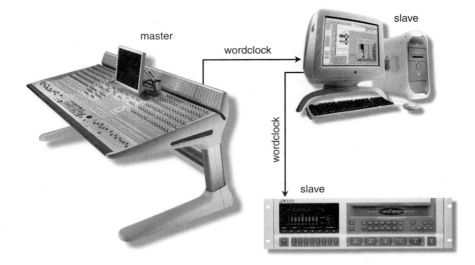

Figure 6.22. *Example of wordclock distribution showing that, within a digital production network, there can be only one master clock.*

timing information is actually embedded within the data bitstream itself. Only when we begin to connect devices that share and communicate digital data will we see the immediate need for wordclock.

It almost goes without saying that there will often be differences in connections and parameter setups, from one system to the next. In addition to proper cabling and impedance termination considerations throughout the network, specific hardware and software setups may be required in order to get all the device blocks to communicate. In order to better understand your particular system's setup (and to keep frustration to a minimum), it's always a good idea to keep all of your device manuals close at hand.

Digital Audio Recording Systems

For the remainder of this section, we'll be looking at the various types of digital audio recording devices that are currently available on the market. From my own personal viewpoint, not only do I find the here and now of recording technology to be exciting and full of cost-effective possibilities . . . I also love the fact that there are lots of recording media and device-type options. In other words, a digital hard- or software system that works really well for me might not be the best and easiest solution for you! As we take an in-depth look at many of these device and system choices, I hope that you'll take the time to learn about each one (and possibly even try your hand at listening to and/or working with each system type). In the earlier years of recording, there were only a few ways that a successful recording could be made. Now, in the age of technological options . . . your mission (should you decide to accept it) is

Figure 6.23. *Detail of digital thin-film head construction. (Courtesy of Sony Professional Audio, www.sony.com/proaudio.)*

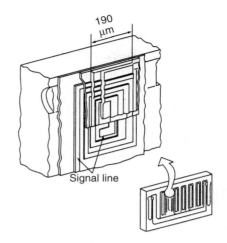

to research and test-drive devices and/or production systems to find the one that best suits your needs, budget, and personal working style.

The Fixed-Head Digital Audio Recorder

The *fixed-head digital audio recorder* is a reel-to-reel system that often emulates its analog counterpart in form and function, although most other similarities end there. These recording systems use digital audio conversion and special data encoding structures to store digital audio data onto specially formulated digital audio tape, using state of-the-art, thin-film heads (Figure 6.23). Although two-channel fixed-head recorders do exist, the vast majority of these systems are multitrack. Of these, the most commonly found recorders are 24- and 48-track recorders that use the Digital Audio Stationary Head (DASH) format, which was ratified and jointly established by Sony, Willi Studer AG, and Matsushita Electric Industries.

The DASH Format

DASH was developed to help ensure standardization between different generations and manu-facturers of digital, fixed-head recorders. The standard provides for three data storage densities (fast, medium, and slow), with the choice being determined by the tape speed of the recorder. Using this system, data isn't encoded onto a single track but actually spread over several interleaved data tracks in order to achieve the high data densities that are required to record digital audio onto longitudinal tape.

Error Correction

The operation of a DASH encoder is based on the cross interleave code (CIC), with increased interleaving being made between the even- and odd-numbered words (allowing up to three consecutive words to be corrected). This interleaving makes it possible for the tape to be spliced, thereby allowing physical tape edits to be made. The correctability of burst (large-scale) errors is determined by the encoders and is the same for all three-speed versions. Error correction encoding and decoding are done independently for each track. For example, if excessive errors appear on one of the recorded tracks (as might occur with a dropout), the correction capabilities on other tracks won't be affected (a feature that safeguards the program material under adverse conditions). In addition to allowing the tape to be physically spliced (using a corrective cross-fade function), the tracks on a DASH tape can be punched in and out and electronically edited (using a special edit control system).

DASH Recording Systems

Recorders using the DASH format are commonly available in 24- and 48-track configurations. With the development of new large-scale integrated circuits (LSIs), smaller and lighter recorders are available that are lower in cost and power consumption than their first-generation models. Both Sony and Studer have developed a series of popular multichannel DASH recorders. The Sony 24-track PCM-3324S and the Studer D 827 Mark II MCH use 1/2-inch tape that runs at a speed of 30 ips and operate using 16- or 24-bit wordlengths at sample rates of 48, 44.1, and 44.056 kHz. Using thin-film head technology (a technology borrowed from integrated circuit fabrication, which is detailed in Figure 6.24), the digital tracks are longitudinally spread across the width of the tape, along with two outside analog tracks and an additional control or external data track. The error detection circuitry of both of these recorders allows splice edits to be made by creating a seamless 90° cross-fade that interpolates the data before and after the splice.

Figure 6.24. Headblock mechanism of the Sony 3324A. (Courtesy of Sony Professional Audio, www.sony.com/proaudio.)

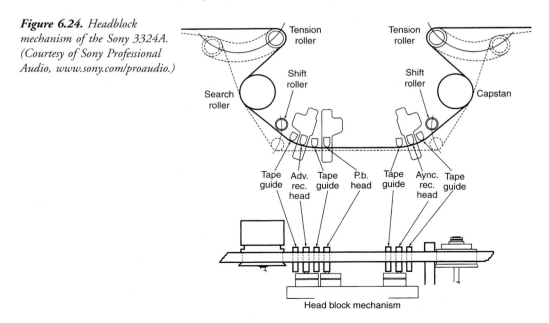

Figure 6.25. The Sony PCM3348 digital
audio multitrack recorder. (Courtesy of Sony
Professional Audio, www.sony.com/
proaudio.)

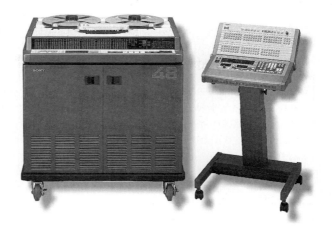

Figure 6.26. The Studer D 827 Mark II MCH
24-/48-track digital audio multitrack recorder.
(Courtesy of Studer Professional Audio AG,
www.studer.ch.)

Both the Sony PCM-3348 (Figure 6.25) and the Studer D 827 Mark II MCH (Figure 6.26) DASH recorders have the added bonus of being able to record 48 tracks of digital audio that's fully compatible with the 24-track DASH format. Using a unique system, tapes that were previously recorded on a 24-track DASH machine can be recorded and reproduced with no signal degradation using tracks 1 to 24, with tracks 25 to 48 being available on the same tape with no problems in compatibility.

The Rotating-Head Digital Audio Recorder

Rotating-head digital recorders fall into several categories, including DAT (digital audio tape) systems and MDM (modular digital multitrack) recording systems.

The Rotary Head

Because of the tremendous amount of data density that's required to record/reproduce PCM digital audio (approximately 2.77 million bits/second), the recording of data onto a single, linear tape track is extremely impractical. In order to get around this bandwidth limitation, when using a narrow-width tape, a rotating-head *helical scan* path is used to effectively increase the overall head-to-tape contact area at a given tape speed. This process (which is also used in VHS videotape transport technology) creates a slanted tape path that wraps around a rotating drum that has two or more magnetic heads built into its scan surface. As the head rotates . . . numerous closely spaced tape "scan" paths are consecutively recorded along the tape path at a slanted angle. These repeated scans result in an effective recorded track that's much longer in length than would be possible with a linear track . . . resulting in a system that can record wider frequency bandwidths at a low physical tape speed. Examples of a helical transport that employ rotary head technology are shown in Figure 6.27.

Digital Audio Tape (DAT) System

The digital audio tape, or DAT recorder, is a compact, dedicated PCM digital audio format (Figure 6.28) that displays a wide dynamic range, low distortion, and virtually immeasurable amounts of wow and flutter. Given the fact that the DAT recorder was initially designed for the consumer market, its professional specs have resulted in its being widely adopted in

Figure 6.27. *Helical scan path: (a) tape path around drum mechanism; (b) recording of slanted tracks.*

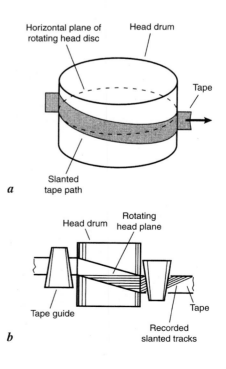

Figure 6.28. *Tascam DA-P1 portable DAT recorder. (Courtesy of Tascam, www.tascam.com.)*

numerous studio and on-location applications. DAT technology makes use of an enclosed compact cassette that's even smaller than a compact audio cassette. Equipped with both analog and digital input/outputs, the DAT format can record and play back at three standard sampling frequencies: 32, 44.1, and 48 kHz (although sample-rate capabilities and system features may vary from one recorder to the next). Current DAT tapes offer running times of up to 2 hours when sampling at 44.1 and 48 kHz, and the standard is capable of three record/reproduce modes for the 32-kHz (broadcast) sampling rate:

◆ Option 1 provides 2 hours of maximum recording time with 16-bit linear quantization.

◆ Option 2 provides up to 4 hours of recording time with 12-bit nonlinear quantization.

◆ Option 3 (which is rarely, if ever, used) allows for the recording of four-channel, nonlinear, 12-bit audio.

DAT Tape/Transport Format

The actual track width of the DAT format's helical scan can range downward to about 1/10th the thickness of a human hair (which allows for a recording density of 114 million bits per square inch . . . the first time such a density has been achieved using magnetic media). To assist with tape tracking and to maintain the highest quality playback signal, a sophisticated tracking correction system is used to center the heads directly over the tape's scan path. The head assembly of a DAT recorder uses a 90° half-wrap tape path. Figure 6.29 shows a two- head assembly in which each head is in direct contact with the tape 25% of the time. Such a discontinuous signal requires that the digital data be communicated to the heads in digital "bursts," which necessitates the use of a digital buffer. On playback, these bursts are again smoothed out into a continuous datastream. This encoding method has the following advantages:

◆ Only a short length of tape is in contact with the drum at one time. This reduces tape damage and allows high-speed searches to be performed while the tape is in contact with the head.

◆ A low tape tension can be used to ensure a longer head life.

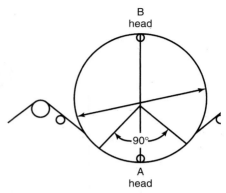

Figure 6.29. *A half-wrap tape path, showing a 90° amount of tape-to-head contact.*

The high-speed search function (approaching 300 times normal speed) is a key feature of the DAT format. In addition to this, the format makes provisions for non-audio information to be written directly into the digital stream's subcode area. This subcode area serves as a digital identifier (just as a compact disc uses subcodes for selection and timing information) and can be written as any one of three data types:

- Start ID, which is used to indicate the beginning of a selection
- Skip ID, which indicates that a selection should be skipped over
- Program number, which makes it easier to identify selections when searching at high speeds

This subcode information can be written, erased, and rewritten at any time without altering the audio program. In addition, the subcode area can be used to encode SMPTE time code for use with audio for film or video, as well as for synchronous music production.

The Modular Digital Multitrack (MDM)

The device that single-handedly ushered professional and project studio production out of the analog age and into affordable digital multitrack audio is the modular digital multitrack system, or MDM (Figure 6.30). MDMs are compact multitrack digital audio recorders that are capable of recording eight tracks of digital audio onto videotape-grade cassette cartridges that can often be bought at your favorite neighborhood drugstore. These recorders are said to be "modular" because several of them can be linked together (in groups of eight) in a synchronous fashion that allows them to work in tandem as a large-scale multitrack recording system (often being configured in the 24- and 32-track range). Another important aspect of the MDM revolution was their price. Given the fact that a tape-based digital multitrack (*i.e.*, DASH) could easily sell for more than $60,000, it's not hard to understand why the (then) average price of $2000 sparked a revolution in production technology.

(*Note:* Just as video killed the radio star . . . DAWs have put a serious dent in the MDM market. As a result, these devices can be had for a virtual song on the used market. Be wary when

Figure 6.30. *MDM-based studio.*
(Courtesy of Tascam,
www.tascam.com.)

Figure 6.31. *Alesis XT20 20-bit*
modular digital multitrack recorder.
(Courtesy of Alesis Studio Electronics,
Inc., www.alesis.com.)

considering a used MDM, as parts are now harder to get for certain machines and replacing a worn head will often cost more than what you paid for the unit.)

ADAT MDM Format

The ADAT standard, which was created by the Alesis Corporation (Figure 6.31) and is no longer available on the new market, is a family of rotary-head, eight-track modular digital multitrack recorders that use standard S-VHS videotape. This format includes such features as 16-/20-bit wordlength capabilities, professional sample-rate standards (including varispeed over a wide range), autolocate functions, and autoloop/rehearse functions (that work in both the play and record punch modes). When using a standard 120-minute S-VHS tape at the highest sampling rate, a recorder that conforms to the ADAT standard can yield recording times of slightly over 40 minutes (or 53 minutes when a 160-minute tape is used).

Analog input/output (I/O) connections vary between models and can include 1/4-inch unbalanced/ balanced jacks and RCA or XLR connections (in either –10 or +4 dBm configurations). Many models also allow multiple I/O connections to be made via a single, standard 56-pin Elco connector at professional +4-dBm reference levels.

Remote ADAT functions are carried out through the use of the LRC (little remote control), BRC (big remote control), or (in certain cases) MIDI machine control. The LRC was shipped

Figure 6.32. Alesis BRC (Big Remote Control) unit for the ADAT. (Courtesy of Alesis Corporation, www.alesis.com.)

with every new ADAT and contains all the device's basic front panel transport controls on a single, palm-sized remote. The optional BRC (Figure 6.32) is capable of acting as a full-featured remote control, digital editor, expanded-feature autolocator, and synchronizer on a single tabletop or free-standing surface. It's capable of controlling up to 16 remote ADAT units and can bounce data from one track to another in the digital domain (while also allowing tracks to be shifted in location). An example of track location shifting is its ability to copy a set of background vocals in a song's bridge to another point near the end of the song, without the use of a DAW. All that's needed is to define a track and location source, as well as its destination, and the BRC will take care of all transport, record, and audio functions.

Digital I/O is transmitted (in a single direction) through the use of a Toslink fiberoptic cable line that's capable of simultaneously carrying all eight channels. This link is used to connect an ADAT to a growing number of digital peripheral I/O devices (such as a digital audio interface, mixer, effects processor, and even certain synthesizers).

On the sync front, since only audio data passes through the optical Toslink cable, it's necessary to interconnect all of the devices together by chaining the sync ports in an in/thru fashion using a 9-pin sync cable. These synchronization ports are used to transmit a proprietary code for interlocking the devices together, with near sample accuracy. From a user's perspective, this code will be displayed as SMPTE and, through the use of a special conversion device or system that includes an ADAT sync interface, this code can often be easily converted to/from SMPTE and/or MIDI time code.

DTRS MDM Format

Although the rotary-head Digital Tape Recording System (DTRS) was created by Tascam (Figure 6.33), several manufacturers offer MDM recorder systems that adhere to this standard. These 8-track modular recorders are capable of recording up to 108 minutes of digital audio onto a standard 120-minute Hi-8-mm video tape, and (like the ADAT format) can be combined with other DTRS recorders to create a system that has 24, 32, or more tracks. Digital I/O is made through the proprietary TDIF connection, which uses a special

Figure 6.33. *Tascam DA-98HR modular digital multitrack recorder. (Courtesy of Tascam, www.tascam.com.)*

25-pin D-sub connector to link to other DRTS recorders, digital mixers, hard disk recorders, or external accessories. The TDIF digital interconnection comes in two flavors:

- TDIF-1, which is capable of transmitting and receiving all eight channels of digital audio (in a bidirectional I/O fashion) over a single cable
- TDIF-2, which, in addition to the features of TDIF-1, is also capable of transmitting sync without the need for an external sync connection

As with all MDMs, the sync capabilities of a DRTS recorder will vary between models and manufacturers. Like the ADAT, systems with TDIF-1 must be synced together through the use of a 15-pin D-sub cable, which will need to be chained between the various devices. This proprietary sync code can also be converted to MTC or SMPTE time code, when using conversion boxes or other special interface options. Autolocators (such as the Tascam RD-848) can be used to remote control many of the basic transport, track arming, and autolocation functions from a single desktop surface.

Third-Party Developments and Accessories for the MDM

Years after MDMs were introduced, an ever-growing number of accessories still pop up on the market for both the ADAT and DTRS MDM formats. These accessories can be used to link the digital audio and possibly the synchronization capabilities to an external device in order to interface and/or synchronize an MDM system to a host of possible applications. For example, it's not uncommon to find a computer-based, multichannel digital audio interface that has ADAT and/or TDIF digital I/O ports. Often, these ports are a cost-effective way to add on extra I/O channels to a host DAW system. When using a system in this manner, be sure that the MDM's wordclock is set to follow the interface (or whatever device is set to generate the master wordclock).

The list of manufacturers and third-party companies is far too large and varied to provide here, as the number of new and innovative applications and accessories for integrating these devices into present-day systems continues to grow. The best way to keep on top of recent developments is to contact the manufacturers, search the Web, and keep reading the trade magazines.

MiniDiscs

Since its introduction by Sony as a consumer device in 1992, the MiniDisc (MD) has actually grown in popularity as a medium for recording and storing CD, MP3 tracks, and original recordings (Figure 6.34). Based on established rewritable magneto-optical (MO) technology, the 64-mm disc system has a standard recording capacity of up to 74 minutes (effectively the same record/play time as the CD, at one-quarter the size). This extended record/play time is due to a compression codec known as Adaptive Transform Acoustic Coding (ATRAC), which makes use of an encoding scheme that reduces data by eliminating or reducing data that has been deemed to be imperceptible by the average listener (in much the manner as the MP3 and WMA). Probably the biggest selling point of the MiniDisc recording medium is its portability (Figure 6.35). Units that are small enough to fit in the palm of your hand are being used to capture song ideas, lectures, instrumental samplefiles, and even live bootleg concerts. These and other advantages help to keep the MiniDisc in the pockets of many music and recording professionals.

Figure 6.34. Tascam MD 801R mkII MiniDisc recorder. (Courtesy of Tascam, www.tascam.com.)

Figure 6.35. Portable MiniDisc recorder.

Hard-Disk Recording

Over the history of disk- and disc-based systems, the style, form, and function of hard disk recording have changed to meet the challenges of faster processors, bigger drives, improved hardware systems, and the ongoing push of marketing forces to sell, sell, sell! As a result, there are numerous hard-disk system types that are designed for various purposes, budgets, and production styles. As new technologies and programming techniques continue to turn out new products on a staggeringly regular basis, many of the long-held limitations and distinctions between system types and even operating systems have gone by the wayside or have been blurred. Although the field of hard-disk recording has matured into a technology that's become pervasive in all forms of media production, it's still in a continual form of evolution as hardware, software, and personal working styles change. As with all evolutionary revolutions, it's always a good idea to keep abreast of these changes by reading the trade magazines, searching the Web, and keeping your eyes and ears open. For the remainder of this chapter, we'll be looking at the various types of hard disk systems that are currently available on the market . . . as well as their basic principles of operation.

Advantages of Hard-Disk Recording

With the advent of the drum machine (the first practical digital playback system) . . . it was soon discovered that advances in digital and semiconductor technology were opening the doors for recording audio as longer and longer samplefiles into random access memory. It wasn't long after this that computer technology advanced to the point of being affordable to the average user . . . and through the use of specialized hardware, software, and I/O interfacing, digital audio could be recorded to, edited on, and played back from a computer's hard disk. Thus, the concept of the hard-disk recorder was born. As most people are aware, there are numerous advantages to using a hard-disk recording system in an audio production environment:

- The ability to handle long samplefiles—Hard-disk recording time is often limited only by the size of the disk itself.
- Random-access editing—Once audio (or any type of data) is recorded onto a disk, any point within the program can be instantly accessed at any time, regardless of the order in which it was recorded.
- Nondestructive editing—This process allows audio segments (often called *regions*) to be placed back in any context and/or order within a program without changing or affecting the originally recorded soundfile in any way. Once edited, these edited tracks and/or segments can be reproduced to create a single, cohesive program.
- DSP—Digital signal processing can be performed on a soundfile and/or segment in either real time or non-real time (often in a nondestructive fashion).

Add to this the fact that computer-based digital audio devices can integrate many of the tasks that are related to both digital audio and MIDI production in a unified fashion that's often easy to use, cost effective, and time effective . . . and you have a system that offers the artist and engineer an unprecedented degree of production power.

Hard-Disk Multitrack Recorders

Once the modular digital multitrack (MDM) came onto the production scene, the expensive analog multitrack and more expensive digital multitrack recorder became less and less of an option for the average project and professional studio budget. As technology marched on, many professionals have chosen not to produce audio using either a tape-based recording system or DAW . . . but have instead opted in favor of a multitrack hard-disk recorder (Figures 6.36 and 6.37).

Unlike software DAWs that use the graphic user interface (GUI) of a personal computer, these dedicated hardware systems mimic the basic transport, operational, and remote controls of a traditional multitrack recorder. Often incorporating one or more removable hard-drive bays and file compatibility with existing multitrack DAW software . . . the basic allure of such a system (which is often designed around a PC motherboard and hard drive) is the fact that it offers a simple, dedicated multitrack hardware interface that's capable of providing the speed, flexibility, and connectivity benefits of a hard-disk recorder.

Hard-Disk and Flash Memory Portable Studios

Another type of dedicated hardware recording system is the modern-day portable studio that's capable of recording to hard disk, MD, or solid-state flash memory cards. These all-in-one

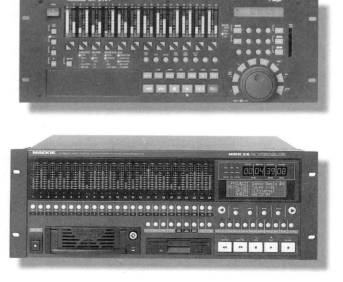

Figure 6.36. Tascam MX-2424 PB multitrack hard-disk recorder. (Courtesy of Tascam, www.tascam.com.)

Figure 6.37. Mackie HDR 24/96 multitrack hard disk recorder. (Courtesy of Loud Technologies, Inc., www.mackie.com.)

Figure 6.38. *Yamaha AW4416HD 16-track hard drive digital audio workstation with CD/RW. (Courtesy of Yamaha Pro Audio, www.yamaha.com.)*

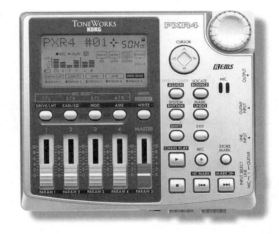

Figure 6.39. *PXR4 digital mini recorder. (Courtesy of Korg USA, www.korg.com/pxr4.)*

systems include all of the hardware and the control system's interface to record, edit, mix down, and play back an on-the-spot composition...virtually anywhere when using an AC adapter or batteries (Figure 6.38). These systems, which range in size, features, and track offerings, are often the system of convenience for musicians, as they include all of the necessary mic preamps, mixing surface controls, effects, and integrated CD burning capabilities. In certain cases, these devices can actually get so small as to become a virtual studio in a pocket...offering all of the recording and overdub features that you might expect, plus surprising ones like flash memory, USB data transfer, effects, built-in stereo mic, and built-in rhythm machine (Figure 6.39).

The Digital Audio Workstation

In recent years, the term *digital audio workstation* (*DAW*) has increasingly come to signify an integrated computer-based hard disk recording system that commonly offers such

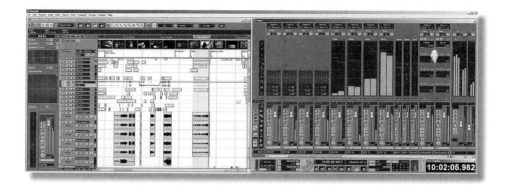

Figure 6.40. *Nuendo Version 2 Media Production System for the Mac or PC. (Courtesy of Steinberg Media Technologies GMBH, www.steinberg.net.)*

features as:

◆ Advanced multitrack recording, editing, and mixdown capabilities

◆ MIDI sequencing, edit, and score capabilities

◆ Integrated and plug-in signal processing support

◆ Support for integrating software plug-in instruments (VSTi) and/or peripheral music programs (ReWire)

◆ Integration of peripheral hardware devices such as controllers and audio and MIDI interface devices

Truth of the matter is…by offering a staggering amount of production power for the buck these software-based programs (Figures 6.40 and 6.41) and their peripherally connected devices have revolutionized the faces of professional, project, and personal studios in a way that touches almost every life within the audio and music production communities.

Integration Now…Integration Forever!

Throughout the history of music and audio production, we've become used to the idea that certain devices were only meant to perform a single task: A recorder records and plays back, a limiter limits, and a mixer mixes. Fortunately, the age of the microprocessor has totally broken down these traditional lines…in a way that has created a breed of digital chameleons that can change their functional colors to match the necessary task at hand. Along these same lines, the digital audio workstation isn't so much a device as a systems concept that can perform a wide range of multichannel audio production tasks with ease and speed. Some of the characteristics that can (or should be) displayed by a DAW include:

◆ *Integration*—One of the major functions of a workstation is its ability to provide centralized control over the digital audio recording, editing, processing, and signal

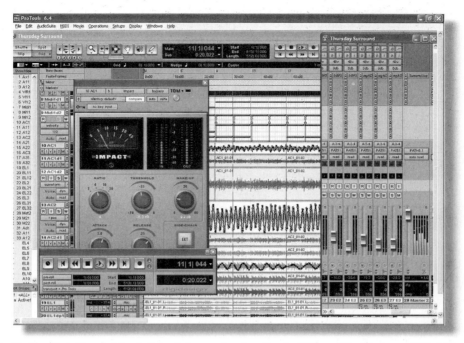

Figure 6.41. *Pro Tools Ver. 6.4 hard-disk editing workstation for the Mac or PC. (Courtesy of Digidesign, a division of Avid, www.digidesign.com.)*

routing functions, as well as to provide transport and/or time-based control over MIDI/electronic music systems, external tape machines, and video recorders.

◆ *Communication*—A DAW should be able to communicate and distribute pertinent audio-, MIDI-, and automation-related data throughout the connected network system. Digital timing (wordclock) and synchronization (SMPTE time code and/or MTC) should also be supported.

◆ *Speed and flexibility*—These are probably a workstation's greatest assets. After you become familiar with a particular system, most production tasks can be tackled in far less time than would be required using similar analog equipment. Many of the extensive signal processing, automation, and system's communications features would simply be next to impossible to accomplish in the analog domain.

◆ *Automation*—Because all of the functions are in the digital domain, the ability to instantly recall a session and to undo a performed action becomes a relatively simple matter.

◆ *Expandability*—Most DAWs are able to integrate new and important hardware and/or software components into the system with little or no difficulty.

◆ *User-friendly operation*—An important element of a digital audio workstation is its ability to communicate with its central interface unit . . . you! The operation of a workstation should be relatively intuitive and shouldn't obstruct the creative process by speaking "computerese."

From the above, I'm sure that you've gathered that a software system (and its associated hardware) that's capable of integrating audio, video, and MIDI under a single, multifunctional umbrella can be a major investment, both in financial terms and in the time that's spent to learn and master the overall program environment. When choosing a system for yourself or your facility, be sure to take the above considerations into account. Each system has its own strengths, weaknesses, and particular ways of working. When in doubt, it's always a good idea to research the system as much as possible before committing to it. Feel free to contact your local dealer for a salesroom test drive. As with a new car, purchasing a digital audio workstation can be an expensive proposition that you'll probably have to live with for a while. Once you've taken the time to make the right choice . . . you can get down to the business of making music.

DAW Hardware

In keeping step with the modern-day truism "technology marches on" . . . the hardware and software specs of a computer and the connected peripherals continue to change at an ever-increasing pace. This is usually reflected as general improvements in such areas as their:

◆ Need for speed
◆ Increased computing power
◆ Increased disk and RAM memory and speed
◆ Operating system (OS) and peripheral integration
◆ General connectivity (networking and the Web)

In this section, we'll be taking a look at many of the hardware devices and connected peripheral devices that help to make a DAW work.

The Desktop Computer

Desktop computers are often (but not always) too large and cumbersome to lug around. As a result, these systems are most often found as a permanent install in the professional, project, and home studio (Figures 6.42 through 6.44) One of the most commonly asked questions is "Which one . . . Mac or PC?" The actual answer to what OS to invest in actually depends upon:

◆ Your preference
◆ Your needs
◆ The kind of software you currently have
◆ The kind of computer platform and software your working associates have

Figure 6.42. *Apple G5 computer in a 19-inch rackmount. (Courtesy of Marathon Computer, www.marathoncomputer.com.)*

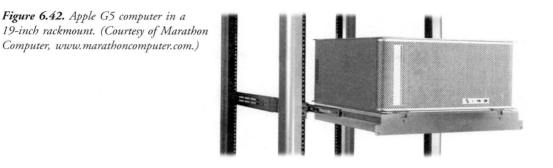

Figure 6.43. *MusicXPC Small Form Factor media production computer. (Courtesy of MusicXPC, www.musicxpc.com.)*

Figure 6.44. *The Audio PC. (Courtesy of Carillon USA, www.carillonusa.com.)*

Truth of the matter is, beyond these important questions, the choice it strictly up to you. Once you've decided which side of the platform tracks that you'd like to live on, the more important questions that you should be asking are:

◆ Is my computer fast and powerful enough for the tasks at hand?

◆ Does it have enough hard disks that are large and fast enough for my needs?

Figure 6.45. Pacific Pro Audio 911 portable FireWire hard drive. (Courtesy of Pacific Pro Audio, www.pacificproaudio.com.)

◆ Is there enough RAM memory?

◆ Do I have enough monitor space (real estate) to see the important things at a glance?

On the "need for speed" front, it's always a good idea to buy (or build) a computer at the top of its performance range at any given time. Keeping in mind that technology marches on, the last thing that you'll want to do is buy a new computer only to soon find out that it's underpowered for the tasks ahead.

There's never been a better time for choices on the hard-disk front. With today's faster and higher capacity IDE, serial ATA, and SCSI drives, it's a simple matter to install cost-effective drives, each with a capacity of hundreds of gigabytes. In addition, portable drive cases (Figure 6.45) can be plugged into either a FireWire or USB2 port (and in some cases both)...making it easy to take your own personal drive with you into the studio.

Speed can be an issue when buying a hard drive for either audio or video applications. The speed at which the disc platters turn will often affect a drive's access time. Modern drives that "spin" at 7200 or higher are often preferable...and it should be noted that production drives often work best when connected directly to the system's motherboard. On-board buffer memory can also be helpful in transferring data and in freeing up the system for other processing functions. It should be mentioned that the portable FireWire and USB2 drives mentioned above will often have reduced access times (over their internal counterparts), often making them a better medium for backing up data...although, with the introduction of FireWire 800 and the onward march of technology, this could change at a moment's notice.

Regarding random access memory...it's always a good idea to use as much (and as fast) RAM as you can muster. If a system doesn't have enough RAM, data will often have to be "swapped" to the system's hard drive...often slowing things down and affecting overall performance. When dealing with video and digital images, having a sufficient amount of RAM becomes even more of an issue.

Just like there never seems to be enough space around the house or apartment, having a single, undersized monitor can leave you feeling cramped for visual "real estate." For starters,

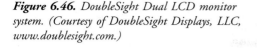

Figure 6.46. *DoubleSight Dual LCD monitor system. (Courtesy of DoubleSight Displays, LLC, www.doublesight.com.)*

a sufficiently sized monitor (either LCD or CRT) that's capable of working at higher resolutions will greatly increase the size of your visual desktop; however, if one is a good thing, two can be better! Both Windows XP and newer OS versions for the Mac offer support for *dual monitors* (Figure 6.46). Through the addition of a "dual head" video card or by simply adding another video card, these systems can be easily configured so that the two monitors will literally double your working space for less bucks than you might think. In short, it's truly a joy to have your edit window, mixer, effects sections, and transport controls in their own place . . . all in plain and accessible view.

The Laptop Computer

One of the most amazing characteristics of the digital age is miniaturization. At the forefront of the studio-a-go-go movement is the laptop computer (Figures 6.47 and 6.48). Out of the advent of smaller, lighter, and more powerful notebooks has come the technological Phoenix of the portable DAW. With the advent of USB and FireWire audio interfaces, controllers, and other peripheral devices, these systems are now capable of handling most (if not all) of

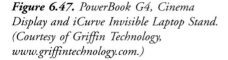

Figure 6.47. *PowerBook G4, Cinema Display and iCurve Invisible Laptop Stand. (Courtesy of Griffin Technology, www.griffintechnology.com.)*

Figure 6.48. A PC laptop in action.

the edit and processing functions that can be handled in the studio. In fact, these AC/battery-powered systems have become powerful enough to handle advanced DAW edit/mixing functions, as well as being able to happily handle a wide range of plug-in effects and virtual instruments . . . all in the comfort of . . . anywhere!

That's the good news! Now, the downside of all this portability is the fact that, since laptops are optimized to run off of a battery with as little power drain as possible, their:

◆ Processors will often run slower.

◆ BIOS (the important subconscious brains of a computer) might be different (especially with regards to battery-saving features).

◆ Hard drives might not spin as fast.

◆ Video display capabilities are sometimes limited when compared to a desktop.

◆ Internal audio interface usually isn't so great.

Although the central processing unit (CPU) will often run slower (often to reduce power consumption in the form of heat), most modern laptops are more than powerful enough to perform on the road. For this reason, it's always best to get a system with the fastest CPU that you can afford.

Most often, the primary problems in a laptop often rest with the basic BIOS and OS battery-saving features. When it comes to making music with a laptop, there is actually a real difference between the Mac and a PC. Basically, there's little difference between a Mac laptop and a Mac tower, as the BIOS's are virtually identical. Conversely, the BIOS of a laptop PC is often limited in power and functional capabilities when compared to a desktop. When shopping for a PC laptop, it's often good to research how a particular BIOS chipset will work for music . . . particularly with regard to certain audio interface devices (some interfaces won't work well or at all with certain chipsets).

Both the PC and Mac laptops often have an automatic power-saving feature (respectively called "speed step" and "processor cycling") that changes the CPU's speed in much the same way that a vehicle changes gears in order to save energy. Often these gear changes will wreak havoc on many of the DSP functions of an audio workstation. Turning them off will greatly improve performance . . . at the expense of reduced battery life.

Hard-drive speeds on a laptop are often limited, when compared to a desktop computer . . . resulting in slower access times and fewer track counts on a multitrack DAW. Even though these speeds are often more than adequate for general music applications, these speed issues can often be improved through the use of an external FireWire drive.

Again, when it comes to RAM, it's often a good idea to pack as much into the laptop as you can. This will reduce data swapping to disk with larger audio applications (especially software synths and samplers). Within certain systems, the laptop's video card capabilities will run off of the system's RAM (which can severely limit audio processing functions in a DAW). Memory-related problems can also crop up when using a motherboard to run a dual-monitor (LCD and external monitor) setup.

It almost goes without saying that the internal audio quality of most laptops ranges from being okay to abysmal. As a result, about the only true choice is to find an external audio interface that works best for you and your applications (Figures 6.49 and 6.50). Fortunately, there are a ton of audio interface choices for either FireWire or USB . . . ranging from a simple stereo I/O device to those that include multitrack audio, MIDI, and controller capabilities in a small, on-the-go package.

In addition to being a studio-on-the-go, the laptop can act as an "expansion module" to a desktop setup. You could trigger a loaded software synth or sampler via MIDI or through the use of VST System Link (a Steinberg product that allows multiple computers to act as a single, connected system via a network connection), and any number of effects, instruments, and VST software devices can work in unison with the main DAW.

Figure 6.49. Tascam US-224 computer interface and control surface. (Courtesy of Tascam, www.tascam.com.)

Figure 6.50. *M-Audio Ozone USB Audio/MIDI mobile workstation. (Courtesy of M-Audio, www.m-audio.com.)*

System Interconnectivity

In the not-too-distant past, installing a device into a computer or connecting between computer systems could be a major hassle. With the development of the USB and FireWire protocols (as well as the improved general programming of hardware drivers), hardware devices such as mice, keyboards, cameras, soundcards, modems, MIDI interfaces, CD and hard drives, MP3 players . . . even portable fans . . . can be plugged into an available port, installed, and up and running in no time, (usually) without a hassle.

With the development of a standardized network protocol, it's now possible to link computers together in a way that allows for the fast and easy sharing of data throughout a connected system. Using such a system, whole businesses are able to share files with a computer across an administered network on the other side of the country, studios can share soundfiles and videofiles throughout an entire production, and houses can share files and a single high-speed Internet connection with relative ease.

USB

In recent computer history, few protocols for interconnecting devices to a host computer have affected our lives like the Universal Serial Bus (USB). In short, USB is an open specification for connecting external hardware devices to the personal computer, as well as a special set of protocols for automatically recognizing and configuring them. The first of the following two speeds are supported by USB 1.0, while all three are supported by USB 2.0:

◆ USB 1.0 (1.5 Mbits/second)—A low speed for the attachment of low-cost peripherals (such as a joystick or mouse).

◆ USB 1.0 (12 Mbits/second)—For the attachment of devices that require a higher throughput (such as data transfer, soundcards, digitally compressed video cameras, and scanners)

◆ USB 2.0 (480 Mbits/second)—For high-throughput and fast transfer of the above applications

The basic characteristics of USB include the following:

◆ Up to 127 external devices can be added to a system without having to open up the computer. As a result, the industry is moving toward a "sealed case" or "locked-box" approach to computer hardware design.

◆ Newer operating systems will often automatically recognize and configure a basic USB device that's shipped with the latest device drivers.

◆ Devices are "hot pluggable," meaning that they can be added (or removed) while the computer is on and running.

◆ The assignment of system resources and bus bandwidth is transparent to the installer and end user.

◆ USB connections allow data to flow bidirectionally between the computer and the peripheral.

◆ USB cables can be up to 5 meters in length (up to 3 meters for low-speed devices) and include two twisted pairs of wires, one for carrying signal data and the other pair for carrying a DC voltage to a "bus-powered" device. Those that use less than 500 milliamps (1/2 amp) can get their power directly from the USB cable's 5-V DC supply, while those having higher current demands will need to be externally powered.

◆ Standard USB cables have two types of connectors at each end. For example, a cable between the PC and a device would have an "A" plug at the PC (root) connection and a "B" plug for the device's receptacle.

◆ Cable distribution and "daisy-chaining" are done via a data "hub" (Figure 6.51). These devices act as a traffic cop, in that they cycle through the various USB inputs in a sequential fashion . . . routing the data into a single data output line. It should be noted that not all hubs are created equal. In certain situations, the chipset that's used within a hub might not be compatible with certain MIDI and audio interface

Figure 6.51. The Griffin 4-Port USB audio hub. (Courtesy of Griffin Technology, www.griffintechnology.com.)

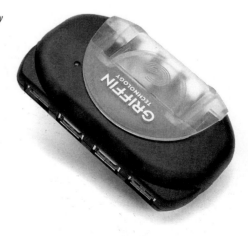

systems. . . . Should a connection problem arise, contact the manufacturer of your audio-related device for advice.

FireWire

Originally created in the mid-1990s by Apple (and later standardized as IEEE-1394), the FireWire protocol is similar to the USB standard in that it uses a twisted-pair wiring to communicate bidirectional, serial data within a hot-swappable, connected chain. Unlike USB (which can handle up to 127 devices per bus), up to 63 devices can be connected within a connected FireWire chain. FireWire supports two speed modes:

◆ FireWire 400 or IEEE-1394a (400 Mbits/second)—Capable of delivering data over cables up to 4.5 meters in length, FireWire 400 is ideally for communicating large amounts of date to such devices as hard drives, video camcorders, and audio interface devices.

◆ FireWire 800 or IEEE-1394b (800 Mbits/second)—FireWire 800 is capable of communicating large amounts of data over cables up to 100 meters in length. When using fiberoptic cables, lengths in excess of 90 meters can be achieved in situations that require long-haul cabling (such as sound stages and studios).

Unlike USB, compatibility between the two modes is mildly problematic, as FireWire 800 ports are configured differently from their earlier predecessor . . . and therefore require adapter cables to ensure compatibility.

Networking

Beyond the concept of connecting external devices to a single computer, another concept hits at the heart of the connectivity age . . . *networking*. The ability to set up and make use of a *local area network* (LAN) can be extremely useful in the home, studio, and/or office, in that it can be used to link multiple computers with various data, platforms, and OS types. In short, a network can be set up in a number of different ways . . . with varying degrees of complexity and administrative levels; however, basically there are two common ways that data can be handled over a LAN (Figure 6.52):

◆ The first is a system whereby the data that's shared between linked computers resides on the respective computers and is communicated back and forth in a decentralized manner.

◆ The second makes use of a centralized computer (called a *server*) that's basically an array of high-capacity hard drives that is used to store "all" of the data that relates to the everyday production aspects of a facility. Often, such a system will have a redundant set of drives that actually clone the entire system on a moment-to-moment basis . . . as a safety backup procedure. Alternatively (and in some cases, in addition to), a set of backup tapes may be made on a daily basic for extra insurance and archival purposes.

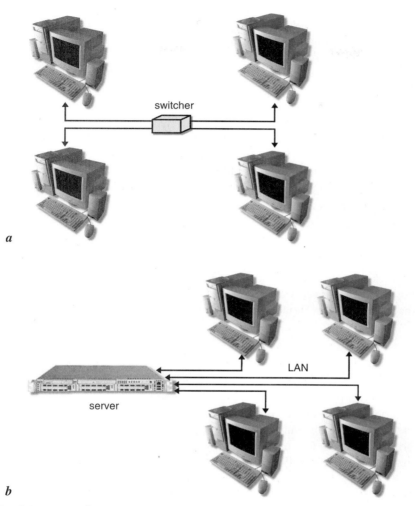

Figure 6.52. *Local Area Network (LAN) connections. (a) Data may be shared between independent computers in a home or workplace LAN environment. (b) Computer terminals may be connected to a centralized server, allowing data to be stored, shared, and distributed from a central location.*

No matter what level of complexity is involved, some of the more common uses for working with a network connection include:

◆ *Sharing files*—Within a connected household, studio or business, a LAN can be used to share files, soundfiles, video images . . . virtually anything, throughout the connected facility. This means that various productions rooms, studios, and offices can simultaneously share and swap data and/or mediafiles in a way that's often transparent to the users.

◆ *Shared Web connection*—One of the cooler aspects of using a LAN is the ability to share an Internet connection over the network from a single, connected computer or server. The ability to connect from any computer with ease is just another reason why you should strongly consider wiring your studio and/or house with LAN connections.

◆ *Archiving and backup*—In addition to the benefits of archiving and backing up data with a server system ... even the simplest LAN can be a true lifesaver. For example, let's say that we need to make a backup DVD of a session but don't have the time to tie up our production DAW. In this situation, we could simply burn the disc on another computer that's connected to the system ... and continue working away, without interruption.

◆ *Accessing soundfiles and sample libraries*—It goes without saying that sound- and samplefiles can be easily accessed from any connected computer ... Hey! If you're wireless (or have a long enough cable), go out to the pool and soak up the sun while you're working!

On a final note, those who are unfamiliar with networking are urged to learn about this powerful and easy-to-use data distribution and backup system. For a minimal investment in cables, hubs, and educational reading, you might be surprised at the time- and trouble-saving benefits that will be almost instantly realized.

The Audio Interface

An important device that deserves careful consideration when putting together a DAW-based production system is the *digital audio interface*. These devices can have a single, dedicated purpose, or they might be multifunctional in nature ... in either case, their main purpose in the studio is to act as a connectivity bridge between the outside world of analog audio and the computer's inner world of digital audio (Figures 6.53 through 6.57). Audio interfaces

Figure 6.53. Mackie Spike USB audio/MIDI interface. (Courtesy of Loud Technologies, Inc., www.mackie.com.)

Figure 6.54. *Universal Audio 2192 master audio interface. (Courtesy of Universal Audio, www.uaudio.com.)*

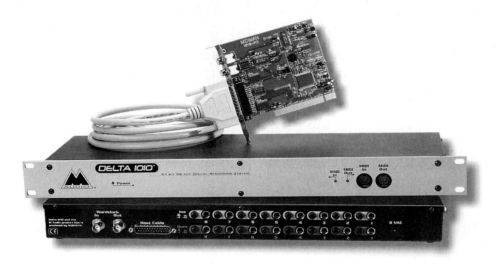

Figure 6.55. *M-Audi Delta 1010 10 in/10 out PCI/rack digital interface. (Courtesy of M-Audio, www.m-audio.com.)*

Figure 6.56. *Mark of the Unicorn 828 mkII FireWire interface. (Courtesy of MOTU, Inc., www.motu.com.)*

come in all shapes, sizes, and functionalities; for example:

◆ Built into a computer (although, more often than not, these devices are often limited in quality and functionality)

◆ A simple, two-I/O audio device

◆ Multichannel offering eight analog I/Os and numerous I/O expansion options

◆ Fitted with one or more MIDI I/O ports

◆ Offering digital I/O, wordclock, and sync options

Figure 6.57. *Digidesign Digi 002 rack audio interface. (Courtesy of Digidesign, a division of Avid,*
www.digidesign.com.)

◆ Fitted with a controller surface (with or without motorized faders) that provides for
hands-on DAW operation

These devices may be designed as hardware cards that fit directly into the computer, or
they might plug into the system via USB or FireWire. They may offer a limited number of
sample-rate and bit-depth options . . . or might be capable of handling rates up to 96 kHz/24
bits or higher. Unless you buy a system that has been designed to operate with a specific
piece of hardware (most notably, Digidesign users), you should weigh the vast number of
interface options and capabilities with patience and care—the system you might save could
be your own.

Native *Versus* Non-Native Processing

In the past, computer technology was far too slow to handle the intense number of processing
and I/O functions that were required for multichannel DAW production; however, with the
introduction of faster CPUs, hard disks, and memory, many computer systems are now able
to handle the workload of dealing with basic program operation, signal processing, and passing
audio to multiple I/O ports, without the need for special, dedicated hardware. As such,
newer software systems that work in the *native processing environment* have now begun to place
the burden of processing and I/O signal routing entirely upon the personal computer, its
operating system, and the DAW software. In short, a DAW that has been programmed to work
in a native environment is able to route all of its processing, file structure, and I/O functions
through the computer's CPU and operating system. The advantage to such a system is that any
audio and interface (and many other devices) that can be seen by the computer's OS can
be accessed by the DAW or host application. That's to say . . . if you already have an XYZ
interface, chances are it'll work with your system. If your friend has a better one for a specific
session . . . no problem.

Non-native devices, on the other hand, have been designed to work with a specific
piece of supporting interface hardware . . . period. If you already have an interface, you won't
be able to use it with that particular DAW software. If a new system comes out that requires

a hardware upgrade, chances are your hardware, software, and possibly your plug-ins will have to be sold on the used market and then be replaced by the new workstation's hardware. Although a non-native system has the appeal of being a complete WYSIWYG ("what you see is what you get"), all-in-one package, you should be aware of the differences and personal options... should you wish to upgrade or add extra functions to your system later.

Audio Driver Protocols

Audio driver protocols are software programs that set standards for allowing allow data to be communicated between the system's software and hardware. A few of the more common protocols are:

◆ *WDM*—The "Windows Driver Model" is a robust driver implementation that's directly supported by Microsoft's Windows. Software and hardware that conform to this basic standard can communicate audio to and from the computer's basic audio ports.

◆ *ASIO*—The "Audio Stream Input/Output" architecture forms the backbone of VST. It does this by supporting variable bit depths and sample rates, multichannel operation, and synchronization. This commonly used protocol offers low latency, high performance, easy set up, and stable audio recording within VST.

◆ *MAS*—The "MOTU Audio System" is a system extension for the Mac that uses an existing CPU to accomplish multitrack audio recording, mixer, bussing, and real-time effects processing. In addition to working with popular interface systems, such as the MOTU 2408 and 1224, this system works with a growing number of interfaces, including Digidesign Audiomedia II/III, Sonorus StudI/O, any PCI-based ProTools system via Direct I/O, Yamaha DSP Factory, Event Layla, and the Korg 1212.

◆ *CoreAudio*—This Digidesign driver allows compatible single-client, multichannel applications to record and play back through most Digidesign audio interfaces on Mac OSX. It supports full-duplex recording and playback of 16-/24-bit audio at sample rates up to 96 kHz (depending on your Digidesign hardware and CoreAudio client application).

Of course, the above listing is far from complete, and further reading can be found from the respective companies. In most circumstances, it won't be necessary for you to be familiar with the protocols... simply that your software and hardware are compatible for use with a driver protocol that works best for you.

Latency

Quite literally, *latency* refers to the buildup of delays (measured in milliseconds) in an audio signal, as they pass through the audio circuitry of the audio interface, CPU, internal mixing structure, and I/O routing chains. When monitoring a signal directly through a computer's signal path, latency can be experienced as short delays between the input and monitored signal. If the delays are excessive, they can be unsettling enough to throw a performer off time. For example, when recording a synth track, you might actually hear the delayed monitor sound shortly after hitting the keys (not a happy prospect). With the advent of faster computers, improved audio drivers, and better programming, latency has now been reduced to levels that are so small as to not be unnoticeable. For example, the latency of a standard Windows audio driver can be truly pitiful (upward to 500 ms). By switching to a supported ASIO driver and by reducing the audio interface (and possibly the DAW) buffers to their lowest operating size (without causing stuttering) . . . these delay values could easily be reduced down to an unnoticeable range.

DAW Controllers

Often, one of the more common complaints that some people have against the digital audio editor and workstation environment (particularly when relating to the use of on-screen mixers) is the lack of a hardware controller that gives the user access to hands-on controls. In recent years, this has been addressed by major manufacturers and third-party companies in the form of a hardware *DAW controller interface* (Figures 6.58 through 6.61). These controllers generally mimic the design of an audio mixer in that they offer slide or rotary gain faders, pan pots, solo/mute, and channel select buttons . . . with the added bonus of a full transport remote. A *channel select button* is used to actively assign a specific channel to a section that contains a series of grouped pots and switches that relate to EQ, effects, and dynamic functions. Often, these controllers offer direct mixing control over eight input strips at a time. By switching between the banks in groups of 8 (1–8, 9–16, 17–24, . . .), any number of the grouped inputs can be access by the virtual mixer. These devices will also often include software function keys that can be programmed to give quick and easy access to the DAW's more commonly used program keys.

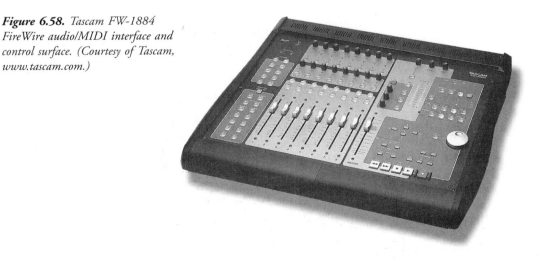

Figure 6.58. *Tascam FW-1884 FireWire audio/MIDI interface and control surface. (Courtesy of Tascam, www.tascam.com.)*

Figure 6.59. *Mackie Control Universal. (Courtesy of Loud Technologies, Inc., www.mackie.com.)*

Figure 6.60. *Digidesign Pro Control. (Courtesy of Digidesign, a division of Avid Technology, Inc., www.digidesign.com.)*

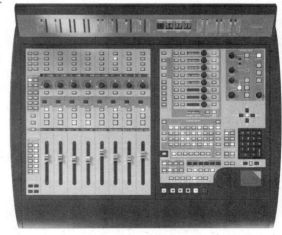

Controller commands are most commonly transmitted between the controller and audio editor via device-specific MIDI System-Exclusive messages. As such, in order to be able to integrate a controller into your system, the DAW's current version must be specifically programmed to accept the control codes from a particular controller... unless the DAW and controller make use of a new plug-in architecture that allows compatible devices to freely connected. Most controller surfaces communicate these messages to the DAW host via the easy-to-use USB or FireWire protocols.

Certain controllers also offer all-in-one capabilities that can be straightforward and cost-effective devices for first-time buyers. Often, these devices include a multichannel audio interface, MIDI interface port(s), monitor capabilities, and full controller functions. Others may already have an existing digital mixer that can actually be used as a fully functional

Figure 6.61. *Digidesign ICON integrated console.*
(Courtesy of Digidesign, a division of Avid
Technology, Inc., www.digidesign.com.)

controller (and in certain circumstances, as a multichannel audio interface) when connected to a DAW host program. For these and other reasons, taking the time to research your needs and current equipment capabilities can save time and money . . . or, at the worst, can simply be educational.

Soundfile Formats

An amazingly varied number of soundfile formats exist within audio and multimedia production. Here are the most commonly used audio production formats that don't use data compression:

◆ *Wave (.wav)*—The Microsoft Windows format supports both mono and stereo files at a variety of resolutions and sample rates. WAV files contain PCM coded audio (uncompressed Pulse Code Modulation formatted data) that follows the Resource Information File Format (RIFF) spec, which allows extra user information to be embedded and saved within the file itself.

◆ *Broadcast wave (.wav)*—In terms of audio content, broadcast wave files are the same as regular wave files; however, text strings for supplying additional information can be imbedded in the file according to a standardized data format.

◆ *Wave64 (.w64)*—This proprietary format was developed by Sonic Foundry, Inc. (now operating under the Sony name). In terms of audio quality, Wave64 files are identical to wave files . . . except that their file headers use 64-bit values (instead of Wave's 32-bit values). As a result, Wave64 files can be considerably larger than standard wave files, and this format is a good choice for long recordings (*e.g.*, surround files and file sizes over 2 GB).

◆ *Apple AIFF (.aif or .snd)*—This standard soundfile format from Apple supports mono or stereo, 8-bit or 16-bit audio at a wide range of sample rates. Like broadcast wave files, AIFF files can contain embedded text strings.

◆ *Sound Designer I & II (.sd and .sd2)*—The Sound Designer is used by Digidesign as a soundfile format for the Mac. SDI was first released in 1985 and can still be found on many CD-ROM and soundfile discs; it was primarily used to store 16-bit, mono samples of short duration (often on the order of seconds). As a later incarnation, SDII can now encode 16- or 24-bit soundfiles of any practical length at a variety of sample rates.

Soundfile Sample Rates

The *sample rate* of a recorded bitstream directly relates to the resolution at which a recorded sound will be digitally captured. Just as with a moving image . . . if you take more "samples" of the image as it moves through time, you'll have a more accurate representation of that recorded image. If the number of samples are too low, the resolution will be substandard and "lossy." On the other hand, too high of a rate might result in a recorded bandwidth that's so high that the gain in resolution is lost on the audience's ability to discriminate it . . . or the storage requirements might become so great that the files become inordinately large. Beyond the basic adherence to certain industry sample rate standards . . . such are the choices and personal decisions that must be made regarding which is the best sample rate to use on a project. Although other sample-rate standards exist, the following are the most commonly used in the professional, project, and general audio production community:

◆ 32 k—This rate is often used by broadcasters to transmit/receive digital data via satellite. With it's overall 15-kHz bandwidth and reduced data requirements, it is also used by certain devices in order to conserve on memory. Although this rate isn't generally used by the procommunity, it's surprising just how good a sound can be captured at 32 k, given a high-quality converter.

◆ 44.1 k—The long-time standard of consumer and pro audio production, 44.1 is the chosen rate of the CD-audio standard. With its overall 20-kHz bandwidth, the 44.1-k rate is generally considered to be the minimum sample rate for professional audio production. Assuming that high-quality converters are used, this rate is capable of recording lossless audio, while conserving on memory storage requirements.

◆ 48 k—This standard was adopted early on as a standard sample rate for professional audio applications (particularly when referring to hardware digital audio devices).

◆ 96 k—With the onset of 24-bit recording capabilities, higher-rate and bit-rate recordings have made it feasible for recordings to be encoded at 96 kHz and higher rates (*e.g.*, 24/96). 96 kHz is also the accepted rate for DVD audio production.

◆ 192 k—This is also an accepted rate for DVD audio production.

Soundfile Bit Rates

The *bit rate* of a digitally recorded soundfile directly relates to the number of quantization steps that are encoded into the bitstream. As a result, the bit rate (or bit depth) is directly correlated to the:

♦ Accuracy at which a sampled level (at one point in time) is to be encoded

♦ Signal-to-error figure . . . and thus the overall dynamic range of the recorded signal

If the number of encoded bits are too low to accurately encode the sample, the resolution will be substandard (*i.e.*, distorted). On the other hand, too high of a bit depth might result in a resolution that's so high that the resulting gain in resolution is lost on the audience's ability to discriminate it . . . or the storage requirements might become so high that the files become inordinately large.

Although other bit-rate standards exist, the following are the most commonly used in the pro, project, and general audio production community:

♦ 16 bits—The long-time standard of consumer and professional audio production, 16 bits is the chosen bit depth of the CD-audio standard (offering a theoretical dynamic range of 97.8 dB). It is generally considered to be the minimum depth for high-quality professional audio production. Assuming that high-quality converters are used, this rate is capable of recoding lossless audio, while conserving on memory storage requirements.

♦ 20 bits—Before the 24-bit rate came onto the scene, 20 bits was considered to be the standard for high-bit-depth resolution. Although it is used less commonly, it can still be found in high-definition audio recordings (offering a theoretical dynamic range of 121.8 dB).

♦ 24 bits—Offering a theoretical dynamic range of 145.8 dB, this standard bit rate is often used in professional-audio, high-definition, and DVD-audio applications.

Format Interchange and Compatibility

At the soundfile level, most software editors and DAWs are able to read a wide range of uncompressed and compressed formats, which can then be saved into a new format. At the session level, there are several standards that allow for the exchange of data for an entire session, from one platform, OS, or hardware device to another. These include:

♦ *Open Media Framework Interchange* (OMFI) is a platform-independent session file format intended for the transfer of digital media between different DAW applications; it is saved with an .omf file extension. OMF (as it is commonly called) can be saved in either of two ways: (1) "export all to one file," when the OMF file includes all of the soundfiles and session references that are included in the session (be prepared for this file to be extremely large), and (2) "export media file references," when the OMF file will not contain the soundfiles themselves but will contain all of the session's region, edit, mix settings; effects (relating to the receiving DAW's available plug-ins and ability to translate effects routing); and I/O settings. This second type of file will be small by comparison; however, the original soundfiles must be transferred into the session folders.

◆ Developed by the Audio Engineering Society, the *AES31 standard* is an open file interchange format that was designed to overcome format incompatibility issues between different software and hardware systems. Transferred files will retain event positions, mix settings, fades, etc. AES31 makes use of Microsoft's FAT32 file system with broadcast wave as the default audio file format. This means that an AES31 file can be transferred to any DAW that supports AES31, regardless of the type of hardware and software used, as long as the workstation can read the FAT32 file system, broadcast wave, or regular wave files.

◆ *OpenTL* is a file exchange format that was developed for Tascam hard disk recording systems. An imported OpenTL project file will contain all audio files and edits that were made within the Tascam system, with all events positioned correctly in the Project window. Conversely, a session can be edited and then exported to a disk in the OpenTL format, making it possible to transfer all edits and audio files back to the Tascam hard-disk device.

DAW Software

Probably one of the strongest playing cards in the modern digital audio deck is the digital audio workstation. By their very nature, DAWs (Figures 6.62 through 6.65) are software programs that integrate with computer hardware and functional applications to create

Figure 6.62. Sonar 3 studio edition. (Courtesy of Twelve Tone Systems, www.cakewalk.com.)

Figure 6.63. *Mark of the Unicorn Digital Performer 4.1. (Courtesy of MOTU, Inc., www.motu.com.)*

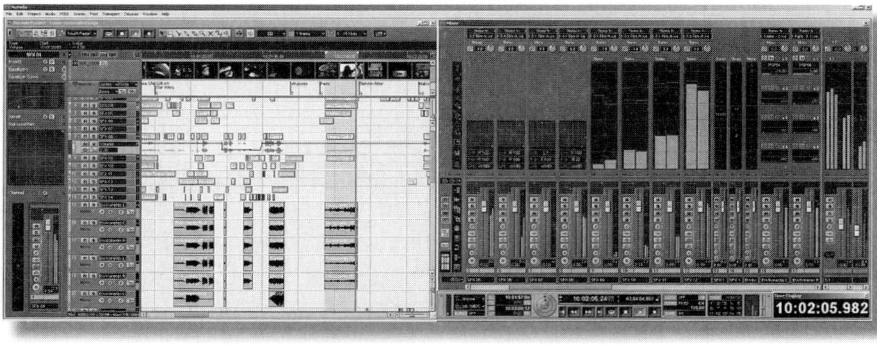

Figure 6.64. *Nuendo 2.0 media production DAW. (Courtesy of Steinberg Media Technologies GMBH, www.steinberg.net.)*

a powerful and flexible audio production environment. These programs commonly offer extensive record, edit, and mixdown facilities through the use of such production tools as:

◆ Extensive soundfile recording, edit, and region definition and placement

◆ MIDI sequencing and scoring

◆ Real-time, on-screen mixing

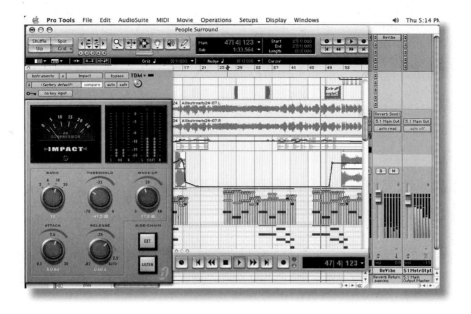

Figure 6.65. *Pro Tools HD hard-disk recording software for the Mac or PC. (Courtesy of Digidesign, a division of Avid Technology, Inc., www.digidesign.com.)*

- ◆ Real-time effects
- ◆ Mixdown and effects automation
- ◆ Soundfile import/export and mixdown export
- ◆ Support for video/picture synchronization
- ◆ Systems synchronization
- ◆ Audio, MIDI, and sync communications with other audio programs (*i.e.*, ReWire)
- ◆ Audio, MIDI, and sync communications with other software instruments (*i.e.*, VST technology)

The above list is but a smattering of the functional capabilities that can be offered by an audio production DAW.

Suffice it to say that these powerful software production tools are extremely varied in their form and function. Even with their inherent strengths, quirks, and complexities...the basic look, feel, and operational capabilities have, to some degree, become unified between the major DAW competitors. Having said this, it goes without saying that there are enough variations in features, layout, and basic operation that individuals, from aspiring beginner to seasoned professional, will have their favorite DAW make and model. With the growth of the DAW and computer industries, people have begun to customize their computers with features, added power, and peripherals that rival their love for souped-up cars and motorcycles. In the end, though...as with many things in life...it's doesn't matter which type of DAW you use—it's how you use it that counts!

For the rest of this section on DAWs, we'll be taking a look at that various functional aspects of the digital audio workstation (and hard disk recorders, in general). Of course, there's no way that all of the general features of a DAW can be covered (let alone specific features of a particular workstation); for that, I'll refer you to the tons of books and manuals that have been written on each DAW. These writings can be amazing gems that often offer specific insights into production tools and techniques that can fine-tune your production habits.

Soundfile Recording, Editing, Region Definition, and Placement

Hard-disk recorders are capable of recording mono, interleaved stereo (where the L/R or multichannel data is alternately encoded within a single file) and multitrack soundfiles directly to disk in a graphic working environment.

Do-It-Yourself Tutorial: Recording a Soundfile to Disk

- Consult your editor's manual regarding recording a soundfile to disk.
- Assign the track to an interface input sound source.
- *Name the track!* It's almost always best to name the track (or tracks) before going into record. In this way, the file will be saved to disk within the session folder under a descriptive name instead of an automatically generated filename (*e.g.*, killerkick.wav . . . instead of track16-01.wav).
- Save the session and assign the input to another track and overdub a track along with the previously recorded track.

Most hard-disk recording systems graphically display soundfile information within the created tracks of a main graphic window (Figure 6.66). These tracks contain drawn waveforms that graphically represent the amplitude of a soundfile over time in a WYSIWYG fashion Depending on the system type, soundfile length, and the degree of zoom, the entire waveform

Figure 6.66. *Graphic display of a soundfile region.*

can be shown on the screen, or only a portion will be shown with it continuing to scroll off one or both sides of the screen.

Graphic editing differs greatly from the "razor blade" approach that's used to cut analog tape in that the waveform gives us both visual and audible cues as to where a precise edit point should be. Using this common display technique, any position, cut/copy/paste, gain, and time changes to the waveform will be instantly reflected on the screen. Usually, these edits are *nondestructive* (a process whereby the original file isn't altered . . . only the way that the region in/out points are accessed or the file is processed as to gain, spectrum, etc.).

Only when a waveform is zoomed-in fully is it possible to see the individual waveshapes of a soundfile (Figure 6.67). At this zoom level, it becomes simple to locate zero-crossing points (points where the level is at the "0," center-level line). In addition, when a soundfile is zoomed-in to a level that shows individual sample points, the program might allow the sample points to be redrawn, in order to remove potential offenders (such as clicks and pops) or to smooth out amplitude transitions between loops or adjacent regions.

When working in a graphic editing environment, regions can usually be defined by positioning the cursor over the waveform, pressing and holding the mouse or trackball button, and then dragging the cursor to the left or right, which highlights the selected region for easy identification. After the region has been defined, it can be edited, marked, named, maimed, or otherwise processed.

As one might expect, the basic cut and paste techniques used in hard-disk recording are entirely analogous to those used in a word processor or other graphics-based programs:

◆ *Cut*—Places the highlighted region into memory and deletes the selected data (Figure 6.68)
◆ *Copy*—Places the highlighted region into memory and doesn't alter the selected waveform in any way (Figure 6.69)
◆ *Paste*—Copies the waveform data that's within the system's clipboard memory into the soundfile beginning at the current cursor position (Figure 6.70)

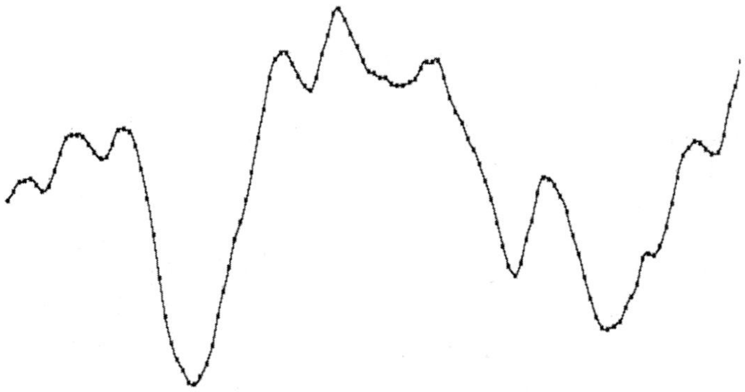

Figure 6.67. *Zoomed in area of a soundfile showing sample edit points.*

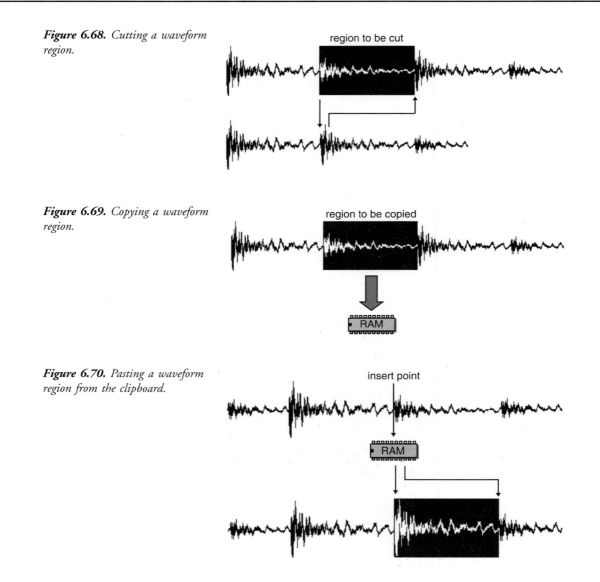

Figure 6.68. *Cutting a waveform region.*

region to be cut

Figure 6.69. *Copying a waveform region.*

region to be copied

RAM

Figure 6.70. *Pasting a waveform region from the clipboard.*

insert point

RAM

Do-It-Yourself Tutorial: Copy and Paste

◆ Consult your editor's manual regarding basic cut and paste commands.

◆ Open a soundfile and define a region that includes a musical phrase or sentence.

◆ Cut the region and try to paste it into another point in the soundfile in a way that makes sense (musical or otherwise).

◆ Feel free to cut, copy, and paste to your heart's desire to create an interesting or totally wacky soundfile.

Besides basic cut and paste techniques, processing the amplitude of a signal is one of the most common types of changes that are likely to be encountered. These include such processes as gain changing, normalization, and fading. *Gain changing* relates to the altering of a region or track's overall amplitude level, such that a signal can be proportionally increased or reduced to a specified level (often in dB or percentage value). In order to increase a soundfile or region's overall level, a function known as *normalization* can be used. Normalization (Figure 6.71) refers to an overall change in a soundfile or defined region's signal level, whereby the file's greatest amplitude will be set to 100% (or a set percentage level), with all other levels in the soundfile or region being proportionately changed in gain level. The *fading* of a region (either in or out) is accomplished by increasing or reducing a signal's relative amplitude over the course of a defined duration. For example, fading in a file (Figure 6.72a) proportionately increases a region's gain from infinity (zero) to full gain. Likewise, a fade-out (Figure 6.72b) has the opposite effect of creating a transition from full gain to infinity. These DSP functions have the advantage of creating a much smoother transition than would otherwise be humanly possible when performing a manual fade. A *cross-fade* (or X-fade) is often used to smooth the transition between two audio segments that either are sonically dissimilar or don't match in amplitude at a particular edit point (a condition that would otherwise lead to an audible "click" or "pop"). This functional tool basically overlaps a fade-in and fade-out between the two waveforms to create a smooth transition from one segment to the next (Figure 6.73). Technically, this process averages the amplitude of the signals over a user-definable length of time in order to mask the offending edit point.

MIDI Sequencing and Scoring

Most DAWs include extensive support for MIDI (Figure 6.74), allowing electronic instruments, controllers, effects devices, and electronic music software to be integrated with

Figure 6.71. Original signal and normalized signal level.

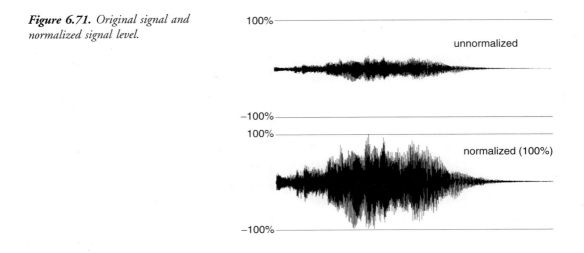

100%

unnormalized

−100%

100%

normalized (100%)

−100%

Figure 6.72. *Examples of various fade curves: (a) fade-in; (b) fade-out.*

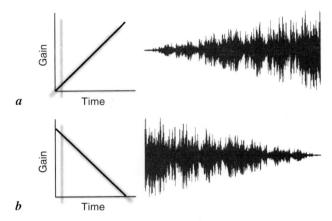

Figure 6.73. *Example of a cross-faded soundfile. (Courtesy of Steinberg Media Technologies GMBH, www.steinberg.net.)*

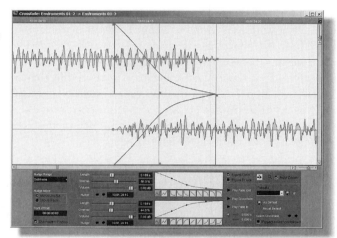

Figure 6.74. *MIDI edit window within the Cubase SE audio production software. (Courtesy of Steinberg Media Technologies GMBH, www.steinberg.net.)*

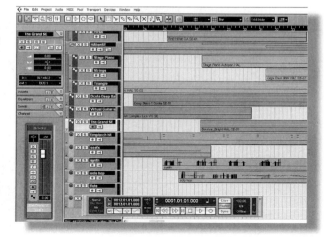

multitrack audio and video tracks. This important feature often includes the full implementation for:

◆ MIDI sequencing, processing and editing
◆ Score editing and printing
◆ Drum pattern editing
◆ MIDI signal processing
◆ Support for linking the timing and I/O elements of an external music application (ReWire)
◆ Support for software instruments (VSTi)

Further reading about the wonderful world of MIDI can be found within Chapter 7.

Real-Time, On-Screen Mixing

In addition to their abilities to offer extensive region edit and definition, one of the most powerful, cost- and time-effective features of a digital audio workstation is its ability to offer on-screen mixing (Figure 6.75). Essentially, most DAWs include a digital mixer interface that offers most (if not all . . . or more) of the capabilities that are offered by larger analog and/or digital consoles that are far more cost and space prohibitive. In addition to the basic input strip fader, pan, solo/mute and select controls . . . most DAW software mixers offer extensive support for EQ, effects plug-ins (offering a staggering amount of DSP flexibility that will be covered later in the chapter), spatial positioning (pan and possibly surround-sound positioning), total automation (both mixer and plug-in automation), external mix, function and transport control from a supported external hardware controller, support for exporting a mixdown to a file . . . the list goes on and on and on and . . .

Figure 6.75. *Nuendo 2.0 on-screen mixer. (Courtesy of Steinberg Media Technologies GMBH, www.steinberg.net.)*

DSP Effects

In addition to being able to cut, copy, and paste regions within a soundfile, it's also possible to alter a soundfile, track, or segment using digital signal processing techniques. In short, DSP works by directly altering the samples of a soundfile or defined region according to a program algorithm (a set of programmed instructions), so as to achieve a desired result. These processing functions can be performed either in real time or non-real time:

◆ *Real-time DSP*—Commonly used in most modern-day DAW systems, this process makes use of the computer's CPU or additional acceleration hardware to perform complex DSP calculations during actual playback. Because no calculations are written to disk in an off-line fashion, significant savings in time and disk space can be realized when working with productions that involve complex or long processing events. In addition, the automation instructions for real-time processing are imbedded within the saved session file, allowing any effect or set of parameters to be changed, undone, and redone . . . without affecting the original soundfile.

◆ *Non-real-time DSP*—Using this method, signal processing (such as changes in level, EQ, dynamics or reverb) that is too calculation intensive to be carried out during playback will be calculated (in an off-line fashion). In this way, the newly calculated file (containing the effect, sub-mix, etc.) will be played back, without having to use up the extra resources that are now available to the CPU for other functions. DAWs will often have a specific term for tracks or processing functions that have been written to disk . . . such as "locking" or "freezing" a file. When DSP is performed in non-real time, its almost *always* wise to save both the original and the effected soundfiles . . . just in case you need to make changes at a later time.

Most DAWs offer an extensive array of DSP options, ranging from options that are built into the basic I/O path of the input strip (*e.g.*, basic EQ and gain-related functions) . . . to DSP effects and plug-ins that come bundled with the DAW package . . . to third-party effects plug-ins that can be either inserted directly into the signal path (direct insertion) or offered as a master effect path that numerous tracks can be assigned to and/or mixed into (side chain).

Although the way that effects are implemented into a DAW will vary from one make and model to the next, the basic fundamentals will be much the same. The following discussion describes but a few of the possible effects that can be "plugged" into the signal path of DAW; however, further reading on effects processing can be found in Chapter 12 (Signal Processing) and Appendix A (DSP Basics).

◆ *Equalization*—EQ is, of course, a feature that's often implemented at the basic level of a virtual input strip (Figures 6.76 and 6.79). Most systems give full parametric control over the entire audible range . . . offering overlapping control over several bands, with a variable degree of bandwidth control (*Q*). Beyond the basic EQ options, many third-party EQ plug-ins are available on the market that vary in complexity, musicality, and market appeal (Figures 6.81 and 6.82).

Figure 6.76. *Nuendo's EQ within the Channel Settings window. (Courtesy of Steinberg Media Technologies GMBH, www.steinberg.net.)*

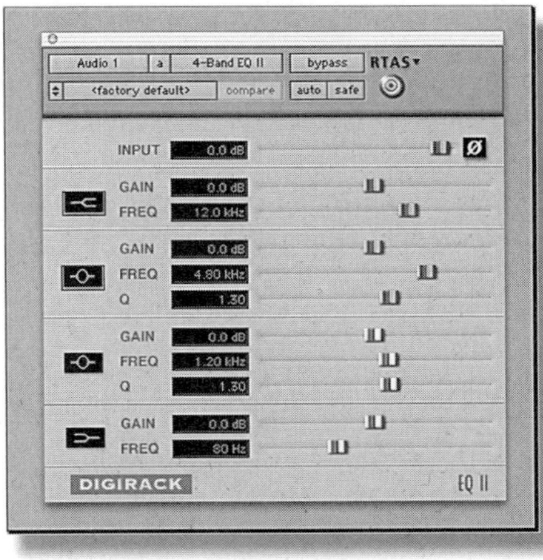

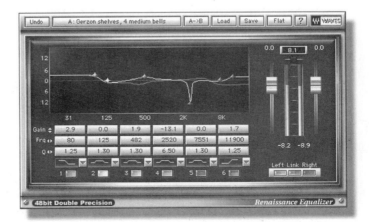

Figure 6.77. *Pro Tools EQ screen. (Courtesy of Digidesign, a division of Avid Technology, Inc., www.digidesign.com.)*

Figure 6.78. *Waves Renaissance equalizer plug-in. (Courtesy of Waves, Ltd., www.waves.com.)*

Figure 6.79. *Universal Audio Cambridge EQ plug-in for the UAD-1. (Courtesy of Universal Audio, www.uaudio.com.)*

Figure 6.80. *Universal Audio Pultec EQ plug-in for the UAD-1. (Courtesy of Universal Audio, www.uaudio.com.)*

Figure 6.81. *Waves Renaissance compressor plug-in. (Courtesy of Waves, Ltd., www.waves.com.)*

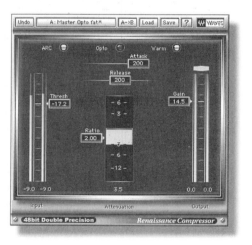

♦ *Dynamic range*—Dynamic range processors (Figures 6.80 and 6.81) can be used to change the signal level of a program. Processing algorithms are available which emulate a compressor (a device that reduces gain by a ratio that's proportionate to the input signal), limiter (reduces gain at a fixed ration above a certain input threshold), or expander (increase the overall dynamic range of a program). These gain changers can be inserted directly into a track, used as a grouped master effect, or inserted into the final output path for use as a master gain processing block.

In addition to the basic complement of dynamic range processors, wide assortments of multi-band dynamic plug-in processors (Figure 6.82) are available for general and mastering DSP

Figure 6.82. *The tc electronic Master X³ multiband dynamic plug-in processor. (Courtesy of tc electronic, www.tcelectronic.com.)*

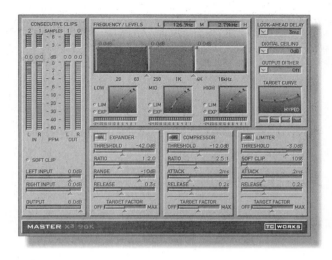

Figure 6.83. *Waves Super Tap Delay plug-in. (Courtesy of Waves, Ltd., www.waves.com.)*

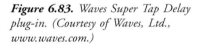

applications. These processors (which are further covered within Chapter 16, Mastering) allow the overall frequency range to be broken down into various frequency bands. For example, a plug-in such as this could be inserted into a DAW's main output path, which allows the lows to be compressed, while the mids are lightly limited and the highs are de-essed to reduce sibilance.

◆ *Delay*—Another important effects category that can be used to alter and/or augment a signal revolves around delays and regeneration of sound over time. These time-based effects use *delay* (Figures 6.83 and 6.84) in order to add a perceived depth to a signal or change the way that we perceive the dimensional space of a recorded sound.

Figure 6.84. *Steinberg's Double Delay bundled plug-in. (Courtesy of Steinberg Media Technologies GMBH, www.steinberg.net.)*

A wide range of time-based plug-in effects exist that are all based upon the use of delay (and/or regenerated delay) to achieve such results as:

◆ Delay

◆ Chorus

◆ Flanging

◆ Reverberation

As was stated toward the beginning of this section, further reading on the subject of delay (and the subject of signal processing in general) can be found in Chapter 12 (Signal Processing).

◆ *Pitch and Time Change*—Pitch change functions make it possible to shift the relative pitch of a defined region or track either up or down by a specific percentage ratio or musical interval. Most systems can shift the pitch of a soundfile or defined region by determining a ratio between the current and the desired pitch and then adding (lower pitch) or dropping (raise pitch) samples from the existing region or soundfile (Figure 6.85).

In addition to raising or lowering a soundfile's relative pitch, most systems can combine variable sample rate and pitch shift techniques to alter the duration of a region or track. These pitch- and time-shift combinations make it possible for such changes as:

◆ *Pitch shift only*—A program's pitch can be changed while recalculating the file so that its length remains the same.

◆ *Change duration only*—A program's length can be changed while shifting the pitch so that it matches that of the original program.

◆ *Change in both pitch and duration*—A program's pitch can be changed while also having a corresponding change in length.

When combined with shifts in time (delay), changes in pitch make it possible for a multitude of effects to be created (such as *flanging*, which results from random fluctuations in

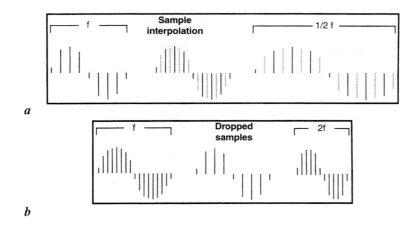

Figure 6.85. *Basic disk-based pitch shifting techniques. (a) Digital audio can be shifted downward by interpolating and adding intermediate samples to the original data and then playing the soundfile back at its original rate. (b) Digital audio can also be shifted upward by dropping samples from the original data and then playing the soundfile back at its original rate.*

delay and time shifts that are mixed with the original signal to create an ethereal "phasey" kind of sound).

DSP Plug-Ins

Workstations often offer a number of DSP effects that come bundled with the program; however, a staggering range of third-party plug-in effects can be inserted into a signal path which perform functions for any number of tasks ranging from the straightforward to the wild 'n' zany (Figures 6.86 through 6.88). These effects can be programmed to seamlessly integrate into a host DAW application that conform to such plug-in platforms as:

◆ *DirectX*—A DSP platform for the PC that offers plug-in support for sound, music, graphics (gaming), and network applications running under Microsoft Windows (in its various OS incarnations)

◆ *AU* (Audio Units)—Developed by Apple for audio and MIDI technologies in OSX; allows for a more advanced GUI and audio interface

◆ *VST* (Virtual Studio Technology)—A native plug-in format created by Steinberg for use on either a PC or Mac; all functions of a VST effect processor or instrument are directly controllable and automatable from the host program

◆ *MAS* (MOTU Audio System)—A real-time native plug-in format for the Mac that was created by Mark of the Unicorn as a proprietary plug-in format for Performer and Digital Performer; MAS plug-ins are fully automatable and do not require external DSP in order to work with the host program

Figure 6.86. *Universal Audio RealVerb Pro reverb plug-in for the UAD-1. (Courtesy of Universal Audio, www.uaudio.com.)*

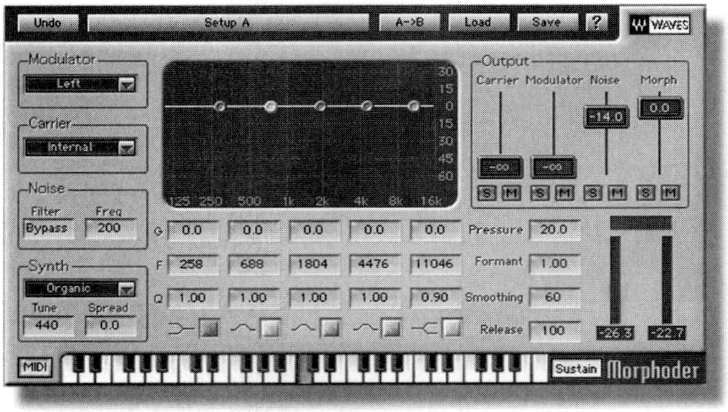

Figure 6.87. *Waves Morphoder Vocoder plug-in. (Courtesy of Waves, Ltd., www.waves.com.)*

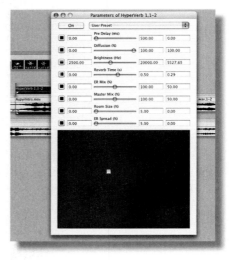

Figure 6.88. *Hyperprism Gold OSX signal processing plug-in in HyperVerb mode. (Courtesy of Arboretum Systems, Inc., www.arboretum.com.)*

◆ *AudioSuite*—A file-based plug-in that destructively applies an effect to a defined segment or entire soundfile . . . meaning that a new, effected version of the file is rewritten in order to conserve on the processor's DSP overhead; when applying AudioSuite, it's often wise to apply effects to a copy of the original file so as to allow for future changes

◆ *RTAS* (Real-Time Audio Suite)—A fully automatable plug-in format that was designed for Digidesign's Pro Tools LE; available on Digi ToolBox and Digi 001 (any system with Pro Tools LE) and runs on the power of the host CPU (host-based processing) on either the Mac or PC

◆ TDM (Time Domain Multiplex)—A plug-in format that can only be used with Digidesign Pro Tools systems (Mac or PC) that are fitted with Digidesign Farm cards; this 24-bit, 256-channel path integrates mixing and real-time digital signal processing into the system with zero latency and under full automation

These popular software applications (which are being programmed by major manufacturers and third-party startups alike) have helped to shape the face of hard-disk recording, by allowing us to pick and choose those plug-ins that best fit our personal production needs. As a result, new companies, ideas, and task-oriented products are constantly popping up on the market . . . literally on a monthly basis.

Accelerator Cards

In most circumstances, the CPU of a host DAW program will have sufficient power and speed to perform all of the DSP effects and processing needs of a project. Under extreme production conditions, however, the CPU might run out of computing steam and choke during real-time playback. Under these conditions, two choices could be made in order to reduce the workload on a CPU: On the one hand, the track(s) could be "frozen," meaning that the processing functions would be calculated in non-real time and then written to disk as a separate file. On the other hand, an *accelerator card* (Figure 6.89) could be placed into the system that's capable of adding an extra CPU into the circuit, giving the system extra processing power to perform the necessary effects calculations.

Figure 6.89. *The tc electronic PowerCore FireWire rack-mountable effects accelerator. (Courtesy of tc electronic, www.tcelectronic.com.)*

Figure 6.90. *Rubberband controls allow a wide range of automated parameters to be varied over time.*

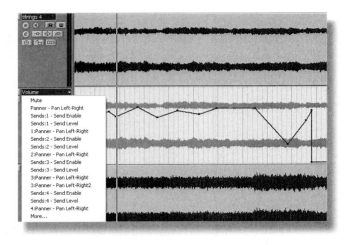

Mixdown and Effects Automation

One of the great strengths of the digital age is how easily all of the mix and effects parameters can be automated and recalled within a mix. A DAW is particularly strong in this area. The ability to change levels, pan, and virtually control any parameter within a project makes it possible for a session to be written to disk, saved, and recalled at a seconds notice. In addition to grabbing a control and moving it (either on-screen or from a physical controller)...''rubberband'' controls let you view, draw, and edit various parameters as a representative graphic line that details the various parameter moves over time (Figure 6.90). Generally, the edit moves that have been made within a mix can be undone, redone, or recalled back to a specific point in the edit stage. Often (but not always), the moves within a mix can't be "undone" and reverted back to a specific point in the mix. Obviously, one of the best ways to save (and revert to) a particular version of a mix in progress (or various versions of an alternate mix) is simply to save the version under a unique (and descriptive) session file title...and then keep on working.

Exporting a Mixdown to File

Most DAWs systems are able to export (print) part or an entire session to a single file or set of soundfiles (Figure 6.91). The former refers to multiple channels that are interleaved together in a L–R–L–R...or multichannel fashion, while the latter renders the channels as individual channel files. Often, the session can be exported in non-real time (a faster than real-time process that can include all mix, plug-in effects, automation, and virtual instrument calculations) or in real-time (a process that's capable of sending and receiving real-time analog signals through the audio interface so as to allow for the insertion of external effects devices, etc.). Usually, a session can be mixed down to a number of final soundfile and bit-/sample-rate formats. Certain DAWs might also allow third-party plug-ins to be inserted into the final (master) output section...allowing for the export of a session to a specific output

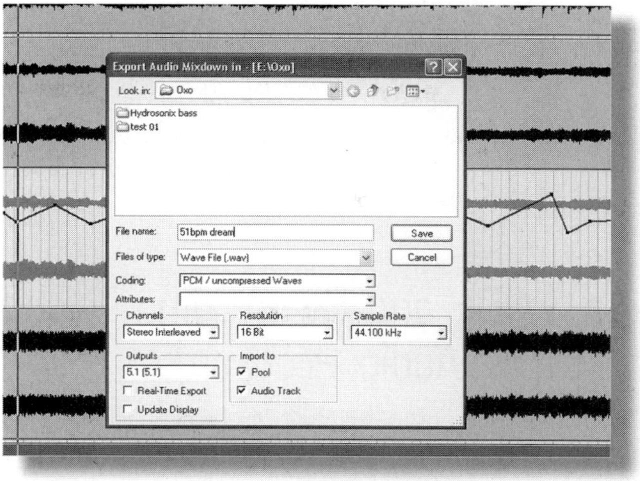

Figure 6.91. *Many DAWs are capable of exporting a session soundfiles, effects, and automation to a final set of mixdown tracks.*

file format. For example, a discrete surround mix could be folded down into a two-channel Dolby ProLogic surround-sound file, or the same file could be rendered as a Dolby Digital 5.1 file for insertion into a DVD video soundtrack.

Support for Video and Picture Sync

Speaking of video . . . most high-end DAWs include support for displaying a video track within a session, both as a video window that can be displayed on the monitor desktop as well as in the form of a video thumbnail track that will often appear in the track view as a linear guide track. Both of these provide important visual cues for tracking live music, sequencing MIDI tracks, and accurately placing effects (sfx) at specific hit points within the scene (Figure 6.92).

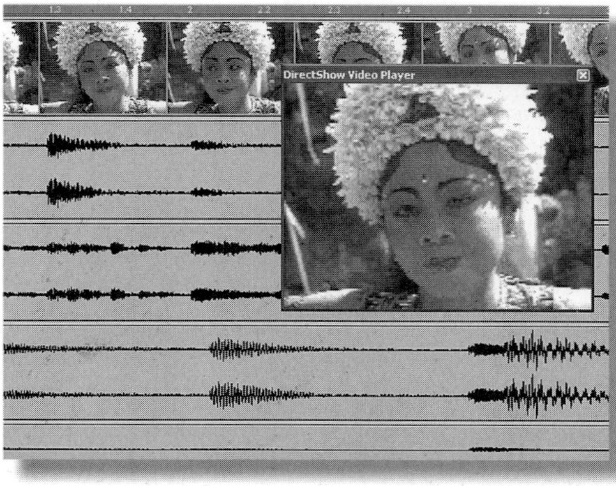

Figure 6.92. *Most high-end DAW systems are capable of importing a videofile directly into the project session window.*

Systems Synchronization

Through the use of SMPTE time code, MTC, and wordclock, the timing elements of a DAW can be locked to various media devices within a studio or media production house. Transport control over external media devices can also be accomplished through the use of Sony 9-pin and MMC (MIDI machine control).

Audio, MIDI, and Sync Communications with Other Audio Programs

Through the use of a standard protocol (such as Propellerhead's ReWire technology), many DAWs are able to communicate audio, MIDI, and timing information between one or more independent music programs to the host DAW. In this way, a music program that was designed to perform to a specific media task could be linked to the DAW's I/O mix and transport functions... effectively allowing multiple programs to work in tandem as a unified production system.

Loop-Based Audio Editors

Loop-based audio editors are groove driven music programs (Figures 6.93 and 6.94) that are designed to let you drag and drop prerecorded or user-created loops and audio tracks into a graphic multitrack production interface. With the help of custom, royalty-free loops (available from the manufacturer and/or third-party companies), users can quickly and easily experiment with setting up grooves, backing tracks, and creating a sonic ambience by simply dragging the loops into the program's main soundfile view... where they can be arranged in a multitrack or special GUI as a session file that can be saved to disk.

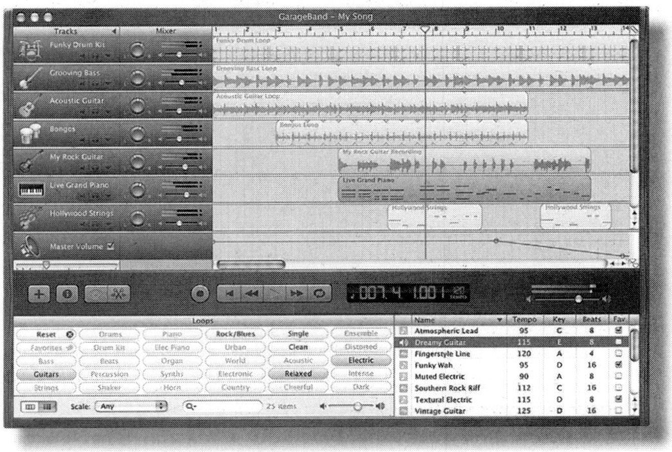

Figure 6.93. Apple GarageBand. (Courtesy of Apple Computers, www.apple.com.)

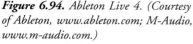

Figure 6.94. *Ableton Live 4. (Courtesy of Ableton, www.ableton.com; M-Audio, www.m-audio.com.)*

One of the most interesting aspects of these editors is their ability to match the tempo of a specially created loop soundfile to the tempo of the current session. Amazingly enough, this process isn't that difficult to perform, as the program extracts the length, native tempo, and pitch information form the imported file's header and (using various digital time and/or pitch change techniques) adjusts the loop to fit the native time/pitch parameters of the current session. This means that loops of various tempos and musical keys can be automatically adjusted in length and pitch so as to fit in time with previously existing loops (a process that would otherwise take a great deal of patience to manually pull off).

These shifts in time to match a loop to the session's native tempo can actually be performed in a number of ways. For example, using basic DSP techniques to time-stretch and pitch-shift a recorded loop will often work well over a given plus-or-minus percentage range (which is often dependant upon the quality of the program algorithms). Beyond this range, the loop will often begin to distort and become jittery. At such extremes, other beat slice detection techniques can be used to make the loop sound more natural. For example, drums or percussion can be stretched in time by adding additional silence between the various hit-points within the loop, at precisely calculated intervals. In this way, the pitch will remain the same while the length is altered. Of course, such a loop would sound choppy and broken up when played on its own; however, when buried within a mix, it might work just fine . . . it's all up to you and the current musical context.

Preprogrammed loops that will work with a number of groove editors can be obtained from any number of sources, such as:

◆ The Web (both free and for purchase)
◆ Commercial CDs
◆ Rolling your own (creating your own loops can add a satisfying and personal touch)

It's important to note that at any point during the creation of a composition, audio and/or MIDI tracks (such as vocals or played instruments) can easily be recorded into a loop session in order to give the performance a fluid and more dynamic feel. It's even possible to record a live instruments into a session with a defined tempo... and then edit these tracks into defined loops that can be dropped into the current and future sessions to add a live touch.

As these programs often include many of the features that you'll find in a full-featured DAW (including real-time effects, mixing, multiple I/O, and synchronization). ... the completed session can be exported as a mixdown file. However, in addition to acting as a stand-alone music application, the I/O, transport, and timing elements of many of these programs can be fully integrated into a DAW (most often through the use of ReWire technology). In this way, groove tracks can be integrated into a master session, where they can be processed and mixed down as a master file for mastering to its intended media.

Beat Slicing

As we now know, it's often necessary to alter the beats per minute (bpm) of one loop to that of another. A loop-based audio editor (and/or a DAW that includes provisions for intelligent beat matching) will be able to use a variety of time-stretching, pitch-shifting, and formant-shifting algorithms. Another method, called *beat slicing*, actually breaks an audiofile into a number of small segments (often called *slices*). Such a system can be used to preserve the pitch, timbre, and sound quality of a file by altering the time between these slices rather than by changing the speed and pitch at which it's played. This process can be done by slicing the loop into equal segments that are based upon its original tempo (*e.g.*, cutting the loop into 8th or 16th intervals)... or by detecting the transient events within the loop and placing the slices at the appropriate points (often according to user-definable sensitivity and detection controls). If the slices are accurately detected to correspond to the "hits" within the loop (*e.g.*, occurring on the various beats of a drum loop), timing modifications can be made in order to extend or shorten its length (while the pitch remains unaltered) to match the session's current tempo. Given that the loop is now divided into smaller segments (that might be small enough so that each segment is made up of one or more instruments... such as an individual kick):

- Each of those segments could actually be replaced with another instrument (*e.g.*, a different kick).
- The various segments could be shuffled in a directed or random order to create a totally new pattern.
- The beat's emphasis could be shifted to create a new "feel" (*e.g.*, make/remove a shuffle groove).
- Different slices could be assigned to different tracks in order to add different effects to various, individual slices, etc.

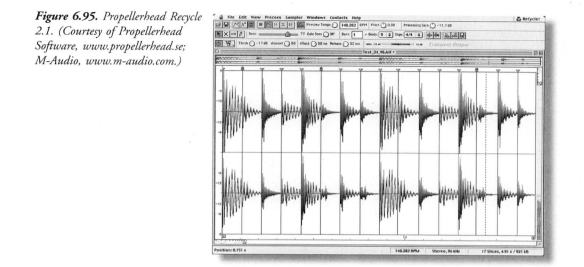

Figure 6.95. Propellerhead Recycle 2.1. (Courtesy of Propellerhead Software, www.propellerhead.se; M-Audio, www.m-audio.com.)

Certain programs are able to match and/or distort the beat timing of a loop by assigning the various slices to a MIDI note scale (Figure 6.95). By cycling through (and altering the tempo, notes, and velocities of) the scale, the loop can be changed, replaced, and otherwise mutilated in any number of imaginable ways.

DJ Software

In addition to music production software, there's a growing number of software players, loopers, groovers, effects, and digital devices on the market for the digital DJ of the 21st century. These hardware/software devices make it possible for digital grooves to be created from a laptop, controller, specially fitted turntable, or digital turntable (jog/scratch CD player) with an unprecedented amount of preprogrammable and/or live performance interactivity that can be used on the floor, on stage, or in the studio (Figure 6.96).

Backup and Archive Strategies

The phrase "nothing lasts forever" is often especially true in the digital domain of lost 1's and 0's, damaged media, dead hard drives, and lost data...you know, the "oops factor." It's a basic fact that you never quite know what lies around the techno bend...and, of course, it's extremely important that you protect yourself as much as is humanly possible against the inevitable. Of course, the answer to most digital dilemmas is to back up your data in the most reliable (or redundant) way. Hardware and program software can (usually) be replaced; on the other hand, when valuable session soundfiles are lost...they're lost!

Figure 6.96. *Tractor DJ Studio 2.0 (www.native-instruments.com).*

Backing up a session can be done in several ways, depending upon the level of assured security that the files will be able to be played in the future with a minimum of hassle and potential problems. Here are a few tips on this important topic:

◆ As you might expect, the most straightforward backup system is to copy the session data, in its entirety, to the most appropriate media. This works well in the short run, as it assumes that all of your program and plug-in data is loaded and up to date.

◆ In the longer run (5+ years), the most ironclad way to back up the track data of a session is to print each track as its own .wav, .aif, or .sd2 file. This file should always be recorded or exported to a file as a contiguous file that flows from the beginning of the session (00:00:00:00 or appropriate begin point) to the end of that particular track. In this way, the individual track files could be loaded into any type of DAW, at the beginning point, for processing and mixdown.

◆ In such a track-by-track safety restoration situation, you might want to save two copies of a track that has a particular effect ... one that contains the original and effected sound and another that simply contains the original, unaltered sound.

◆ For those that might want additional protection against the degradation of unproven digital media, you might also want to back each track (or group of tracks) to the individual tracks of an analog recorder.

◆ For those sessions that contain MIDI tracks, you should always keep these tracks within the session (*i.e.*, don't delete them). These tracks might come in very handy during a remix or future mixdown.

◆ Speaking of MIDI ... It's always wise to export all of the MIDI tracks/data within a session as standard MIDI file. You should at least save all of the tracks as a type 1 file (where all of the multichannel track/data information is left intact ... and whenever possible save it as both a type 0 and a type 1 (you never know what file format obstacles might haunt you in the future). For those wanting additional safety, it might be wise to create a contiguous soundfile (in the same manner as above) that's a recorded file of each MIDI instrument track...just in case the instrument is no longer available for any reason.

◆ Whenever possible, make multiple backups and store them in separate locations. Having a backup copy in your home as well as the studio can save your proverbial butt in case of a fire or unforeseen situation.

As you might expect, all of these tracks can add up to a ton of data. For example, a 3-minute song that's been recorded onto 24 tracks at 16 bit/4.1 kHz would add up to 720 Mb (10 Mb min/track × 3 minutes × 24 tracks = 720 Mb). When you add up the sessions tracks from multiple songs, original DAW session data, and alternate track takes...you could easily end up backing up huge amounts of data onto such backup media as:

◆ *Hard disk*—Large amounts of session/track data can be backed to a large removable drive (or duplicated onto multiple drives).

◆ *DVD+R or DVD-R*—This removable media lets you back up to 5 Gb (gigabytes) of data onto a single disc. (My good buddy, Craig Anderton, recommends that you individually back each song onto its own disc...In case something happens to the media, just that song would be lost, not *all* of them.)

◆ *CD-R*—This removable media can hold up to 800 Mb of data and is reliable assuming that proper care is taken in labeling the disc with a water-based CD marker and that proper storage and handling precautions are taken.

Just think of how you would feel if all of your most precious, original session material would be lost...forever! All of the frustration and deep sense of loss can be avoided (or at least cushioned) by thinking ahead and saving the day...by backing up your data.

Session Documentation

Closely related to backing up your session data is the need for comprehensive session documentation. There are few things that are more frustrating than going back to an archived session and finding that no information exists on what instrument patch, mic type, or outboard effect was used on a DAW session (or any session, for that matter). The importance of documenting a session within a separate written document and/or within the notepad apps within a DAW can't be overemphasized (Figure 6.97). Just a few of the important information and useful parameters that should be included in your session documentation include:

◆ Session tempo

◆ Participants in the project and important dates (for future credit references)

◆ Mic choice and placement (for future overdub reference)

◆ Outboard equipment type and their settings

Figure 6.97. Nuendo Notepad apps. (Courtesy of Steinberg Media Technologies GMBH, www.steinberg.net.)

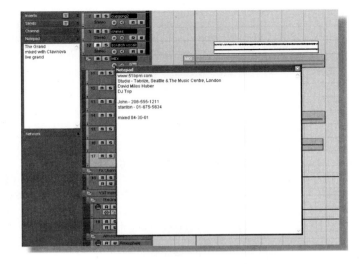

◆ Plug-in effects and their settings or general descriptions (you never know if they'll be available at a future time, so a description can help you to duplicate it with another app)

In short, the more information that can be archived with a session (and its backups) . . . the better chance that you'll be able to duplicate the session in greater detail, at some point in the future. Just remember that it's up to us to save and document the music of today for the fans and future playback/mix technologies of tomorrow.

CHAPTER

MIDI and Electronic Music Technology

Today, professional and nonprofessional musicians are using the language of the *musical instrument digital interface* (MIDI) to perform an expanding range of music and automation tasks within audio production, audio for video and film post, stage production, etc. This industry-wide acceptance can, in large part, be attributed to the cost-effectiveness, power, and general speed of MIDI production. Once a MIDI instrument or device comes into the production picture, there's often less need (if any at all) to hire outside musicians for a project. This alluring factor allows a musician to compose, edit, and arrange a piece in a multichannel electronic music environment that's extremely flexible. By this, I'm not saying that MIDI replaces (or should replace) the need for acoustic instruments, microphones, and the traditional performance setting. In fact, it is a powerful production tool that helps countless musicians create music and audio productions that are innovative and highly personal. In short, MIDI is all about control, repeatability, flexibility, cost-effective production power and fun.

The affordable potential for future expansion and increased control over an integrated production system has spawned the growth of an industry that allows an individual to cost-effectively realize a full-scale sound production, not only in his or her own lifetime...but in a relatively short time. For example, much of modern-day film music owes its very existence to MIDI. Before this technology, composers were forced to compose without the benefits of hearing their composition or by creating a reduction score that could be played on a piano or small ensemble. With the help of MIDI, composers can hear their work in real time, make any necessary changes, print out the scores...and, finally, take a full orchestra into the studio to replace some or all of the parts into a final recording. At the other end of the spectrum, MIDI can be an extremely personal tool that lets you "lay down," edit, and layer synthesized and/or sampled instruments to create a song that helps you to express yourself to the masses...all within the comfort of your own home or personal project studio. The moral to the story is that today's music industry would look and sound very different if it weren't for this powerful four-letter word.

MIDI Production Environments

One of the better aspects of MIDI production is that a system can be designed to handle a wide range of tasks with a degree of flexibility and ease that best suits an artist's main instrument, playing style, and even the user's personal working habits. By opening up almost any industry-related magazine, it's easy to see that a vast number of electronic musical instruments, effects devices, computer systems, and other MIDI-related devices are currently available on the new and used electronic music market. MIDI production systems exist in any number of shapes and sizes and can be designed to match a wide range of production and budget needs. For example, working and aspiring musicians commonly install digital audio and MIDI systems in their homes (Figures 7.1). These production environments can range from ones that take up a corner of an artist's bedroom to larger systems that are integrated into a dedicated project studio. Systems such as these can be specially designed to handle a multitude of applications and have the important advantage of letting artists produce their music in a comfortable environment...whenever the creative mood hits. Newer, laptop-based systems (I call 'em studio-a-go-gos) can also let you make music "wherever" the creative mood hits. Such production luxuries, which would have literally cost an artist a fortune in the not-too-distant past, are now within the reach of almost every musician.

Figure 7.1. *MIDI in the project studio. (Courtesy of Loud Technologies, Inc., www.mackie.com; B.O.M.B. Faktory, www.bombfaktory.com. Photo credit: Christopher Buttner.)*

MIDI has also dramatically changed the sound, technology, and production habits of the professional recording studio. Before MIDI and the concept of the home project studio, the recording studio was one of the only production environments that allowed an artist or composer to combine instruments and sound textures into a final recorded product. With the advent of MIDI, electronic musicians were given the tools to record, overdub, mix, and play back their performances in a working environment that loosely resembles the multitrack recording process. Once MIDI has been mastered, its repeatability and edit control offer production challenges and possibilities that can stretch beyond the capabilities and cost-effectiveness of the traditional multitrack recording environment. When combined with digital audio workstations (DAWs) and modern-day recording technology, much of the music production process can be preplanned and rehearsed before you step into the studio. . . . In fact, recorded tracks are often laid down before they ever see the hallowed halls of the professional studio (if they see them at all). In business jargon, this luxury has reduced the number of billable hours to a cost-effective minimum.

Since its inception, electronic music has been an indispensable tool for the scoring and audio postproduction of television and radio commercials, industrial videos, and full-feature motion picture sound tracks (Figure 7.2). For productions that are on a budget, an entire score can be created in the artist's project studio using MIDI, hard-disk tracks and digital recorders . . . all at a mere fraction of what it might otherwise cost to hire the musicians and rent out a studio.

Electronic music production and MIDI are also very much at home on the stage. In addition to using synths, samplers, DAWs, and drum machines on the stage, most or all of a MIDI instrument and effects device parameters can be controlled from a presequenced or real-time controller source. This means that all the necessary settings for the next song (or section of a song) can be automatically called up before being played. Once underway, various instrument patch and controller parameters can also be changed during a live performance from a stomp box controller, keyboard controller, or active sequencer.

Figure 7.2. James Newton Howard's Studio, Los Angeles, California. (Courtesy of Bryston, Ltd, www.bryston.ca.)

One of the media types included in the notion of multimedia is definitely MIDI. With the advent of General MIDI (GM, a standardized spec that allows any sound card or GM-compatible device to play back a score using the originally intended sounds and program settings), it's possible (and common) for MIDI scores to be integrated into multimedia games, Web sites, and CD-ROMs.

Since MIDI is simply a series of performance commands (unlike digital audio, which actually encodes the audio information itself), only a small amount of data is communicated and processed at any point in time. This means that almost no processing power is required to play MIDI, making it ideal for playing real-time music scores while browsing text, graphics, or other media over the Web. . . . Truly, when it comes to weaving MIDI into the various media types, your imagination is the limit—not the system's processing and bandwidth power.

What's a MIDI?

Simply stated, MIDI is a digital communications language and compatible hardware specification that allows multiple electronic instruments, performance controllers, computers, and other related devices to communicate with each other throughout a connected network (Figure 7.3). It is used to translate performance- and/or control-related events (such as playing a keyboard, selecting a patch number, or varying a modulation wheel) into equivalent digital messages that can be transmitted to other MIDI-capable devices for such purposes as:

◆ Controlling sound generation (synthesis)
◆ Controlling digital audio playback (sample playback)
◆ Transport control
◆ Transmitting time-related information (time code)
◆ Triggering events

The true beauty of MIDI lies in its ability to transmit data that can be easily understood and/or recorded by hardware devices and software programs for the purpose of digitally communicating the real-time language of music and system-related control . . . all throughout a connected network.

Figure 7.3. MIDI allows electronic instruments, digital audio devices, effects processors, and other equipment to communicate performance- related data within a connected audio production network.

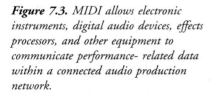
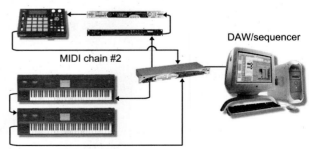

In artistic terms, this digital language is an important medium that lets artists create with a degree of expression and control that was, before its inception, often not possible on an individual level. It does this by offering complete control over a wide palette of sound types, their timbre (sound and tonal quality) and overall blend (level, panning, tonal layering, etc.) ... all in a digital environment that can be recorded, edited (using a wide range of music and control parameters), and reproduced in a multitrack fashion. In short, through the transmission of this performance language, an electronic musician can compose and develop a song in a practical, flexible, affordable, and (hopefully) fun production environment.

System Interconnections

Using the actual data communications and hardware links that exist within a connected MIDI network, it's possible for data to be communicated to all the electronic instruments and devices through the transmission of real-time performance and control-related messages. Up to 16 channels of MIDI data can be communicated between devices and/or instruments (or individual sound generators within an instrument) through each cable/port connection. One or more of these connections can be established in a number of ways, including:

◆ A MIDI cable and port connection
◆ Universal Serial Bus (USB)
◆ Network connection

MIDI Cable and Port Connections

MIDI digitally communicates musical performance data through a standard MIDI cable as a string of MIDI messages that are transmitted at a speed of 31.25 kbaud (bits/second). This data can only travel over a single MIDI line in one direction, from a single source to a destination (Figure 7.4a). In order to make two-way communication possible, a second MIDI data line must be used to communicate data back to the first device (Figure 7.4b). A standard MIDI cable (Figure 7.5) consists of a shielded, twisted pair of conductor wires that has a male five-pin DIN plug located at each of its ends. The MIDI specification uses only three of the five pins; pins 4 and 5 are used to conduct MIDI data, and pin 2 is used to connect the cable's shield to equipment ground (the typical MIDI input leaves this pin floating, thereby preventing any potential ground loops between devices in complicated MIDI setups). Pins 1 and 3 are currently not in use but are reserved for possible changes in future MIDI applications. Twisted cable and metal shield groundings are used to reduce outside interference, such as radio frequency (RFI) or electrostatic interference, which can distort or disrupt MIDI message transmissions. The fact that MIDI is optically isolated also helps prevent ground loops.

Prefabricated MIDI cables in lengths of 2, 6, 10, and 20 feet can often be found at your favorite music store or site. 50 feet is the maximum length stated by the MIDI specification in order to reduce the effects of signal degradation and external interference that might

Figure 7.4. *MIDI data can only travel in one direction through a single MIDI cable: (a) data transmission from a single source to a destination; (b) two-way communication using two MIDI cables.*

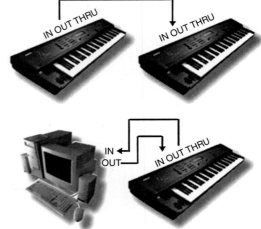

a

b

occur over extended cable runs (although you'll probably have to make your own cable). MIDI cables are connected to an instrument, interface or device at a MIDI port. Three types of MIDI ports are used to connect MIDI devices in a network: MIDI in, MIDI out, and MIDI thru (Figure 7.6):

◆ *MIDI in port*—Receives MIDI messages from an external source and communicates this performance, control, and timing data to the device

◆ *MIDI out port*—Transmits MIDI messages from the device out to another MIDI instrument or device

◆ *MIDI thru port*—Transmits an exact copy of the incoming data at the MIDI in port to another MIDI instrument or device that follows in the connected MIDI data chain

Certain MIDI devices that don't include a MIDI thru port will sometimes offer a software-based "echo" function that can be selected to provide an exact copy of the data received at the MIDI in port and route it out through a MIDI out/echo port.

As a general rule, there are only two valid ways to connect to a MIDI device to another using a standard MIDI cable (Figure 7.7):

◆ Connect the MIDI out port of one device to the MIDI in port of the next device.

◆ Connect the MIDI thru port (or MIDI out/echo port) of one device to the MIDI in port of the next device.

One of the simplest and most common ways to distribute data throughout a MIDI cable network is to use a *daisy chain*. This method relays MIDI from one device to the next in the chain by throughputting the data received at a device's MIDI in port directly out to another device via its MIDI thru (or MIDI echo) port . . . where the chain continues onto and thru the next device. For example, a typical MIDI daisy chain (Figure 7.8a) will flow from the MIDI out of a source device (such as a controller or sequencer) to the MIDI in port of a second device. The MIDI data being received by the second device is relayed out through its

Figure 7.5. *Wiring diagram and picture of a MIDI cable.*

Figure 7.6. *MIDI in, out, and thru ports.*

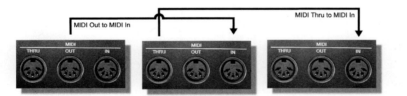

Figure 7.7. *The only two valid ways to connect one MIDI device to another via a standard MIDI cable.*

MIDI thru port, which is then plugged into a third device's MIDI in port. The data received by the third device is then relayed out through its MIDI thru port, which in turn is plugged into the fourth device's MIDI in port...and so on until the final device in the chain is reached. Put in simpler terms, this allows source 1 to be plugged into device 2, which echoes the data thru to device 3, which echoes the data thru to device #4. ...In this way, MIDI data can be communicated to each and every device within a single, connected chain (according to its assigned channel number (more about this in a moment).

A computer can also be (and often is) designated to be the master source in a daisy chain. In this way a MIDI-capable DAW or sequencer can be used to control the playback, channelizing, and signal processing functions of an entire system. In Figure 7.8b, the MIDI out of a master MIDI keyboard controller is routed to the computer. The computer's MIDI out port (or ports) can then relay the performance data thru to a device that's connected via one or more daisy chains.

Figure 7.8. Example of
a connected MIDI
system using a daisy
chain: (a) typical daisy
chain hookup;
(b) example of how a
computer can be
connected into a daisy
chain.

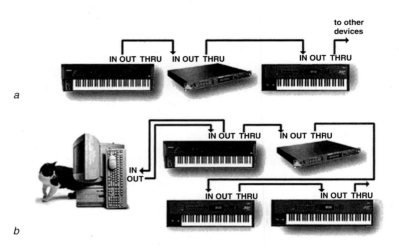

a

b

Figure 7.9. Various I/O
devices can be connected
to the system via USB and/or
FireWire.

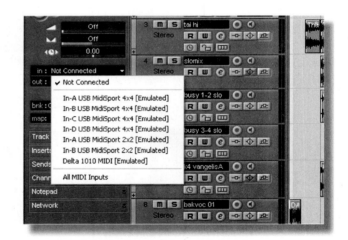

USB and FireWire Connections

Many of the newer peripheral devices that can communicate MIDI data connect to a personal
computer via the USB 1 and 2 protocols and IEEE 1394 (FireWire) protocols. These devices
are able to use these high-speed connections to communicate MIDI spec 1.0 controller
and performance data to the computer and/or to interface various MIDI devices to a host
DAW, sequencer, or program. Often, these devices (and the various MIDI ports that
are designed into them) will show up as a dedicated software port on your computer
(Figure 7.9).

mLAN Network Connections

One of the major problems that faces the electronic musician are cords, cords, and more cords. Given that MIDI and audio are often routed separately in a production system, the resulting mess can easily end up as a complex pile of spaghetti that's not only cumbersome and complex but can also be unreliable and extremely difficult to reconfigure. Although several proprietary network interconnections are beginning to solve this problem, an innovative system from Yamaha (known as *mLAN*) seems to be leading the pack by allowing multichannel digital audio and MIDI music data to be transferred via a single 1394-standard (FireWire) cable. Theoretically, mLAN can transfer approximately 100 channels of digital audio data and up to 256 ports of MIDI data (16 channels × 256 connections) over a single cable in a bidirectional fashion . . . at speeds of 100, 200, or 400 Mbps!

All of this means that an mLAN-based music system could be quickly and easily configured, saved, and recalled, with none of the frustration and downtime that occur when reconfiguring a conventional system. Using the patch bay application that's provided with all mLAN products, devices can be routed and reconfigured in software. Since it is "hot pluggable," devices can be plugged and unplugged without having to power down or reset the system. In addition, mLAN is able to transmit and resolve world clock issues . . . even allowing devices to run at differing sample rates on the same network. Finally, this format can run with or without a computer. Should multiple mLAN hardware devices be connected, the system will establish a network link for easy, on –the-spot connectivity.

Although mLAN was developed by Yamaha, it is available to other manufacturers on a royalty-free (no-cost) basis. As of this writing, over 40 manufacturers had signed on as mLAN licensees.

The MIDI Message

MIDI messages are made up of groups of 8-bit words (known as bytes), which are transmitted in a serial fashion to convey a series of performance and/or control instructions to one or all MIDI devices within a system. Only two types of bytes are defined by the MIDI specification:

◆ Status byte
◆ Data byte

Status bytes are used as an identifier to tell the receiving device which particular MIDI function and channel are being addressed. The *data byte* information is used to encode the actual numeric values that accompany the status byte. Although a byte is made up of 8 bits, the most significant bit (MSB; the leftmost binary bit within a digital word) is used solely to identify the byte type. The MSB of a status byte is always 1, while the MSB of a data byte is always 0 (Figure 7.10). For example, a 3-byte MIDI note-on message (which is used to signal the

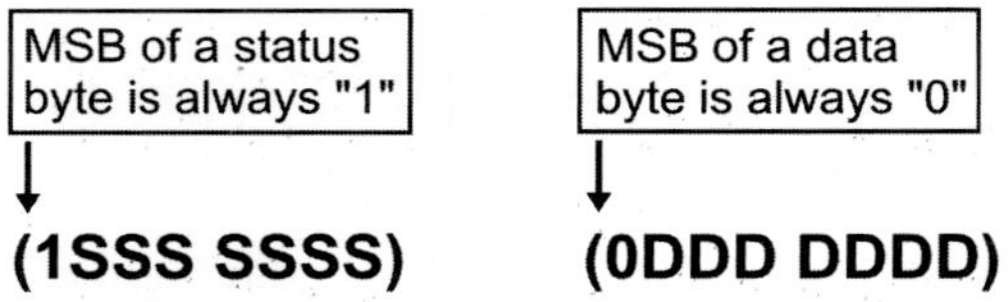

Figure 7.10. *The most significant bit of a MIDI data byte is used to identify between a status byte 1 and a data byte 0.*

MSB of a status byte is always "1"

(1SSS SSSS)

MSB of a data byte is always "0"

(0DDD DDDD)

beginning of a MIDI note) in binary form might read as shown in the following table:

	Status Byte	*Data Byte 1*	*Data Byte 2*
Description	Status/channel no.	Note no.	Attack velocity
Binary Data	(1001 0100)	(0100 0000)	(0101 1001)
Numeric Value	(Note on/channel 5)	(64)	(89)

Thus, a 3-byte note-on message of (10010100) (01000000) (01011001) will transmit instructions that would be read as "transmitting a note-on message over MIDI channel 5 using keynote 64 with an attack velocity (the note's volume level) of 89."

MIDI Channels

Just as it's possible for a public speaker to single out and communicate a message to one individual within a crowd, MIDI messages can be directed to a specific device or range of devices in a MIDI system. This is done by embedding a nibble (four bits) within the status/channel number byte that makes it possible for performance or control information to be communicated to a specific device or one of the sound generators in a device over its own channel. Since the nibble is 4 bits wide, up to 16 discrete MIDI channels can be transmitted through a single MIDI cable (Figure 7.11). Whenever a MIDI device is instructed to respond to a specific channel number, it will only react to messages that are transmitted on that channel and will ignore channel messages that are transmitted on any other channel.

Figure 7.11. *Up to 16 channels can be transmitted through a single MIDI cable.*

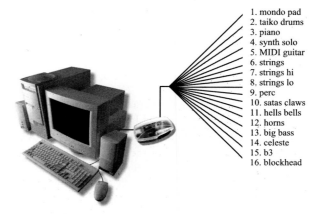

1. mondo pad
2. taiko drums
3. piano
4. synth solo
5. MIDI guitar
6. strings
7. strings hi
8. strings lo
9. perc
10. satas claws
11. hells bells
12. horns
13. big bass
14. celeste
15. b3
16. blockhead

Figure 7.12. *MIDI setup showing a set of channel assignments.*

For example, let's assume that we have a MIDI keyboard controller, a synth, and a sampler that are linked together in a MIDI chain (Figure 7.12). In this instance, we've assigned the synth to respond to data being transmitted over channel 3, while the sampler is set to respond to messages that are sent over channel 5. Setting the controller to transmit over channel 3 will allow the synth to be played when the controller's keys are pressed, while the sampler ignores the messages. Likewise, setting the controller to channel 5 will allow the sampler to be played, while the synth remains silent. Splitting the controller's keyboard (so that the lower octaves transmit on channel 3 while the upper octaves transmit over channel 5) allows us to play different musical parts on both instruments at once.

Using Figure 7.12 as an example, we could create a short song using a keyboard/synth workstation that has a built-in *sequencer* (a device that's capable of recording, editing, and playing back MIDI data), an external synth module, and our handy-dandy sampler. Let's start off by recording a percussion track into the sequencer that's set to transmit on MIDI channel 3 (a channel that's set to trigger a rock drum set on our synth module). After laying down a simple, repeating groove, we can set the controller to transmit notes on channel 5 (a channel that's set to play a mellow electric piano sample)... and begin to lay down a melody track. When these are done, we can begin recording a bass riff onto new track that'll internally transmit to the workstation's synth on channel 1.

When the song is complete, we can sit back and watch the workstation play the finished sequence back in all its glory. We can even have control over each instrument (using a multitude of musical and mixing parameters), in what can be viewed as a virtual multitrack working environment. ... From here, the number of possible channel, track, and/or instrument combinations are limited only by your setup and your imagination.

Channel Messages

Channel messages are used to transmit real-time performance data throughout a connected MIDI system and are generated whenever the controller of a MIDI instrument is played, selected, or varied by the performer. These messages are commonly generated whenever a MIDI instrument is played and controlled during a live or recorded performance. Each channel message contains a MIDI channel number in its status byte and therefore can be addressed by any device that's set to the assigned channel number. There are seven channel voice message types: Note On, Note Off, Polyphonic Key Pressure, Channel Pressure, Program Change, Control Change, and Pitch Bend Change. These message types are explained in the following list.

- ◆ *Note On* indicates the beginning of a MIDI note. This message is generated each time a note is triggered on a keyboard, drum machine or other MIDI instrument (by pressing a key, striking a drum pad, etc.). A Note On message consists of three bytes of information: a MIDI channel number, a MIDI pitch number, and an attack velocity value (messages that are used to transmit the individually played volume levels [0–127] of each note).

- ◆ *Note Off* indicates the release (end) of a MIDI note. Each note played through a Note On message is sustained until a corresponding Note Off message is received. A Note Off message doesn't cut off a sound; it merely stops playing it. If the patch being played has a release (or final decay) stage, it begins that stage upon receiving this message.

- ◆ *Polyphonic Key Pressure* messages are transmitted by instruments that can respond to pressure changes applied to the individual keys of a keyboard. A Polyphonic Key Pressure message consists of three bytes of information: a MIDI channel number, a MIDI pitch number, and a pressure value.

- ◆ *Channel Pressure* (or *Aftertouch*) messages are transmitted and received by instruments that respond to a single, overall pressure applied to the keys. In this way, additional pressure on the keys can be assigned to control such variables as pitch bend, modulation, and panning.

- ◆ *Program Change* messages change the active voice (generated sound) or preset program number in a MIDI instrument or device. Using this message format, up to 128 presets (a user- or factory-defined number that activates a specific sound-generating patch or system setup) can be selected. A Program Change message consists of two bytes of information: a MIDI channel number (1–16) and a program ID number (0–127).

- ◆ *Control Change* messages are used to transmit information that relates to real-time control over a MIDI instrument's performance parameters (such as modulation, main volume, balance, and panning). Three types of real-time controls can be communicated through control change messages: continuous controllers, as shown in Figure 7.13, which communicate a continuous range of control settings, generally in value ranging from 0–127; switches (controls having an ON or OFF state with no intermediate settings); and data controllers, which enter data either through numerical keypads or stepped up/ down entry buttons. A full listing of control-change parameters and their associated numbers can be found in Figure 7.14.

- ◆ *Pitch Bend*—Transmitted by an instrument whenever its pitch bend wheel is moved in either the positive (raise pitch) or negative (lower pitch) direction from its central (no pitch bend) position.

Figure 7.13. *Continuous controller data value ranges.*

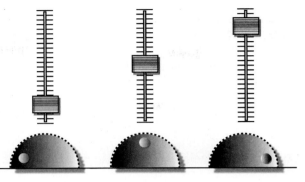

System Messages

As the name implies, system messages are globally transmitted to every device within a MIDI network. This is possible, as MIDI channel numbers aren't addressed in the byte structure of a system message. Consequently, any device will respond to these messages, regardless of what MIDI channel or channels the device is assigned to. System Common messages are used to transmit MIDI Time Code, Song Position pointers, Song Select, Tune Request, and System-Exclusive data throughout the MIDI system or the 16 channels of a specified MIDI port. The following list describes most of the existing System Common messages:

◆ *MIDI Time Code (MTC)* provides a cost-effective and easily implemented way of translating SMPTE time code into a format that conforms to the MIDI 1.0 specification. MTC messages allow time-based code and commands to be distributed throughout the MIDI chain.

◆ *Song Position Pointer (SPP)* allows a sequencer or drum machine to be synchronized to an external source (such as a tape machine) from any measure position within a song. This complex timing protocol isn't commonly used, as most users and design layouts currently favor MTC.

◆ *Song Select Message* uses an identifying song ID number to request a specific song from a sequence or controller source. After being selected, the song responds to MIDI Start, Stop, and Continue messages.

◆ *Tune Request* is used to request that an equipped MIDI instrument initiate its internal tuning routine.

◆ *End of Exclusive (EOX) messages* indicate the end of a System Exclusive message.

System-Exclusive Messages

The system-exclusive (sys-ex) message allows MIDI manufacturers, programmers, and designers to communicate customized MIDI messages between MIDI devices. The purpose of these

14-BIT CONTROLLER MOST SIGNIFICANT BIT			7-BIT CONTROLLERS (continued)		
Controller Hex	Number Decimal	Description	Controller Hex	Number Decimal	Description
.	.	.	.	.	.
00H	0	Underfined	4FH	79	Undefined
01H	1	Modulation Controller	50H	80	General Purpose Controller # 5
02H	2	Breath Controller	52H	81	General Purpose Controller # 6
03H	3	Underfined	52H	82	General Purpose Controller # 7
04H	4	Foot Controller	53H	83	General Purpose Controller # 8
05H	5	Portamento Time	54H	84	Undefined
06H	6	Data Entry MSB	.	.	.
07H	7	Main Volume	.	.	.
08H	8	Balance Controller	5AH	90	Undefined
09H	9	Undefined	5BH	91	External Effects Depth
0AH	10	Pan Controller	5CH	92	Tremolo Depth
0BH	11	Expression Controller	5DH	93	Chorus Depth
0CH	12	Underfined	5EH	94	Celeste (Detune) Depth
			5FH	95	Phaser Depth

14-BIT CONTROLLER MOST SIGNIFICANT BIT (cont.)			PARAMETER VALUE		
0FH	15	Undefined	Controller Hex	Number Decimal	Description
10H	16	General Purpose Controller # 1	60H	96	Data Increment
11H	17	General Purpose Controller # 2	61H	97	Data Decrement
12H	18	General Purpose Controller # 3			

			PARAMETER SELECTION		
13H	19	General Purpose Controller # 4	Controller Hex	Number Decimal	Description
14H	20	Undefined	62H	98	Non-Registered Parameter Number LSB
.	.	.	63H	99	Non-Registered Parameter Number MSB
.	.	.	64H	100	Registered Parameter Number LSB
1FH	31	Undefined	65H	101	Registered Parameter Number MSB

14-BIT CONTROLLER LEAST SIGNIFICANT BIT			UNDEFINED CONTROLLERS		
Controller Hex	Number Decimal	Description	Controller Hex	Number Decimal	Description
20H	32	LSB Value for Controller 0	66H	102	Undefined
21H	33	LSB Value for Controller 1	.	.	.
22H	34	LSB Value for Controller 2	.	.	.
.	.	.	78H	120	Undefined
.	.	.			
3EH	62	LSB Value for Controller 30			
3FH	63	LSB Value for Controller 31			

7-BIT CONTROLLERS			RESERVED FOR CHANNEL MODE MESSAGES		
Controller Hex	Number Decimal	Description	Controller Hex	Number Decimal	Description
40H	64	Damper Pedal (sustain)	79H	121	Reset All Controllers
41H	65	Portamento On/Off	7AH	122	Local Control On/Off
42H	66	Sostenuto On/Off	7BH	123	All Notes Off
43H	67	Soft Pedal	7CH	124	Omni Mode Off
44H	68	Undefined	7DH	125	Omni Mode On
45H	69	Hold 2 On/Off	7EH	126	Mono Mode On (Poly Mode Off)
46H	70	Undefined	7FH	127	Poly Mode On (Mono Mode Off)
.	.	.			
		continues			

Figure 7.14. *Listing of controller ID numbers, outlining both the defined format and conventional controller assignments.*

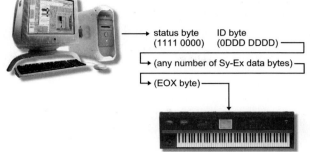

Figure 7.15. *System-exclusive data (one ID byte format.)*

messages is to give manufacturers, programmers, and designers the freedom to communicate any device-specific data of an unrestricted length, as they see fit. Most commonly, sys-ex data are used for the bulk transmission and reception of program/patch data and sample data, as well as real-time control over a device's parameters. The transmission format of a sys-ex message (Figure 7.15), as defined by the MIDI standard, includes a sys-ex status header, manufacturer's ID number, any number of sys-ex data bytes, and an EOX byte. When a sys-ex message is received, the identification number is read by a MIDI device to determine whether or not the following messages are relevant. This is easily accomplished, as a unique 1- or 3-byte ID number is assigned to each registered MIDI manufacturer and make. If this number doesn't match the receiving MIDI device, the subsequent data bytes will be ignored. Once a valid stream of sys-ex data has been transmitted, a final EOX message is sent, after which the device will again begin to respond normally to incoming MIDI performance messages.

In actual practice, the general idea behind sys-ex is that it uses MIDI messages to transmit and receive program, patch, and sample data or real-time parameter information between devices. It's sort of like having an instrument or device that's a musical chameleon. One moment it can be configured with a certain set of sound patches and setup data and then, after it receives a new sys-ex data dump, you could easily end up with an instrument that's literally full of new and exciting (or not-so-exciting) sounds and settings. Here are a few examples of how sys-ex can be put to good use:

◆ Transmitting patch data between synths—Sys-ex can be used to transmit patch and overall setup data between synths of identical make and (most often) model. Let's say that we have a Brand X Model Z synthesizer, and, as it turns out, you have a buddy across town that also has a Brand X Model Z. That's cool, except your buddy's synth has a completely different set of sound patches in her synth . . . and you want them! Sys-ex to the rescue! All you need to do is go over and transfer your buddy's patch data into your synth (to make life easier, make sure you take the instruction manual along).

◆ Backing up your current patch data—This can be done by transmitting a sys-ex dump of your synth's entire patch and setup data to disk, to a sys-ex utility program (often shareware), or to your DAW/MIDI sequencer. This is important: *Back up your factory preset or current patch data before attempting a sys-ex dump!* If you forget and download a sys-ex dump, your previous settings will be lost until you contact the manufacturer, download the dump from their Web site, or take your synth back to your favorite music store to reload the data.

◆ Getting patch data from the Web—One of the biggest repositories of sys-ex data is the Internet. To surf the Web for sys-ex patch data, all you need to do is log on to your favorite search engine site and enter the name of your synth. You'll probably be amazed at how many hits will come across the screen, many of which are chock-full of sys-ex dumps that can be downloaded into your synth.

◆ Varying sys-ex controller or patch data in real time—Patch editors or hardware MIDI controllers can be used to vary system and sound generating parameters, in real time. Both of these controller types can ease the job of experimenting with parameter values or changing mix moves by giving you physical or on-screen controls that are often more intuitive and easier to deal with than programming electronic instruments that'll often leave you dangling in cursor and 3-inch LCD screen hell.

Before moving on, I should also point out that sys-ex data grabbed from the Web, disk, disc, or any other medium will often be encoded using several sys-ex file format styles (unfortunately, none of which are standardized). Unfortunately, sequencer Y might not recognize a sys-ex dump that was encoded using sequencer Z. For this reason, dumps are often encoded using easily available, standard sys-ex utility programs for the Mac or PC.

Finally, a single unified standard has begun to emerge from the fray that's so simple it's amazing that it wasn't universally adopted from the start. This system simply records a sys-ex dump as a single MIDI file. Before recording a dump to a sequencer track, you may need to consult the manual to make sure that sys-ex filtering is turned off . . . once done, simply place the track into record, initiate the dump and save the track in an all-important sys-ex dump directory. Using this approach, it would also be possible to:

◆ Import the appropriate sys-ex dump track (or set of tracks) into the current working session so as to automatically program the instruments before the sequence is played back.

◆ Import the appropriate sys-ex dump track (or set of tracks) into separate MIDI tracks that can be muted or unassigned. Should the need arise, the track(s) can be activated and/or assigned in order to dump the data into the appropriate instruments.

MIDI Machine Control

MIDI Machine Control (MMC) is a protocol that's been designed to allow DAWs, MIDI sequencers, hard-disk recorders, tape and video transports, and other recording systems to be remotely controlled from a hardware device or computer program via MIDI. This works via the transmission of specially designated system-exclusive messages throughout the MIDI system to devices that can respond to MMC (Figure 7.16). An increasing number of devices have begun to support this protocol (including stage lights and other nonmusical equipment).

Figure 7.16. *Example of a MIDI machine control (MMC)-equipped system.*

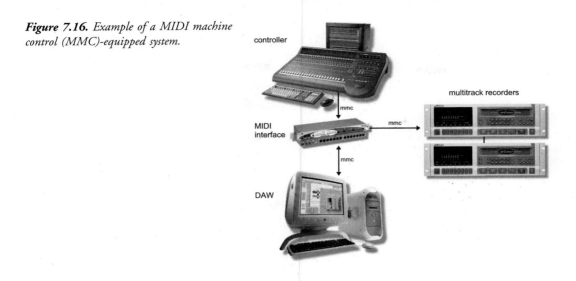

MMC control messages are able to communicate with individual devices within the connected network by assigning ID numbers to each relevant device. For example, a hard-disk recorder might have an ID of 1, while a MIDI sequencer would have an ID of 2, etc. As with all synchronous systems, any number of slaves can be controlled from one master device or program.

The MIDI Interface

Although computers and electronic instruments communicate using the digital language of 1's and 0's, computers simply don't understand the language of MIDI without the use of a device that translates these serial messages into a data structure that it can understand. Such a device is known as a *MIDI interface*. A wide range of MIDI interfaces currently exist that can be used with most computer systems and OS platforms. Here are some of the interface types that you can expect to find:

◆ USB interfaces are external devices that plug into a computer's USB port. The simpler devices range from being 1 × 1 (1 in and 1 out port) to designs that offer 8 × 8 port capabilities (Figure 7.17).

◆ Newer audio interface, instrument, and controller interface systems (Figure 7.18) often include a 1 × 1, 2 × 2, or even a 4 × 4 MIDI I/O interface, directly within its design.

◆ When talking about MIDI interfaces, who could pass up the one design that surpasses all others in numbers (by the millions)? I'm referring, of course, to all of those SoundBlaster PC sound cards and their look-alikes that have a 1 × 1 interface built right into them. Almost every sound card, no matter how inexpensive, will have a 15-pin connector that can be used either as a game joystick port or (with an optional cable) as a MIDI in, out, and thru port.

Figure 7.17. *USB Midisport 1×1 MIDI interface. (Courtesy of M-Audio, www.m-audio.com.)*

Figure 7.18. *Tascam US-428 computer interface and control surface. (Courtesy of Tascam, www.tascam.com.)*

The Multiport Interface

As DAW and software capabilities become more powerful, much of the MIDI routing and processing is handled from within the host software; however, for those requiring added routing and synchronization capabilities, the interface of choice for most professional electronic musicians is the external multiport MIDI interface (Figure 7.19). These rack-mountable devices also often plug into the computer's USB port to provide eight independent MIDI in and outs that can easily distribute MIDI data through separate lines over a connected network. As an added note, multiport interfaces often include a software-controlled patch bay for routing and processing MIDI data between ports and throughout the MIDI network. For example, it could be used to merge together several MIDI inputs (or outputs) into a single data stream, filter out specific MIDI message types (used to block out unwanted commands that might adversely change an instrument's sound or performance), or rechannel

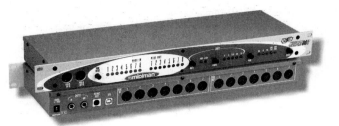

Figure 7.19. USB Midisport 8×8/s
MIDI interface. (Courtesy of M-Audio,
www.m-audio.com.)

data being transmitted on one MIDI channel to another channel that can be recognized by an instrument or device.

Another important function that can be handled by most multiport interfaces is *synchronization.* Synchronization (*sync,* for short) allows other, external devices (such as DAWs, MDMs, video decks, and other media systems) to be simultaneously played back using the same timing reference. Interfaces that includes sync features will often read and write SMPTE time code, convert SMPTE to MIDI time code (MTC), and allow recorded time-code signals to be cleaned up when copying code from one analog device to another (jam sync). Further reading on synchronization can be found in Chapter 9.

Electronic Musical Instruments

Since their inception in the early 1980s, MIDI-based electronic instruments have played a central and important role in the development of music technology and production. These devices (which fall into almost every instrument category), along with the advent of cost-effective digital and analog recording systems, have helped to shape the music production industry into what it is today. In fact, it's the combination of these technologies that have made the personal project studio into an important driving force behind modern-day music.

Inside the Toys

Although electronic instruments often differ from one another in looks, form, and function, they almost always share a standard set of basic building-block components (Figure 7.20). These include the following:

◆ *Central processing units (CPU)*—One or more dedicated processors (often in the form of a specially manufactured chip) contain all the necessary brains to control the hardware, voice data, and sound-generating capabilities of an instrument or device.

◆ *Performance controllers*—These include keyboards, drum pads, wind controllers, etc., which transmit performance/controller data directly to the processor and then to the voice circuitry or out of the instrument as MIDI messages. Not all instruments have a built-in controller. These devices (commonly known as *modules*) contain all the

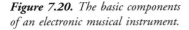

Figure 7.20. *The basic components of an electronic musical instrument.*

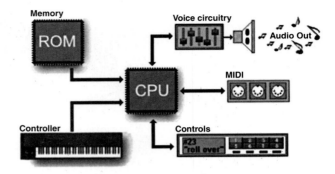

necessary processing and sound-generating circuitry; however, they save space by eliminating redundant keyboards or other controller surfaces.

◆ *Control panel*—This all-important human interface of data-entry controls and display panels lets you select and edit sounds, route and mix output signals, and control the instrument's basic operating functions.

◆ *Memory*—Used for storing important internal data (such as patch and setup configurations and/or digital waveform data), this information can be stored in ROM (read-only memory) or RAM (random-access memory) or on a factory-encoded chip, cartridge, or CD-ROM.

◆ *Voice circuitry*—Depending upon the device type, this section can either generate analog sounds (voices) or can be used to instruct digital samples that are permanently recorded into memory to be played back according to a set of parameters (which are often user defined). In short, it's used to generate or reproduce a sound patch, which can then be reproduced and recorded.

◆ *Auxiliary controllers*—These are external controlling devices that can be used in conjunction with an instrument or sequencer. Examples of these include foot pedals (providing continuous-controller data), breath controllers, and pitch-bend/modulation wheels. Certain controllers may only have two or a limited number of switching states (such as a foot controller, sustain pedals, or vibrato switches).

◆ *MIDI communications ports*—These are used to transmit and/or receive MIDI data.

Most (if not all) of an instrument's component blocks can be accessed or communicated between several devices (in one form or another) via MIDI. In fact, there are many ways in which MIDI can be used to communicate performance, patch, system setup, and even audio data between devices. For example, a played synth keyboard could transmit performance messages to a synth module . . . or all of an instrument's sound-patch data settings could be transmitted to or received from a sequencer or other instrument as MIDI system-exclusive messages (sys-ex) . . . or sampled audio could be exchanged between samplers using the MIDI sample dump standard or SCSI (a computer protocol that communicates data at higher speeds).

For the remainder of this section, we'll be discussing the various types of MIDI instruments and controller devices that are currently available on the market. These instruments can be

Figure 7.21. *Korg Triton Extreme keyboard workstation. (Courtesy of Korg USA, www.korg.com.)*

grouped into such categories as keyboards, percussion, controlling devices, MIDI guitars and strings, and sequencers.

Keyboards

By far the most commonly encountered instrument within almost any MIDI production facility belongs to the keyboard family. This is due, in part, to the fact that keyboards were the first electronic music devices to gain wide acceptance and that MIDI was initially developed to record and control many of their performance and control parameters. The two basic keyboard-based instruments are the *synthesizer* and the *digital sampler*.

The Synthesizer

A *synthesizer* (Figure 7.21) is an electronic instrument that uses multiple sound generators to create complex waveforms that can be combined (using various waveform synthesis techniques) into countless sonic variations. These synthesized sounds have become a basic staple of modern music and vary from sounding "cheesy"... to closely mimicking traditional instruments... to generating rich, otherworldly sounds that literally defy classification. Synthesizers (also known as *synths*) generate sounds and percussion sets using a number of different technologies or program algorithms. The earliest synths were analog in nature and generated sounds using additive or subtractive synthesis: *Frequency modulation* (FM) synthesis initially appeared in 1982 in the form of the first completely digital synthesizer... the Yamaha DX-7. The FM process usually involves the use of at least two signal generators (commonly referred to as *operators*) to create and modify a voice. Often, this is done through the analog or digital generation of a signal that modulates or changes the tonal and amplitude characteristics of a base carrier signal. More sophisticated FM synths can use up to four or six operators per voice and also often use filters and variable amplifier types to alter the signal's characteristics into a sonic voice that either roughly imitates acoustic instruments or creates sounds that are totally unique.

Another technique that's used to create sounds is *wavetable synthesis*. This technique works by storing small segments of digitally sampled sound into a ROM chip. Various sample-based synthesis techniques use sample looping, mathematical interpolation, pitch shifting, and digital filtering to create extended and richly textured sounds that use a very small amount of sample memory. These sample-based systems are often called wavetable synths because a large number of prerecorded samples are encoded within the instrument's memory and can

Figure 7.22. Orbit Dance Planet synth module. (Courtesy of E-MU/ Ensoniq, www.emu.com.)

be thought of as a table of sound waveforms that can be looked up and utilized when needed. Once selected, a range of parameters (such as wavetable mixing, envelope, pitch, volume, pan, and modulation) can be modified to control an instrument's overall sound characteristics.

Synthesizers are also commonly designed into rack- or half-rack-mountable modules (Figure 7.22) that contain all of the features of a standard synthesizer, except that they don't incorporate a keyboard controller. This space-saving feature means that more synths can be placed into your system and can be controlled from a master keyboard controller or sequencer, without cluttering up the studio with redundant keyboards.

Software Synthesis and Sample Resynthesis

Since wavetable synthesizers derive their sounds from prerecorded samples that are stored in a digital memory media, it logically follows that these sounds can also be stored on hard disk (or any other medium) and loaded into the RAM of a personal computer. This process of downloading wavetable samples into a computer and then manipulating these samples is used to create what is known as a *virtual* or *software synthesizer* (Figure 7.23). In recent years, software synths have grown from being novel and obscure programs that were primarily used by the academic community to their current state of being widely accepted in the production community as a cost-effective musical instrument. These cost-effective software modules can be used in conjunction with a digital audio workstation to offer up a wide range of complex sounds that can mimic traditional instruments, as well as create sonic textures that are both new and interesting.

Figure 7.23. Steinberg xphrase VSTi software synth. (Courtesy of Steinberg Media Technologies GMBH, www.steinberg.net.)

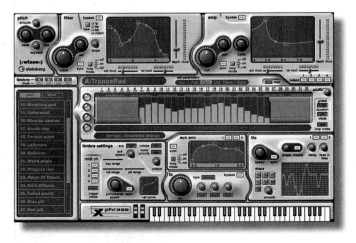

Figure 7.24. *REAKTOR 4 modular sound studio, a tool that lets musicians and engineers design and build their own instruments, samplers, effects, and sound designs. (Courtesy of NATIVE INSTRUMENTS Software Synthesis GmbH, www.native-instruments.com.)*

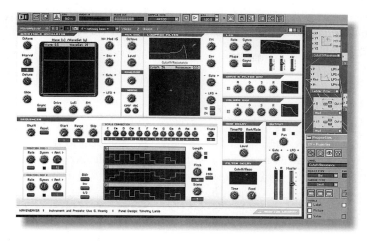

Sample resynthesis software systems (Figures 7.24) are able to take software synthesis to a new level by allowing the user to build, save, and recall sonic patches that can be built from traditional synthesis building blocks (such as oscillators, voltage-controlled amplifiers, voltage-controlled filters, and mixers). In addition to sound generation, digital audio samples can be imported and resynthesized in a way that can create sounds of almost any texture or type that you can possibly imagine. All of these software blocks can be combined in a graphic environment that allows these instruments, textures, and soundscapes to be easily saved to disk for later recall.

Using various internal software data communications protocols, it's possible to communicate MIDI, audio, timing sync, and control data between an instrument (or effect plug-in) and a host DAW program/CPU processor. These plug-in protocols make it possible for much or all of the audio and timing data to be routed through the host audio application, allowing the instrument or application to either integrate into the DAW or application . . . or work in tandem so as to route the audio and performance/control data through the host application with relative ease. A few of these protocols include:

◆ Steinberg's Virtual Studio Technology (VST)
◆ MOTU's Audio System (MAS)
◆ Propellerhead's ReWire

Further reading on virtual and plug-in instruments, plug-in protocols, and applications can be found within Chapter 6.

Sound Card Synths

By far, the greatest number of installed synthesizers have been designed into generic PC sound cards. These devices (which can be found in almost every home) are generally designed into a single chipset and generate sounds using a simple form of digitally controlled FM synthesis.

Figure 7.25. Akai Z4 24/96
hardware sampler. (Courtesy of Akai
Professional, www.akaipro.com.)

Although more expensive sound cards will often use wavetable synthesis to create richer
and more realistic sounds, both card types conform to the General MIDI specification, which
has been universally defined with an overall patch and drum-sound structure that can be played
by all synths with the correct voicings and levels. Further information about General MIDI
can be found in Chapter 8.

Samplers

A *sampler* (Figure 7.25) is a device that can convert audio into a digital form and then import
that data into its internal RAM. Once audio has been sampled or loaded into RAM
(from diskette, disk, or disc), segments of sampled audio can be edited, transposed, processed,
and played in a polyphonic, musical fashion. Basically, a sampler can be thought of as a
wavetable synth that lets you record, edit, and reload the samples into RAM. Once loaded,
these sounds (whose length and complexity are often limited only by memory size and your
imagination) can be looped, modulated, filtered, and amplified (according to user or factory
setup parameters) in a way that allows the waveshapes and envelopes to be modified. Signal
processing capabilities, such as basic editing, looping, gain changing, reverse, sample-rate
conversion, pitch change, and digital mixing can also be altered and/or varied.

A sampler's design will often include a keyboard or set of trigger pads (Figure 7.26) that lets
you polyphonically play samples as musical chords, sustain pads, triggered percussion sounds,
or sound-effect events. These samples can be played according to the standard Western musical

Figure 7.26. Akai MPC-1000
Music Production Center. (Courtesy
of Akai Professional,
www.akaipro.com.)

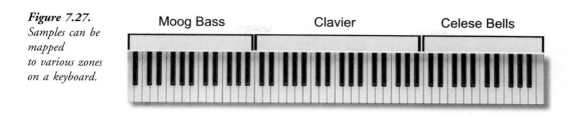

Figure 7.27.
Samples can be mapped to various zones on a keyboard.

Moog Bass   Clavier   Celese Bells

scale (or any other scale, for that matter) by altering the playback sample rate over the controller's note range. For example, pressing a low-pitched key on the keyboard will cause the sample to be played back at a lower sample rate, while pressing a high-pitched one will cause the sample to be played back at rates that would put Mickey Mouse to shame. By choosing the proper sample-rate ratios, sounds can be polyphonically played (whereby multiple notes are sounded at once) at pitches that correspond to standard musical chords and intervals.

A sampler (or synth) with a specific number of *voices*—for example, 64 voices—simply means that up to 64 notes can be simultaneously played on a keyboard at any one time. Each sample in a multiple-voice system can be assigned across a performance keyboard, using a process known as *splitting* or *mapping*. In this way, a sound can be assigned to play across the performance surface of a controller over a range of notes, known as a *zone* (Figure 7.27). In addition to grouping samples into various zones, velocity can enter into the equation . . . by allowing multiple samples to be layered across the same keys of a controller, according to how soft or hard they are played. For example, a single key might be layered so that pressing the key lightly would reproduce a softly recorded sample, while pressing it harder would produce a louder sample with a sharp, percussive attack. In this way, mapping can be used to create a more realistic instrument or wild set of soundscapes that change not only with the played keys but with different velocities as well.

Most samplers have extensive edit capabilities that allow the sounds to be modified in much the same way as a synthesizer, using such modifiers as:

◆ Velocity
◆ Panning
◆ Expression (modulation and user control variations)
◆ Low-frequency oscillation (LFO)
◆ Attack, delay, sustain and release (ADSR) and other envelope processing parameters)
◆ Keyboard scaling
◆ Aftertouch

Many sampling systems will often include such features as integrated signal processing, multiple outputs (offering isolated channel outputs for added mixing and signal processing power or for recording individual voices to a multitrack recording system), as well as integrated MIDI sequencing capabilities.

Figure 7.28. Steinberg's HALion VST software sampler. (Courtesy of Steinberg Media Technologies GMBH, www.steinberg.net.)

Figure 7.29. MOTU MachFive software sampler. (Courtesy of Mark of the Unicorn, www.motu.com.)

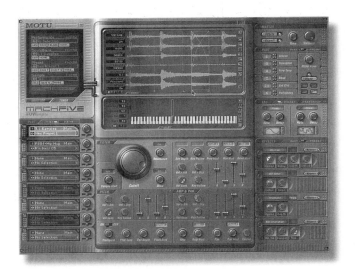

Software Samplers

In addition to hardware sampling systems, a growing number of virtual or *software samplers* exist that use a computer's existing memory, processing, and signal routing capabilities in order to polyphonically reproduce samples in real time. Offering much of the same functionality as their hardware counterparts, these software-based systems (Figures 7.28 and 7.29) are capable of editing, mapping, and splitting sounds across a MIDI keyboard using on-screen graphic controls and DAW integration and have improved to the point of equaling or surpassing their hardware counterparts in cost-effectiveness, power, and ease of use.

As with a software synth, software samplers derive their sounds from recorded and/or imported audio data stored as digital audio data within a personal computer. Using the digital signal processing (DSP) capabilities of today's computers (as well as the recording, sequencing, processing, mixing, and signal routing capabilities of most digital audio workstations), most software samplers are able to store and access samples within the internal memory of a laptop or desktop computer. Using a graphic interface, these sampling systems often allow the user to:

◆ Import previously recorded soundfiles (often in .wav, .aif, and other common formats)
◆ Edit and loop sounds into a usable form
◆ Vary envelope parameters (e.g., dynamics over time)
◆ Vary processing parameters
◆ Save the edited sample performance setup as a file for later recall

As with a virtual synth application, it's possible for a software sampler to communicate MIDI, audio, timing sync, and control data between a software instrument (or effect plug-in) and a host DAW program/CPU processor. Using an internal plug-in communications protocol, it's possible for most or all of the audio and timing data to be routed through the host audio application, allowing the instrument or application either to integrate into the DAW or application . . . or to work in tandem so as to route the audio and performance/control data through the host application with relative ease. A few of these protocols include:

◆ Steinberg's VST
◆ MOTU's MAS
◆ Propellerhead's ReWire

Further reading on virtual and plug-in instruments, plug-in protocols, and applications can be found within Chapter 6.

Sample Editing

Whenever a recorded sound is transferred into a hardware or software sampler, the original source material may contain extraneous sounds, breathing and fidget noises, or other music that occurs both before and after the desired sample (Figure 7.30a). Using the sampler's edit function, these unwanted sounds can be deleted by *trimming* the in and out points to include only the desired sounds. Trimming is accomplished by instructing the system's microprocessor to ignore (not access or reproduce) the samples that exist before a user-defined in point and/or those samples following a desired out point (Figure 7.30b). After trimming, the final sample can be played, looped, dynamically processed (if a fade or other function needs to be performed), and then saved for later recall.

Figure 7.30. *Sample editing: (a) unedited sample; (b) sample that has been trimmed, normalized, and faded at its end.*

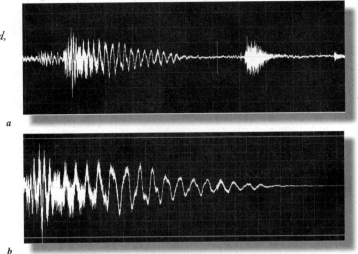

a

b

Sample Looping

Another editing technique that's regularly used to maximize the system's available RAM and disk-based memory is a process known as *looping*. Using this technique, a sample that occupies a finite memory space in RAM can be played in a repeated fashion. In this way, a carefully edited sample can be sustained for long periods of time (well past the length of the original sample) and affected over time simply by continuing to hold down the note key. Once the key is released, the note might continue to die out in a natural fashion over time (according to its parameter settings). Such a loop can be created by defining a segment of sound within a sample that doesn't significantly change in amplitude and composition over time. Once done, the sampler can continually access the segment over time within the RAM (Figure 7.31).

The most challenging part of creating a realistically convincing loop is to make sure that the loop splice point is carefully matched in level and frequency balance. This can be made easier by following this simple rule: Be sure to carefully match the waveform shape and amplitude at the beginning of the loop with the waveform shape and amplitude at the end of the loop. This simply means that the beginning and end amplitudes must match (Figure 7.32). If they don't, the signal levels will vary and an annoying "pop," audible "tick," or discontinuity in the sample will result. Many samplers and sample-editing programs provide a way to automatically search out the closest level match or display the loop crossover points on a screen; however, more often than not the levels will need to be manually finessed at their final crossover points.

In addition to allowing multiple, layered samples to be assigned to a single note, many samplers are able to access multiple loop points within a single sample. This has the effect of making the sample sound less repetitive and more natural and adds to its overall expressiveness

Figure 7.31. *Example of a soundfile with a highlighted sustain loop. (Courtesy of Sony Pictures Digital, Inc., http:// mediasoftware.sonypictures.com.)*

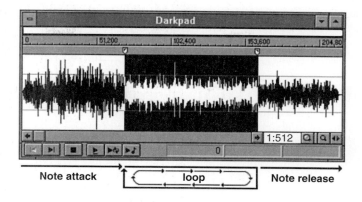

Figure 7.32. *A loop waveform window allows the beginning and end levels of a loop to be manually matched. (Courtesy of Sony Pictures Digital, Inc., http:// mediasoftware.sonypictures.com.)*

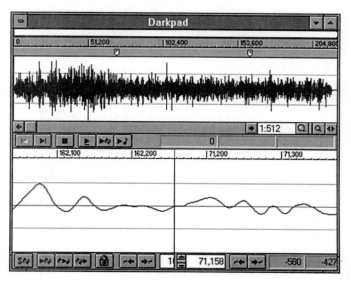

when played on a keyboard. In addition to having multiple sustain loops, a different "release loop" can be programmed to create a unique decay whenever the sample note is released.

Editors for Hardware Sampling Systems

Since MIDI's inception in 1981, several MIDI sample-dump formats have been developed that allow samples to be transmitted to and received from various hardware sampling systems in the digital domain. These sample dumps can be transferred between like samplers or between a personal computer and various samplers. To take full advantage of the latter,

Figure 7.33. Example diagram of a sample-editor network.

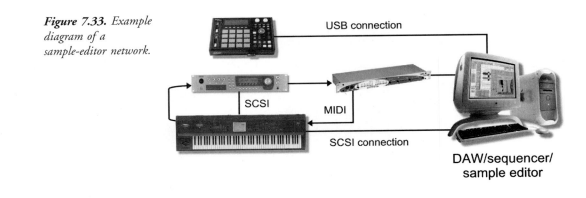

USB connection

SCSI

MIDI

SCSI connection

DAW/sequencer/
sample editor

Figure 7.34. Sample Wrench 24/96. (Courtesy of dissidents software, http://www.dissidents.com.)

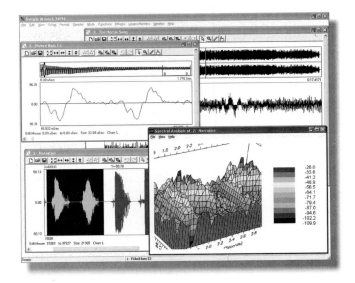

sample-editing software (Figures 7.33 and 7.34) was developed to perform such tasks as:

◆ Loading samples into a computer, where they can be stored to hard disk, arranged into a library that best suits the user's needs, and transmitted to any hardware sampling device in the system

◆ Editing and arranging the sample before copying to disk using standard computer cut and paste edit tool; because most samplers support multiple loops, segments of a sample can be repeated to save valuable RAM in the hardware sampler

◆ Digitally processing a signal to alter or mix a samplefile and using such functions as gain changing, mixing, equalization, inversion, reversal, muting, fading, crossfading, and time compression

In all fairness, it should be noted that with the proliferation of software-based sampling systems and their integration into the DAW environment, the use of software sample editors for transmission and storage between hardware samplers is on the decline. Quite simply, communications in the computer-based sample and network environment are simpler and much less cumbersome.

Distribution of Sampled Audio

Distributing data to and from a sampler can be carried out in many ways. Almost every sampler has a floppy, fixed/removable hard drive or external data port for saving waveform and system data to a computer, external drive, or CD-R. Distribution between samplers or to/from a sample editing program can be handled through the use of the slower MIDI sample-dump standard, a high-speed small computer system interface (SCSI) port, or FireWire protocol. Within a hardware-based MIDI setup, it's important that sampled audio be distributed using a method that's as fast and as painless as possible; therefore, standards have been adopted that allow samplers and related software programs to communicate, categorize, and store sample-based digital audio. It should be noted that samplers might or might not support these protocols. As always, consulting the manuals for each device before attempting a data transfer will most likely reduce stress, frustration, and wasted time.

MIDI Sample-Dump Standard

The MIDI sample-dump standard (SDS) was developed and proposed by the MIDI Manufacturers Association as a protocol for transmitting sampled digital audio and sustain-loop information between sampling devices. This data is transmitted over regular MIDI lines as a series of MIDI system-exclusive (sys-ex) messages, which are unspecified in length and data structure. Although samplers of different manufacture and model type can be used to perform similar musical functions, the inner electronics and the way that data is internally structured often vary widely from device to device. As a result, most samplers communicate using their own unique system- exclusive data structure (as identified by a unique manufacturer and device ID number). In order for data to be successfully transmitted between samplers, they must be of the same or compatible manufacture and design. If this isn't the case, a computer-based program (such as a sample editor) must be used to translate from one sampler's data format and structure into one that can be understood by another make or model. Samplers that communicate sample data via SDS have the distinct disadvantage of being rather slow, since the digital audio data are transmitted over standard MIDI lines at the 31.25-kbaud rate. When transmitting anything more than a short sample, be prepared to take a coffee break.

SCSI Sample Dump Formats

A number of computer-based digital audio systems and professional samplers are capable of transmitting and receiving sampled audio via SCSI, which is a bidirectional communications line that's commonly used by personal computers to exchange digital data at high speeds. When used in digital audio applications, it provides a direct parallel data link for transferring soundfiles at a rate of 16 MB/sec or higher (literally hundreds of times faster than MIDI). Although the data format will change from one device to the next (meaning that data can only be transmitted between like devices or via specific device/computer system combinations), SCSI still wins out as a fast and straightforward way to transfer data to and from an editing program, hard disk, or CD-ROM sample library.

SMDI

The SCSI Musical Data Interchange (SMDI) was developed as a standardized, non-device-specific format for transferring digitally sampled audio between SCSI-equipped samplers and computers at speeds up to 300 times faster than MIDI's transmission rate of 31.25 kB per second. Using this format, all you need to transfer digital audio directly from one supporting sampler to another is to connect the SMDI ports by way of a standard SCSI cable and follow the steps for transferring the sample. Although SMDI is loosely based on MIDI SDS, it has more advantages over its slower cousin than just speed. For example, SMDI can distribute stereo or multichannel samplefiles. Also, it isn't limited to files that are less than two megawords in length and can transmit associated file information (such as filename, pitch values, and sample number range). Sound patch and device-specific setup parameters can also be transmitted and received over SMDI lines via standard system-exclusive (sys-ex) messages.

Commercial and Personal Sample Libraries

As with MIDI sys-ex data sound patches, a wide range of previously recorded samplefiles can be purchased as a collection on CD (in edited form on a CD-ROM or unedited on a CD-Audio disc) or on the Web. The edited files might exist in your sampler's native format or might be in another format that can be imported, converted, and then saved into your sampler's native format. These commercially available libraries are often the mainstay of many electronic musicians and visual postproduction facilities. Of course, commercially available sample libraries need not be the only option. Professional and nonprofessional artists will often record and edit their own samples for their own personal library. These sounds can be created from original acoustic or electronically generated sound sources, in addition to lifting soundclips from previously recorded source material on CD, television, records, and videotapes (an area that requires careful attention to copyright issues).

Percussion

Historically speaking, one of the first applications of sample technology made use of digital audio to record and playback drum and percussion sounds. Out of this virtual miracle sprang a major class of sample and synthesis technology that lets an artist (mostly keyboard players) add drum and percussion to their own compositions with relative ease. Over the years, MIDI has brought sampled percussion within the grasp of every electronic musician, whatever

Figure 7.35. *Zoom MRT-3B Micro Rhythm Trak Drum Machine. (Courtesy of Samson Technologies Corp., www.samsontech.com/zoom.)*

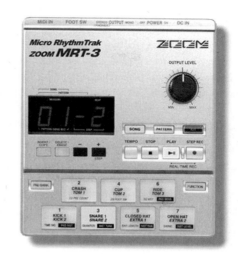

Figure 7.36. *Groove Agent VST virtual drummer. (Courtesy of Steinberg Media Technologies GMBH, www.steinberg.net.)*

skill level—from the frustrated drummer to professional percussionist/programmers . . . all of whom use their skills to perform live or to sequenced compositions.

The Drum Machine

The drum machine (Figures 7.35 and 7.36) is most commonly a sample-based digital audio device that can't record audio into its internal memory. Instead, these hardware or software systems use ROM-based, prerecorded samples to reproduce high-quality drum sounds. These factory-loaded sounds often include a wide assortment of drum sets, percussion sets, rare and wacky percussion hits, and effected drum sounds (e.g., reverberated, gated). Who knows, you might even encounter "Hit me!" screams from the venerable King of Soul—James Brown.

Most hardware drum machines allow prerecorded samples to be assigned to a series of playable keypads that are often located on the machine's top face. This provides a straightforward controller surface that usually includes velocity and aftertouch dynamics. Drum voices can be assigned to each pad and edited using such control parameters as tuning, level, output assignment, and panning position. Multiple outputs are often provided, enabling individual or groups of voices to be routed to a specific output on a mixer or console. This feature allows isolated voices to be individually processed or recorded onto separate tracks of a DAW or tape machine.

The assigned percussion sounds can be played live or from a programmed sequence track, although most drum machines have a built-in sequencer that's been specifically designed to arrange drum/percussion sounds into a rhythmic sequence (known as a *drum pattern*). These patterns often consist of basic variations on a rhythmic groove or can be built from patterns that are taken from a variety of playing styles (such as rock, country, or jazz). Drum machines that have this feature will often let you chain patterns together into a continuous song.

Although a number of hardware drum machine designs include a built-in sequencer, it's more likely that these workhorses will be triggered from a MIDI sequencer. This lets us take full advantage of the real-time performance and editing capabilities that a sequencer has to offer. For example, sequenced patterns can easily be created in step time (where notes are entered and assembled into a rhythmic pattern one note at a time) and can then linked together into a song that's composed of several rhythmic patterns. Alternatively, performing into a sequencer on the fly can help create a live feel, or you can combine step- and real-time tracks to create a human-sounding composite rhythm track. In the final analysis, the style of and approach to composition are entirely up to you.

Alternative Percussion Voices

In addition to the sounds that can be found in a drum machine, a virtually unlimited number of percussion sounds can be obtained from other sources. For example, most synth wavetables will include several drum and/or percussion setups that are often mapped over the entire keyboard surface. Sampler libraries will almost always include a never-ending range of percussion instruments and drumsets. Soundfiles can be loaded into a hard-disk editor to be built up as rhythm tracks, or you can lift percussion loops from the inexhaustible number of loop libraries that are available on CD and the Web. Heck, you could even be bold and record your own samples!

Performance Controllers

MIDI *performance controllers* are used to translate the voicings and expressiveness of a musical performance into MIDI data. These can fall into such categories as keyboard, percussion, wind, and guitar controllers.

Figure 7.37. *M-Audio Oxygen 8 USB keyboard controller. (Courtesy of M-audio, www.m-audio.com.)*

Figure 7.38. *The Korg microKontrol USB MIDI controller. (Courtesy of Korg USA, www.korg.com.)*

Figure 7.39. *Roland A-37 MIDI keyboard controller. (Courtesy of Roland Corp. U.S., www.rolandus.com.)*

Keyboard Controllers

MIDI keyboard controllers (Figures 7.37 to 7.39) are keyboard devices which are expressly designed to control hard/software synths, samplers, modules, and other devices within a connected MIDI system. They contain no internal tone generators or sound-producing elements; instead, their design includes a performance keyboard and controls for handling MIDI performance, control and device switching events.

Percussion Controllers

MIDI percussion controllers are used to translate the voicings and expressiveness of a percussion performance into MIDI data. These devices are great for capturing the feel of a live performance while giving you the flexibility of recording and automating a performance within

Figure 7.40. Kits such as the V-Drums from Roland provide standard playing surfaces that let you be the drum "machine." (Courtesy of Roland Corp. U.S., www.rolandus.com.)

Figure 7.41. Yamaha WX5 MIDI wind controller. (Courtesy of Yamaha Corp., www.yamaha.com.)

a DAW/sequencer environment. These controllers vary over a wide range from being a simple and cost-effective setup (i.e., using the pads on a drum machine, keys on a keyboard surface, or pads on an entry-level drum controller) to a full-blown drum kit that mimics its acoustic cousin (Figure 7.40).

Wind Controllers

MIDI wind controllers (Figure 7.41) are expressly designed to bring the breath and key articulation of a woodwind or brass instrument to a MIDI performance. These controller types are used because many of the dynamic- and pitch-related expressions (such as breath and controlled pitch glide) simply can't be communicated from a standard music keyboard. In these situations, wind controllers can often help create a dynamic feel that's more in keeping with their acoustic counterparts by using an interface that provides special touch-sensitive keys, glide- and pitch-slider controls, and real-time breath sensors for controlling dynamics.

MIDI Guitars

Guitar players often work at stretching the vocabulary of their instruments beyond the traditional norm. They love doing nontraditional gymnastics using such tools of the trade as distortion, phasing, echo, feedback, etc. Due to advances in guitar pickup and microprocessor technology, it's also possible for the notes and minute inflections of guitar strings to be accurately translated into MIDI data. With this innovation, many of the capabilities that MIDI has to offer are available to the electric (and electronic) guitarist. For example, a guitar's natural sound can be layered with a synth pad that's been transposed down, giving it a rich, thick sound that just might shake your boots. Alternatively, recording a sequenced guitar track into a session would give a producer the option of changing and shaping the sound...later in mixdown! On-stage program changes are also a big plus for the MIDI guitar, allowing the player to radically switch between guitar voices from the guitar or sequencer or by stomping on a MIDI foot controller.

Sequencers

Apart from electronic musical instruments, one of the most important tools that can be found in the modern-day project studio is the *MIDI sequencer*. Basically, a sequencer is a digital device that's used to record, edit, reproduce, and distribute MIDI messages in a sequential fashion. Most sequencers function using a traditional track-based interface, separating different instruments, voices, beats, etc. in a way that makes it easier for us humans to view MIDI data as though they were linear tracks on a DAW or tape machine. These virtual "tracks" contain MIDI-related performance and control events that are made up of such channel and system messages as note on/off, velocity, modulation, aftertouch, and program/ continuous-controller messages. Once a performance has been recorded into a sequencer's memory, these events can be graphically (or audibly) edited into a musical performance, played back and saved to a digital storage media for recall at any time.

Integrated Sequencers

Some of the newer and more expensive keyboard synth and sampler designs include a built-in sequencer. These portable keyboard workstations have the advantage of letting you take both the instrument and sequencer on the road without having to drag a computer-based system along. Integrated sequencers are designed into an instrument for the sole purpose of sequencing MIDI data and include integrated controls for performing sequence-specific functions. Ease of use and portability are the main advantages of a hardware sequencer, most of which are designed to emulate the basic functions of a tape transport (record, play, start/stop, fast forward, and rewind). These devices generally offer a moderate amount of editing features, including note editing, velocity and other controller messages, program change, cut-and-paste and track-merging capabilities, tempo changes, etc. Programming, track, and edit information is commonly viewed on a liquid-crystal display (LCD) that's often

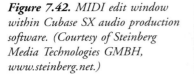

Figure 7.42. *MIDI edit window within Cubase SX audio production software. (Courtesy of Steinberg Media Technologies GMBH, www.steinberg.net.)*

limited in size and resolution and generally limits information to a single parameter or track at a time. These sequencers often don't offer a wide range of editing tools beyond standard transport functions: punch-in/out commands and other basic edit tools. However, they're often more than adequate for capturing and reproducing a performance and can be integrated with other instruments that are connected in a MIDI chain.

Software Sequencers

By far, the most common sequencer type is the *software MIDI sequencer* (Figures 7.42). These programs or integrated components of a digital audio workstation, for both the Mac and the PC, take advantage of the hardware and software versatility that a computer can offer in the way of speed, flexibility, digital signal processing, memory management, and signal routing.

Computer-based sequencers offer numerous functional advantages over their hardware counterparts. Among these are increased graphic capabilities (which often offer extensive control over track- and transport-related functions), standard computer cut-and-paste techniques, an on-screen graphic environment (allowing easy manipulation of program and edit-related data), routing of MIDI to multiple ports in a connected system, and the graphic assignment of instrument voices via program change messages (not to mention the ability to save and recall files using standard computer memory media). . . . Now, let's take a look into how these devices function.

A Basic Intro to Sequencers

When dealing with any type of sequencer, one of the most important concepts to grasp is the fact that these devices don't store sound directly—instead, they encode MIDI messages that instruct an instrument to play a particular note, over a certain channel, at a specific velocity and with any optional controller values. In other words, a sequencer stores music-related data commands that follow in a sequential order which then tell instruments and/or devices how their voices are to be played and/or controlled. This simple (but important) fact means that the amount of encoded data is far less memory-intensive than its hard-disk audio or video-recording counterparts and that the data overhead that's required by MIDI is very small. In short, a computer-based sequencer can simultaneously operate in a digital audio, digital video, and processing environment...without placing an added load on a computer's CPU. As you might expect, many types of sequencers are currently on the market, each offering its own set of advantages and disadvantages. It's also true that each sequencer has its own basic operating feel, thus choosing the best tool and toy for the job or studio is totally up to you.

Recording

Commonly, a sequencer is used as a digital workspace for creating personal compositions in environments that range from the bedroom to more elaborate project studios. Whether they're hard- or software-based, most sequencers use a working interface that's designed to emulate the traditional multitrack recording environment. A tape-like transport lets you move from one location to the next using standard Play, Stop, Fast Forward, Rewind, and Record command buttons. Beyond using traditional record-enable button(s) to arm selected recording track(s), all you need to do is select the MIDI input (source) and outputs (destination) ports, instrument/voice MIDI channel, instrument patch, and other setup information... press the Record button, and start playing. Once you've finished laying down a track, you can jump back to the beginning of a sequence and listen to your original track while continuing to lay down additional tracks until the song begins to form.

Almost all sequencers are capable of punching in and out of record mode while playing a sequence. This common function lets you drop in and out of record on a track (or tracks) in real time. Although punch in/out points can often be manually performed on the fly, most sequencers can perform a punch automatically, once the in/out measure numbers have been graphically or numerically entered. The sequence can then be rolled back a few measures and the artist can play along, while the sequencer automatically performs the necessary switching functions (usually with multiple-take and full undo capabilities).

In addition to recording a performance in a track-based environment, most sequencers let you enter note values into sequence one note at a time. This feature (known as *step time*) lets you give the sequencer a basic tempo and note length (e.g., quarter note, sixteenth note) and then manually enter the notes from a keyboard or other controller. This data entry style is often (but not always) used with fast, high-tech and dance styles, where a real-time performance just isn't possible or accurate enough for the song.

Whether you're recording a track in real-time or in step-time, it's often best to select the proper song tempo before recording a sequence. I bring this up because most sequencers are able to output a click track that can be used as an accurate, audible guide for keeping in time

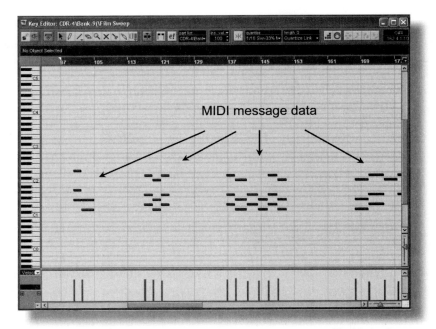

Figure 7.43. *The presence of Midi message data will often appear as a series of highlighted areas within a sequence track or a window.*

with the song's selected tempo... it's also critical that the tempo be accurate when trying to sync groove loops and rhythms to a sequence via VST or MIDI time code.

Editing

One of the more important features that a sequencer (or MIDI track within a DAW) has to offer is its ability to edit sequenced tracks or blocks within a track. Of course, these editing functions and capabilities often vary from one sequencer to another. The main track window of a sequencer or MIDI track on a DAW is used to display such track information as the existence of track data, track names, MIDI port assignments for each track, program change assignments, volume controller values, and other transport commands. Depending on the sequencer, the existence of MIDI data on a particular track at a particular measure point (or over a range of measures) is indicated by the highlighting of a track range in a way that's highly visible. For example, in Figure 7.43, you'll notice that a number of bars within the MIDI track contain highlighted areas. This means that these measures contain MIDI messages, while the other non-highlighted areas don't. By navigating around the various data display and parameter boxes, it's possible to use cut-and-paste and/or edit techniques to vary note values and parameters for almost every facet of a section or musical composition. For example, let's say that we really screwed up a few notes when laying down an otherwise killer bass riff. With MIDI, fixing the problem is totally a no-brainer—simply highlight each fudged note and drag it to it's proper note location... we can even change the beginning and end points in the process. In addition, tons of other parameters can be changed, including velocity, modulation and pitch bend, note and song transposition, quantization, and humanizing (factors that eliminate or introduce human timing errors that are generally present in a live performance); in addition to full control over program and continuous controller messages . . . and the list goes on.

Playback

Once a composition is complete, all of the MIDI tracks in a project can be transmitted through the various MIDI ports and channels to the instruments or devices for playback. Since the data exists as encoded real-time control commands, you can listen to the sequence and make changes at any time. For example, you could change instruments (by changing or editing patch voices), alter volume and other mix changes, or experiment with such controllers as pitch bend, modulation, or aftertouch . . . even change the tempo and key signature. In short, this medium is infinitely flexible with regard to how a performance and/or set of parameters can be created, saved, folded, spindled, and mutilated until you've arrived at the sound and feel that you want.

Another of the beauties of MIDI production is its ability to be altered at any later point in time. For example, let's say that 5 years ago you laid down a killer synth riff in a song that made it onto the charts. A couple of weeks ago a producer came to you in hopes of collaborating on a remix. Of course, technology marches on and your studio has improved over time. First off, even though a lot of the setup parameters have been saved with the original sequence, let's assume that you were smart enough to keep really good setup notes. One big change, however, is that you have a new software synth which has a patch that totally runs rings around the original patch. Since the remix is to be used in an upcoming film track . . . we'll tweak things up a bit by splitting the riff into two parts: one that contains the lower notes and another the highs. By sending the lows to one patch on the synth and the highs to another, not only have you improved the overall sound but you've also expanded the soundfield in a way that thickens it up and/or allows it to be panned into a surround mix environment. Without MIDI, you'd have to arrange for a new session and hope that all goes well; with MIDI, the performance is exactly the same and improvements can be made in a no-brainer environment . . . this is what MIDI's all about: performance, repeatability, easy editing, and cost-effective power!

Other Software Sequencing Applications

In addition to DAW and sequencing packages that are designed to handle most of the day-to-day production needs of the musician, other types of software tools and applications exist that can help to carry out specialized tasks. A few of these packages include drum pattern editors, algorithmic composition programs, patch editors, and music printing programs.

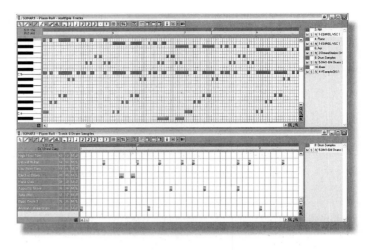

Figure 7.44. *Sonar 3 Studio Edition, showing the MIDI piano roll and drum grid window. (Courtesy of Twelve Tone Systems, Inc., www.cakewalk.com.)*

Drum-Pattern Editor/Sequencers

At any one time, a handful of companies have software or hardware devices that are specifically designed to create and edit drum patterns. In addition, most of the higher-end DAW audio production systems also include a drum pattern editor app that relies on user input and quantization to construct and chain together any number of user-created percussion grooves. More often than not, these editors use a grid pattern that displays drum-related MIDI notes or subpatterns along the vertical axis, while time is represented in metric divisions along the horizontal axis (Figure 7.44). By clicking on each grid point with a mouse or other input system, individual drum or effect sounds can be built into rhythmic patterns. Once created, these and other patterns can be linked together to create a partial or complete song. These editors commonly offer such features as the ability to change MIDI note values (thereby changing drum voices), note length, quantization, and humanization, as well as making adjustments to note and pattern velocities. Once completed, the sequenced drum track (or chained patterns) can be imported into a sequence, saved, and/or exported.

Algorithmic Composition Programs

Algorithmic composition programs (Figure 7.45) are interactive sequencers that directly interface with MIDI controllers or imported MIDI files to generate a performance in real time according to user-programmed computer parameters. In short, once you give it a few basic musical guidelines, it can begin to generate performances or musical parts on its own in order to help you gain new ideas for a song, create an automatic accompaniment, make improvisational exercises, create special performances, or just plain have fun. This type of sequencer can be programmed to control the performance according to musical key, generated notes, basic order, chords, tempo, velocity, note density, rhythms, accents, etc. Alternatively, an

Figure 7.45. Jammer Professional 5 algorithmic composition program. (Courtesy of Soundtrek, www.soundtrek.com.)

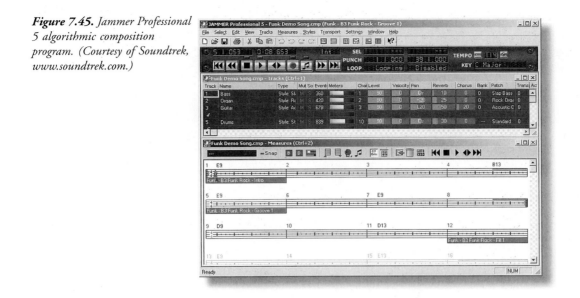

existing standard MIDI file can be imported and further manipulated in real time according to new parameters that can be varied from a computer keyboard, mouse, or controller. Often such interactive sequencers will accept input from multiple players, allowing it to be performed as a collective jam. Once a composition has been satisfactorily generated, a standard MIDI file can be created and exported into any sequencer.

Patch Editors

The vast majority of MIDI instruments and devices store their internal patch data within RAM. Synths, samplers, or other devices contain information on how the oscillators, amplifiers, filters, tuning, and other presets are to be configured in order to create a particular sound timbre or effect. In addition to controlling sound patch parameters, a unit's internal memory can also store such setup information as effects processor settings, keyboard splits, MIDI channel routing, controller assignments, etc. These settings, which can be accessed by from the device's preset buttons, alpha dial, or keypad entry, can be manually edited from the device's front panel. Another (and sometimes more straightforward) way to gain real-time control over the parameters of a specific instrument or wide range of MIDI devices is through the use of a *patch editor* (Figure 7.46). A patch editor is a software package that's used to provide on-screen controls and graphic windows for emulating and varying an instrument's parameter controls in real time. Direct communication between the software/computer system and a device's microprocessor commonly occurs through the use of MIDI sys-ex messages. Almost all popular voice and setup editing packages include provisions for receiving and transmitting bulk patch data in this way. This makes it possible to save and organize large numbers of patch data files, vary settings in real time, and print out patch parameter settings. In addition to software editing packages, there are also hardware solutions for gaining

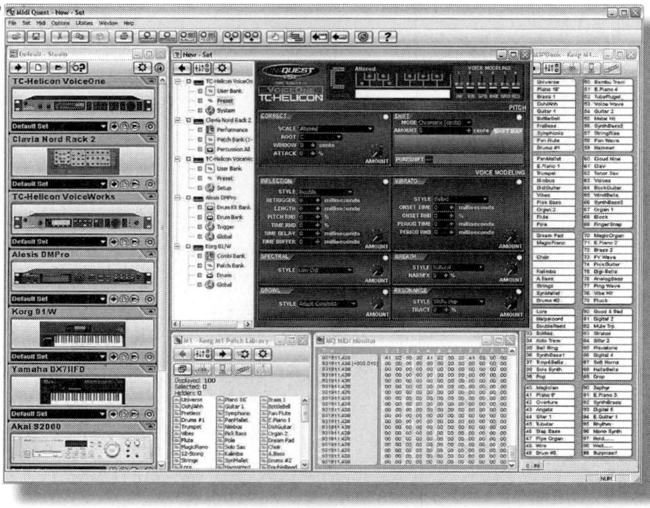

Figure 7.46. *Midi Quest software patch editor/librarian. (Courtesy of Sound Quest, Inc., www.squest.com.)*

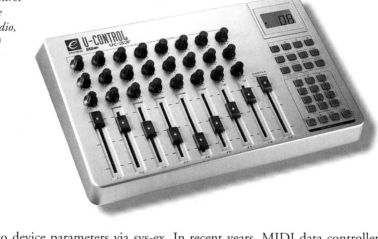

Figure 7.47. *Evolution U-control UC-33 USB MIDI hardware controller. (Courtesy of M-Audio, www.m-audio.com/evolution.)*

quick and easy access to device parameters via sys-ex. In recent years, MIDI data controllers (Figure 7.47) have sprung onto the market that can control a wide range of instruments using data faders, and "soft" buttons vary patch, system, and performance parameters in real time. In many situations, these controllers can also be used to the control volume and mix parameters of a DAW.

Music-Printing Programs

In recent years, the field of transcribing musical scores onto paper has been strongly affected by the computer, the DAW and MIDI technology. This process has been enhanced through the use of newer generations of software that make it possible for music notation data to be

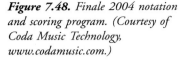

Figure 7.48. Finale 2004 notation and scoring program. (Courtesy of Coda Music Technology, www.codamusic.com.)

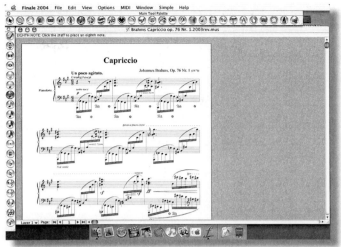

entered into a computer either manually (by placing the notes onto the screen via keyboard and/or by mouse movements) or via direct MIDI input. Once entered, these notes can be edited in an on-screen environment using a *music printing program* (or notation app within a DAW) that lets you change and configure a musical score or lead sheet using standard cut- and- paste edit techniques. In addition, most printing programs can play the various instruments in a MIDI system directly from the score. A final and important program feature is their ability to print out hard copies of a score or lead sheets in a wide number of print formats and styles.

These programs or DAW program apps (Figures 7.48 and 7.49) allow musical data to be entered into a computerized score in a number of manual and automated ways (often with varying degrees of complexity and ease). Although scores can be manually entered, most music-transcription programs will generally accept direct MIDI input, allowing a part to be played directly into a sequence. This can be done in real time (by playing a MIDI instrument or finished sequence into the program), in step time (entering the notes of a score one note at a time from a MIDI controller), or from an existing standard or program-specific MIDI file.

Another way to enter music into a score is through the use of an optical recognition program (Figure 7.50). These programs let you place sheet music or a printed score onto a standard flatbed scanner, scan the music into a program, and then save the notes and general layout as a Notation Interchange File Format (NIFF) file. One of the biggest drawbacks to automatically entering a score via MIDI (either as a real-time performance or from a MIDI file) is the fact that music notation is a very interpretive art. "To err is human," and it's commonly this human touch that gives music its full range of expression. It is very difficult, however, for a program to properly interpret these minute, yet important imperfections and place the notes into the score exactly as you want them. (For example, it might interpret a held quarter note as either a dotted quarter note or one that's tied to a 32nd note.)

Figure 7.49. Score application within Nuendo 2.0. (Courtesy of Steinberg Media Technologies GMBH, www.steinberg.net.)

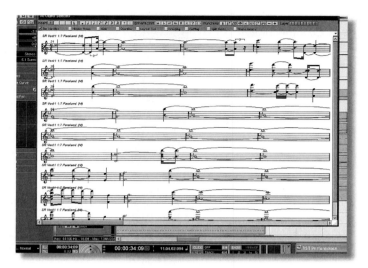

Figure 7.50. SmartScore Piano Edition 3 Music Scanning Program. (Courtesy of Music Imaging Technologies, www.musitek.com.)

Even though these computer algorithms are getting better at interpreting musical data and quantization can be used to tell a computer to round a note value to a specified length...a score will still often need to be manually edited to correct for misinterpretations.

Mixing in the MIDI Environment

With the advent of the digital audio workstation and its integration of sequencing into the production environment, MIDI-based mixing has made steady inroads with regard to

Figure 7.51. *System-wide mixing via MIDI channel messages.*

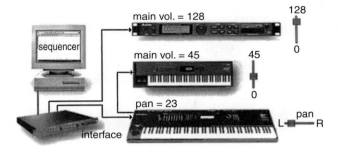

ease of use, flexibility, incorporation of signal processing, and direct control over parameters. These improvements are largely due to the integration of MIDI into the DAW's on-screen software mixer...and its ability to be remotely controlled from a hardware controller (Figure 7.51). For example, by adding a MIDI track into a DAW session, it's often a simple matter to assign that track to a hardware or virtual MIDI instrument, arm the track, and begin recording. Once done, the mixer and/or controller can be used to change volume, pan, and other parameter controls—using the DAW's graphic user interface (GUI) or hardware controller...in as simple a manner as controlling audio. There's no hidden cost and no additional hardware...just pure production power that's waiting to be put to use. This degree of automation is accomplished by transmitting MIDI channel messages directly to the receiving instruments or devices in such a way as to provide extensive, real-time control over such parameters as note volume, main (track) volume, pan position, timbre, effects, program changes, etc. This is commonly accomplished through the transmission of continuous controller messages that range in value from 0 (minimum) to 127 (maximum).

Console Automation via MIDI

Although MIDI isn't generally implemented into larger, professional analog consoles, steps have been taken to include basic mix functions into certain project-based analog mixers. This implementation can range from having simple MIDI control over mute functions to the creation of static snapshots for reconfiguring user-defined mixer settings...all the way to complete dynamic control over basic mix functions via MIDI. Since the audio, routing, and processing functions of a digital console are handled in the digital domain, it's often a simple matter for the console mixing surface and automation functions to transmit and respond to MIDI controller messages. Commonly, the digital console or mixer is used to transmit sys-ex messages that can be interpreted by a DAW for use as an external controller surface. If you check your DAW's manual for compatibility, you just might find that your mixer can actually be used as a bonus controller surface!

Effects Automation in MIDI Production

In the same way that electronic instruments can store sound patches into their memory-location registers for later recall, a number of hardware effects and multi-effects devices store patch-related data within RAM. Since many of these devices include a MIDI port, it's often possible to recall and/or automate effects in real time through the use of program change and continuous-controller messages. For example, a MIDI stomp box on stage could be used to transmit controller 63 out to an effects box, which will promptly switch it to the "Killer Verb-Box from Hell" patch setting. . . . Alternatively, a sequencer could send out the same message in a studio to call up the same effect at any point in the song. In many cases, a program change that relates to the desired effects patch can be inserted at the beginning of a sequenced track so that the proper effect will be automatically recalled once the sequence is opened. . . . Basically, it's an easy way to automatically recall all (or some) of your outboard effects settings.

In addition to all this, many effects boxes are capable of communicating with an external MIDI controller via sys-ex so as to easily fine-tune many of the parameter settings. Once edited, the patches could be saved to the device's patch RAM and as a sys-ex data dump for archival and recall purposes. Again, remember to save the factory settings as a sys-ex dump before making changes.

In closing, mixing in the MIDI environment isn't only fun, it's often a lesson in understanding the power and flexibility of automated mixing within a DAW or analog environment. In short, once you've assigned a MIDI channel and MIDI port for each instrument and/or effects device, you can then go about the business of assigning sounds, editing a sequence, mixing via MIDI, synchronizing the various media and devices in your system . . . and, most important of all, making music. As we all know, technology isn't always transparent or bug free. There's a lot to learn, but once you've begun to understand the basics of MIDI and electronic music production, your studio will be an even better playground. . . . Just hit the big red Record button and start playing!

CHAPTER 8

Multimedia and the Web

It's no secret that modern-day computers have gotten faster, sleeker, and sexier in their overall design. In addition to its ability to act as a multifunctional production workhorse, one of the crowning achievements of the modern computer is the degree of media and networking integration that has worked its way into our collective consciousness in a way that has come to be universally known by what is now known as the household buzzword . . . *multimedia*.

The combination of working and/or playing with multimedia has found its way into modern computer culture through the use of various hardware and software systems that work in a multitasking environment and combine to bring you a unified experience that seamlessly involves such media types as:

- ◆ Text
- ◆ Graphics
- ◆ Audio and music
- ◆ Computer animation
- ◆ Musical instrument digital interface (MIDI)
- ◆ Video

The obvious reason for integrating and creating these media types is the human desire to create content with the intention of sharing and communicating one's experiences with others. This has been done for centuries in the form of books and more recently by movies and television. In the here and now, the Web has been added to the communications list . . . in that it has created a vehicle that allows individuals (and corporate entities alike) to communicate a multimedia experience to millions and then allows each individual to manipulate that experience, learn from it, and even respond in an interactive fashion. The Web has indeed unlocked the potential for experiencing multimedia events and information in a way that makes each of us a participant . . . not just a passive spectator. To me, this is the true revolution occurring at the dawn of the twenty-first century!

The Multimedia Environment

When you get right down to it, multimedia is nothing more than a unified programming and operating system (OS) environment that allows multiple forms of program data and content media to simultaneously stream and be routed to the appropriate hardware ports for output, playback, and/or processing (Figure 8.1). The two most important concepts behind this environment are:

◆ Multitasking
◆ The device driver

Basically, *multitasking* can be thought of as a modern-day form of illusion. Just as a magic trick can be quickly pulled off with sleight of hand or a film that switches frames 24 times each second can create the illusion of continuous movement . . . the multimedia environment deceives us into thinking that all of the separate program and media types are working at the same time. In reality, the computer uses multitasking to quickly switch from one program to the next in a cyclic fashion. Similar to the film example, newer computer systems have gotten so lightning fast at cycling between programs and applications that they give the illusion that all are running at the same time.

Another central concept to multimedia is that of the *device driver*. Briefly, a driver acts as a device-specific software patch cord that routes media data from the source application to

Figure 8.1. *Example of a multimedia program environment.*

Figure 8.2. Basic interaction between a software app and a hardware device via the device driver.

software application   driver   hardware I/O

the appropriate hardware output device (also known as a port)...and/or from a port back to the application's input (Figure 8.2). Thus, whenever any form of media playback is requested, it will be recognized as being a particular data type that will be routed to the appropriate device driver and finally sent out to the selected output port (or ports).

From all of this, a multimedia computer can be seen as being a device that operates and switches between multiple programs and media players and then (on outputting or receiving any supported media type) routes the data to or from the appropriate hardware ports; all in a manner that's so fast as to be virtually seamless to our human senses. Pretty amazing, huh?

Delivery Media

Although media data can be stored and/or transmitted over a wide range of media storage devices, the most commonly found delivery media at this time of writing are the:

◆ CD
◆ DVD
◆ Web

The CD

One of the most important developments in the mass marketing and distribution of large amounts of media and entertainment information to the masses is the CD (Compact Disc)...both in the form of the CD-Audio and the CD-ROM. As most are aware, the CD-Audio disc is capable of storing up to 74 minutes of audio at a rate of 44.1 kHz/16 bits. Its close optical cousin, the CD-ROM, is capable of storing up to 700 Mb of graphics, video, digital audio, MIDI, text, and raw data. Consequently, these premanufactured and/or user-encoded media are able to store large amounts of music, text, video, graphics, etc., to such an interactive extent that this medium has become an important driving force among all communications media. Table 8.1 details the various CD standards that are currently in use.

Table 8.1. CD format standards.

Format	Description
Red Book	Audio-only standard; also called CD-A (Compact Disc Audio)
Yellow Book	Data-only format; used to write/read CD-ROM data
Green Book	CD-I (Compact Disc Interactive) format; never gained mass popularity
Orange Book	CD-R (Compact Disc Recordable) format
White Book	VCD (Video Compact Disc) format for encoding CD-A audio and MPEG-1 or MPEG-2 video data; used for home video and Karaoke
Blue Book	Enhanced Music CD format (also known as CD Extra or CD+) can contain both CD-A and data
ISO-9660	A data file format that's used for encoding and reading data from CDs of all types across platforms
Joliet	Extension of the ISO-9660 format that allows for up to 64 characters in its file name (as opposed to the 8 file + 3 extension characters allowed by MS-DOS)
Romeo	Extension of the ISO-9660 format that allows for up to 128 characters in the file name
Rock Ridge	Unix style extension of the ISO-9660 format that allows for long file names
CD-ROM/XA	Allows for extended usage for the CD-ROM format—Mode-1 is strictly Yellow Book, while Mode-2 Form-1 includes error correction and Mode-2 Form-2 doesn't allow for error correction; often used for audio and video data
CD-RFS	Incremental packet writing system from Sony that allows data to be written and rewritten to a CD or CD-RW (in a way that appears to the user much like the writing/retrieval of data from a hard drive)
CD-UDF	UDF (Universal Disc Format) is an open incremental packet writing system that allows data to be written and rewritten to a CD or CD-RW (in a way that appears to the user much like the writing/retrieval of data from a hard drive) according to the ISO-13346 standard
HDCD	The High-Definition Compatible Digital system adds 6 dB of gain to a Red Book CD (when played back on an HDCD-compatible player) through the use of a special compansion mastering technique
Macintosh HFS	An Apple file system that supports up to 31 characters in a file name; includes a data fork and a resource fork that identify which application should be used to open the file

The DVD

Similar to their illustrious optical cousins, the DVD (which, after a great deal of industry deliberation, actually stands for ... DVD) can contain any form of data. Unlike the CD, these discs are capable of storing a whopping 4.7 gigabytes (GB) within a single-sided disc and 9.4 GB on a double-layered disc. This capacity makes the DVD the perfect delivery medium for encoding video in the MPEG-2 encoding format, data-intensive games, DVD-Audio, and numerous DVD-ROM titles. The increased demand for multimedia games, educational products, etc., has spawned the computer-related industry of CD and DVD-ROM authoring. The term *authoring* refers to the creative, design, and programming aspects of putting together a CD/DVD project. At its most basic level, a project can be authored, mastered, and "burned" to disc from a single commercial authoring program. Whenever the stakes are higher, trained professionals and expensive systems are often called in to assemble, master, and produce the final disc for mass duplication and sales.

The Web

One of the most powerful aspects of multimedia is the ability to communicate experiences either to another individual or to the masses . . . for this, you need some kind of a network connection. The largest and most common network that can be found in the home, studio, office, classroom (you name it) . . . is a connection to the Internet. Here's the basic gist of how the Internet works:

◆ The Internet (Figure 8.3) can be thought of as a communications network that allows your computer (or connected network) to be connected to an Internet Service Provider (ISP) server (a specialized computer or cluster of ISP computers designed to handle, pass, and route data between large numbers of user connections).

◆ These ISPs are connected (through specialized high-speed connections) to an interconnected network of network access points (NAPs) . . . which essentially forms the connected infrastructure of the World Wide Web (WWW).

◆ In its most basic form the WWW can be thought of as a unified network of networks.

Internet browsers transmit and receive information on the Web via a Uniform Resource Locator (URL) address. This address is then broken down into three parts: the protocol (*e.g.*, http), the server name (*e.g.*, www.modrec.com), and the requested page or file name (*e.g.*, index.htm). The connected server is able to translate the server name into a specific Internet Provider (IP) address that's used to connect your computer with the desired server . . . after which the requests to receive or send data are communicated and the information is passed to your desktop.

E-mail works in a similar data transfer fashion, with the exception that an e-mail isn't sent to or requested from a specific server (which might be simultaneously connected to a large group of users); rather, it's communicated from one specific e-mail address (*e.g.*, myname@ myprovider.com) to a destination address (*e.g.*, yourname@yourprovider.com).

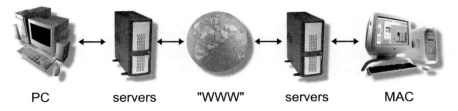

PC   servers   "WWW"   servers   MAC

Figure 8.3. *The Internet works by communicating requests and data from a user's PC to a single server that's connected to other servers around the world, which are likewise connected to other users' PCs.*

Delivery Formats

When creating content for the various media systems, it's extremely important that you match the media format and bandwidth requirements to the content delivery system that's being used. In other words... it's always smart to maximize the efficiency of the message (media format and required bandwidth) to match (and not alienate) your intended audience. The following section outlines many standard and/or popular formats for delivering media to a target audience.

Digital Audio

Digital audio is obviously a component that adds greatly to the multimedia experience. It can augment a presentation by adding a dramatic music soundtrack, help us to communicate through speech, or give realism to a soundtrack by adding sound effects. Because of the large amounts of data required to pass video, graphics, and audio from a CD-ROM, the Internet, or other media, the bit- and sample-rate structure of an "uncompressed" audio file is usually limited compared to that of a professional-quality sound file. The general accepted sound file standard for multimedia production is either 8-bit or 16-bit audio at a sample rate of 11.025 or 22.050 kHz. This standard has come about mostly because older single- and two-speed CD-ROMs generally couldn't pass the professional 44.1-kHz rate. With the introduction of faster processing systems and better hardware, these limitations have generally been lifted to include 44.1-kHz/16-bit audio and compressed data formats that offer CD quality and surround sound playback capabilities. In addition, there are obvious limitations to communicating uncompressed professional-rate sound files over the Internet... or from a CD or DVD disc that is also streaming full-motion video. Fortunately, with improvements in codec (encode/decode) techniques, hardware speed, and design, the overall sonic and production quality of compressed audio data has greatly improved in audience acceptance.

Uncompressed Sound File Formats

Although several formats exist for encoding and storing sound file data, only a few have been universally adopted by the industry. These standardized formats make it easier for files to be exchanged between compatible media devices. By far the most common file type is the Wave (or .wav) format. Developed for the Microsoft Windows format, this universal file type supports both mono and stereo files at a variety of uncompressed resolutions and sample rates. Wave files contain PCM coded audio (pulse code modulation formatted data) that follows the Resource Information File Format (RIFF) spec, which allows extra user information to be embedded and saved within the file itself. Another file format that's most commonly used with Mac computers is the Audio Interchange File (.aif) format (AIFF). As with Wave files, AIFF files support mono or stereo, 8-bit or 16-bit audio at a wide range of sample rates. Like Broadcast Wave files, AIFF files can also contain embedded text strings. Table 8.2 details the differences between uncompressed file sizes as they range from

Table 8.2. Audio bit rate and file sizes.

PCM Audio File Sizes

Sample Rate	Bit Rate	Channel No.	Data Rate (Bbps)	File Size	MB/min	MB/hour
192	32	2	1536	92,160	92.16	5529.6
192	32	1	768	46,080	46.08	2764.8
192	24	2	1152	69,120	69.12	4147.2
192	24	1	576	34,560	34.56	2073.6
96	32	2	768	46,080	46.08	2764.8
96	32	1	384	23,040	23.04	1382-4
96	24	2	576	34,560	34.56	2073.6
96	24	1	288	17,280	17.28	1036.8
48	32	2	384	23,040	23.04	1382.4
48	32	1	192	11,520	11.52	691.2
48	24	2	288	17,280	17.28	1036.8
48	24	1	144	8640	8.64	518.4
48	16	2	192	11,520	11.52	691.2
48	16	1	96	5760	5.76	345.6
44.1	32	2	352	21,120	21.12	1267.2
44.1	32	1	176	10,560	10.56	633.6
44.1	24	2	264	15,840	15.84	950.4
44.1	24	1	132	7920	7.92	475.2
44.1	16	2	176	10,560	10.56	633.6
44.1	16	1	88	5280	5.28	316.8
32	16	2	128	7680	7.68	460.8
32	16	1	64	3840	3.84	230.4
22	16	2	88	5280	5.28	316.8
22	16	1	44	2640	2.64	158.4
22	8	1	22	1320	1.32	79.2
11	16	2	44	2640	2.64	158.4
11	16	1	22	1320	1.32	79.2
11	8	1	11	660	0.66	39.6

professional 32-bit/192-kHz rates that are used to encode DVD-Audio sound all the way down to voice-quality 8-bit/11-kHz files.

Compressed Codec Sound File Formats

As was mentioned earlier, high-quality uncompressed sound files often present severe challenges to media delivery systems that are restricted in terms of bandwidth, download times, and/or memory storage. Although the streaming of audio data from disk or disc and through various high-bandwidth networks (including the Web) has improved over the years, memory storage space and other limitations have led to the popular acceptance of audio-related data formats (codecs) that can encode audio data in a manner that reduces data file size and bandwidth requirements . . . and then decode the information upon playback using a system known as *perceptual coding*.

Perceptual Coding

The central concept behind perceptual coding is based upon the psycho-acoustic principle that the human ear will not always be able to hear all of the information that's present in a recording. In Chapter 2, we found that louder noises will often mask sounds that are both lower in level and relatively close to another louder signal. These perceptual coding schemes take advantage of this masking effect by filtering out noises and sounds that can't be detected and removing them from the encoded audio stream.

The perceptual encoding process is said to be lossy or destructive . . . as once the filtered data has been eliminated, it can't be replaced or introduced back into the file. For reasons of audio quality, the level of just how much audio is to be removed from the data can be varied within an audio codec. Higher bandwidth compression rates will remove less data from a stream (resulting in a reduced amount of filtering and higher audio quality), while low bandwidth rates will greatly reduce the data stream (resulting in smaller file sizes, increased filtering, increased artifacts, and lower audio quality). The amount of filtering that is to be applied to a file will depend upon the intended audio quality and the delivery medium's bandwidth limitations. Due to the lossy character of these encoded files, it's always a good idea to keep a copy of the original, uncompressed sound file as a data archive backup . . . should changes in content or future technologies occur (never underestimate Murphy's Law).

The most commonly used perceptual coding schemes that are in use today are:

◆ MP3
◆ MWA
◆ AAC
◆ Ogg Vorbis
◆ Real Audio

Many of the listed codecs are capable of encoding and decoding audio using a constant bit rate (CBR) and variable bit rate (VBR) structure:

◆ CBR encoding is designed to work effectively in a streaming scenario where the end user's bandwidth is a consideration. With CBR encoding, the chosen bit rate will remain constant over the course of the file or stream.

◆ VBR encoding is designed for use when you want to create a downloadable file that has a smaller file size and bit rate without sacrificing sound and video quality. This is carried out by detecting which sections will need the highest bandwidth and adjusting the encode process accordingly. When lower rates will suffice, the encoder will adjust the process to match the content. Under optimum conditions, you might end up with a VBR-encoded file that has the same quality as a CBR-encoded file, but with only half the file size.

MP3

MPEG (which is pronounced "M-peg" and stands for the Moving Picture Experts Group [www.mpeg.org]) is a standardized format for encoding digital audio and video into a compressed format for storage to various media and for transmission over the Web. As of this writing, the most popular format is the ISO-MPEG Audio Level-2 Layer-3, commonly referred to as MP3 (Figure 8.4). Developed by the Fraunhofer Institute (www.iis.fhg.de) and Thomson Multimedia in Europe, MP3 has advanced the public awareness and acceptance of compressing and distributing digital audio by creating a codec that can compress audio by factors of 10:1 or greater, while still maintaining quality levels that approach those of a CD (depending on which compression levels are used). In fact, compression ratios of up to 24 times can be attained without seriously degrading the sound quality, and even higher levels of compression can be used for nonmusical tracks (such as voice).

Although faster Web connections are capable of streaming MP3 in real time, this format is most often downloaded to the end consumer for storage to disk, disc, and/or memory media for the storage and playback of downloaded songs. Once saved, the data can then be transferred to solid-state playback devices (such as portable MP3 players, PDAs, cell phone players . . . you name it!). In fact, over one billion music tracks are currently downloaded every month on the Internet using MP3, practically every personal computer contains licensed MP3 software, virtually every song has been MP3 encoded, and some 150 million MP3 players are soon expected to reach the global market, currently making it the Web's most popular audio compression format by far.

In 2001, MP3 Pro was introduced to the public as an encoding system for enhancing sound quality and improving the compression scheme. MP3 Pro works by splitting the encoding process into two parts. The first analyzes the low-frequency band information and encodes it into a normal MP3 stream (which allows for complete compatibility with existing MP3 players). The second analyzes the high-frequency band information and encodes it in a way that helps preserve the high-frequency content. When combined, the MP3 Pro codec creates a file that's more compact than original MP3 files, with equal or better sound quality and complete backward and forward compatibility. Although MP3 Pro professes to offer 128-kbps

Figure 8.4. A large number of soft-, share-, and freeware players are available to play MP3 files from the hard disk or any number of portable and/or desktop memory sources. (Courtesy of Nullsoft, Inc., www.winamp.com.)

performance at a 64-kbps encoding rate (effectively doubling the digital music capacity of flash memory and compact discs), the general download community has been slow to adopt this new codec.

WMA

Developed by Microsoft as their corporate response to MP3, Windows Media Audio (WMA) allows for compression rates that can encode high-quality audio at low bit-rate and file-size settings. Designed for "ripping" (encoding audio from audio CDs) and sound file encoding/playback from within the popular Window's Media Player (Figure 8.5), this format has grown in its general acceptance and popularity. In addition to its high quality at low bit rates, WMA also has the advantage of allowing for real-time streaming over the internet (as witnessed by the large amount of radio and Internet stations that stream to the Windows Media Player at various bit-rate qualities). Finally, content providers often favor MWA over MP3 as it is able to provide for a degree of content copy protection through its incorpora-tion of DRM (Digital Rights Management) coding. With the introduction of the Windows Media Player Version 9, WMA is also capable of encoding and delivering audio in discrete surround sound. Using various high-end workstations and encoders...it is now possible to deliver surround audio to an ever-increasing number of homes whose computers are equipped with surround-sound playback (Figure 8.6).

Ogg Vorbis

Developed in 1993 under the name of "Squish," the Ogg Vorbis codec has begun to grow in its general use. Originally designed as a substitute for MP3 and WMA, the strength of this format is that it circumvents any required royalty fees that must be paid by equipment

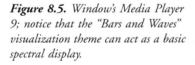

Figure 8.5. *Window's Media Player 9; notice that the "Bars and Waves" visualization theme can act as a basic spectral display.*

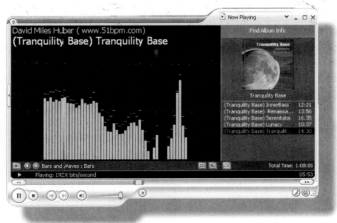

Figure 8.6. Windows XP speaker setup box for stereo, surround, and other playback schemes.

manufacturers for the use of the media patents. In addition to offering a high-quality alternative to the major players, Ogg Vorbis is capable of delivering audio in a number of channel formats at both constant and variable bit-rates.

AAC

Jointly developed by Dolby Labs, Sony, ATT, and the Fraunhofer Institute, the Advanced Audio Coding (AAC) scheme is touted as a "multichannel-friendly format for secure digital music distribution over the Internet." Stated as having the ability to encode CD-quality audio at lower bit rates than other coding formats, AAC is not only capable of encoding 1, 2, and 5.1 surround-sound files but can also encode up to 48 channels within a single bitstream at bit/sample rates of up to 24/96. This format is also SDMI compliant (see Secure Digital Music Initiative section, below), allowing copyrighted material to be protected against unauthorized copy and distribution.

RealAudio

With the introduction of their RealPlayer server application, RealNetworks (www.realnetworks.com) became one of the first companies to provide real-time streaming over the Web. From a technical point of view, RealAudio data is transmitted using any of more than 12 proprietary encoding levels that range from transmission rates of 8 kbps (low-fidelity mono voice quality over a 56-k modem) to speeds that range above the 1.5 Mbps point. Although there are several compression levels to choose from, the most common "music mode" type compresses data in a way that doesn't introduce extreme artifacts over a wide dynamic range, thereby creating an algorithm that can faithfully reproduce music with near-FM quality over 56-k or faster lines. At the originating Internet site, the RealAudio server can automatically recognize which modem, cable, or network connection speed is currently in use and then transmit the data in the best possible audio format. This reduced data throughput ultimately means that the RealPlayer will take up very little of your computer's resources, allowing you to keep on working while audio is being played.

Tagged MetaData

Within certain types of multimedia file formats it's possible to imbed a wide range of data into the file header that can identify and provide extensive information that relates to the content of the file. For example, within the Windows OS file structure, right-clicking on a file name within the Windows Explorer menu and selecting Properties/Summary (or Control-clicking/ Get Info on a Mac) ... will bring up a set of extensive metadata file tags that can be used to enter and identify such media information as artist, album title, year, track number, genre, lyrics (yes, the entire set), title, comments, etc. (Figure 8.7). It should be pointed out that this meta tag info can also be entered within many digital audio workstations and editors ... and will often transfer from one format type to another upon conversion.

MIDI

One of the unique advantages of MIDI, as it applies to multimedia, is the rich diversity of musical instruments and program styles that can be played back in real time while requiring almost no overhead processing from the computer's CPU. This makes MIDI a perfect candidate for playing back soundtracks from multimedia games or over the Internet. It's interesting to note that MIDI has taken a back seat to digital audio as a serious music playback format for multimedia. Most likely, this is due to several factors, including:

◆ A basic misunderstanding of the medium
◆ The fact that producing MIDI content requires a basic knowledge of music

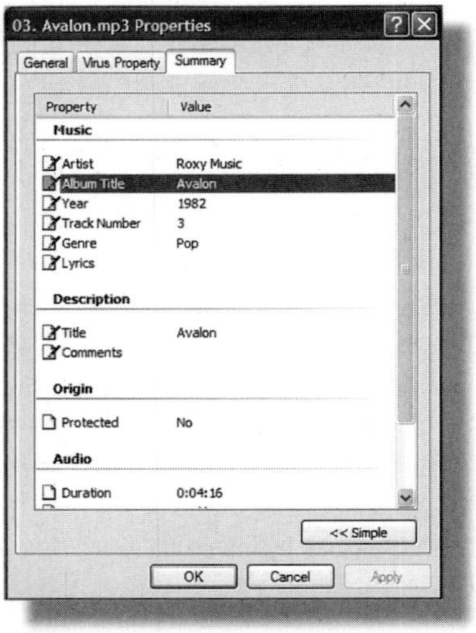

Figure 8.7. Embedded metatag file tags can be added to a media file within Windows Explorer.

◆ The frequent difficulty of synchronizing digital audio to MIDI in a multimedia environment

◆ The fact that soundcards often include poorly designed FM synthesizers (although most operating systems now include a higher quality software synth).

Fortunately, companies such as Microsoft have taken up the banner of embedding MIDI within their media projects and have helped push MIDI a bit more into the Web mainstream. As a result, it's becoming more common for your PC to begin playing back a MIDI score, on its own or perhaps in conjunction with a more data-intensive program or game.

Standard MIDI Files

The accepted format for transmitting files or real-time MIDI information in multimedia (or between sequencers from different manufacturers) is the standard MIDI file. This file type (which is stored with a .mid or .smf extension) is used to distribute MIDI data, song, track, time signature, and tempo information to the general masses. Standard MIDI files can support both single and multichannel sequence data and can be loaded into, edited, and then directly saved from almost any sequencer package. When exporting a standard MIDI file, keep in mind that they come in two basic flavors: type 0 and type 1:

◆ Type 0 is used whenever all of the tracks in a sequence need to be compressed into a single MIDI track. All of the original channel messages still reside within that track; however, the data will have no definitive track assignments. This type might be the best choice when creating a MIDI sequence for the Internet (where the sequencer or MIDI player application might not know or care about dealing with multiple tracks).

◆ Type 1, on the other hand, will retain its original track information structure and can be imported into another sequencer type with its basic track information and assignments left intact.

General MIDI

One of the most interesting aspects of MIDI production is the absolute setup and patch uniqueness of each professional and even semipro project studio. In fact, no two studios will be alike (unless they've been specifically designed to be the same or there's some unlikely coincidence). Each artist will be unique in having his or her own favorite equipment, supporting hardware, favorite way of routing channels and tracks, and assigning patches. The fact that each system setup is unique and personal has placed MIDI at odds with the need for system's compatibility in the world of multimedia. For example, after importing a standard MIDI file over the Net and loading it into a sequencer, you might hear a song that's being played with a totally irrelevant set of sound patches (it might sound interesting, but it won't sound anything like it was originally intended). If the MIDI file is loaded into a new computer, the sequence will again sound completely different, with patches that are so irrelevant that the guitar track might sound like a bunch of machine-gun shots from the planet Glob.

In order to eliminate (or at best reduce) the basic differences that exist between systems, a patch and settings standard known as General MIDI (GM) was created. In short, General MIDI assigns a specific instrument patch to each of the 128 available program change numbers. Since all electronic instruments that conform to the GM format must use these patch assignments, placing GM program change commands at the header of each track will automatically configure the sequence to play with its originally intended sound. As such, no matter what sequencer is used to play the file back, as long as the receiving instrument conforms to the GM spec the sequence will be heard using its intended instrumentation. Tables 8.3 and 8.4 detail the program numbers and patch names that conform to the

Table 8.3. GM nonpercussion instrument patch map.

1. Acoustic Grand Piano	33. Acoustic Bass	65. Soprano Sax	97. FX 1 (rain)
2. Bright Acoustic Piano	34. Electric Bass (finger)	66. Alto Sax	98. FX 2 (soundtrack)
3. Electric Grand Piano	35. Electric Bass (pick)	67. Tenor Sax	99. FX 3 (crystal)
4. Honky-tonk Piano	36. Fretless Bass	68. Baritone Sax	100. FX 4 (atmosphere)
5. Electric Piano 1	37. Slap Bass 1	69. Oboe	101. FX 5 (brightness)
6. Electric Piano 2	38. Slap Bass 2	70. English Horn	102. FX 6 (goblins)
7. Harpsichord	39. Synth Bass 1	71. Bassoon	103. FX 7 (echoes)
8. Clavi	40. Synth Bass 2	72. Clarinet	104. FX 8 (sci-fi)
9. Celesta	41. Violin	73. Piccolo	105. Sitar
10. Glockenspiel	42. Viola	74. Flute	106. Banjo
11. Music Box	43. Cello	75. Recorder	107. Shamisen
12. Vibraphone	44. Contrabass	76. Pan Flute	108. Koto
13. Marimba	45. Tremolo Strings	77. Blown Bottle	109. Kalimba
14. Xylophone	46. Pizzicato Strings	78. Shakuhachi	110. Bag pipe
15. Tubular Bells	47. Orchestral Harp	79. Whistle	111. Fiddle
16. Dulcimer	48. Timpani	80. Ocarina	112. Shanai
17. Drawbar Organ	49. String Ensemble 1	81. Lead 1 (square)	113. Tinkle Bell
18. Percussive Organ	50. String Ensemble 2	82. Lead 2 (sawtooth)	114. Agogo
19. Rock Organ	51. SynthStrings 1	83. Lead 3 (calliope)	115. Steel Drums
20. Church Organ	52. SynthStrings 2	84. Lead 4 (chiff)	116. Woodblock
21. Reed Organ	53. Choir Aahs	85. Lead 5 (charang)	117. Taiko Drum
22. Accordion	54. Voice Oohs	86. Lead 6 (voice)	118. Melodic Tom
23. Harmonica	55. Synth Voice	87. Lead 7 (fifths)	119. Synth Drum
24. Tango Accordion	56. Orchestra Hit	88. Lead 8 (bass + lead)	120. Reverse Cymbal
25. Acoustic Guitar (nylon)	57. Trumpet	89. Pad 1 (new age)	121. Guitar Fret Noise
26. Acoustic Guitar (steel)	58. Trombone	90. Pad 2 (warm)	122. Breath Noise
27. Electric Guitar (jazz)	59. Tuba	91. Pad 3 (polysynth)	123. Seashore
28. Electric Guitar (clean)	60. Muted Trumpet	92. Pad 4 (choir)	124. Bird Tweet
29. Electric Guitar (muted)	61. French Horn	93. Pad 5 (bowed)	125. Telephone Ring
30. Overdriven Guitar	62. Brass Section	94. Pad 6 (metallic)	126. Helicopter
31. Distortion Guitar	63. SynthBrass 1	95. Pad 7 (halo)	127. Applause
32. Guitar harmonics	64. SynthBrass 2	96. Pad 8 (sweep)	128. Gunshot

Table 8.4. GM percussion instrument patch map (Channel 10).

35. Acoustic Bass Drum	51. Ride Cymbal 1	67. High Agogo
36. Bass Drum 1	52. Chinese Cymbal	68. Low Agogo
37. Side Stick	53. Ride Bell	69. Cabasa
38. Acoustic Snare	54. Tambourine	70. Maracas
39. Hand Clap	55. Splash Cymbal	71. Short Whistle
40. Electric Snare	56. Cowbell	72. Long Whistle
41. Low Floor Tom	57. Crash Cymbal 2	73. Short Guiro
42. Closed Hi-Hat	58. Vibraslap	74. Long Guiro
43. High Floor Tom	59. Ride Cymbal 2	75. Claves
44. Pedal Hi-Hat	60. Hi Bongo	76. Hi Wood Block
45. Low Tom	61. Low Bongo	77. Low Wood Block
46. Open Hi-Hat	62. Mute Hi Conga	78. Mute Cuica
47. Low-Mid Tom	63. Open Hi Conga	79. Open Cuica
48. Hi Mid Tom	64. Low Conga	80. Mute Triangle
49. Crash Cymbal 1	65. High Timbale	81. Open Triangle
50. High Tom	66. Low Timbale	

Note: In contrast to Table 8.3, the numbers in Table 8.4 represent the percussion keynote numbers on a MIDI keyboard, not program change numbers.

GM format (Table 8.3 for nonpercussion and Table 8.4 for percussion instruments). These patches include sounds that imitate synthesizers, ethnic instruments, and/or sound effects that have been derived from early Roland synth patch maps. Although the GM spec states that a synth must respond to all 16 MIDI channels, the first nine channels are reserved for instruments, while GM restricts the percussion track to MIDI channel 10.

Graphics

Graphic imaging occurs on the computer screen in the form of pixels. These are basically tiny dots that blend together to create color images in much the same way that dots are combined to give color and form to your favorite comic strip. In much the same way that word length affects the overall amplitude range of a digital audio signal, the number of bits in a pixel's word will affect the range of colors that can be displayed in a graphic image. For example, a 4-bit word only has 16 possible combinations. Thus, a 4-bit word will allow your screen to have a total of 16 possible colors; an 8-bit word will yield 256 colors; a 16-bit word will give you 65,536 colors; and a 24-bit word will yield a whopping total of 16.7 million colors! (See Figure 8.8.)

The methods of displaying graphics onto a screen can be broken down into several categories:

◆ *Raster graphics*—In raster graphics, each image is displayed as a series of pixels. This image type is what is used when a single graphic image is used (*i.e.*, bitmap, JPEG, GIF, or TIFF format). The sense of motion can only come from raster images by successively

Figure 8.8. *The Windows Display Properties Box is used to change screen sizing and pixel resolution.*

stepping through a number of changing images every second (which is how a standard video image creates the sense of motion).

◆ *Vector graphics*—This process often creates a sense of motion by projecting a background raster image and then overlaying one or more objects that can be animated according to a series of programmable vectors. By instructing each object to move from point A to point B then point C, according to a defined script, a sense of animated motion can be created without the need to project separate images for each frame. This script form reduces a file's data size dramatically and is used with several image animation programs (including MacroMedia's Flash, Shockwave, and Director).

◆ *Wireframe animation*—This form of animation uses a computer to create a complex series of wireframe image vectors of a real or imaginary object. Once programmed, these stick-like objects can be filled in with any type of skin, color, shading, etc., and then programmed to move with a staggering degree of realism. Obviously, with the increased power of modern computers and supercomputers, this graphic art form has attained higher degrees of artistry and/or realism within modern-day film, video, and desktop visual production.

Desktop Video

With the proliferation of digital VCRs, video interface hardware, and video editing software systems... *desktop and laptop video* has begun to play an increasingly important role in multimedia production and content (Figure 8.9). Video is encoded into a datastream as a continuous series of successive frames, which are refreshed at rates that vary from 12 or fewer frames/second (fr/sec) to the standard video rate of 30 fr/sec. As with graphic files, a single full-sized video frame can be made up of a gazillion pixels... which are themselves encoded as a digital word of *n* bits. Multiply these figures by nearly 30 frames and you'll come up with a rather impressive data file size.

Figure 8.9. *Playback of a digital video clip using the Windows Media Player.*

Obviously, it's more common to find such file sizes and data throughput rates on higher-end desktop systems and professional video editing workstations; however, several options are available to help bring video down to data rates that are suitable for multimedia and even the Internet:

◆ *Window size*—The basics of making the viewable picture smaller is simple enough . . . reducing the frame size will reduce the number of pixels in a video frame, thereby reducing the overall data requirements during playback.

◆ *Frame rate*—Although standard video frame rates run at around 30 fr/sec (United States and Japan) and 25 fr/sec (Europe), these rates can be lowered to 12 fr/sec in order to reduce the encode file size and/or throughput.

◆ *Compression*—In a manner similar to that for audio, codecs can be applied to a video frame to reduce the amount of data necessary to encode the file by filtering out and smoothing over pixel areas that consume data. In situations where high levels of compression are needed, it's common to accept degradations in the video's resolution in order to reduce the file size and/or data throughput to levels that are acceptable to a restrictive medium (*e.g.*, the Web).

From all of this, you can see that there can be many options for encoding a desktop video file. When dealing with video clips, tutorials, and the like, it's common for the viewing window to be medium in size and encoded at a medium to lower frame rate. This middle ground is often chosen to accommodate the standard data throughput that can be streamed off of most CD-ROMs. These files are commonly encoded in Microsoft's Audio-Video Interleave (AVI) format, which uses little to no compression; QuickTime, a common codec that was developed by Apple and can be played by either a Mac or PC; or MPEG I or II, codecs that vary in level from those for use with multimedia to higher resolutions that are used to encode DVD movies. Both the Microsoft and Macintosh OS platforms include built-in or easily obtained applications that allow all or most of these file types to be played without additional hardware or software.

Multimedia and the Web in the "Need for Speed" Era

The household phrase "surfin' the Web" has become synonymous with jumping onto the Net, browsing the sites, and grabbin' onto all of those hot songs, videos, and graphics that might wash your way. Dude, with improved audio and video codecs and faster data connections (Table 8.5), the ability to search on any subject, download files, and stream audio or radio stations (Table 8.6) from any point in the world has definitely changed our perception of modern-day communications.

Table 8.5. Internet connection speeds.

Connection	Speed (bps)	Description
56k dial-up	56 kbps (usually less)	Common modem connection
ISDN	128 kbps; older technology	
DSL	384 kbps or higher high-bandwidth phone line technology	
Cable	384 kbps and higher high-bandwidth cable technology	
T1	1.5 Mbps	
T3	45 Mbps	
OC-1	52 Mbps	Optical fiber
OC-3	155 Mbps	Optical fiber
OC-12	622 Mbps	Optical fiber
OC-48	2.5 Gbps	Optical fiber
Ethernet	10 Mbps	Local area network (LAN), not an Internet connection
Fast Ethernet	100 Mbps	Local area network (LAN), not an Internet connection

Table 8.6. Streaming data file sizes.

Data Rate (Kbps)	File Size (kB/min)	File Size (Mb/hr)	Minutes on a 650-Mb CD	Hours on a 650-Mb CD
20	150	9	4333	72
64	480	28.8	1354	23
96	720	43.2	903	15
128	960	57.6	677	11
160	1200	72	542	9
256	1920	115.2	339	6
384	2880	172.8	226	4

Thoughts on Being and Getting Heard in Cyberspace

Most of us grew up in the age of the supermarket . . . where everything is wholesaled, processed, packaged, and distributed to a single clearinghouse that's gotten so big that older folks can only shop with the aid of a motorized shopping cart. For more than six decades, the music industry has largely worked on a similar principle: Find artists who'll fit into an existing marketing formula (or, more rarely, create a new marketing image), produce and package them according to that formula, and put tons of bucks behind them to get them heard and distributed. Not a bad thing in and of itself; however, for independent artists the struggle has been and continues to be one of getting themselves heard, seen, and noticed—without the aid of the well-oiled megamachine. With the creation of cyberspace, not only are established record industry forces able to work their way onto your desktop screen (and into your multimedia speakers), but independent artist also have a new medium for getting heard. Through the creation of a dedicated Web site, search engines, links from other sites, and independent music dotcoms, as well as through creative gigging and marketing . . . new avenues have begun to open up to the Web-savvy independent artist.

Uploading to Stardom!

If you build it, they will come! . . . This early-1900s concept definitely doesn't apply to the Web. With an ever-increasing number of dot-whatevers going online every month, the idea of expecting people to come to your site "just because it's there" simply isn't realistic. Like anything that's worthwhile, it takes connections, persistence, a good product, and good ol'-fashioned dumb luck to be seen as well as heard! If you're selling your music, T-shirts, or whatever at gigs, on the streets, and to family and friends . . . cyberspace can help increase sales by making it possible (and even easy) to get your band, music, or clients onto several independent music Web sites that offer up descriptions, downloadable samples, direct sales, and a link that goes directly to your main Web site. Such a site could definitely help to get the word out to a potentially new public . . . and help clue your audience in to what you and your music are all about.

Cyberproducts can be sold and shipped via the traditional mail or phone-in order channels; however, it's long been considered hip in the Web world to flash the silver, gold, or platinum credit card to make your purchase. Due to the fact that attaining your own credit card processing and authorization system can be costly, a number of cyber companies have sprung up that offer secure credit card authorization, billing, and artist payment plans for an overall sales percentage fee (Figure 8.10).

The preview and/or distribution format choice for releasing all or part of your music to the listening audience will ultimately depend on you and the format/layout style that's been adopted by the hosting site. For example, you could do any or all of the following:

◆ Place short, low-fidelity segments onto a site that can be streamed and/or downloaded to entice the listener to buy.

Figure 8.10. *CD Baby independent music site. (Courtesy of CD Baby, www.cdbaby.com.)*

- ◆ Provide free access to the entire project (at low or medium fidelity) while encouraging the listener to buy the CD.
- ◆ Place several high-fidelity cuts on your site for free as a teaser or as a gift to your fan base.
- ◆ Place the music on a secure site that's SDMI compliant, using the pay-per-download system.
- ◆ Sell the completed CD package on the site.
- ◆ Create a fanzine to keep your fans up to date on goings-on, upcoming releases, diaries, etc.

No matter what or how many cyber distribution methods you choose to get your music out, always take the time to read through the contractual fine print. Although most are above board and offer to get your music out on a nonexclusive basis . . . *caveat emptor* (let the buyer [or seller] beware)! In your excitement to get your stuff out there, you might not realize that you are signing away the rights for free distribution of that particular project (or worse). This hints at the fact that you're dealing with the music "business" . . . and, as with any business, you should always tread carefully in the cyber jungle.

Copyright Protection: Wanna Get Paid?

I hope that by the time you read this edition that many of the problems that currently exist with "ripping" off an artist from CD to a compressed file format and then posting that music onto the Web for everyone to download for free will be carefully considered and dealt with in a way that's fair to all those involved. With the shutdown of illegal music download sites and peer-to-peer networks . . . and the risen-from-the ashes version of newer pay-per-download sites (such as www.iTunes.com and the reborn www.napster.com), many of the major labels and larger independent artists are beginning to see a light at the end of the online tunnel. Even so, the problems with sharing both major label and independent music online are still fraught with problems that can lead to lost revenues.

Secure Digital Music Initiative

With the vast number of software (and hardware) systems that are able to "rip" CDs to MP3 (or any other format), MP3s back to audio and CD-R...the powers that be in the recording industry have grown increasingly fearful of the rising prevalence of copyright infringement. Although many online music sites legally use these formats to allow potential buyers to preview music before buying or to simply put unreleased cuts onto the Web as a freebie...a number of sites still exist that can connect online users with databases of music that have been illegally ripped off of CDs without paying royalties to the record company or independent artist.

As a result, the Recording Industry Association of America (RIAA), major record labels, and industry organizations have helped to form the Secure Digital Music Initiative (SDMI, www.sdmi.org). SDMI is an independent forum that brings together the worldwide recording, consumer electronics, and information technology industries to develop open specifications for protecting digital music distribution. As a result of these efforts, the Digital Rights Management (DRM) system was developed as a secure and encrypted means of solving the issue of unauthorized copying by digitally locking the content and limiting its distribution to only those who pay for the content. In short, DRM acts as a digital "watermark" that identifies the copyright owner and provides for an electronic "key" that allows access to the music or information once the original copy has been legally purchased.

One such DRM-compliant online and digital distribution system is Weed (www.weedshare.com). This system, which was developed primarily as a distribution tool for the independent artist, allows the end-user to listen to a song three times for free. After three listens, the built-in DRM will alert the user that it's time to pay up (an amount that's set by the artist or copyright owner). Once the payment is made into a PayPal account, the artist receives 50% of the sale. If that end-user shares the Weed file with a friend who also buys the song...the artist gets 50% of that sale and the original end-user gets a percentage of the sale...and so on. In the end, the artist, Weed, and those down the musical food chain get paid (Figure 8.11).

Figure 8.11. The Weed distribution/payment model. (Courtesy of Shared Media Licensing, Inc., www.weedshare.com.)

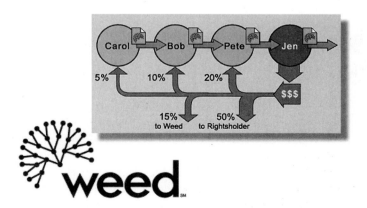

Figure 8.12. *Web site for*
Capital Radio, London.
(Courtesy of Capital Radio plc,
www.capitalfm.co.uk.)

Internet Radio

Due to the increased bandwidth of many Internet connections and improvements in
audio streaming technology, many of the world's radio stations have begun to broadcast on the
Web. In addition to offering a worldwide platform for traditional radio listening audiences,
a large number of corporate and independent "Web" radio stations have begun to spring
up which can help to increase the fan and listener base of musicians and record labels
(Figure 8.12). Go ahead, get on the Web and listen to your favorite St-st-st-station *en vivo en*
Mexico City . . . the latest dance craze from London . . . or Jamaican reggae streaming in from
the islands.

The Virtual "E-Dub"

In addition to providing another vehicle for getting an artist's music out to the public at large,
the Internet is making it easier for artists to e-collaborate over cyberspace. One approach
is to "E-dub" across the Web. Using this approach, you can create a rough session mix of
all or a portion of a song and export it to a medium-resolution MP3 (or preferred codec).
This file (along with sheet music jpgs, descriptive docs, etc.) could be sent to a collaborative
buddy across the street or across the world . . . who will then load the file into his or her
DAW. A track or set of tracks can then be overdubbed to the original mix in the traditional

fashion. The resulting file (or files) should then be encoded without compression or at a high resolution (at least 192 kbps) and then emailed back to the artist or producer for re-import back into the original session at the proper time and position. (If your DAW doesn't automatically convert the file back into the session's native file format, you'll need to manually convert the file yourself.) Using this system, the world could be your cost-effective E-dub oyster.

On a Final Note

One of the most amazing things about multimedia, cyberspace, and their related technologies is the fact that they're ever changing. By the time you read this book, many new developments will have occurred. Old concepts fade away; new and possibly better ones will take over and then begin to take on a new life of their own. Although I've always had a fascination with crystal balls and have often had a decent sense about new trends in technology... there's simply no way to foretell the many amazing things that lie before us in the fields of music, music technology, multimedia... and especially cyberspace. As with everything techno, I encourage you to read the trades, check out the Webzines, and surf the Web to keep abreast of the latest and greatest tools that have recently arrived or are about to rise on the horizon.

CHAPTER

Synchronization

With the dawning of DVD surround-sound movie production, high-definition television production, 3D computer games, and blazing Internet/network connections, the field of audio no longer has to take the back seat . . . as it now plays an increasingly important role in the entertainment and multiple media markets. As a result, it's become commonplace for the technologies of audio, video, film, multimedia, and the electronic music arts to coexist within a production. Within video postproduction, for example, audio and video transports, digital audio workstations, automated console systems, and electronic musical instruments routinely work together in tandem to help one or more operators to create and refine a video soundtrack into its final form (Figure 9.1). The technology that allows multiple audio and visual media to operate in tandem so as to maintain a direct time relationship is known as *synchronization*, or *sync*.

Strictly speaking, synchronization occurs when two or more events happen at precisely the same time. With respect to analog audio and video systems, sync is achieved by interlocking the transport speeds of two or more machines. For computer-related systems (such as digital audio, MIDI, and digital video), synchronization between devices is often achieved through the use of a timing clock that can be fed through a separate line or can be directly embedded within the digital data line itself. Frequently, it's necessary for analog and digital devices to be synchronized together . . . as a result, a number of ingenious forms of systems communication and data translation have been developed. In this chapter, the various forms of synchronization used for both analog and digital devices are discussed, as well as current methods for maintaining sync between media types.

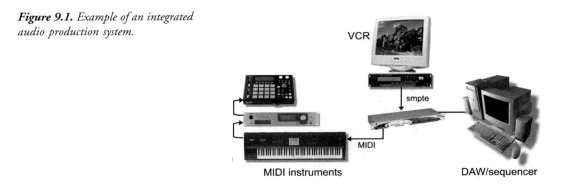

Figure 9.1. *Example of an integrated audio production system.*

Synchronization between Transports

Maintaining relative sync between analog tape transports doesn't require that all transport speeds involved in the process be constant; however, these devices must maintain the same relative speed at all points over the course of a program. Because of differences in mechanical design, voltage fluctuations, and tape slippage, it's a simple fact of life that analog tape devices aren't able to maintain a constant playback speed...not even over relatively short durations. For this reason, sync between two or more machines without some form of timing "lock" would be impossible over any reasonable program length. It therefore quickly becomes clear that if production is to utilize multiple forms of media and record/playback systems, a method for maintaining a synchronous lock is essential.

Time Code

The standard method of interlocking audio, video, and film transports makes use of a code that was developed by the Society of Motion Picture and Television Engineers (SMPTE). This *time code* (or *SMPTE time code*) identifies an exact position on a tape or within a media program by assigning a digital address that increments along its length. This address code can't slip, always retains its original location, and allows for the continuous monitoring of tape position to an accuracy of between 1/24th and 1/30th of a second (depending on the media type and frame standards being used). The specified tape segments are called *frames*, a term taken from film production. Each audio or video frame is tagged with a unique identifying number, known as a *time code address*. This eight-digit address is displayed in the form 00:00:00:00, in which the successive pairs of digits represent hours:minutes:seconds:frames— HH:MM:SS:FF (Figure 9.2).

The recorded time code address is used to locate a position on magnetic tape (or any other recorded media) in much the same way that a letter carrier uses an address to deliver the mail to a specific residence (Figure 9.3a). Suppose that a time-encoded videotape begins at time 00:01:00:00 and ends at 00:28:19:05 and contains a specific cue point (such as a glass shattering) at 00:12:53:18 (Figure 9.3b). By monitoring the time code readout, it's a simple

Figure 9.2. *Readout of a SMPTE time code address in HH:MM:SS:FF (www.smpte.org).*

00:01:20:05

Figure 9.3. *Location of relative addresses: (a) postal address; (b) time code addresses and a cue point on longitudinal tape.*

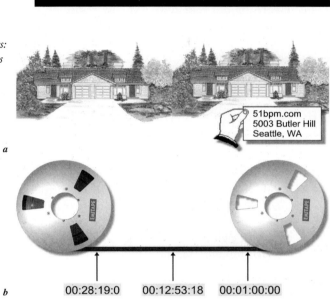

a

51bpm.com
5003 Butler Hill
Seattle, WA

b   00:28:19:0   00:12:53:18   00:01:00:00

matter to locate the precise position that corresponds to the cue point on the tape and then perform whatever function is necessary, such as inserting an effect into the sound track at that specific point... CRASH!

The standard method for encoding time code in analog audio production is to record (stripe) an open audio channel (usually the highest available track) with SMPTE time code that can then be read directly from the track in either direction and at a wide range of transport tape speeds. As we'll see later, digital audio production has settled on a rather straightforward system for distributing time code between CPU-based devices via MIDI.

Time Code Word

The total of all time-encoded information that's encoded into each audio or video frame is known as a *time code word*. Each word is divided into 80 equal segments, called *bits*, which are numbered consecutively from 0 to 79. One word covers an entire audio or video frame, such that for every frame there is a unique and corresponding time code address. Address information is contained in the digital word as a series of bits that are made up of binary 1's and 0's, which (in the case of an analog SMPTE signal) are electronically encoded in the form

Figure 9.4. *Biphase modulation encoding.*

of a modulated square wave. This method of encoding information is known as *biphase modulation*. Using this code type, a voltage transition in the middle of a half-cycle of a square wave represents a bit value of 1, while no transition within this same period signifies a bit value of 0 (Figure 9.4). The most important feature about this system is that detection relies on shifts within the pulse and not on the pulse's polarity. Consequently, time code can be read in either the forward or reverse direction, as well as at fast or slow shuttle speeds.

Do-It-Yourself Tutorial: SMPTE Time Code

◆ Go to the "Tutorial section" of www.modrec.com, click on "Ch. 9—SMPTE Audio Example" and play the time code sound file. Not my favorite tune, but it's a useful one!

◆ The 80-bit time code word is subdivided into groups of 4 bits (Figure 9.5), whereby each grouping represents a specific coded piece of information. Each 4-bit segment represents a binary-coded decimal (BCD) number that ranges from 0 to 9. When the full frame is scanned, all eight of these 4-bit groupings are read out as a single SMPTE frame number (in hours, minutes, seconds, and frames).

User-Information Data

An additional 32 bits (called *user bits*) join the 26 bits that make up the time code address. This extra information has been set aside for time code users to enter their own ID information. The SMPTE standards committee has placed no restrictions on the use of this "slate code," which can contain such information as date of shooting, shot or take ID, reel number, and so on.

Sync Information Data

Another form of information that's encoded into the time code word is *sync data*. This information exists as 16 bits at the end of the time code word which are used to define the

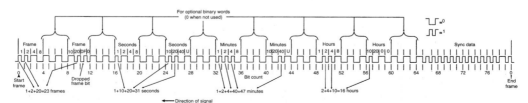

Figure 9.5. *Biphase representation of the SMPTE time code word.*

end of each frame. Because time code can be read in either direction, sync data is also used to tell the device which direction the tape or digital device is moving.

Time Code Frame Standards

In productions using time code, it's important that the readout display be directly related to the actual elapsed time of a program, particularly when dealing with the exacting time requirements of broadcasting. For this reason, time code frame rates may vary from one medium, production house, or country of origin to another:

♦ 30 fr/sec (monochrome U.S. video)—In the case of a black and white (monochrome) video signal, a rate of exactly 30 frames per second (fr/sec) is used. If this rate (often referred to as non-drop code) is used on a black and white program, the time code display, program length, and actual clock-on-the-wall time would all be in agreement.

♦ 29.97 fr/sec (drop-frame time code for color NTSC video)—This simplicity was broken, however, when the National Television Standards Committee (NTSC) set the frame rate for the color video signal in the United States and Japan at 29.97 fr/sec. Thus, if a time code reader that's set up to read the monochrome rate of 30 fr/sec were used to read a color program, the time code readout would pick up an extra 0.03 frame for every second that passes. Over the duration of an hour, the time code readout would differ from the actual elapsed time by a total of 108 frames (or 3.6 seconds). In order to correct for this difference and bring the discrepancy between the time code readout and the actual elapsed time back into agreement, a series of frame adjustments were introduced into the code. Because the goal is to drop 108 frames over the course of an hour, the code used for color has come to be known as *drop frame code*. In this system, two frame counts for every minute of operation are omitted, with the exception of minutes 00, 10, 20, 30, 40, and 50. This has the effect of adjusting the frame count so that it agrees with the actual elapsed duration of a program.

♦ 29.97 fr/sec (non-drop time code)—In addition to the color 29.97 drop frame code, a 29.97 non-drop-frame color standard can also be found in video production. When using non-drop time code, the frame count will always advance one count per frame, without any drops. As you might expect, this mode will result in a disagreement between the frame count and the actual clock-on-the-wall time over the course of the program. Non-drop, however, has the distinct advantage of easing the time calculations that are often required in the video editing process (because no frame compensations need to be taken into account).

♦ 25-fr/sec EBU (standard rate for PAL video)—Another frame rate format that's used throughout Europe is the European Broadcast Union (EBU) time code. EBU utilizes SMPTE's 80-bit code word but differs in that it uses a 25-fr/sec frame rate. Because both monochrome and color video EBU signals run at exactly 25 fr/sec, an EBU drop frame code isn't necessary.

♦ 24-fr/sec (standard rate for film work)—The medium of film differs from all of these as it makes use of an SMPTE time code format that runs at 24 fr/sec.

From the above, it's easy to understand why confusion often exists around what frame rate is to be used on a project. Basically, if you are working on an in-house project that doesn't incorporate time-encoded material that comes from the outside world . . . you should choose a rate that both makes sense for you and is likely to be compatible with an outside facility (should the need arise). For example, American electronic musicians who are working in-house will often choose to work at 30 fr/sec. Those in Europe have it easy, as 25 fr/sec is the logical choice for all music and video productions. On the other hand, those who work with projects that come through the door from other production houses will need to take special care to reference their time code rates to those used by the originating media house. This can't be stressed enough: If care isn't taken to keep your time code references at the proper rate, while keeping degradation to a minimum from one generation to the next . . . the various media might have trouble syncing up when it comes time to put the final master together . . . and that can spell BIG trouble.

LTC and VITC Time Code

Currently, two major systems exist for encoding time code onto magnetic tape:

- ◆ Longitudinal time code (LTC)
- ◆ Vertical interval time code (VITC)

Time code recorded onto an analog audio or video cue track is known as *longitudinal time code* (*LTC*). LTC encodes a biphase time code signal onto the analog audio or cue track in the form of a modulated square wave at a bit rate of 2400 bits/second. The recording of a perfect square wave onto a magnetic audio track is, under the best of conditions, difficult. For this reason, the SMPTE standard has set forth an allowable rise time of 25 ± 5 microseconds for the recording and reproduction of valid code. This tolerance requires a signal bandwidth of 15 kHz, which is well within the range of most professional audio recording devices. Variable-speed time code readers are often able to decode time code information at shuttle rates ranging from 1/10th to 100 times normal playspeed. This is effective for most audio applications; however, in video postproduction it's often necessary to monitor a videotape at slow or still speeds. As LTC can't be read at speeds slower than 1/10th to 1/20th normal playspeed, two methods can be used for reading time code. The first of these uses a character generator to "burn" time code addresses directly into the video image of a worktape copy. This superimposed readout allows the time code to be easily identified, even at very slow or still picture shuttle speeds (Figure 9.6). In most situations, LTC code is preferred for audio, electronic music, and mid-level video production, as it's a more accessible and cost-effective protocol.

A second method, which is used by major video production houses, is the *vertical interval time code* (*VITC*). VITC makes use of the same SMPTE address and user code structure as LTC, but is encoded onto videotape in an entirely different manner. VITC actually encodes the time code information into the video signal itself . . . inside a field (known as the vertical blanking interval) that's located outside the visible picture scan area. Because the time code data is encoded into the video signal itself, professional helical scan video recorders are able to read

Figure 9.6. *Video image showing burned-in time code window.*

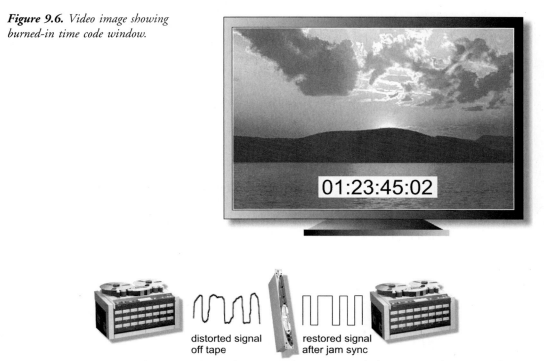

01:23:45:02

distorted signal
off tape

restored signal
after jam sync

Figure 9.7. *Jam sync is used to restore distorted SMPTE when copying code from one machine to another.*

time code at very slow and even still-frame speeds. Since time code is encoded in the video signal, an additional track can be opened up on a video recorder for audio or cue information, while also eliminating the need for a burned window dub.

Time Code Refreshment/Jam Sync

Longitudinal time code operates by recording a series of square-wave pulses onto magnetic tape. As you now know, it's somewhat difficult to record a square waveform onto analog magnetic tape without suffering moderate to severe waveform distortion (Figure 9.7). Although time code readers are designed to be relatively tolerant of waveform amplitude fluctuations, such distortions are severely compounded when code is copied from tape to tape by one or more generations. For this reason, a *time code refresher* has been incorporated into most time code synchronizers and MIDI interface devices with sync capabilities. Basically, this process reads the degraded time code information from a previously recorded track and then amplifies and regenerates the square wave back into its original shape so it can be freshly recorded to a new track or read by another device.

Should the quality of a SMPTE signal degrade to the point where the synchronizer can't differentiate between the pulses, the code will disappear and the slave(s) will stop unless the system includes a feature known as *jam sync*. Jam sync refers to the synchronizer's ability to output the next time code value even though one has not appeared at its input. The generator is said to be working in a "freewheeling fashion" since the generated code may not agree with the actual recorded address values—however, if the dropout occurs for only a short period, jam syncing works well to detect and/or refresh the signal. (This process is often useful when dealing with dropouts or undependable code from VHS audio tracks.) Two forms of jam sync options are available:

◆ Free-wheeling
◆ Continuous

In the *free-wheeling* mode, the receipt of time code causes the generator's output to initialize when a valid address number is detected. The generator then begins to count in an ascending order on its own, ignoring any deterioration or discontinuity in code...producing fresh, uninterrupted SMPTE address numbers. *Continuous jam sync* is used in cases where the original address numbers must remain intact and shouldn't be regenerated as a continuously ascending count. After the reader has been activated, the generator updates the address count for each frame in accordance with incoming address numbers and outputs an identical, regenerated copy.

Synchronization Using SMPTE Time Code

In order to achieve a frame-by-frame time code lock between multiple audio, video, and film analog transports, it's necessary to use a device or integrated system that's known as a *synchronizer* (Figure 9.8). The basic function of a synchronizer is to control one or more tape, computer-based, or film transports (designated as slave machines) so that their speeds and relative positions are made to accurately follow one specific transport (designated as the master). Although the lines of distinction often break down, synchronization as a whole can be divided into two basic system types: those that are used in project or electronic music production facilities and those that can be found in larger audio and video production and postproduction facilities. The greatest reason for this division is not so much a system's performance, as it is its price and device types that are used.

The use of a synchronizer within a project studio environment often involves a multiport MIDI interface that includes provisions for locking an analog audio or video transport to a digital audio, MIDI, or electronic music system by translating LTC SMPTE code into MIDI time code (more on this later in the chapter). In this way, one simple device can cost-effectively serve multiple purposes to achieve lock with a high-degree of accuracy.

Systems that are used in video production and higher levels of production will often require a greater degree of control and remote-control functions throughout the studio or production facility. Such a setup will often require a more sophisticated device, such as a control synchronizer or an edit decision list (EDL) controller.

Control synchronizers are used to gain both manual and automated control over all of the transports that are connected within a larger system. By using a central CPU/controller

Figure 9.8. *The synchronizer in time code production: (a) simple MIDI interface synchronizer within a project studio; (b) more sophisticated control synchronizer, which may be required for a larger production facility. (Courtesy of Timeline Vista, Inc., www.timelinevista.com.)*

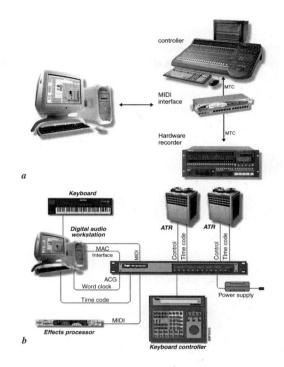

(which can either exist as a stand-alone device or can be integrated into a recording console's design), a control synchronizer provides such options as:

- *Machine selection*—Lets us choose which machines are to be connected in the control slave chain, as well as allowing for the selection of a designated master.
- *Transport control*—Provides conventional remote control functions over any or all machines in the system.
- *Locate*—A transport command that causes all selected machines to automatically locate to a specific time code address
- *Loop*—Enters the system into a continuous repeat cycle (play, rewind, play again) between any two address cue points that are stored in memory
- *Offset* – Allows for the correction of any time code discrepancy that might exist between two or more transports (lets you adjust relative time by $\pm x$ frames to achieve or improve sync)
- *Event points*—A series of cues that can be entered into memory for triggering event(s) at specific time code addresses (*e.g.*, triggering a sound effect or changing console snapshots from one mix scene to another)
- *Record punch in/out*—Allows the synchronizer to take control over transport record/edit functions, enabling tight record in/out points that can be repeated with frame accuracy

An edit decision list (EDL) controller/synchronizer uses a process that has evolved from the online video editing process and is most commonly found in larger video and audio-for-video postproduction suites. This synchronizer type goes one step further, in that it uses a user-built list of edit commands to electronically control, synchronize, and switch all video and audio transports and then exert control over their edit in/out points, tape positions, time code offsets, and so on to assemble all of the various takes and scenes of a project into a finished, assembled master . . . all under automated computer control.

SMPTE Offset Times

In the real world of audio production, programs or songs don't always begin at 00:00:00:00. Let's say that you were handed an ADAT tape that needed a synth track laid down onto track 7 of a song that goes from 00:11:24:03 to 00:16:09:21. Instead of inserting more than 11 minutes of empty bars into a MIDI track on your synched digital audio workstation (DAW), you could simply insert an offset start time of 00:11:24:03. This means that the sequenced track will begin to increment from measure 1 at 00:11:24:03 and will maintain relative offset sync throughout the program. Offset start times are also useful when synchronizing devices to an analog or videotape source that doesn't begin at 00:00:00:00. As you're probably aware, it takes a bit of time for an analog audio transport to settle down and begin playing (this wait time often quadruples whenever videotape is involved). If a program's time code were to begin at the head of the tape, it's extremely unlikely that you would want to start a program at 00:00:00:00 (since playback would be delayed and extremely unstable at this time). Instead, most programming is striped with an appropriate pre-roll of anywhere from 10 seconds to 2 minutes. Such a pre-roll gives all of the transports ample time to begin playback and sync up to the master TC source. In addition, it's often wise to start the actual production or first song at an offset time of 01:00:00:00 (some facilities begin at 00:01:00:00). This minimizes the possibility that the synchronizer will become confused by "rolling over midnight"; that is, if the content starts at 00:00:00:00, the pre-roll would be in the 23:59:00:00 range and the synchronizer would try to rewind to zero (rolling the tape off the reel) instead of rolling forward . . . Not always fun in the heat of a production!

Loops

SMPTE can be set up to play, record, and/or punch-in at a specific address and then loop back to the beginning and start over. This process is common in studio and audio for visual production, as well as in theme parks where a short demonstration or show is repeatedly played under time code lock in a continuous loop.

Distribution of SMPTE Signals

In a basic audio production system, the only connection that's usually required between the master machine and a synchronizer is the LTC time code track. Generally, when connecting analog slave devices, two connections will need to be made between each transport

Table 9.1. Optimum time code recording levels.

Tape Format	Track Format	Optimum Recording Level
ATR	Edge track (highest number)	−5 to −10 VU
3/4-inch VTR	Audio 1 (L) track or time	−5 to 0 VU code
1-inch VTR	Cue track or audio 3	−5 to −10 VU
MDM	Highest number track	−20 dB

Note: If the VTR is equipped with automatic gain compensation (AGC), override the AGC and adjust the signal gain controls manually.

and the synchronizer. These include lines for the time code reproduce track and the control interface (which often uses the Sony 9-pin remote protocol for giving the synchronizer full logic transport and speed-related feedback information). LTC signal lines can be distributed throughout the production system in the same way that any other audio signal is distributed. They can be routed directly from machine to machine or patched through audio switching systems via balanced, shielded cables or unbalanced cables, or a combination of both. Because the time code signal is biphase or symmetrical, it's immune to cable polarity problems.

Time Code Levels

One problem that can plague systems using time code is crosstalk. Such problems arise from having a high-level time code signal leak into adjacent signal paths or analog tape tracks. Currently, no industry standard levels exist for the recording of time code onto magnetic tape or digital tape track; however, the levels shown in Table 9.1 can help you get a good signal level while keeping distortion and analog crosstalk to a minimum.

MIDI-Based Synchronization

Just as sync is routinely used within larger audio and video facilities, the wide acceptance of MIDI and digital audio within media production has created the need for cost-effective synchronization in project studio and mid-sized production environments. Over the years, devices such as digital audio workstations, MIDI sequencers, digital mixing consoles, and effects devices have become increasingly integrated and networked together. Advances in design have fashioned these technologies into an integrated system that makes extensive use of synchronization and time code in a manner that's both easy-to-use and inexpensive to implement—all through the use of MIDI (Figure 9.9). The following sections outline the various forms of synchronization that are often encountered in a MIDI-based sync and production environment.

MIDI Real-Time Messages

Although not related to SMPTE time code or any external reference, it's important to know that MIDI has a built-in (and often transparent) protocol for synchronizing all of the tempo

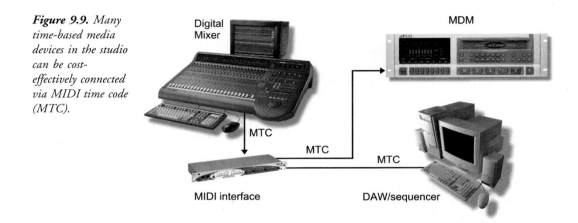

Figure 9.9. *Many time-based media devices in the studio can be cost-effectively connected via MIDI time code (MTC).*

and timing elements of each MIDI device in a system to a master clock. This protocol operates by transmitting real-time messages to the various devices through standard MIDI cables, USB, and internal CPU paths. Although these relationships are often automatically defined within a system setup, one MIDI device must be designated as the master device in order to provide the timing information to which all other slaved devices are locked. *MIDI real-time messages* consist of four basic types that are each 1 byte in length:

◆ *Timing clock*—A clock timing that's transmitted to all devices in the MIDI system at a rate of 24 pulses per quarter note (ppq). This method is used to improve the system's timing resolution and simplify timing when working in nonstandard meters (*e.g.*, 3/8, 5/16, 5/32, . . .).

◆ *Start*—Upon receipt of a timing clock message, the start command instructs all connected devices to begin playing from the beginning of their internal sequences. Should a program be in mid-sequence, the start command repositions the sequence back to its beginning, at which point it begins to play.

◆ *Stop*—Upon the transmission of a MIDI stop command, all devices in the system stop at their current positions and wait for a message to follow.

◆ *Continue*—Following the receipt of a MIDI stop command, a MIDI continue message instructs all instruments and devices to resume playing from the precise point at which the sequence was stopped. Certain older MIDI devices (most notably drum machines) aren't capable of sending or responding to continue commands. In such a case, the user must either restart the sequence from its beginning or manually position the device to the correct measure.

MIDI Time Code

MIDI time code (MTC) was developed to allow electronic musicians, project studios, video facilities, and virtually all other production environments to cost-effectively and easily translate time code into time-stamped messages that can be transmitted via MIDI. Created by

Chris Meyer and Evan Brooks, MIDI time code enables SMPTE-based time code to be distributed throughout the MIDI chain to devices or instruments that are capable of synchronizing to and executing MTC commands. MIDI time code is an extension of MIDI 1.0, which makes use of existing message types that were either previously undefined or were being used for other, nonconflicting purposes. Since most modern recording devices include MIDI in their design, there's often no need for external hardware when making direct connections. Simply chain the MIDI cables from the master to the appropriate slaves within the system (via physical cables, USB, or virtual internal routing). Although MTC uses a reasonably small percentage of MIDI's available bandwidth (about 7.68% at 30 fr/sec), it's customary (but not necessary) to separate these lines from those that are communicating performance data when using MIDI cables. As with conventional SMPTE, only one master can exist within an MTC system, while any number of slaves can be assigned to follow, locate, and chase to the master's speed and position. Because MTC is easy to use and is often included free in many system and program designs, this technology has grown to become the most common and most straightforward way to lock together such devices as digital audio work-stations, modular digital multitracks, and MIDI sequencers, as well as analog and videotape machines (by using a MIDI interface that includes a SMPTE-to-MTC converter).

MIDI Time Code Control Structure

The MIDI time code format can be divided into two parts:

◆ Time code
◆ MIDI cueing

The time code capabilities of MTC are relatively straightforward and allow devices to be synchronously locked or triggered to SMPTE time code. MIDI cueing is a format that informs a MIDI device of an upcoming event that's to be performed at a specific time (such as load, play, stop, punch in/out, reset). This protocol envisions the use of intelligent MIDI devices that can prepare for a specific event in advance and then execute the command on cue.

MIDI Time Code Messages

MIDI time code is made up of three message types: quarter-frame messages, full messages, and MIDI cueing messages.

◆ *Quarter-frame messages*—These are transmitted only while the system is running in real or variable speed time, in either forward or reverse direction. True to its name, four quarter-frame messages are generated for each time code frame. Since 8 quarter-frame messages are required to encode a full SMPTE address (in hours, minutes, seconds, and frames—00:00:00:00), the complete SMPTE address time is updated once every two frames. In other words, at 30 fr/sec . . . 120 quarter-frame messages would be transmitted per second, while the full time code address would be updated 15 times in the same period. Each quarter frame message contains 2 bytes. The first byte is "F1," the quarter-frame common header, while the second byte contains a nibble (four hits) that represents the message number (0 through 7) and a nibble for encoding the time field digit.

◆ *Full messages*—Quarter-frame messages are not sent in the fast-forward, rewind, or locate modes, as this would unnecessarily clog a MIDI data line. When the system is in any of these shuttle modes, a full message is used to encode a complete time code address within a single message. After a fast shuttle mode is entered, the system generates a full message and then places itself in a pause mode until the time-encoded slaves have located to the correct position. Once playback has resumed, MTC will again begin sending quarter frame messages.

◆ *MIDI cueing messages*—MIDI cueing messages are designed to address individual devices or programs within a system. These 13-bit messages can be used to compile a cue or edit decision list, which in turn instructs one or more devices to play, punch in, load, stop, and so on . . . at a specific time. Each instruction within a cueing message contains a unique number, time, name, type, and space for additional information. At the present time, only a small percentage of the possible 128 cueing event types has been defined.

SMPTE/MTC Conversion

A SMPTE-to-MIDI converter is used to read incoming SMPTE time code and convert it into MIDI time code (and *vice versa*). These conversion systems are available as a stand-alone device or as an integrated part of a multiport MIDI interface/patch bay/synchronizer system (Figure 9.10). Certain analog and digital multitrack systems include a built-in MTC port within their design, meaning that the machine can be synchronized to a DAW/sequencing system (with a MIDI interface) without the need for any additional hardware.

Proprietary Synchronization Systems for Modular Digital Multitrack Recorders

Modular digital multitrack (MDM) recorders, such as the Tascam DA-98HR and Alesis ADAT, encode a proprietary form of time-encoded sync data onto tape, along with the audio information. This sync coding can be used to lock several MDMs together in order to give us more tracks. It can also be used to lock these devices to an external SMPTE source via an interface

Figure 9.10. *SMPTE time code can be easily converted to MTC for distribution throughout a production system.*

analog multitrack   SMPTE/MIDI interface   DAW/sequencer

that can translate MTC or SMPTE into the MDM's sync code (and *vice versa*). In this way, one or more digital multitracks can be easily rigged to act as a master or slave within an MDM- or DAW-based system.

Digital Audio's Need for a Stable Timing Reference

The process of maintaining a synchronous lock between digital audio devices or between digital and analog systems differs fundamentally from the process of maintaining relative speed between analog transports. This is due to the fact that a digital system generally achieves synchronous lock by adjusting its playback sample rate (and thus its speed and pitch ratio) so as to precisely match the relative playback speed of the master transport. Whenever a digital system is synchronized to a time-encoded master source, a stable timing source is extremely important. Such a reference is preferable in order to keep jitter (in this case, an increased distortion due to rapid pitch shifts) to a minimum. In other words, the source's program speed should vary as little as possible to prevent any degradation in the digital signal's quality. For example, all analog tape machines exhibit speed variations, which are caused by tape slippage and transport irregularities (a basic fact of analog life known as *wow and flutter*). If we were to synchronize a digital device to an analog master source that contains excessive wow and flutter, the digital system would be required to constantly speed up and slow down to precisely match the transport's speed fluctuations. One way to avoid such a problem would be to use a source that's more stable, such as a video deck, DAW, or MDM.

Video's Need for a Stable Timing Reference

Whenever a video signal is copied from one machine to another, it's essential that the scanned data (containing timing, video, and user information) be copied in perfect sync from one frame to the next. Failure to do so will result in severe picture breakup or, at best, the vertical rolling of a black line over the visible picture area. Copying video from one machine to another generally isn't a problem, as the VCR or VTR that's doing the copying obtains its sync source from the playback machine. Video postproduction houses, however, often simultaneously use any number of video decks, switchers, and edit controllers during the production of a single program. Mixing and switching between these sources will usually result in nonsynchronous chaos, with the end result being a very unhappy client. Fortunately, referencing all of the video, audio, and timing elements to an extremely stable timing source (called a *black burst* or *house sync* generator) will generally resolve this sync nightmare. This reference clock serves to synchronize the video frames and time code addresses that are received and/or transmitted by every video-related device in a production facility, so that the leading frame edge of every video signal occurs at exactly the same instant in time (Figure 9.11). By resolving all video and audio devices to a single black burst reference, you're assured that relative frame transitions and speeds throughout the system will be consistent and stable.

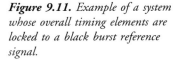

Figure 9.11. *Example of a system whose overall timing elements are locked to a black burst reference signal.*

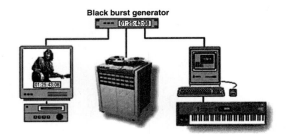

This even holds true for slaved analog machines, as their transport's wow and flutter can be smoothed out when locked to such a stable timing reference.

Real-World Sync Applications for Using Time Code and MIDI Time Code

Before we delve into the many possible ways that a system can be set up to work in a time code environment, it needs to be understood that each system will often have its own particular "personality" and that the connections, software, and operation of one system might differ from that of another. This is often due to factors such as system complexity and the basic hardware types that are involved, as well as the type of hard/software systems that are installed in a DAW. Larger, more expensive setups that are used to create television and film soundtracks will often involve extensive time code and system interconnections that can easily get complicated. Fortunately, the use of MIDI time code has greatly reduced the cost and complexity of connecting and controlling a synchronous electronic music and project studio down to levels that can be easily managed by both experienced and novice users. Having said these things, I'd still like to stress that solving synchronization problems will often require as much intuition, perseverance, insight, and art as it will technical skill. For the remainder of this chapter, we'll be looking into some of the basic concepts and connections that can be used to get your system up and running. Beyond this, the next best course of action will be to consult your manual(s), seek help from an experienced friend, or call the tech department for the particular hardware or software that's giving both you and your system the willies.

Master/Slave Relationship

Since synchronization is based upon the timing relationship between two or more devices, it follows that the logical way to achieve sync is to have one or more devices (known as *slaves*) follow the relative movements of a single transport or device (known as the *master*). The basic rule to keep in mind is that there can be only one master in a connected system; however, any number of slaves can be set to follow the relative movements of a master transport or device

Figure 9.12. *There can be only one master in a synchronized system; however, there can be any number of slaves.*

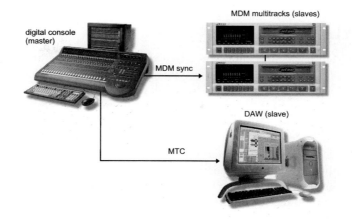

(Figure 9.12). Generally, the rule for deciding which device will be the master in a production system can best be determined by asking a few questions:

◆ What media is the master time code media recorded on?

◆ Which device will easily serve as the master?

◆ Which device will provide the most stable timing reference?

If the master comes to you from an outside source, the first question will most likely solve your dilemma. If the project is in-house and you have total say in the matter, you might want to research your options more fully. The following sections can help give you insights into which devices will want to serve as the master within a particular system.

Audio Recorders

In many audio production situations, whenever an analog tape recorder is connected in a time code environment, this machine will want to act as the master as costly hardware is often required to lock an analog machine to an external time source. This is due to the fact that the machine's speed regulator (generally a DC capstan servo) must be connected into a special feedback control loop in a way that allows it to continuously adjust its speed by comparing its present location with that of the master SMPTE time code. As a result, it is often far easier and less expensive to set the analog device as the master . . . especially if the slave device is a DAW or other digital device. When starting a new session, the course of action is to stripe the highest track on a clean roll of tape (with ascending code that continues from the tape's beginning to its end). Once done, the reproduced code can be routed to the SMPTE input on your MIDI interface or synchronizer. If you don't have a multiport interface or if your interface doesn't have a SMPTE input, you'll need to get hold of a box that converts SMPTE to MTC . . . which can then be plugged into a MIDI IN port for assignment to a DAW or sequencer device.

VCRs

Since video is often an extremely stable timing source, an analog VCR or VTR should almost invariably act as a system master. In fact, without expensive hardware, a VCR can't easily be set to act as a slave at all, since the various sync references within the machine would be thrown off and the picture would immediately break up or begin to roll. From a practical standpoint, locking other devices to a standard VCR is done in much the same way as with an analog tape machine. Professional video decks generally include a separate track that's dedicated to time code (in addition to other tracks that are dedicated to audio). As with the earlier analog scenario, the master time code track must be striped with SMPTE before beginning the project. This process shouldn't be taken lightly, as the time code must conform to the time code addresses on the original video master or working copy (see the discussion of jam sync). Basically, the rule of thumb is *if you're working on a project that was created "out of house" always use the code that was provided by the original production team.* Striping your own code or erasing over the original code with your own would render the timing elements useless, as the new code wouldn't relate to the original addresses or include any variations that might be a part of the original master source. In short, make sure that your working copy includes SMPTE that's a regenerated copy of the original code! Should you overlook this, you can expect to run into timing and sync troubles, either immediately or later in the postproduction phase, while assembling the music or dialog back together with the final video master . . . factors that'll definitely lead to premature hair and client loss.

MDMs

As a digital device, a modular digital multitrack machine is also an extremely stable timing reference and often works well as a master in an audio setting. Because of extensive sync and pitch shifting technology, these devices can also be slaved within a system without too much difficulty. As with an analog machine, it's possible to record SMPTE onto the highest available track and then route this track to a valid SMPTE sync input; however, if you don't feel like losing a physical track to SMPTE, you might want to pick up a sync interface that can translate the MDM's proprietary sync code into SMPTE or MTC.

Software Applications

In general a MIDI sequencer will be programmed to act as a slave device. This is due to its ability to chase a master MTC source by easily changing its timing position.

Digital Audio Workstations

A computer-based DAW can often be set to act as either a master or slave. This will ultimately depend on the software, as most professional workstations can be set to chase (or be triggered by) a master time code source.

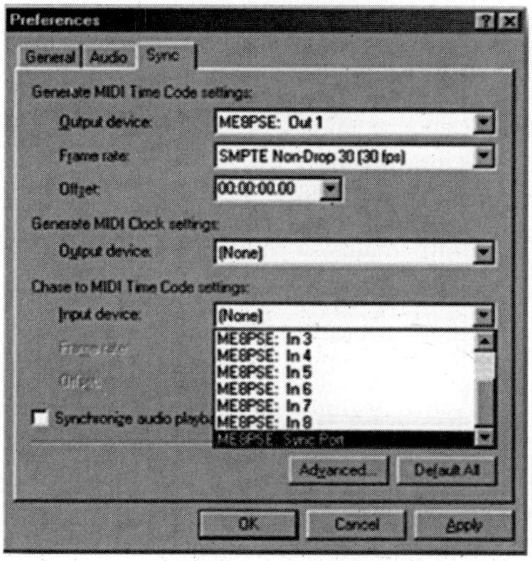

Figure 9.13. Sync Preferences dialog box.
(Courtesy of Sony Pictures Digital, Inc.,
http://www.sony.com/mediasoftware.)

Routing Time Code to and from Your Computer

From a connections standpoint, most DAW, MIDI, and audio application software packages are flexible enough to let you choose from any number of available sync sources (whether connected to a hardware port, MIDI interface port, or virtual sync driver). All you have to do is assign all of the slaves within the system to the device driver that's generating the system's master code (Figure 9.13). In most cases, where the digital audio and MIDI sequencing applications are operating within the same computer, it's generally best to have your DAW or editor generate the master code for the system. From time to time, you might run into an application or editor that's unable to generate time code in any form. When faced with such an all-slave software environment, you'll actually need a physical time code master that can be routed to your editor, MIDI sequencer, etc. In practice, this source could be an MDM or analog recorder, but if you simply want to sync the two pieces of software together without a tape machine, the easiest solution is to use a multiport MIDI interface that includes a software applet for generating time code. In such a situation, all you need to do is to select the interface's sync driver as your sync source for all slave applications. Pressing the Generate SMPTE button from the interface's application window or from its front panel will lock the software to the generated code, beginning at 00:00:00:00 or at any specified offset address.

As more and more DAWs and digital mixing boards become locked to SMPTE/MTC, the issue of locking the word clock of a digital device directly to the SMPTE time code stream will become more and more important. For example, slaving a DAW to external time code under a full time code lock (as opposed to a triggered free-run start) will usually require that specialized sync hardware be used to maintain a frame-by-frame lock.

Keeping Out of Trouble

Here are a few guidelines that can help save your butt when using SMPTE and other time code translations during a project:

◆ When in doubt about frame rates, special requirements, or literately anything (no matter how small you think it might be) . . . ask! You and your client will be glad you did.

◆ Fully document your time code settings, offsets, start times, etc.

◆ If the project isn't to be used in-house, ask the producer what the proper frame rate should be. Don't assume or guess it.

◆ When beginning a new session (when using a tape-based device), always stripe the master contiguously from the beginning to end *before* the session begins. It never hurts to stripe an extra tape, just in case. This goes for both analog and digital MDM devices.

◆ Start generating new code at a point 1 to 2 minutes before 01:00:00:00 or 00:01:00:00 (to allow for a pre-roll). If the project isn't to be used in-house, ask the producer what the start times should be. Don't assume or guess it.

◆ Never dub time code directly. Always make a refreshed copy of the original time code *before* the session begins.

◆ Never use slow videotape speeds. In EP mode, a VHS deck runs too slowly to record or reproduce a reliable code.

◆ Disable noise reduction on audio tracks (on both audio and video decks).

◆ Work with copies of the original production video, and make a new one when sync troubles appear.

◆ It's not unusual for the time code to be read incorrectly (when short dropouts occur on the track . . . usually on videotape). When this happens, you might set the synchronizer to "free wheel" once the transports have initially locked.

In closing, I'd like to point out that synchronization can be a simple procedure or it can be a fairly complex one (depending on your experience and the type of equipment that's involved). A number of books and articles have been written on the subject. If you're serious about production, I'd suggest that you do your best to keep up on the subject. Although the fundamentals often stay the same, new technologies and techniques are constantly emerging. As always, the best way to learn is simply by reading, jumping in, and then doing it.

CHAPTER 10

◆

Amplifiers

In the world of audio, amplifiers have many applications. They can be designed to amplify, equalize, combine, distribute, or isolate a signal. They can even be used to match the signal impedance between devices. At the heart of any *amplifier* (*amp*) system is either a vacuum tube or a semiconductor-type transistor device. Just about everyone knows about these regulating devices, but not everyone has a grasp on how they operate. Here are a few, simple insights into these electronic wonders.

Amplification

In order to best understand how the theoretical process of amplification works, let's draw an analogy to a valve (a term that was originally used for the vacuum tube and is still in use in England and other Commonwealth countries). Let's begin by assuming that we have a high-pressure fire hose. Connected to this hose is a valve that, when turned, can control large amounts of water pressure with very little effort (Figure 10.1). Using a small amount of energy, we can then turn a trickle of water into a high-powered gusher and back again. In practice, both the vacuum tube and the transistor work much like this valve. For example, a vacuum tube operates by placing a DC current across its plate and a heated cathode element (Figure 10.2). A wire mesh grid separating these two elements acts like a control valve for allowing electrons to pass from the plate to the cathode. By introducing small, varying amounts of voltage onto the tube's grid, a much larger electrical current can be correspondingly passed between the plate and the cathode (Figure 10.3).

Figure 10.1. *Current through a vacuum tube or transistor is controlled in a manner that's similar to the way that a valve tap can control water pressure through a water pipe.*

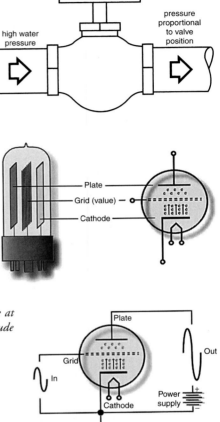

Figure 10.2. *An example of a triode vacuum tube.*

Figure 10.3. *A schematic showing how small changes in voltage at the tube's grid can produce a much larger, corresponding amplitude changes between its cathode and plate.*

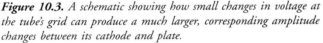

Although the transistor (a term derived from "trans-resistor," referring to a device that can easily change resistance) operates under a different electrical principle, the valve analogy is still relevant. Figure 10.4 shows a basic amplifier schematic with a DC power source that's been set up across the transistor's collector and emitter points. Now, let's revisit the valve analogy. By presenting a small control voltage at the transistor's base, it's possible for a corresponding change in the resistance to be introduced between the collector and emitter. This allows a much larger and corresponding voltage/current to be passed from the supply source through to the device's output.

As a device, the transistor isn't inherently linear; that is, applying a signal to the base won't always produce a corresponding output change. The linear operating region of a transistor lies between the device's lower-end cutoff region and an upper saturation point (Figure 10.5). Within this operating region, changes in the base current will produce a corresponding change

Figure 10.4. A simple schematic showing how small changes in current at the transistor's base can produce much larger, corresponding amplitude changes through the emitter and collector to the output.

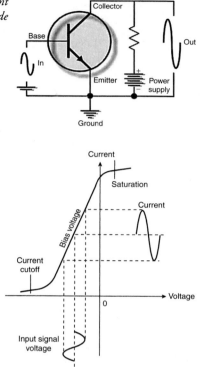

Figure 10.5. Operating region of a transistor.

in the collector current and voltage. When operating near the cutoff or saturation points, the base current lines won't be linear and the output will be distorted. In order to keep the signal within this linear operating range, a DC bias signal is applied to the base of the transistor (for much the same reason a high-frequency bias signal is applied to an analog recording head). After a corrective bias voltage has been applied and sufficient amplifier design characteristics have been met, the amp's dynamic range will be limited by only two factors: noise (which results from thermal electron movement within the transistor and other circuitry) and saturation.

Amplifier *saturation* results when the input signal is so large that its DC supply output voltage isn't large enough to produce the required output current. Overdriving an amp in such a way will produce a mild to severe waveform distortion effect known as *clipping* (Figure 10.6). For example, if an amp having a supply voltage of +24 volts (V) is operating at a gain ratio of 30:1, an input signal of 0.5 V will produce an output of 15 V. Should the input be raised to 1 V, the required output level would be increased to 30 V. However, since the maximum output voltage is limited to 24 V, wave excursions of greater levels will be chopped off or "clipped" at the upper and lower ends of the waveform, until the signal level falls below the maximum 24-V supply level. Whenever a transistor clips, severe odd-order harmonics are introduced that are immediately audible with transistors and most integrated circuit designs. Although clipping can be a sought-after part of a musical instrument's sound (tube

Figure 10.6. *A clipped waveform.*

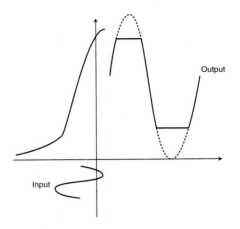

amps tend to have a more musical, even-order harmonic aspect when clipping), it's rarely a desirable effect in quality studio and monitoring gear.

The Operational Amplifier

The *operational amplifier* (*op-amp*) is a stable, high-gain, high-bandwidth amp that has a high-input impedance and a low-output impedance. These qualities allow op-amps to be used as a basic building block for a wide variety of audio and video applications, simply by tagging additional components onto the basic circuit to fit the design needs. Figure 10.7 shows a typical op-amp design that's used for basic amplification. In order to reduce an op-amp's output gain to more stable, workable levels, a negative feedback loop is often required. Negative feedback is a technique that applies a portion of the output signal through a limiting resistor (which determines the gain) back into the negative or phase-inverted input terminal. By feeding a portion of the amp's output back into its input out of phase, the device's output signal

Figure 10.7. *Basic operational amp configuration.*

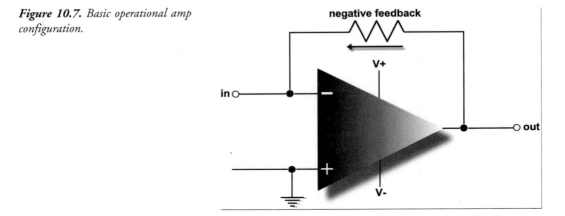

level will be reduced. Controlling the gain in this fashion has the added effect of stabilizing the amp and further reducing distortion.

Preamplifiers

One of the mainstay amplifier types found at the input section of most professional mixer, console, and outboard devices is the *preamplifier (preamp)*. This amp type is often used in a wide range of applications, such as boosting a mic's signal to line level, providing variable gain for various signal types, or isolating signals from extraneous input interference and improper grounding or signal voltage conditions, as well as equalization . . . just to name a few. Preamps are an important component in audio engineering as they can often set the "tone" of how a device or system will sound. Just as a microphone has its own sonic character, preamp designs will often have their own "sound." Questions such as "Are the op-amp blocks designed from quality components?" "Do they use tubes or transistors?" "Are they quiet or noisy?" are all important considerations to the overall sound of a device. This holds especially true when choosing a mic preamp (a device that boosts a mic's low output signal into the line-level range), because dynamic ranges in excess of 120 dB are often required for high-quality pickup conditions.

Equalizers

Basically, an *equalizer* is nothing more than a frequency-discriminating amplifier. In most analog designs, equalization (EQ) is achieved through the use of resistive/capacitive networks that are located in an op-amp's negative feedback loop (Figures 10.8 and 10.9) in order to boost (amplify) or cut (attenuate) certain frequencies in the audible spectrum. By changing the circuit design and/or parameters, any number of EQ curves can be achieved.

Figure 10.8. *Low-frequency equalizer circuit.*

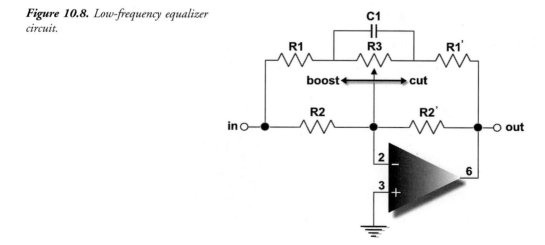

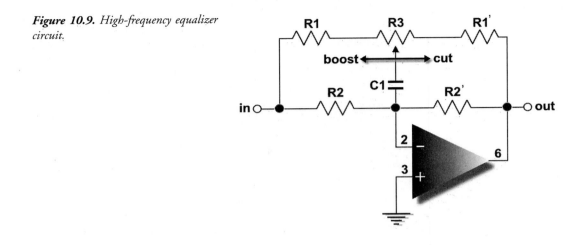

Figure 10.9. *High-frequency equalizer circuit.*

Summing Amplifiers

A *summing amp* (also known as an active combining amplifier) is designed to combine any number of discrete inputs into a single signal bus, while providing a high degree of isolation between them (Figure 10.10). The summing amplifier is an important component in analog console design because the large number of internal signal paths require a high-degree of isolation in order to prevent signals from inadvertently leaking into other input/output paths.

Isolation Amplifiers

An amplifier that's used to isolate combined signals (such as a summing amp) might also be used to prevent unwanted electrical and ground potentials at a device's input from reaching its

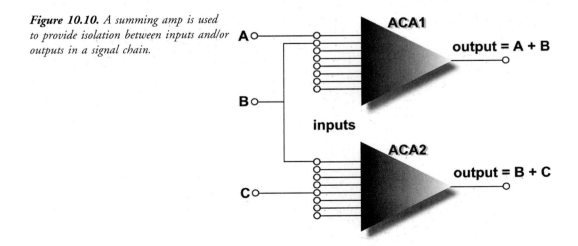

Figure 10.10. *A summing amp is used to provide isolation between inputs and/or outputs in a signal chain.*

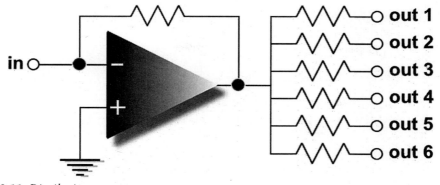

Figure 10.11. *Distribution amp.*

output. Such a device is called an *isolation amplifier* (*iso-amp*). An example of such an iso-amp is an active direct box, which, in addition to reducing a signal's impedance, can prevent spurious ground and voltage potentials of the output of an electric instrument or amplifier from being injected into a console or mic preamp input.

Distribution Amplifiers

Often, it's necessary for audio signals to be distributed from one device to several other devices or signal paths within a recording console or music studio. In this situation, a *distribution amp* isn't used to provide gain but instead will amplify the current (power) that's delivered to one or more signal loads (Figure 10.11). Such an amp, for example, could be used to boost the signal so that a single headphone feed could be distributed to a large number of musicians during a string session.

Impedance Amplifiers

Amps can also be used to change the impedance (resistive load) of a signal. A good example of this is the preamp in a condenser mic, which is used to reduce the mic capsule's impedance (which can have a rating in excess of a billion ohms) down to a workable impedance of around 200 ohms.

Power Amplifiers

As you might expect, *power amplifiers* (Figures 10.12 and 10.13) are used to boost the current of a signal to a level that can drive one or more loudspeakers at their rated volume levels. Although these are often reliable devices, power amp designs have their own special set of

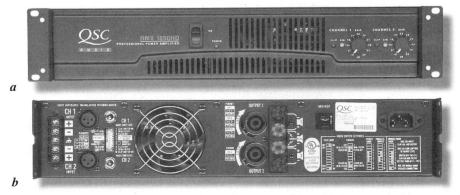

Figure 10.12. *QSC Audio's RMX 1850HD professional power amplifier: (a) front panel; (b) rear panel. (Courtesy of QSC Audio Products, Inc., www.qscaudio.com.)*

Figure 10.13. *Bryston 9B SST professional five-channel amplifier. (Courtesy of Bryston, Ltd., www.bryston.ca.)*

inherent problems. This includes the fact that transistors don't like to work at the high temperatures that can be generated during continuous, high-level operation. Such temperatures can also result in changes in the unit's response and distortion specifications . . . or outright failure, requiring that protective measures (such as fuse and thermal protection) be taken. Many of the newer amplifier models offer protection under a wide range of circuit conditions (such as load shorts, mismatched loads, and even open [no-load] circuits) and are usually designed to work with speaker impedance loads ranging between 4 and 16 □ (with most speaker models being designed to present a nominal load of 8 □). When matching amplifier

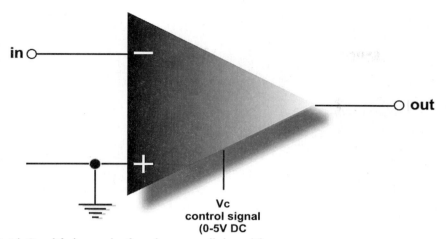

Figure 10.14. *Simplified example of a voltage-controlled amplifier.*

and speaker combinations, it's important that the amp be capable of delivering sufficient power to properly drive the speaker system. If the speaker's sensitivity rating is too low or the power rating too high for what the amp can deliver, there could be a tendency to "overdrive" the amp at levels that could cause the signal to be clipped. In addition to sounding distorted, clipped signals will often contain a high-level DC component that could potentially damage the speaker's voice coil drivers.

Voltage- and Digitally-Controlled Amplifiers

Up to this point, the discussion has largely focused on analog amps whose output levels are directly proportional to the signal level that's present at the input. Several exceptions to this principle are the *voltage-controlled amplifier* (*VCA*) and the *digitally-controlled amplifier* (*DCA*). In the case of the VCA, the overall output gain is a function of an external DC voltage (generally ranging from 0–5 V) that's applied to the device's control input (Figure 10.14). As the control voltage is increased, the analog signal will be proportionately attenuated. By inserting a digital-to-analog (D/A) converter at this point, a digitally controlled external voltage could be used to change the amp's overall gain. Certain older console automation systems, automated analog signal processors, and even newer digital console designs often make use of VCA technology. With the wide acceptance of digital technology in the production studio, it's now far more common to find devices that use digitally controlled amplifiers (also known as a digitally controlled attenuators) to control the gain of an audio signal. Although most digital devices change the gain of a signal directly in the digital domain (see the Digital Signal Processing/Multiplier section in Appendix A), it's also possible to change the gain of an analog signal using an external digital source. Much like the VCA, the overall gain of an analog amp can be altered by placing a series of digitally controlled step resistors into its negative feedback loop and digitally switching the amount of inserted resistance to achieve the desired gain.

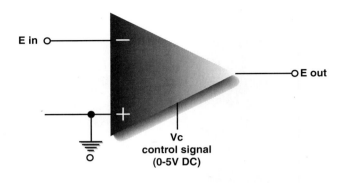

Figure 10.15. *The Analog Devices SSM2163 digitally controlled integrated circuit makes use of DCAs to control up to eight analog input signals. (Courtesy of Analog Devices, Inc., www.analogdevices.com.)*

An example of an eight-channel mixing chip that uses DCAs is the Analog Devices SSM2163 (Figure 10.15). This circuit can mix up to eight analog inputs under digital control to a stereo output bus. Each input can be attenuated up to 63 dB in 1-dB intervals and can be assigned to or panned between either of the stereo outputs.

CHAPTER

The Audio Production Console

The basic purpose of an *audio production console* (Figure 11.1) is to give us full control over volume, tone, blending, and spatial positioning for any or all signals that are applied to its inputs from microphones, electronic instruments, effects devices, recording systems, and other audio devices. An audio production console (which also goes by the name of *board*, *desk*, or *mixer*) should also provide a straightforward way to quickly and reliably route these signals to any appropriate device in the studio or control room so they can be recorded, monitored, and/or mixed into a final product. By analogy, a console can be likened to an artist's palette in that it provides a creative control surface that allows an engineer to experiment and blend all the possible variables onto a sonic canvas.

Before the introduction of multitrack recording, all the sounds and effects of a recording were mixed together at one time during a live performance. If the recorded blend (or *mix*, as it is called) wasn't satisfactory or if one musician made a mistake, the selection had to be performed over until the desired balance and performance was obtained. However, with the introduction of multitrack recording, the production phase of a modern recording session has radically changed into one that generally involves three stages:

- ◆ Recording
- ◆ Overdubbing
- ◆ Mixdown

Figure 11.1. *The final mix of Sting's album "Brand New Day" was mixed at Mega Studio, Paris, using a Solid State Logic SL 9000 J console. The majority of the album was recorded in Tuscany, Italy, using Sting's portable SL 4064 G+ console, which was designed to be easily dismantled and flight-cased. The console has been used on all of Sting's albums since 1993's "Ten Summoner's Tales." (Courtesy of Solid State Logic, www.solid-state-logic.com.)*

Recording

The *recording* phase involves the physical process of capturing live or sequenced instruments onto a recorded medium (tape, disk or whatever). Logistically, this process can be carried out in a number of ways:

◆ All the instruments to be used in a song can be recorded in one live pass.

◆ Live musicians can be used to lay down the basic foundation tracks (usually rhythm) of a song. Other instruments, vocals, etc., can then be added at a later time during the overdub phase.

◆ Electronic instruments, which were previously arranged and sequenced to form the basic foundation of a song, can be recorded onto the various tracks of recorder or digital audio workstation (DAW) in such a way that other live instruments, vocal tracks, and so on, can be added at a later time.

The last two of these procedures are most commonly encountered in the recording of popular music. The resulting foundation tracks (to which other tracks can be added at a later time) are called *basic*, *rhythm*, or *bed* tracks. These consist of instruments that provide the rhythmic foundations of a song and often include drums, bass, rhythm guitar, and keyboards (or any combination thereof). An optional vocal guide (scratch track) can also be recorded at this time to help the musicians and vocalists capture the proper tempo and that all-important feel of a song.

When recording popular music, each instrument is generally recorded onto separate tracks of a tape or DAW recorder (Figure 11.2). This is accomplished by plugging each mic into an input strip on the console (either directly into the mixer itself or into an appropriate input on a mic panel that's located in the studio), setting the gain throughout the input strip's signal path to its optimum level, and then assigning each signal to an appropriate console

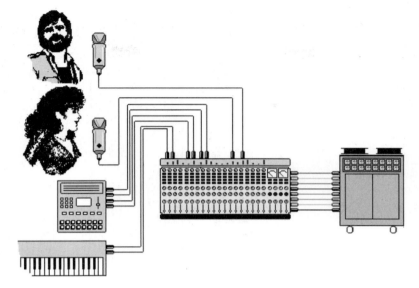

Figure 11.2. *When recording popular music, each instrument is generally recorded onto a separate track (or stereo tracks) of a multitrack recorder.*

output... which is finally routed to a desired track on the recording device. Although monitoring is important during this phase, the beauty behind this process is that the final volume, effects, and placement changes can be made at a later time—during the mixdown stage.

Should the need arise, an instrument or group of instruments that require multiple mics can be recorded onto a single track (or a stereo pair of tracks) by assigning the various input strips on a console to the same output bus (in a process known as *grouping*). These combined signals can then be balanced in level, panned, equalized, and processed by monitoring the grouped outputs at the console or recording device (Figure 11.3). Whenever multiple sources are recorded in a grouped fashion onto a single track or tracks, a great deal of care should be taken when setting the various volumes, tonalities, and placements. As a general rule, it's much more difficult to make changes to recorded grouped tracks, as changes to one instrument will almost always directly affect the overall group mix.

In the conventional one-instrument-per-track setting, each signal should be recorded at the highest possible level without overloading the analog or digital track. Recording at the highest level will result an optimum signal-to-noise ratio, so that tape hiss, preamp noise, or other artifacts won't impair the final product. Although digital tracks are more forgiving (due to the increased headroom...especially at higher bit rates), it's always a good idea to record signals to tape or disk at recommended levels (often peaking at 12 dB below the maximum overload point).

The recording stage is vitally important to the outcome of the overall project. The name of the game is to capture the best performance with the highest possible quality...and at optimum signal levels (often without regard to level balances on other tracks). In other words, a chanted whisper from a vocalist might easily be boosted to recorded levels that are equal to those of

Figure 11.3.
Several instruments can be "grouped" onto a single track (or stereo pair of tracks) by assigning each signal's input strip to the same console output bus.

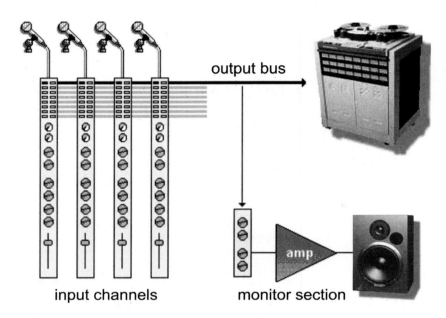

output bus

input channels   monitor section

the electric guitar (which ideally has been placed in another, separate room). From this, you might guess that the process of monitoring recorded signals is extremely important.

When you ask producers about the importance of preparation, they will most likely place it near the top of the list for capturing a project's sound, feel, and performance. The rhythm tracks (*i.e.*, drums, guitar, bass, and possibly piano) are often the driving backbone of a song, and recording them improperly can definitely get the project off to a bad start. Beyond making sure that the musicians are properly prepared and that the instruments are tuned and in top form (both of these being primarily the producer and/or band's job), it's the engineer's job to help capture a project's sound to tape, disk, or disc during the recording phase. Recording the best possible sound (both musically and technically), without having to excessively rely on the "fix it in the mix" approach, will definitely start the project off on the right track.

Monitoring

Since the instruments have been recorded at levels that don't relate to the program's final balance, a separate mix must be made in order for the artists, producer, and engineer to hear the instruments in their proper musical perspective . . . for this, a separate mix is often set up for *monitoring*. As you'll learn later in this chapter, a multitrack performance can be monitored in several ways. No particular method is right or wrong; rather, it's best to choose a method that matches your own personal production style. No matter which monitoring style is chosen, the

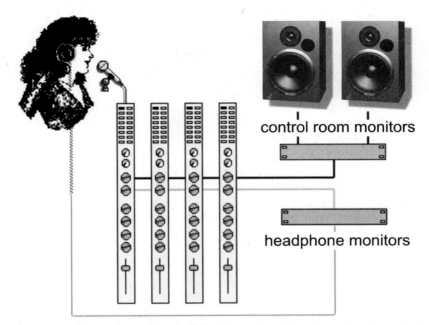

Figure 11.4. *During recording, each signal can be fed to the monitor mix section, where the various instruments can be mixed and then fed to the control room's main speakers and/or the performer's headphones.*

overall result will generally be as follows:

◆ When using a console or mixer during the recording process, each signal being fed to a track on the multitrack recorder or DAW will be fed to a studio monitor mix section (Figure 11.4). This submixer section is used to mix the individual inputs and instrument groups (with regard to level, panning, effects, etc.) into a musical balance that is then fed to the control room's main monitor speakers. It should be noted that when using a DAW's software mixer, the track balance levels are often independent of the interface inputs so the program's overall balance can be adjusted without fear of changing the record levels . . . a feature that greatly simplifies the control room monitoring process.

◆ When using a console or DAW, a separate monitor mix can often be created that can be heard by musicians in the studio over headphones. In fact, two or more separate "cue" mixes are often available (depending upon each musician's individual listening needs).

Overdubbing

Instruments that aren't present during the original performance can be added to the existing multitrack project during a process known as *overdubbing*. At this stage, musicians listen to the previously recorded tracks over headphones and then play along while recording new and separate tracks. A new "take" can be laid down to tape or disk if a musician has made minor mistakes during an otherwise good performance or if various instruments need to be added to the basic tracks at a later time in order to finish the project. These new performances are recorded in sync with the original performances and are laid onto unrecorded tracks or previously recorded tracks that are no longer needed. When overdubbing tracks onto an analog multitrack recorder, it's important to remember to place the tracks that are to be played back into the sync mode (a process whereby the record head is temporarily used as a playback head in order to maintain a proper time relationship during playback). Most modern analog multitracks can be placed into a master sync mode, which automatically switches the machine between the input and sync monitor modes (thereby eliminating the need for manual switching). For more information on sync playback, refer to Chapter 5 (The Analog Tape Recorder).

Mixdown

Once all the musical parts have been performed and recorded to everyone's satisfaction, the *mixdown* or *mix* stage can begin. At this point, playback outputs of a multitrack recorder (or recorders) are fed to the console's line inputs. When using a traditional console layout, this is done by switching the console to the mixdown mode or by changing the appropriate input switches to the line or Tape position. Most DAW software mixers need no switching or preparation. The master tape is then repeatedly played while adjustments in level, panning, EQ, effects, etc., are made for each track and/or track grouping. Throughout this artistic process, the individually recorded signals are blended into a composite stereo, surround, or mono signal that's fed from the main output busses to the master mixdown recorder (or are internally mixed within a DAW's software mixer). When a number of mixes have been made and a single version has been approved, this recording (called the *master* or *final mix*) can be mastered to its intended medium and/or assembled (along with other programs in the project) into a final product.

The Professional Analog Console

Although their numbers are dwindling as the digital age reaches adolescence, the vast majority of analog audio consoles used in professional recording studios (Figures 11.5 through 11.7) are laid out using controls that are traditional in both form and function. Their surfaces and functions basically differ only in appearance, location of controls, signal processing features, and routing abilities... as well their use of automation and recall features (if any exist).

Before we delve into the details of how a console works, it's important to take a look at one of the most important concepts in all of audio technology: the *signal chain* (also known as the *signal path*). As is true with literally any audio system, the recording console can be broken down into functional components that are chained together into a larger (and hopefully manageable) number of signal paths. By identifying and examining the individual components

Figure 11.5. *Mackie 8-bus with expansion sidecars. (Courtesy of Loud Technologies, Inc., www.mackie.com.)*

Figure 11.6. *Neve 88RS. (Courtesy of AMS Neve plc, www.ams-neve.com.)*

that work together to form this chain, it becomes easier to understand the basic layout of any console, no matter how large or complex.

In order to better understand a console or mixer system, let's start with the concept that it is built from numerous building-block components, each having an output (source) and an input (destination). In such a signal flow chain, the output of each source device must be connected to the input of the following device...whose output must be connected to the input of the following device...and so on, until the end of the audio path is reached. Whenever a link in this source-to-destination path is broken, no signal will pass. Although this might seem like a simple concept, keeping it in mind can save your sanity when paths,

Figure 11.7. *Solid State Logic SL 9000 K. (Courtesy of Solid State Logic, www.solid-state-logic.com.)*

devices, and cables that look like tangled piles of spaghetti get out of hand. It's as basic as the age-old knitting adage: knit one, purl two... meaning that it's important to be patient and keep your wits about you.

The next concept to grasp is that the signal for each input of a recording console or mixer will almost always flow vertically through a series of components in the signal chain, known as an *input strip* or *I/O module* (Figure 11.8). I/O (input/output) is so named because all of the associated electronics for a single track/output bus combination are often designed into a single circuit board that's physically attached to the input strip controls. Since the electronics of an I/O module are self-contained, they can be fitted into a modular mainframe in a number of channel and layout configurations (so as to better match the current and future production needs of a particular studio). This plug-in nature also makes them easily interchangeable and removable for service. The following sections describe the various I/O stages of a professional audio production console. Although consoles tend to vary in layout, this introduction will give you a better understanding of basic I/O strip design.

Channel Input

The channel input (Figure 11.9) serves as a preamp section to optimize the signal gain levels at the input of an I/O module before the signal is processed and routed. Either mic or line inputs can be selected and are continuously varied in gain by shared or independent level controls (often called *gain trims*). Although these values vary between designs, *mic trims* are typically capable of boosting a signal over a range of +20 to +70 dB, while a *line trim* can be varied in gain over a range of −15 (15-dB pad) to +45 dB. Gain trims are a necessary component in the signal path, as the output level of a microphone is typically very

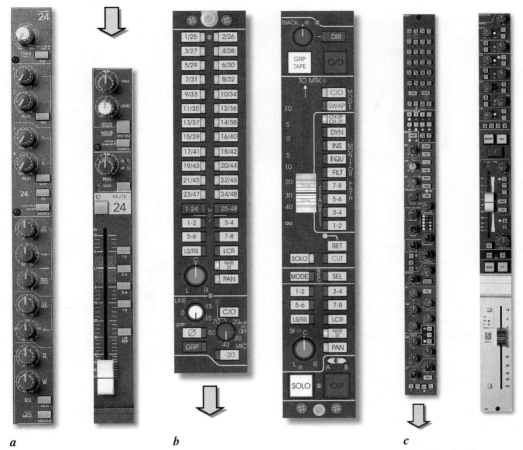

Figure 11.8. *Console I/O module input strips (side view): (a) Mackie 8-bus (courtesy of Loud Technologies, Inc., www.mackie.com); (b) Neve 88RS (courtesy of AMS Neve plc, www.ams-neve.com); (c) Solid State Logic XL 9000 K (courtesy of Solid State Logic, www.solid-state-logic.com).*

low (−45 to −55 dB) and requires that a high-quality, low-noise amp be used to raise and/or match the various mic levels in order for the signal to be passed throughout the console at an optimum level (as determined by the console's design and standard operating levels).

Whenever a mic or line signal is boosted to levels that cause the preamp's output to be driven above +28 dBm, severe clipping distortion will almost certainly occur. In order to avoid the dreaded LED overload light, the input gain must be reduced (by simply turning down the gain trim or by inserting an attenuation pad into the circuit). Conversely, signals that are too low in level will unnecessarily add noise into signal path. Finding the right levels is often a matter of knowing your equipment, watching the meter/overload displays, and using your experience.

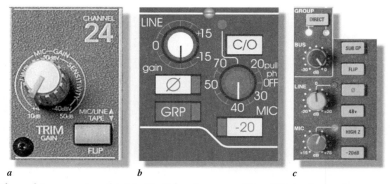

Figure 11.9. *Channel input section: (a) Mackie 8-bus (courtesy of Loud Technologies, Inc., www.mackie.com); (b) Neve 88RS (courtesy of AMS Neve plc, www.ams-neve.com); (c) Solid State Logic XL 9000 K (courtesy of Solid State Logic, www.solid-state-logic.com).*

Figure 11.10. *Although a console's signal path generally flows vertically from top to bottom, an "aux" sends path flows in a horizontal fashion . . . in that the various channel signals are mixed together to feed a mono or stereo output bus. This mix can be used for virtually any purpose (effects send, headphone cue, broadcast feed . . . you name it).*

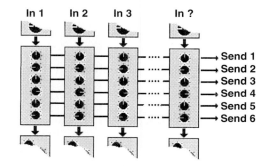

Input attenuation pads that are used to reduce a signal by a specific amount (*e.g.*, −10 or −20 dB) may be inserted ahead of the preamp, in order to prevent input overload. On many consoles, the preamp outputs may be phase-reversed, via the "Ø" button. This is used to change the signal's phase by 180° in order to compensate for polarity problems in the mic cable or during placement. High- and low-pass filters may also follow the preamp, allowing such extraneous signals as tape hiss or subsonic floor rumble to be filtered out.

Auxiliary Send Section

The *auxiliary sends* (*aux sends*) are used to route and mix signals from an input strip to the various effects output sends and monitor/headphone cue sends of a console. These sections are used to create a *submix* of any (or all) of the various console input signals to a mono or stereo send, which can then be routed to any destination (Figures 11.10 and 11.11). Commonly, up to eight individual aux sends can be found on an input strip. An auxiliary send can serve many purposes. For example, one send could be used to drive a reverb unit, signal processor, etc., while another could be used to drive a speaker that's placed in that great-sounding bathroom down the hall. A pair of sends (or a stereo send) could be used to provide a headphone mix for several musicians in the studio, while another send could feed

Figure 11.11. *Auxiliary send section: (a) Mackie 8-bus (courtesy of Loud Technologies, Inc., www.mackie.com); (b) Solid State Logic XL 9000 K (courtesy of Solid State Logic, www.solid-state-logic.com).*

a   b

a separate mix to the hard-of-hearing drummer. Hopefully, you get the idea that a send can be used for virtually any task that needs to be handled. Setting up a special mix for a satellite feed to Moscow or driving a speaker that's in the producer's motorcycle saddlebag shouldn't be a problem . . . how you deal with sends is literally up to you, your needs, and your creativity.

Equalization

The *equalization* (EQ) section (Figure 11.12), like the auxiliary sends, derives its feed directly from the channel input section. An input strip's equalizer is used to compensate for variations or discrepancies in frequencies that are present in the audio signal. Although the specifics vary from console to console, larger parametric EQ designs often include four continuously variable, overlapping, frequency-control bands, each having a variable bandwidth (Q) and a boost or cut control (often having values that can be boost or cut over a ±18 dB range). Most designs include an EQ in/out button that lets the engineer silently switch the equalizer in or out (bypass) of circuit. A complete explanation of equalization can be found in Chapter 12 (Signal Processing).

Insert Point

Many mixer and console designs provide a break in the signal chain that follows after the equalizer. At this point, a *direct send/return* or *insert point* (often referred to simply as *direct* or *insert*) can be used to send the strip's line level audio signal out to an external processing

Figure 11.12. Equalization section: (a) Mackie 8-bus (courtesy of Loud Technologies, Inc., www.mackie.com); (b) Solid State Logic XL 9000 K (courtesy of Solid State Logic, www.solid-state-logic.com).

a   b

or recording device. The external device's output signal can then be inserted back into the strip's signal path, where it can be routed and/or mixed back into the program. It's important to note that plugging a dynamics, EQ, or effects processor into an insert point will only affect the signal that passes through the selected I/O channel. Should you wish to affect a number of channels, the auxiliary send or group outputs can be used to process a combined set of input channels. Physically, the send and return jacks of a console or mixer can be accessed as two separate jacks on the top or back of a mixer/console (Figure 11.13a), as a single, stereo jack . . . in the form of a tip–ring–sleeve (TRS) jack that carries the send, return, and common signals (as shown in Figure 11.13b), or as access points on a console's patch bay.

A number of mixers and console manufacturers (such as Mackie and Tascam) use unbalanced, stereo TRS jacks to directly insert a signal that can be accessed in several interesting ways (Figure 11.14; please consult the manual for your system's jack layout):

- Inserting a mono send plug to the first click will connect the cable to the direct out signal, without interrupting the return signal path.
- Inserting a mono send plug all the way in will connect the cable to the direct out signal while interrupting the return signal path.
- Inserting a stereo TRS plug all the way in allows the cable to be used as its intended send/return loop.

Dynamics Section

Many top-of-the-line analog consoles offer a *dynamics section* on each of their I/O modules (Figure 11.15). This allows individual signals to be dynamically processed more easily, without

Figure 11.13. *Direct send/ return signal paths: (a) Two jacks can be used to send signals to and return signals from an external device. (b) A single TRS (stereo) jack can be used to insert an external device into an input strip's path. (c) Cable wiring diagram for a TRS send/return signal path. (Courtesy of Loud Technologies, Inc., www.mackie.com.)*

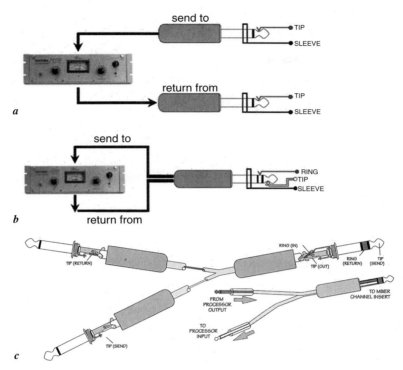

Figure 11.14. *Various insert positions for an unbalanced TRS send/return loop: (a) direct out with no signal interruption to a mixer's main output path—insert only to first click; (b) direct out with an interruption to the mixer's main output path—insert all the way to the second click; (c) for use as an insert loop (tip = send to external device, ring = return from external device, sleeve = common circuit ground). (Courtesy of Loud Technologies, Inc., www.mackie.com.)*

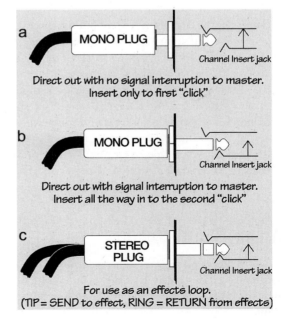

Figure 11.15. *Dynamics section of the Solid State Logic SL 9000 K.*
(Courtesy of Solid State Logic, www.solid-state-logic.com.)

the need to scrounge up tons of outboard devices. Often a full complement of compression, limiting, and expansion (including gating) is provided. A complete explanation of dynamic control can be found in Chapter 12 (Signal Processing).

Monitor Section

Since all the signals have commonly been recorded to tape or DAW at their optimum levels (without regard to the relative musical balance on other tracks), a means for creating a separate monitor mix in the control room is necessary in order to hear a musically balanced version of the production. As a result, a separate *monitor section* is usually designed into the console to provide control over each input's level, pan, effects, etc., and then route this mix to the control room's mono, stereo, or surround speakers. The approach and techniques of monitoring tracks during a recording will often vary from console to console (as well as among individuals). Again, no method is right or wrong compared to another. It simply depends on what type of equipment you're working with, as well as your personal working style. The following sections briefly describe a few of the most common approaches to monitoring.

In-Line Monitoring

Many newer console designs incorporate an I/O small fader section (Figure 11.16), which can be used to directly feed the recorded signal that's being fed to either the multitrack recorder or the monitor mixer (depending on its selected operating mode). In the standard monitor mix mode (Figure 11.17a), the small fader is used to adjust the monitor level for the associated recording track. In the *flipped* mode (Figure 11.17b), the small fader is used to control

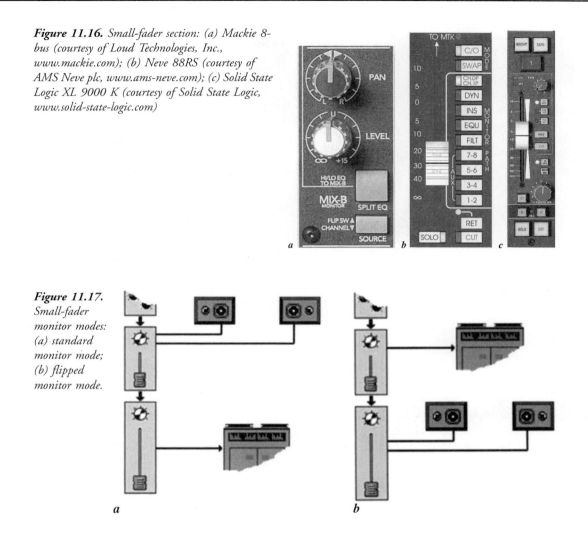

Figure 11.16. *Small-fader section: (a) Mackie 8-bus (courtesy of Loud Technologies, Inc., www.mackie.com); (b) Neve 88RS (courtesy of AMS Neve plc, www.ams-neve.com); (c) Solid State Logic XL 9000 K (courtesy of Solid State Logic, www.solid-state-logic.com)*

Figure 11.17. *Small-fader monitor modes: (a) standard monitor mode; (b) flipped monitor mode.*

the signal level that's being sent to the recording device, while the larger, main fader is used to control the monitor mix levels. This function allows multitrack record levels (which aren't often changed during a session) to be located out of the way, while the more frequently used monitor levels are assigned to the more accessible master fader position.

Separate Monitor Section

Certain British consoles (particularly those of older design) incorporate a separate mixing section that's dedicated specifically to the task of mixing the monitor feed. Generally located on the console's right-hand side (Figure 11.18), the inputs to this section are driven by

Figure 11.18. *Older style English consoles may have a separate monitor section (circled), which is driven by the console's multitrack output and/or tape return buses. (Courtesy of Buttermilk Records, www.buttermilkrecords.com.)*

the console's multitrack output and tape return buses . . . and offer level, pan, effects, and "foldback" (an older British word for headphone monitor control). During mixdown, this type of design has the distinct advantage of offering a large number of extra inputs that can be assigned to the main output buses during mixdown for use with effects returns, electronic instrument inputs, and so on. During a complex recording session, this monitoring approach will often require an extra amount of effort and concentration to avoid confusing the inputs that are being sent to tape or DAW with the corresponding return strips that are being used for monitoring (which is probably why this design style has fallen out of favor).

Direct Insertion Monitoring

If a console doesn't have any of the preceding monitor facilities (or even if it does), a simple and effective third option is still available to the user. This approach uses the direct send/returns of each input strip to insert the recorder directly into the input strip's signal path. Using this approach, the direct send for each of the associated tracks (which can be either before or after the EQ section) is routed to the associated track input on a multitrack recording device (Figure 11.19). The return signal is then routed from the recorder's output back into the same input strip's return path (where it injects the signal back into the strip's effects send, pan, and main fader sections for monitoring). Using this approach, the input signal directly following the mic/line preamp will be fed to the recorder (with record levels being adjusted by the preamp's gain trim). The return path (from the recorder) is then fed back into the input strip's signal path so it can be mixed (along with volume, pan, effects, sends, etc.) without regard for the levels that are being recorded to tape, disk, or other medium. Playing back the track or placing any track into the sync mode won't affect the overall monitor mix at all as

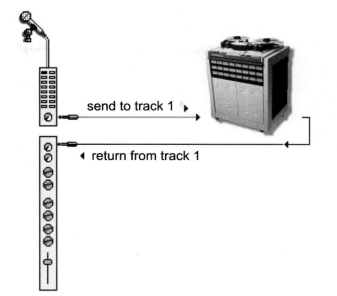

Figure 11.19. *By directly inserting the tape send and return into the signal path, a recorded track can be easily monitored.*

send to track 1 ▶

◀ return from track 1

the recorder's outputs are already being used to drive the console's monitor mix signals. This approach can be used effectively on mixers and larger consoles alike, with the only potential drawback being the large number of patch cords that might be required to patch the inputs and outputs to and from the multitrack recorder.

Output Fader

Each input strip contains an associated *output fader* (which determines the strip's bus output level) and *pan pot* (which is often designed into or near this section and determines the signal's left/right placement in the stereo and/or surround field). Generally, this section (Figures 11.20 and 11.21) includes a Solo/Mute feature, which performs the following functions:

◆ *Solo*—When pressed, the monitor outputs for all other channels will be muted, allowing the listener to monitor only the selected channel (or soloed channels) without affecting the multitrack or main stereo outputs during the recording and mixdown process. In certain designs, this function can be programmed to mute all non-soloed tracks from the main mix bus (thereby reducing inadvertent noise or hiss that might be coming from other tracks).

◆ *Mute*—This function basically works opposite to the solo button, in that when it is pressed, the selected channel is muted or cut from the monitor outputs. This function can often be programmed to mute the signal in the main bus outputs (thereby reducing inadvertent noise or hiss that might be coming from the track or selected tracks).

Figure 11.20. Output fader section of the Solid State Logic SL 9000 K. (Courtesy of Solid State Logic, www.solid-state-logic.com.)

Figure 11.21. Pan pot configurations: (a) stereo pan left/right control; (b) surround pan control/joystick.

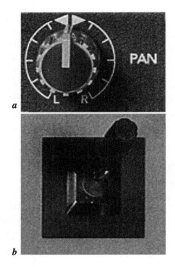

Output Bus

In addition to the concept of the signal chain, one other important signal path concept should be understood: the *output bus*. From the discussion of the input strip, we've seen that a channel's audio signal, by and large, follows down from the top to the bottom of the strip; however, as we take the time to follow this path, it's easy to spot where audio is routed off the strip and onto a horizontal output path. Conceptually, we can think of this path (or bus) as a single electrical conduit that runs the horizontal length of a console or mixer (Figure 11.22). In fact, this path is often a heavy copper wire or a single wire on a ribbon connector cable that runs the entire width of the console. It can be thought of as an electrical

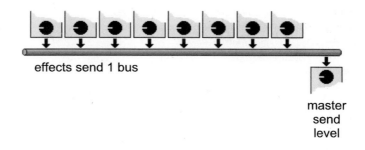

Figure 11.22. *Example of an effects send bus, whereby multiple inputs are mixed and routed to a master send output.*

effects send 1 bus

master
send
level

junction point that allows any number of signals to be injected into the bus (where it can be mixed in with other signals that are present on the line). Signals can also be routed off the bus to one or more output destinations (such as a console output, track output, or auxiliary effects send). Much like a city transit bus, this signal path follows a specific route and allows audio signals to get on or off the line at any point along its path.

Aux sends, monitor sends, channel assignments, and main outputs are all examples of signals that are injected into buses for routing to one or more output destinations. For example, it's easy to see that the aux send controls are horizontally duplicated across a console's surface. In fact, these gain controls are physically tied to an auxiliary send bus that routes the mixed levels to an output destination. An example of a stereo bus is the console's main stereo output. Following the strip fader and pan pot positioning, the relative output is injected into the console's main left/right mix buses . . . which are then combined with the various effects return signals and routed to a master recording device.

Channel Assignment

After the main output fader, the signal is often routed to the strip's track assignment matrix (Figure 11.23), which is capable of distributing the signal to any or all tracks of a connected multitrack recorder. Although this section electrically follows either the main or small fader section (depending on the channel's monitor mode), the track assign buttons will often be located either at the top of the input strip or designed into the main output fader (often being placed at the fader's right-hand side). Functionally, pressing any or all assignment buttons will route the input strip's main signal to the corresponding track output buses. For example, if a vocal mic is plugged into channel 14, the engineer might assign the signal to track 14 by pressing (you guessed it) the "14" button on the matrix. If a quick overdub on track 15 is also needed, all the engineer has to do is unpress "14" button and reassign the signal to track 15.

Many newer consoles offer only a single button for even- and odd-paired tracks, which can then be individually assigned by using the strip's main output pan pot. For example, pressing the button marked "5/6" and panning to the left routes the signal only to output bus 5 . . . while panning to the right routes it to bus 6. This simple approach accomplishes two things:

◆ Fewer buttons need to be designed into the input strip (lowering production cost and reducing the number of moving parts).

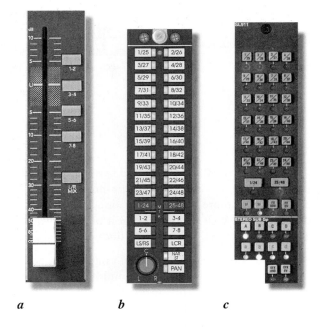

Figure 11.23. *Channel assignment section: (a) Mackie 8-bus (courtesy of Loud Technologies, Inc., www.mackie.com); (b) Neve 88RS (courtesy of AMS Neve plc, www.ams-neve.com); (c) Solid State Logic XL 9000 K (courtesy of Solid State Logic, www.solid-state-logic.com).*

◆ Stereo submixes can easily be built up by panning instruments within a stereo soundfield and then assigning their outputs to a pair of tracks on the multitrack recorder.

Master Output/Group Faders

Many mixer and console designs include a *master output fader* section (Figure 11.24) that allows the overall levels of an output bus or individual group bus levels to be trimmed. Basically, a master output fader or rotary pot serves as a convenient point for controlling the overall mono/stereo/surround mix signal level that's being sent to the master recording device. Individual group bus faders often have two functions:

◆ Vary the overall level of a grouped signal that's being sent to a recorded track.
◆ Vary the overall submix level of a grouped signal that's being routed to the master output bus.

In the latter instance, it would be much easier (and would cause less disruption to level balances) to alter the level of a bus fader than it would be to try to proportionately turn down all of the gain trims or channel faders on each strip.

During mixdown, the output groups are often invaluable in making relative volume changes to instrument sections or ensembles. By assigning any number of input channels to a single set of grouped outputs, a lot of time and frustration can be saved when changing the level of a master group fader within a mix. For example, let's say that we're in the middle of a mix that includes 10 drum tracks, piano, bass, vocals, full string section, and small choir. Instead of

Figure 11.24. *Mackie 8-bus master output fader section. (Courtesy of Loud Technologies, Inc., www.mackie.com.)*

assigning each track to the main L/R or surround bus, let's assign the drums to its own subgroup, the strings to another, and the choir to another. Once all of the relative balance, pan, and effects levels have been set on each input strip, the relative balance of each section can be changed by varying its associated group fader. That is, instead of turning down all the inputs (and risk losing the balance), we can simply turn down the group faders . . . which are, of course, being fed to the main mix output bus.

Monitor Level Section

Most console and mixing systems include a central *monitor section* that controls levels for the various monitoring functions (such as control room level, studio level, headphone levels, and talkback). This section (Figure 11.25) often makes it possible to easily switch between multiple speaker sets and can also provide switching between the various input and recording device sources that are found in the studio (*e.g.*, surround, stereo, and mono output buses; tape returns; aux send monitoring; solo monitoring).

Patch Bay

A *patch bay* (Figure 11.26) is a panel that (under the best of conditions) contains accessible jacks that correspond to the various inputs and outputs of every access point within a mixer or recording console. Most professional patch bays (also known as patch panels) offer centralized I/O access to most of the recording, effects, and monitoring devices or system blocks within the production facility (as well as access points that can be used to connect between different production rooms).

Figure 11.25. *Monitor level section for the Mackie 8-bus. (Courtesy of Loud Technologies, Inc., www.mackie.com.)*

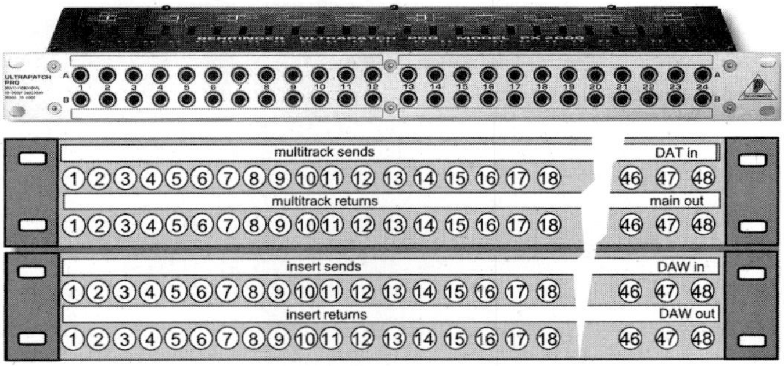

Figure 11.26. *The patch bay: (a) Ultrapatch PX2000 patch bay (courtesy of Behringer International GMBH, www.behringer.de); (b) rough example of a labeled patch bay layout.*

Patch bay systems come in a number of plug and jack types as well as wiring layouts. For example, prefabricated patch bays are available using tip–ring–sleeve (balanced), or tip–sleeve (unbalanced) 1/4-inch phone configurations, as well as RCA (phono) connections. These models will often place interconnected jacks at the panel's front and rear so that studio users can reconfigure the panel simply by rearranging the plugs at the rear access points. Other professional systems using the professional telephone-type (TT or mini [Bantam/TT]) plugs often require that you hand-wire the connections in order to configure or reconfigure a bay (usually an amazing feat of patience, concentration, and stamina).

Patch jacks can be configured in a number of ways to allow for several signal connection options among inputs, outputs, and external devices (Figure 11.27):

◆ *Open*—When no plugs are inserted, each I/O connection entering or leaving the panel is independent of the other and has no electrical connection.

◆ *Half-normalled*—When no plugs are inserted, each I/O connection entering the panel is electrically connected (with the input being routed to the output). When a jack is inserted into the top jack, the in/out connection is still intact, allowing you to tap into the signal path. When a jack is inserted into the bottom jack, the in/out connection is broken, allowing only the inserted signal to pass to the input.

◆ *Normalled*—When no plugs are inserted, each I/O connection entering the panel is electrically connected (with the input routing to the output). When a jack is inserted into the top jack, the in/out connection is broken, allowing the output signal to pass to the

Figure 11.27. Typical patch bay signal routing schemes. (Courtesy of Behringer International GMBH, www.behringer.de.)

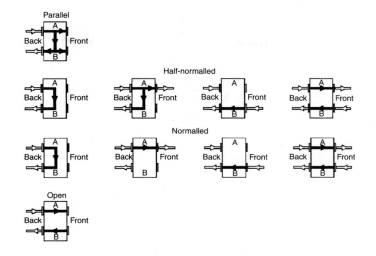

cable. When a jack is inserted into the bottom jack, the in/out connection is broken, allowing the input signal to pass through the inserted cable connection.

◆ *Parallel*—In this mode, each I/O connection entering the panel is electrically connected (with the input routing to the output). When a jack is inserted into either the top or bottom jack, the in/out connection will still be intact, allowing you to tap into both the signal path's inputs and outputs.

Breaking a normalled connection allows an engineer to patch different or additional pieces of equipment into a circuit that's normally connected. For example, a limiter might be temporarily patched between a mic preamp output and an equalizer input. The same preamp/EQ patch point could be used to insert an effect or other device type. These useful access points can also be used to bypass a defective component or to change a signal path order.... Versatility is definitely the name of the game here!

Metering

The input, bus outputs, and other level points of a console (such as an effects send) are often measured by a meter that displays signal level strength (Figure 11.28). Meter and indicator types will often vary from system to system. For example, banks of readouts that indicate console bus output and tape return levels might use VU metering, peak program meters

Figure 11.28. A set of LED, light-bar, and VU meter displays.

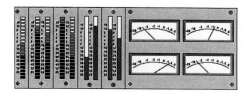

(PPMs, found in European designs), or LED/liquid crystal displays. It's also not uncommon to find LED overload indicators on an input strip's preamp which give quick and easy peak indications as to whether you've approached or reached the component's headroom limits (a sure sign to back off your levels).

The basic rule regarding levels isn't nearly as rigid as you might think and will often vary depending on whether the device or recording medium is analog or digital. In short, if the signal level is too low, tape, system, and even digital noise could be a problem, as the levels throughout the signal chain will probably not be optimized. If the level is too high, overloaded preamps, saturated tape, or clipped digital converters will often result in a distorted signal. Here are a few rules of thumb:

♦ In analog recording, the proper recording level is achieved when the highest reading on the meter is near the zero level, although levels slightly above or below this might not be a problem, as shown in Figure 11.29. In fact, (slightly) overdriving some analog devices and tape machines will often result in a sound that's "rough" and "gutsy."

♦ When recording digitally, noise is often less of a practical concern (especially when higher bit rates are used). It's often a good idea to keep levels as high as possible, while keeping a respectful distance from the dreaded clip or "over" indicator Unlike analog, digital is usually very unforgiving of clipped signals and will generate grunge that's guaranteed to make you cringe! As there is no real standard for digital metering levels beyond these guidelines, it's best to get specifics about levels from the device's manual (if one is provided).

Gain Level Optimization

Before leaving our discussion of the analog console, it's extremely important that we touch base on the concept of signal flow or *gain level optimization*. In fact, the idea of optimizing levels as they pass from one functional block in an input strip to the next or from one device to

Figure 11.29. *VU meter readings for analog devices: (a) too low; (b) too high; (c) just right.*

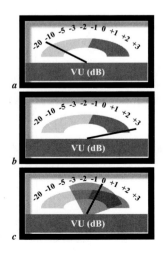

another is one of the more important concepts to be grasped in order to create professional-quality recordings. Although it's possible to go into a great deal of math in this section, I feel that it's far more important that you understand the underlying principles of level optimization, internalize them in everyday practice, and let common sense be your guide. For example, it's easy to see that if a mic that's plugged into an input strip is overdriven to the point of distortion that the signal following down the entire path will be distorted. By the same notion, driving the mic preamp at too low a signal will require that it be excessively boosted at a later point in the chain . . . resulting in increased noise. From this, it follows that the best course of action is to optimize the signal levels at each point in the chain (regardless of whether the signal path is within an input strip, internal to a specific device, or pertains to I/O levels as they pass from one device to another throughout the studio).

From a practical standpoint, level adjustments usually begin at the mic preamp. Although many of these points are situational and even debatable, one straightforward approach to setting proper gains on an input strip is to set the gain on the main strip fader (and possibly the master and group output faders) to 0 dB (unity gain). While monitoring levels for that channel or channel grouping, turn the mic preamp up until an acceptable gain is reached. Should the input overload LED light up, back off on the input level and adjust the output gain structure accordingly. Care should be taken when inserting devices into the signal chain at a direct insert point, making sure that the in and out signals are at or near their optimum level. In addition, the EQ section can also cause level overload problems whenever a signal is overly boosted within a frequency range.

Digital Console Technology

Console design, signal processing, and signal routing technology have undergone a tremendous design revolution with the advent of digital audio. As we move into the new millennium, digital consoles, digital mixers, software DAW mixers, and controller designs are finding their ways into professional, project, and audio production facilities at an amazing pace. These systems use large-scale integrated circuits and central processors to convert, process, route, and interface to external audio and computer-related devices with relative ease. In addition, this technology makes it possible for many of the costly and potentially faulty discrete switches and level controls that are required for such functions as track selection, gain, and EQ to be replaced by assignable digital switches and control networks. The big bonus, however, is that since routing and other processing control functions are digitally encoded, it becomes a simple matter for routing and dynamic automation settings to be saved in computer memory for instantaneous recall at any time. One of the other surprising aspects of this technology is the fact that digital console design is often more cost-effective than their analog counterparts.

Digital Hardware Consoles

In a digital console design (Figures 11.30 through 11.32), input signals are converted from analog into digital data or are directly inserted into the console's signal chain as digital

Figure 11.30. *Mackie dXb Digital Console. (Courtesy of Loud Technologies, Inc., www.mackie.com.)*

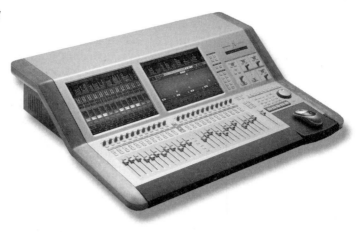

Figure 11.31. *Sony DMX-R100 Digital Console. (Courtesy of Sony Electronics, Inc., www.sony.com/proaudio.)*

Figure 11.32. *SSL C200 digital production console. (Courtesy of Solid State Logic, www.solid-state-logic.com.)*

data. Once done, these signals are thereafter distributed, routed, and processed entirely in the digital domain. At the console's various aux sends, main, and/or group outputs, the signal may either be decoded back into analog form or remain in the digital domain for distribution to a digital recording, processing, or transmission device. Since all audio, routing, and processing data is digitally controlled, many (if not all) of the system's mix, processing, and output moves can be recorded directly into the system's memory.

Central Control Panel

From a functional standpoint, the basic signal flow of a digital console or mixer's input strip is similar to that of an analog console, in that the input will first need to be boosted in level by a mic/line preamp (where it will be converted into digital data). From this point, the signals can pass through EQ, dynamics, and other signal processing blocks; various effects and monitor sends, volume, routing, and other sections that you might expect... such as main and group output fader controls. From a control standpoint, however, sections of a digital console might be laid out in a manner that's completely different. The main difference is a fundamental and philosophical change in the concept of the input strip, itself.

By the very nature of its design, all of the controls on an analog input strip must be duplicated for each strip along the console. This is the single-most factor that adds to the cost and reliability problems in traditional analog design. Since most (if not all) of the functions of a digital console pass through the device's central processor, this duplication is neither necessary nor cost effective. As a result, designers have often opted to keep the most commonly used controls (such as the pan, solo, mute, and channel volume fader) in their traditional input strip position. However, controls such as EQ, input signal processing, effects sends, monitor levels, and (possibly) track assignment have been designed into a central control panel (Figure 11.33) that can be used to vary a particular channel's setting parameters.

Figure 11.33. *Centralized control panel for the Sony DMX-R100 digital console. (Courtesy of Sony Electronics, Inc., www.sony.com/proaudio.)*

These controls can be quickly and easily assigned to a particular input strip by pressing the "select" button on the relevant input strip channel.

In certain designs, each input strip may be fitted with a virtual pot (V-pot) that can be assigned to a particular parameter for instant access, with position and level readouts being displayed via LEDs around the control. In others, the main parameter panel can be multipurpose in its operation, allowing itself to be reconfigured in a chameleon-like fashion to fit the task at hand. In such a case, touch-screen displays can be used to provide for an infinite degree of user control, while others might opt for software "soft" buttons that can easily reconfigure the form and function of the various buttons and dial controls in the control panel. Finally, console parameters can be controlled by physically placing a readout display at their traditional control points on the input strip. Using such a "knob per function" control/display system you can simply grab the desired parameter and move it to its new position (as though it were an analog console). Generally, this system is more expensive than those that use a central control panel, as more readout indicators, control knobs, and digital control interface systems are required. The obvious advantage to this interface would be instant access (both physically and visually) to any and all these parameters at once, in an analog fashion.

From the above, it's easy to get the idea that digital consoles often vary widely in their layout, ease of operation, and cost effectiveness. As with any major system device, it's often a good idea to familiarize yourself with the layout and basic functions before buying or taking on a project that involves such a beastie. As you might expect, operating manuals and basic tutorials are often available on the various manufacturers' Web sites.

Do-It-Yourself Tutorial: Channel Select

- ◆ Find yourself a digital mixer, console, or DAW with a software mixer. (If one's not nearby, you could visit a friend or a friendly studio or music store.)
- ◆ Feed several tracks into it and create a simple mix.
- ◆ Press the channel select for the first input and experiment with the EQ and/or other controls.
- ◆ Move on to the next channel and have fun building your mix.

Studio Interface

Beyond the human aspect of digital mixing, the interfacing of these devices to the production studio environment is often eased by the fact that they're already fluent in digital—the primary language of DAWs, digital recorders, audio interface systems, digital effects devices, MTC and time code synchronizers . . . as well as external transport controls. Often the console is capable of providing a systemwide word clock, so as to lock all of the digital audio converters and processing functions to a single, master timing element (see Chapter 6 for more information on this important topic). Although digital console systems often provide for analog I/O, digital I/O (such as S/PDIF, TDIF, and/or ADAT lightpipe) may allow for extensive audio, sync, and transport communications with a DAW or digital audio recorders. These I/O interface connections might be part of the console's design, or different I/O options might be available as plug-in cards that can be easily configured by the user. Another cost-effective option that's gaining in popularity allows the I/O and signal routing facilities of a digital console to interface with a DAW via the IEEE 1394 (FireWire) protocol. In this way, the console is able to integrate with the DAW in a way that allows the two to work in tandem as a single I/O, recording, controller, routing, and digital signal processing (DSP) system.

Although certain of the earlier console/mixer designs don't provide for digital audio connectivity, they are able to communicate with a DAW so as to act as a controller surface. In this way, the console is able to directly communicate mix, display, and transport control between the console and DAW via MIDI sys-ex messages. (*Note:* Consult your DAW's manual and Web site to see if your console or mixer is currently supported using the latest software or driver download version.)

DAW Software Mixers and Controllers

The increased power of the computer-based digital audio workstation has brought about a new, powerful and cost-effective audio mixing interface, the DAW software mixer (Figures 11.34 and 11.35). Through the use of a traditional (and sometimes not-so-traditional) user interface, these mixers offer functional on-screen control over levels, panning, EQ, effects, DSP, mix automation, and a host of functions that are too long to list here. Often these software consoles emulate their hardware counterparts by offering basic controls (such as fader, solo, mute, select, and routing) in the virtual input strip. Likewise, pressing the Select button will assign a multitude of virtual parameters on the central control panel to that strip.

When using a software mixer in conjunction with a DAW's waveform/edit window on a single monitor screen, it's easy to feel squeezed by the lack of visual "real estate." For this reason, many opt for a dual-monitor display arrangement. Whether you are working in

Figure 11.34. Cubase SX 2.0 mixer screen. (Courtesy of Steinberg Media Technologies GMBH, www.steinberg.net.)

Figure 11.35. *Pro Tools HD. (Courtesy of Digidesign, a division of Avid Technology, Inc., www.digidesign.com.)*

a PC or Mac environment, this working arrangement is easier and more cost effective than you might think and the benefits will immediately become evident . . . no matter what audio, graphics, or program environment you're working in.

When using a DAW controller (Figures 11.36 and 11.37), the flexibility of the DAW's software environment can be combined with the hands-on control and motorized automation (if available) of a hardware controller surface. More information on the DAW software mixing environment and hardware controllers can be found in Chapter 6 (Digital Audio Technology).

Console Automation

In multitrack recording, the final musical product generally isn't realized until the individual tracks are combined and processed in the mixdown stage. Each console channel that's being fed from the multitrack tape (as well as from other sources) will often have its own particular volume, spatial positioning, EQ, and other settings. Multiply each of these by 24 or more

Figure 11.36. *Tascam US-2400 USB DAW controller. (Courtesy of Tascam Corporation, www.tascam.com.)*

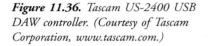

Figure 11.37. Digidesign's Control for Pro Tools TDM system. (Courtesy of Digidesign, a division of Avid, www.digidesign.com.)

tracks, then add the auxiliary effects send and return volumes, compression/limiting, and other signal-processing functions . . . and it quickly becomes obvious that the mixdown process can easily turn into a beast that's hard for an engineer to efficiently handle. As a result, mixdowns often need to be repeatedly rehearsed so the engineer can learn which controls must be operated, how much they should be varied, and at what point in the program any changes should come. It's not uncommon for the engineer, producer, and artists to spend 12 hours mixing a complicated piece of music before an acceptable mix can be obtained. Often mixes must be rejected because the engineer simply forgot to make one simple, but important, control change. Although the producer and engineer know how and when the control settings should be changed, the memory and physical dexterity required to execute them can sometimes simply exceed human abilities.

One solution to this problem is to mix with the aid of *console automation*. With such help, all of the settings and changes can be entered during the mixdown phase and then executed under computer control during a final mix pass. This way of working has the obvious advantages of allowing those involved to continually improve the mix until the desired effect is achieved. In addition to providing an extra set of "hands," an automation system's recall abilities can be a virtual lifesaver should you need to go back into the studio at a later date and make changes to an existing mix . . . or if an alternative mix is needed for a different medium (such as radio, television, music video, or film).

Console Automation in Action

In practice, an automation system can range from being able to sense only the position of a volume fader and level-related switching functions (often called *snapshot automation* because these settings represent mixer or console values at one point in time) to being a fully automated system that can store and recall all the dynamic functions on a production console. This amazing feat is accomplished by a system that continuously scans the various controls of a mixing surface over the course of a mix. Once a control has been altered, the automation system will detect the associated moves and convert them into a series of corresponding digital words that can be stored directly into the system's automation memory.

When dealing with automated system, mix moves are often carried out by converting the positions and states of the various dynamic controls on a console surface (such as volume, pan, EQ, and sends) into a DC voltage that can be translated into digital data for easy storage and processing. Stated more simply, an automated control might not pass audio, but instead would vary a DC voltage whose value would then be converted into digital data. Once in digital form, the automation data can be used to control gain and other analog audio parameters as it passes through the various links in the console chain using any of the following device types:

◆ *Voltage-controlled amplifier (VCA)*—This amplifier type is used to control the gain of a signal as a function of an external DC control voltage. A level control or other parameter can be designed to output a scaled DC voltage (often 0–5 VDC), which is then translated into a digital equivalent that can be stored into memory. . . . Upon playback, the data is converted back into the equivalent voltage, which directly controls the parameter level.

◆ *Digitally controlled amplifier (DCA)*—This amplifier type is used to control the gain of a signal as a function of a scaled set of digital word values. A DC control voltage is converted into equivalent digital data or a digital sensor inputs the data directly. Once in the digital domain, these word values can directly vary the gain of a DCA . . . or they can be used to scale the gain of an encoded signal directly in the digital domain.

◆ *Moving fader*—In an analog system, a fader may have two separate control paths: one to pass audio and one to pass a DC control voltage. Changing the fader position will vary the control voltage, which is then digitally encoded for storage and processing by the system. On playback, the demodulated control voltage is fed to a DC servo motor in the fader . . . which in turn causes the fader to automatically move to the proper fader position. In a digital mixing system, the process is the same, except that no analog is passed through the fader. . . . Quite simply, the scaled DC voltage is converted into digital values that can vary the gain of a DCA or the parameters of a digital processing block. Often, such a system (Figure 11.38) relies less on the traditional write, update, and read modes, as the faders are *touch sensitive*. This means that a fader can easily be programmed to enter into the update mode simply by touching it and moving it to the desired position. Since it's always physically at the current read position, all that's needed to update the levels to a new position is to take hold of the fader and move it to its new location. One thing's for sure . . . this type of automation brings out the kid in almost everybody. It's lots of fun to watch a mixer or console literally make complex moves all by itself!

As we've seen, any of the above gain changing devices can be controlled by converting analog voltage levels into equivalent digital data for storage and/or processing . . . and (in some cases) re-converting the data back into its original analog control form. When purely digital systems are used, it's not uncommon for parameter controllers to be used that output their settings

Figure 11.38. *Motorized moving faders doing what they do best.*

directly in the digital domain...a language that can be directly understood by the device's native processors. Variable encoders and alpha-dials often use optical encoding technology to continuously vary gain and positioning functions. Switching functions are often much easier to encode/decode, as buttons tend to have an off/on (0/1) state that can be easily converted to digital, with the two states being handled by a simple, reliable switching network.

Grouping

Often, console automation systems allow signals to be easily organized into *groups* according to instrument or scene change type. As noted earlier, this feature makes it possible for several instruments to retain their relative balance while offering control over their overall level from a single fader (thereby avoiding the need to change each channel volume individually). For example, 12 tracks of a string ensemble could be easily varied in overall level by assigning them to a single group. From a technical standpoint, grouping is easily accomplished by using a single control voltage or digital value to control the relative balance of several grouped channels and/or tracks (Figure 11.39).

Automation Modes of Operation

Just as different mixer or console designs often vary in design, ability, and function, the approach to automating a mix will also differ in form and sophistication from one system to the next. For example, some analog automation systems might offer only a few functional options, whereas the functionality of most digital consoles, DAWs, and large-scale analog consoles can be entirely automated. Control over these mix automation functions can be carried out in any of these three basic operating modes:

◆ Write mode
◆ Update mode
◆ Read mode

Figure 11.39. A single control voltage or digital value can be used to control the relative balance of several channels that are assigned to the same group.

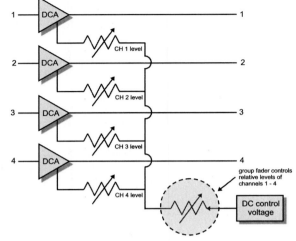

Write Mode

Once the mixdown process is underway, the process of writing the automation data into the system's memory can begin (of course, a basic automation mix can be made during the recording or overdub phase). When in the write mode, the system will begin the process of encoding mix moves for the selected channel or channels in real time. This mode can be used to globally record all of the settings and moves that are made on all the input strips (in essence, storing all of the mix moves, live and in one pass) . . . or on a selected strip or strips (allowing track mixes to be built up individually). The latter approach can help us focus all of our attention on a difficult or particularly important part and/or passage. Once done, another track or group of tracks can then be "written" into the system's memory . . . and then another . . . until an initial mix is built up.

Update Mode

As the name implies, the update mode lets us to go back at a later time and alter mix settings that were originally written into memory. In this mode, the updated settings can be changed by adding or subtracting from the track's control data, rather than by completely rewriting it. An advantage to this is best shown by example. . . . Let's say that we did a really good job at writing several complex and time-consuming moves for a group of tracks. The only problem is that a few of the tracks were simply not loud enough. Instead of taking the time to rewrite all of the tracks from scratch, we can simply go into the update/offset mode and change the relative balance levels of one or more tracks. In this way, the complex moves will remain intact, while the relative volume levels are changed (thereby making the overall passage louder or softer). Certain systems (including many DAW software mixers) don't offer an update mode. Instead, whenever a pass is made over a section that contains previously written automation data, these moves will remain unchanged unless a new mix move is written into memory at that point.

Level Matching

When writing over previous automation data (or when entering into the update mode), the concept of matching current fader level or controller positions to their previously written position often becomes important with automation systems that don't have moving faders. For example, let's say that we needed to redo several track moves that occur in the middle of a song. If the current volume fader positions don't match up with their previously written positions, the mix levels could jump during the transition on playback. In order to manually set the faders to their previous mix positions, some form of indicator is needed. For those who are working with a console with moving faders, there will be no need to manually match the settings as the faders and/or controls will conveniently and automatically move to their current mix level position. However, those who are working in a purely VCA-based system or non-moving DAW controller will often need to match the positions by hand. This is often done with the aid of nulling indicator lights that reads the difference and indicates whether the setting is higher, lower, or equal to the current mix position (Figure 11.40). After a level match is achieved, the engineer can switch between the modes without fear of a sudden jump in level. After the write mode is entered, the controls can be moved at will.

Read Mode

Consoles that are placed into the read mode will play the mix information from the system's automation and convert the data into a form that can be understood by the system's VCA, DCA, moving fader, or digital readout controls. Once the final mix has been achieved, all that's needed is to roll the tape, sit back, and listen to the mix.

Figure 11.40. Readout null displays are used to show current fader positions.

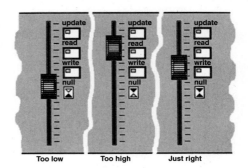

Figure 11.41. *Example of a DAW's rubberband mix interface.*

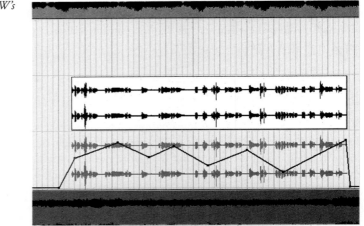

Drawn (Rubberband) Automation

In addition to standard fader modes of operation, some console automation systems and most DAW editing systems offer a graphic form of automation that allows the operator to draw fades and complicated mix moves in an on-screen fashion. This interface is often referred to as *rubberband* automation or mixing, in that the graphic lines (that represent relative fade, volume, pan, and other parameters) can be bent, stretched, and contorted like a rubberband (Figure 11.41). Commonly, all that's needed to define a new mix point is to click on a point on the rubberband (at which point a box "handle" will appear) and drag it to the desired position. Changing a move is simply done by clicking on an existing handle and moving it to its new position.

Other Automation Options

Certain mixer and console designs that are equipped with automation may also allow "snapshots" to be taken of the current mix settings. Just as a camera freezes a Kodak moment in time, a console snapshot can place many or all of the mixing surface's settings into memory for recall at a later time. With a nonautomated mixer or console, one of the more tedious tasks for an assistant engineer or engineer is to "zero" the board ... the process of setting all of the controls to their initial zero, null, or centered state—in other words, one where all the controls are ready for a new session. This process can be a tedious exercise in repetition (particularly if there are 96 inputs and tons of parameter controls). With the help of snapshot automation, this process can automatically take place at the simple touch of a button. Of course, this same kind of recall automation is commonly used to reset a console to its former settings when returning to the studio several weeks after a session.

From all of the above, it quickly becomes obvious that different automation systems will likely work and perform in very different ways. Most large-console systems and DAW mixers offer extensive editing and merging facilities, allowing you to mix short sections of a song, project, or soundtrack and combine the best of these takes into a final mix master. These takes (or different versions of a single or composite mix) could then be saved to disk or disc, for recall at a later time. As with any device and/or system in the studio...which, by its very nature, is flexible and moderately complicated...it's often a good idea to consult the user's manual and take a bit of time to become familiar with its operation. Alternately, you could always have an assistant on hand who can show you the ropes.

MIDI-Based Automation

Many project-oriented mixers and console systems are capable of communicating automation information via MIDI. The immediate advantages to this include the following:

◆ Automation information can be stored into a conventional MIDI sequencer.

◆ Most modern effects devices can be MIDI controlled, allowing effects to respond to program-change messages during a mix.

◆ MIDI data is an established, cost-effective standard in worldwide use for professional and project studio production.

◆ Automation data can be easily synchronized to a project via MIDI time code.

Most digital mixers, controllers, and small console designs can be controlled (at a basic or complex mix level) via MIDI, in either a dynamic or snapshot fashion (in fact, this is how most DAW hardware controllers communicate). These mixer systems include MIDI ports for communicating mix-related data to and from a DAW or MIDI sequencer as a stream of MIDI sys-ex messages. Because of the expense, most analog mixer designs don't include MIDI mixing capabilities (unless they have been specially fitted with a MIDI-controlled VCA-based automation system), although a small number of console designs, however, do include switching and routing systems that can be dynamically or snapshot automated in this manner.

Effects Automation via MIDI

One of the most common ways to automate external hardware devices is through the use of MIDI (from a sequence, parameter controller, or MIDI stomp-box controller). In the same way that electronic instruments can store sound patches into a memory-location register for later recall, most modern effects devices will let you store program patches into memory, where they can be recalled at any time using program change commands. Through the use of program change commands and continuous-controller sys-ex messages, it's possible for signal processing patches and parameters to be altered in real time. By locking a sequence to a mix (*e.g.*, from a DAW MIDI track), it's possible for a series of effect changes to be automatically triggered at the proper time within an automated mix.

The Art of Mixing

Actually, the topic of the art of mixing could easily fill a book (and I'm sure it has); however, I'd simply like to point out the fact that it is indeed an art form...and as such is a very personal process. I remember the first time that I sat down at a console (an older Neve 1604). I was truly petrified and at a loss as to how to approach the making of *a mix*. Am I over-equalizing? Does it sound right? Will I ever get used to this sea of knobs? Well, folks, as with all things...the answers come to you when you simply sit down and mix, mix, mix! It's always a good idea to watch others in the process of practicing their art and take the time to listen to the work of others (both the known and the not so well known). With practice, it's a foregone conclusion that you'll begin to develop your own sense of the art and style of mixing...which, after all, is what it's all about. It's up to you now. Just dive into the deep end—and have fun!

CHAPTER

Signal Processing

Signal processing has become an increasingly important part of modern audio and music production. It's the function of a *signal processor* to change, augment, or otherwise modify an audio signal in either the analog or digital domain. Before we delve into the various types of effects devices and functional applications that are available in audio production, it's important that we first cover a few basic device types and applications that are near and dear to the process of sculpting sound into forms that are subtle, lush, or just plain whimsical and wacky.

The Wonderful World of Analog, Digital, or Whatever

Signal processing devices and their applied practices, come in all sizes, shapes, and flavors. These tools and techniques might be analog, digital, or even acoustic in nature. The very fact that early analog processors have made a serious comeback (in the form of reissued hardware and software plug-in emulations) points to the importance of embracing past tools and techniques while combining them with the technological advances of the day to make the best possible productions.

Although they aren't the first choices that comes to mind, the use of acoustic and ambient mic techniques are often the first line of defense when dealing with the processing of an audio signal. For example, as we saw in Chapter 4, changing a mic or its position might be the best option for changing the

Figure 12.1. *Effects bay at Four Seasons Media Productions, St. Louis, MO. (Courtesy of Russ Berger Design Group, Inc., www.rbdg.com.)*

character of a pickup. Placing a pair of mics out into a room or mixing a guitar amp with a second, distant pickup might fill the ambience of an instrument in a way that a device just might not be able to duplicate. In short, never underestimate the power of your acoustic environment as an effect.

For those wishing to work in the world of analog, an enormous variety of devices can be put to use in a production. Although these generally relate to devices that alter a source's relative volume levels (*e.g.*, equalization and dynamics), there are a number of analog devices that can be used to alter effects that are time based. For example, an analog tape machine can be patched so as to make an analog delay device or a regenerative echo (echo, echo) device. Although they're not commonly used, spring and plate reverb units that add their own distinctive sound can still be found on the used and sometimes new market.

The world of digital audio has definitely set signal processing on fire by offering an almost unlimited range of effects that are available to the musician, producer, and engineer. One of the biggest advantages to working in the digital signal processing (DSP) domain is the fact that software programming can be used to configure a processor in order to achieve a wide range of effects (such as reverb, echo, delay, equalization, dynamics, pitch shifting, or gain changing).

The task of processing a signal in the digital domain is accomplished by combining logic circuits in a building-block fashion. These logic blocks follow basic binary computational rules that operate according to a special program algorithm. When combined, they can be used to alter the numeric values of sampled audio in a highly predictable way. After a program has been configured (from either internal ROM, RAM, or system software), complete control over a program's setup parameters can be altered and output as an effected digital audio stream. Since the process is fully digital, these settings can be saved and precisely duplicated at any time upon recall. Even more amazing is how the overall quality and functionality have steadily increased while at the same time becoming more available and cost effective. It has truly brought a huge amount of production power to the beginner as well as to the pro.

For those wishing to gain a better understanding of the under-the-hood aspects of DSP, I urge you to read (or reread) the section in Chapter 6). The point to this discussion is that there are many ways to accomplish an effect. Acoustic, analog, and digital all have their place in the scheme of sound production. . . . It's all a matter of personal choice and of choosing the right tool and toy for the job.

Plug-Ins

In addition to both analog and digital hardware devices, an ever-growing list of signal processors is available for the Mac and PC platforms in the form of software plug-ins. These software utilities offer virtually every processing function imaginable (often at a fraction of the price of their hardware counterparts) with little or no reduction in quality, capabilities, or automation features. These programs (which are programmed and marketed by large and small companies alike) are designed to be integrated into an editor or DAW production environment in order to perform a particular real- or non-real-time processing function.

Currently, several plug-in standards exist, each of which functions as a platform that serves as a bridge to connect the plug-in—through the computer's operating system (OS)—to the digital audio production software. This means that any plug-in (regardless of its manufacturer) will work with an OS and DAW that's compatible with that platform standard, regardless of its form, function, and/or manufacturer. As of this writing, the most popular standards are VST (PC/Mac), DirectX (PC), AudioSuite (Mac), Audio Units (Mac), MAS (MOTU for PC/Mac), and TDM (Digidesign for PC/ Mac).

By and large, effects plug-ins operate in a native processing environment. This means that the computer's host CPU processor carries out the DSP functions. With the ever-increasing speed and power of modern-day CPUs, this has become less and less of a problem; however, when working on a complex session it's still possible for your computer to run out of DSP steam. This can be dealt with in several ways:

- Your computer or processor can be beefed up in order to take full advantage of your system.
- Many DAWs offer a "freeze" or "lock" function that allows a track or processing function to be written to disc in a non-real-time fashion in order to free up the CPU for other real-time calculations.
- A DSP accelerator card (Figure 12.2) that can be plugged into your computer acts as a dedicated plug-in processor to share the processing workload with the CPU.
- When using Steinberg's VST plug-ins, it's possible to share the CPU workload over several networked computers using their VSTLink (V-Stack) protocol.

Inline vs. Side-Chain Processing

Before we delve into process of effecting and/or altering sound in various ways, we should first take a quick look at an important signal path concept . . . the fact that a signal processing

Figure 12.2. *Universal Audio's UAD-1*
DSP card. (Courtesy of Universal Audio,
www.uaudio.com.)

device can be inserted into an audio chain in one of two ways:

◆ Inline routing
◆ Side-chain routing

Inline Routing

Inline routing is often used to alter a signal. It occurs whenever a processor is inserted directly into a signal path in a serial fashion (Figure 12.3). Using this approach, the signal path passes from the device's input, through the processing element and directly out to another

Figure 12.3. *An example of inline routing, whereby the*
processed signal is inserted directly into the signal path.

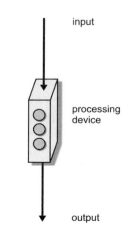

input

processing
device

output

device or point in the chain. This method for inserting a device is generally used for processing a single instrument, voice or grouped signals that are present on that particular line. Often, but not always, this device tends to be a level-based processor (such as an equalizer, compressor, or limiter). In keeping with the "no-rules" concept, time- and pitch-changing devices can also be used to tweak an instrument or voice in the signal chain.

Here are but a few examples of how inline routing can be used:

◆ A device can be plugged into an input strip's insert (direct send/return) point. This approach is often used to insert an outboard device directly into an input strip's signal path.

◆ A console's main output bus could be run through a device (such as a stereo compressor) to affect an overall mix or submix grouping.

◆ An effects stomp box could be placed between a mic preamp and console input to create a grungy distortion effect.

◆ A DAW plug-in could be inserted into an input path to process only the signal on that channel.

Side-Chain Routing

Side-chain routing is often used to augment a signal. It occurs whenever a signal is allowed to pass through the chain while a side signal is simultaneously fed to an effects device... once effected, the signal is then proportionately mixed back in with the original signal to create an effects blend (Figure 12.4). In other words, a portion of the unbroken signal path is fed to an effects device (often via an effects send bus), where it is processed and then mixed back into the original or main output signal path (often via an effects return bus or spare input strip). This approach is commonly used to proportionately feed a group of signals from their respective input strips (using an effects send bus) to a single effects processor. Often, but not always, this device tends to be a time- and pitch-changing processor (e.g., adding reverb, chorus, flanging, etc.).

Figure 12.4. *An example of side-chain routing, whereby the processed signal is proportionately mixed with the signal's main path.*

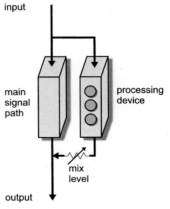

dry/effects mix control

Figure 12.5. *Example of a commonly found mix control that allows an inline device to work in a side-chain fashion.*

Side-chain routing is primarily used by plugging the output of an effects send into a reverb device and then routing its stereo outputs back into a spare input strip. Portions of any signal can be effected by that device simply by turning up the related effects send pot on any strip. It's not at all uncommon for a patch that looks like an inline path to actually be a side-chain effects route. This occurs when the effects device has a *mix* control (Figure 12.5), which allows for a variable mixture between the incoming signal (which passes through the device unaffected) and a side-chain processing path (which affects the signal according to the device's function). In this way, the variations between the uneffected path (0%, dry) and the fully effected (100%, wet) path can be mixed using a simple device routing setup.

From this point on, this chapter will be taking an in-depth look at many of the signal processing devices, applications, and techniques that have traditionally been the cornerstone of music and sound production, including systems and techniques that excerpt an ever-increasing degree of control over:

◆ *The spectral content of a sound*—In the form of equalization or intelligent bandpass filtering

◆ *Amplitude level processing*—In the form of dynamic range processing

◆ *Time-based effects*—Augmentation or re-creation of room ambience, delay, time/pitch alterations, and tons of other special effects that can range from being sublimely subtle to "in yo' face."

Equalization

The most common form of signal processing is equalization (EQ). The audio equalizer (Figure 12.6) is a device or circuit that lets us control the relative amplitude of various

Figure 12.6. *Manley Massive Passive stereo equalizer. (Courtesy of Manley Laboratories, Inc., www.manleylabs.com.)*

frequencies within the audible bandwidth. Put another way, it exercises tonal control over the harmonic or timbral content of a recorded sound. EQ may need to be applied to a single recorded channel, to a group of channels, or to an entire program (often as a step in the mastering process) for any number of other reasons, including:

- To correct for specific problems in a recording or room (possibly to restore a sound to its natural tone)
- To overcome deficiencies in the frequency response of a mic or in the sound of an instrument
- To allow contrasting sounds from several mics or tape tracks to blend together better in a mix
- To increase the separation between mics or recorded audio tracks by seeking to reduce those frequencies that excessively "leak" between channels.
- To alter a sound purely for musical or creative reasons.

Equalization refers to the alteration in frequency response of an amplifier so that the relative levels of certain frequencies are more or less pronounced than others. EQ is specified as either plus or minus a certain number of decibels at a certain frequency. For example, you might want to boost a signal by "+4 dB at 5 kHz." Although only one frequency was specified in this example, in reality a range of frequencies above, below, and centered around the specified frequency will often be affected. The amount of boost or cut at frequencies other than the one named is determined by whether the curve is peaking or shelving, by the bandwidth of the curve (a factor that's affected by the Q settings and determines how many frequencies will be affected around a chosen centerline), and by the amount of boost or cut at the named frequency. For example, a +4 dB boost at 1000 Hz might easily add a degree of boost or cut at 800 and 1200 Hz (Figure 12.7).

Older equalizers and newer "retro" systems often base their design around filters that use passive components (*i.e.*, inductors, capacitors, and resistors) and employ amplifiers only to make up for internal losses in level, called *insertion loss*. Figure 12.8a shows typical signal levels in a passive equalizer that's set for a flat response, while Figure 12.8b shows the signal level structure of an equalizer that has a low-end boost. Most equalization circuits today, however, are of the active filter type that change their characteristics by altering the feedback loop of an operational amp (Figure 12.9). This is by far the most common analog EQ type and is generally favored over its passive counterpart due to its low cost, size, and weight . . . as well as its wide gain range and line-driving capabilities.

Peaking Filters

The most common EQ curve is of the *peaking filter*. As its name implies, a peak-shaped bell curve can either be boosted or cut around a selected center frequency. Figure 12.10 shows the curves for a peak equalizer that's set to boost or cut at 1000 Hz. The *quality factor* (Q) of a peaking equalizer refers to the width of its bell-shaped curve. A curve with a high Q will have a narrow bandwidth with few frequencies outside the selected bandwidth being

Figure 12.7. *Various boost/cut EQ curves centered around 1 kHz: (a) center frequency, 1 kHz bandwidth 1 octave, ±15 dB boost/cut; (b) center frequency, 1 kHz bandwidth 3 octaves, ±15 dB boost/cut. (Courtesy of Mackie Designs, www.mackie.com.)*

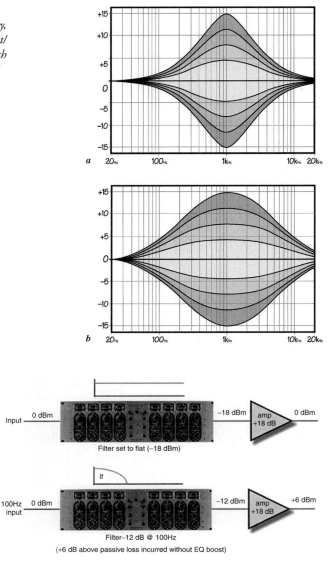

Figure 12.8. *Typical signal levels in a passive equalizer. (a) EQ is set for flat response. (b) EQ filter is set for 6-dB boost at 100 Hz.*

affected, whereas a curve having a low Q is very broadband and can affect many frequencies (or even octaves) around the center frequency. *Bandwidth* is a measure of the range of frequencies that lie between the upper and lower −3 dB (half-power) points on the curve (Figure 12.11). The Q of a filter is an inverse measure of the bandwidth (such that higher Q values mean that fewer frequencies will be affected . . . and *vice versa*). To calculate Q, simply divide the center frequency by the bandwidth. For example, a filter centered at 1 kHz that's a third of an octave wide will have its −3 dB frequency points located at 891 and 1223 Hz, yielding a bandwidth of 232 Hz (1123 − 891). This EQ curve's Q, therefore, will be 1 kHz divided by 232 Hz, or 4.31.

Figure 12.9. *Simplified example of an active equalization circuit.*

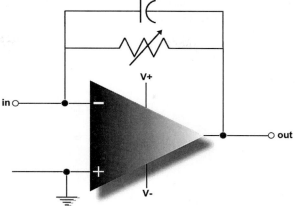

Figure 12.10. *Peaking equalization curves.*

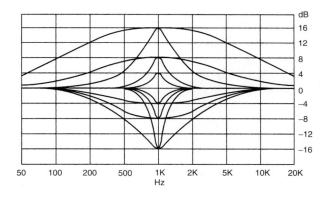

Figure 12.11. *The number of hertz between the two points that are 3 dB down from the center frequency determines the bandwidth of a peaking filter.*

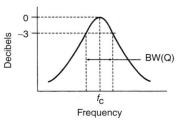

Shelving Filters

Another type of equalizer is the shelving filter. Shelving refers to a rise or drop in frequency response at a selected frequency, which tapers off to a preset level and continues at that level to the end of the audio spectrum. Shelving can be inserted at either the high or low end of the audio range and is the curve type that's commonly found on home stereo bass and treble controls (Figure 12.12).

Figure 12.12. *High/low, boost/cut curves of a shelving equalizer.*

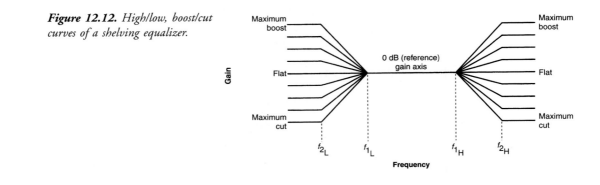

High-Pass and Low-Pass Filters

Equalizer types also include high-pass and low-pass filters. As their names imply, this EQ type allows certain frequency bandwidths to be passed at full level while other sections of the audible spectrum are attenuated. Frequencies that are attenuated by less than 3 dB are said to be inside the *passband*; those attenuated by more than 3 dB are in the *stopband*. The frequency at which the signal is attenuated by exactly 3 dB is called the *turnover* or *cutoff frequency* and is used to name the filter frequency. Ideally, attenuation would become infinite immediately outside the passband; however, in practice this isn't always attainable. Commonly, attenuation is carried out at rates of 6, 12, and 18 dB per octave. This rate is called the *slope* of the filter. Figure 12.13, for example, shows a 700-Hz high-pass filter response curve with a slope of 6 dB per octave, and Figure 12.14 shows a 700-Hz low-pass filter response curve having a slope of

Figure 12.13. *A 700-Hz high-pass filter with a slope of 6 dB per octave.*

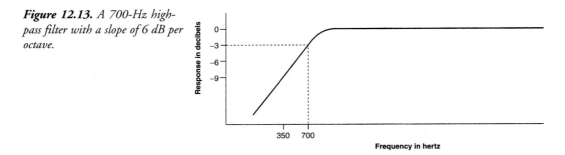

Figure 12.14. *A 700-Hz low-pass filter with a slope of 12 dB per octave.*

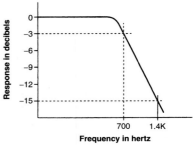

Figure 12.15. *A bandpass filter is created by combining a high- and low-pass filter with different cutoff frequencies.*

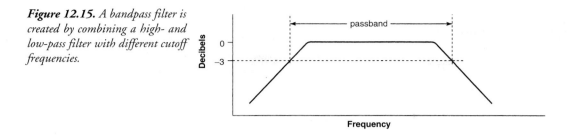

12 dB per octave. High- and low-pass filters differ from shelving EQ in that their attenuation doesn't level off outside the passband. Instead, the cutoff attenuation continues to increase. A high-pass filter in combination with a low-pass filter can be used to create a *bandpass filter*, with the passband being controlled by their respective turnover frequencies and the Q by the filter's slope (Figure 12.15).

Equalizer Types

The four most commonly used equalizer types that can incorporate one or more of the previously described filter types are the:

- ◆ Parametric equalizer
- ◆ Selectable frequency equalizer
- ◆ Graphic equalizer
- ◆ Notch filter

The *parametric equalizer* (Figure 12.16) lets you adjust most or all of its frequency parameters in a continuously variable fashion. Although the basic design layout will change from model-to-model, each band will often have an adjustment for continuously varying the center frequency. The amount of boost or cut is also continuously variable. Control over the center frequency (Q) may be either selectable or continuously variable, although certain manufacturers might not have provisions for a variable Q.

Generally, each set of frequency bands will overlap into the next band section, so as to provide smooth transitions between frequency bands or allow for multiple curves to be placed in nearby frequency ranges. Because of its flexibility and performance, the parametric equalizer has become the standard design for most input strips, digital equalizers, and workstations.

Figure 12.16. *The EQF-100 full range, parametric vacuum tube equalizer. (Courtesy of Summit Audio, Inc., www.summitaudio.com.)*

Figure 12.17. *The selectable frequency equalizer: (a) venerable API 550A switchable equalizer within the 7600 outboard module (courtesy of API Audio, www.apiaudio.com); (b) plug-in emulation of the API 550B for TDM, RTAS, and AudioSuite (courtesy of Unique Recording Software, www.ursplugins.com).*

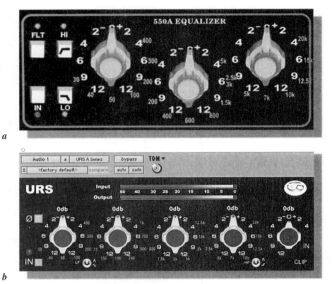

a

b

The *selectable frequency equalizer* (Figure 12.17), as its name implies, has a set number of frequencies from which to choose. These equalizers usually allow a boost or cut to be performed at a number of selected frequencies with a predetermined Q; they are most often found on older console designs, certain low-cost production consoles, and outboard gear.

A *graphic equalizer* (Figure 12.18) provides boost and cut level control over a series of center frequencies that are equally spaced (ideally according to music intervals). An "octave band" graphic equalizer might, for example, have 12 equalization controls spaced at the octave intervals of 20, 40, 80, 160, 320, and 640 Hz and 1.25, 2.5, 5, 10, and 20 kHz, while 1/3-octave equalizers could have up to 36 center-frequency controls. The various EQ band controls generally use vertical sliders that are arranged side by side so that the physical positions of these controls could provide a "graphic" readout of the overall frequency response curve at a glance. This type is often used in applications that can help with the fine-tuning a system to compensate for the acoustics in various types of rooms, auditoriums, and studio control rooms.

Notch filters are often used to zero in on and remove 60- or 50-Hz hum or other undesirable discrete-frequency noises. It uses a very narrow bandwidth to fine-tune and attenuate

Figure 12.18. *Rane GE 130 single-channel, 30-band, 1/3-octave graphic equalizer. (Courtesy of Rane Corporation, www.rane.com.)*

Figure 12.19. *Notch filter response curves. (Courtesy of Orban Associates, Inc., www.orban.com.)*

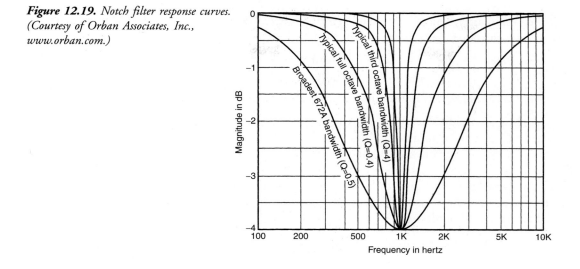

a particular frequency in such a way as to have little effect on the rest of the audio program (Figure 12.19). Notch filters are used more in film location sound and broadcast than in studio recording, as severe narrow-band problems aren't often encountered in a well-designed studio . . . hopefully.

Applying Equalization

When you get right down to it, EQ is all about compensating for deficiencies in a sound pickup or about reducing extraneous sounds that make their way into a pickup signal. To start our discussion on how to apply EQ, let's revisit the *"Good Rule"* from Chapter 4:

> Good musician + good instrument + good performance + good acoustics
>
> + good mike + good placement = good sound.

Let's say that at some point in the "good" chain something falls short—like, a mic was placed in a bad spot for a particular instrument during a session that's still under way. Using this example, we now have two options:

◆ We can change the mic position and overdub the track or rerecord the entire song, or . . .

◆ We can decide to compensate for the positioning by applying EQ.

These choices represent an important philosophy that's held by many producers and engineers (including myself): Whenever possible, EQ should not be used as a Band-Aid. By this, I mean that it's often a good idea to correct for a problem on the spot rather than to rely on the hope that you can "fix it in the mix" at a later time using EQ and other methods.

When in doubt, it's often better to deal with a problem as it occurs. This isn't always possible, however; therefore, EQ is best used in situations where:

- There's no time or money left to redo the track.
- The existing take was simply magical.
- The track was already recorded during a previous session.

According to Keith Medley (a good friend of mine at Mackie, er...Loud Technologies), when EQ is applied to a track, bus, or signal, "the whole idea is to take out the bad and leave the good." If you keep adding EQ to the signal, it'll degrade the gain structure and lead to the "creeps." By this, he means that it's often a good idea to take away a deficiency in the signal, not necessarily to boost the desirable part of the track (which would in effect serve to turn up the overall gain). Such a boost will often throw off a mix's overall balance and reduce its overall headroom. A couple active examples would include:

- Reducing the high-end on a bass guitar instead of boosting it's primary bass notes.
- Using a peak filter to pull out the ring of a snare drum (a perfect example of a problem that should've been corrected during the session).

This may not always be the best course of action; for example, to bring out an upper presence on a recorded vocal track, it might be best to use a peak curve to slightly boost the upper mid-range. Just like life, nothing's ever absolute.

EQ in Action!

Although most equalization is done by ear, it's helpful to have an idea about which frequencies affect an instrument in order to achieve a particular effect. On the whole, the audio spectrum can be divided into four frequency bands: low (20–200 Hz), low-mid (200–1000 Hz), high-mid (1000–5000 Hz), and high (5000–20,000 Hz). When the frequencies in the 20- to 200-Hz (low) range are modified, the fundamental and the lower harmonic range of most bass information will be affected. These sounds often are felt as well as heard, so boosting in this range can add a greater sense of power or punch to music. Lowering this range will weaken or thin out the lower frequency range.

The fundamental notes of most instruments lay within the 200- to 1000-Hz (low-mid) range. Changes in this range often result in dramatic variations in the signal's overall energy and add to the overall impact of a program. Because of the ear's sensitivity in this range, a minor change can result in an effect that's very audible. The frequencies around 200 Hz can add a greater feeling of warmth to the bass without loss of definition. Frequencies in the 500- to 1000-Hz range could make an instrument sound hornlike, while too much boost in this range can cause listening fatigue.

Higher-pitched instruments are most often affected in the 1000–5000 Hz (high-mid) range. Boosting these frequencies often results in an added sense of clarity, definition and brightness. Too much boost in the 1000–2000 Hz range can have a "tinny" effect on the overall sound, while the upper mid-frequency range (2000–4000 Hz) affects the intelligibility of speech.

Table 12.1. Instrumental frequency ranges of interest.

Instrument	*Frequencies of Interest*
Kick drum	Bottom depth at 60–80 Hz, slap attack at 2.5 kHz
Snare drum	Fatness at 240 Hz, crispness at 5 kHz
Hi-hat/cymbals	Clank or gong sound at 200 Hz, shimmer at 7.5 kHz to 12 kHz
Rack toms	Fullness at 240 Hz, attack at 5 kHz
Floor toms	Fullness at 80–120 Hz, attack at 5 kHz
Bass guitar	Bottom at 60–80 Hz, attack/pluck at 700–1000 Hz, string noise/pop at 2.5 kHz
Electric guitar	Fullness at 240 Hz, bite at 2.5 kHz
Acoustic guitar	Bottom at 80–120 Hz, body at 240 Hz, clarity at 2.5–5 kHz
Electric organ	Bottom at 80–120 Hz, body at 240 Hz, presence at 2.5 kHz
Acoustic piano	Bottom at 80–120 Hz, presence at 2.5–5 kHz, crisp attack at 10 kHz, "honky tonk" sound (sharp Q) at 2.5 kHz
Horns	Fullness at 120–240 Hz, shrill at 5–7.5 kHz
Strings	Fullness at 240 Hz, scratchiness at 7.5–10 kHz
Conga/bongo	Resonance at 200–240 Hz, presence/slap at 5 kHz
Vocals	Fullness at 120 Hz, boominess at 200–240 Hz, presence at 5 kHz, sibilance at 7.5–10 kHz

Boosting in this range can make music seem closer to the listener, but too much of a boost often tends to cause listening fatigue.

The 5000- to 20,000-Hz (high-frequency) region is composed almost entirely of instrument harmonics. For example, boosting frequencies in this range will often add sparkle and brilliance to a string or woodwind instrument. Boosting too much might produce sibilance on vocals and make the upper range of certain percussion instruments sound harsh and brittle. Boosting at around 5000 Hz has the effect of making music sound louder. A 6-dB boost at 5000 Hz, for example, can sometimes make the overall program level sound as though it's been doubled in level; conversely, attenuation can make music seem more distant. Table 12.1 provides an analysis of how frequencies and EQ settings can interact with various instruments. (For more information, refer to the Microphone Placement Techniques section in Chapter 4.)

Do-It-Yourself Tutorial: Equalization

1. Solo an input strip on a mixer, console, or DAW . . . experiment with the settings using the previous frequency ranges. Can you improve upon the original recorded track or does it take away from the sound?

2. Using the input strip equalizers on a mixer, console, or DAW . . . experiment with the EQ settings and relative instrument levels of an entire mix using the previous frequency ranges as a guide. Can you bring an instrument out without changing the fader gains? Can you alter the settings of two or more instruments to increase the mix's overall clarity?

3. Plug the main output busses of a mixer, console, or DAW into an outboard equalizer, and change the program's EQ settings using the previous frequency range discussions as a guide.

One way to zero in on a particular frequency using an equalizer (especially a parametric one) is to accentuate or attenuate the EQ level and then vary the center frequency until the desired range is found. The level should then be scaled back until the desired effect is obtained. If boosting in one-instrument range causes you to want to do the same in other frequency ranges, it's likely that you're simply raising the program's overall level. It's easy to get caught up in the "bigger! better! more!" syndrome of wanting an instrument to sound louder. If this continues to happen on a mix, it's likely that one of the frequency ranges of an instrument or ensemble is too dominant and requires attenuation. On the subject of laying down a recorded track with EQ, there are a number of situations and differing opinions regarding them:

◆ Some use EQ liberally to make up for placement and mic deficiencies, whereas others might use it sparingly, if at all. One example where EQ is used sparingly is when an engineer knows that someone else will be mixing a particular song or project. In this situation, the engineer who's doing the mix might have a very different idea of how an instrument should sound. If large amounts of EQ were recorded to tape during the session, the mix engineer might have to work very hard to counteract the original EQ settings.

◆ If everything was recorded flat, the producer and artists might have difficulty passing judgment on a performance or hearing the proper balance during the overdub phase. Such a situation might call for equalization in the monitor mix, while leaving the recorded tracks alone.

◆ In situations where several mics are to be combined onto a single tape track, the mics can be individually equalized only during the recording phase. In situations where a project is to be engineered, mixed, and possibly even mastered, the engineer might want to know in advance the type and amount of EQ that the producer and/or artist would want. Obviously, this comes under the "art" category of recording and comes with experience.

◆ Above all, it's wise that any "sound-shaping" should be determined and discussed with the producer and/or artist before the sounds are committed to tape.

Whether you choose to use EQ sparingly, as a right-hand tool for correcting deficiencies, or not at all . . . there's no getting around the fact that an equalizer is a powerful tool. When used properly, it can greatly enhance or restore the musical and sonic balance of a signal. Experimentation and experience are the keys to proper EQ usage, and no book can replace the trial-and-error process of "just doing it!"

Before moving on, it's important to keep one age-old viewpoint in mind—that an equalizer shouldn't be regarded as a cure-all for improper mic technique; rather, it should be used as a tool for correcting problems that can't be easily fixed on the spot through mic and/or performance adjustments. If an instrument is poorly recorded during an initial recording session, it's often far more difficult and time consuming to "fix it in the mix" at a later time. Getting the best possible sound down onto tape will definitely improve your chances for attaining a sound and overall mix that you can be proud of.

Dynamic Range

Like most things in life that get out of hand from time to time, the level of a signal can vary widely from one moment to the next. For example, if a vocalist gets caught up in the moment and lets out an impassioned scream following a soft whispery passage, you can almost guarantee that the mic's signal will jump from its optimum recording level into severe distortion . . . OUCH! Conversely, if you set an instrument's mic to accommodate the loudest level, its signal might be buried in the mix during the rest of the song. For these and other reasons, it becomes obvious that it's sometimes necessary to exert some form of control over a signal's dynamic range by using various techniques and dynamic controlling devices. In short, the dynamics of an audio program's signal resides somewhere in a continuously varying realm between three categories:

- Saturation
- Average signal level
- System/ambient noise

As you may remember from various chapters in this book, *saturation* occurs when an input signal is so large that an amp's supply voltage isn't large enough to produce the required output current . . . or is so large that a digital converter reaches full scale (where the A/D output reads as all 1's). In either case, the results generally don't sound pretty and should be avoided in the studio's audio chain. The *average signal level* is where the overall signal level of a mix resides. Logically, if an instrument's level is too low, it can get buried in the mix . . . if it's too high, it can unnecessarily stick out and throw the entire balance off. It is here that the art of mixing at an average level that's high enough to stand out in the sonic crowd, while still retaining enough dynamic "life," truly becomes an applied balance of skill and magic. The following sections on readout displays and dynamic range processors can be used in the recording process to tame the audio signal in such a way that instruments, vocals, and the like can be placed squarely in the sonic pocket, at an optimum signal level.

Metering

Amplifiers, magnetic tape, and even digital media are limited in the range of signal levels that they can pass without distortion. As a result, audio engineers need a basic standard to help determine whether the signals they're working with will be stored or transmitted without distortion. The most convenient way to do this is to use a visual level display, such as a *meter*. Two types of metering ballistics (active response times) are encountered in recording sound to either analog or digital media:

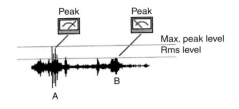

Figure 12.20. *A peak meter reads higher at point A than at point B, even though the average loudness level is the same.*

◆ Average (rms)

◆ Peak

From Chapter 2, we know that the root-mean-square (rms) value was developed to determine a meaningful average level of a waveform over time. Since humans perceive loudness according to a signal's average value (in a way that doesn't bear much relationship to a signal's instantaneous peak level), the displays of many meters will indicate an average signal level readout. The total amplitude measurement of the positive and negative peak signal levels is called the *peak-to-peak value*. A readout that measures the maximum amplitude fluctuations of a waveform is a *peak-indicating meter*.

One could definitely argue that both average and peak readout levels have their own set of advantages. For example, the ear's perception of loudness is largely proportional to the rms (average) value of a signal, not its peak value. On the other hand, a peak readout displays the actual amplitude at a particular point in time and not the overall perceived level. For this reason, a peak meter might show readings that are noticeably higher at a particular point in the program than the averaged rms counterpart (Figure 12.20). Such a reading will alert you to the fact that the short-term peaks are at levels that are above the clipping point, while the average signal is below the maximum limits. Under such conditions (where short duration peaks are above the distortion limit), you might or might not hear distortion as it often depends upon the makeup of the signal that's being recorded; for example, the clipped peaks of a bass guitar will not be nearly as noticeable as the clipped high-end peaks of a cymbal. The recording medium often plays a part in how a meter display will relate to sonic reality; for example, recording a signal with clipped peaks onto a tube analog tape machine might be barely noticeable (as the tubes and the tape medium act to smooth over these distortions)... whereas a DAW or digital recorder might churn out hash that's as ugly as the night (or your current session) is long.

Getting to the heart of the matter, it goes without saying that, whenever the signal is too high (hot), it's an indication for you to grab hold of the channel's mic trim, output fader, or whatever level control is the culprit and turn it down. In doing so, you've actually become a dynamic range-changing device. In fact, the main channel fader (which can be controlling an input level during recording or a tape track's level during mixdown) is by far the most intuitive and most often used dynamic gain changing device in the studio.

In practice, the difference between the maximum level that can be handled without incurring distortion and the average operating level of the system is called *headroom*. Some studio-quality preamplifiers are capable of signal outputs as high as 26 dB above 0 VU and thus are said to have 26 dB of headroom. With regard to analog tape, the 3% distortion level for analog magnetic tape is typically only 8 dB above 0 VU. For this reason, the best recording level for most program material is around 0 VU (although higher levels are possible provided that short-term peak levels aren't excessively high). In some circumstances (*i.e.*, when using higher bias, low-noise/high-output analog tape), it's actually possible to record at higher levels without distortion, as the analog tape formulation is capable of handling higher magnetic flux levels. With regard to digital media, the guidelines are often far less precise and will often depend upon your currently chosen bit rate. Since a higher bit rate (*e.g.*, 24 or 32 bits) directly translates into a wider dynamic range, it's possible to back off from the optimal recording level without fear of incurring additional noise. For example, the optimal average recorded level for a 44.1-k/16-bit recording is often agreed to be around −12 dB. When recording at higher word lengths, the encoded dynamic range is often so wide that the average level could easily be reduced by 10 or more dB. Reducing the recorded levels beyond this could have the unintended side effect of unnecessarily increasing distortion due to quantization noise (although this usually isn't an overriding factor at higher bit rates).

Now that we've gotten a few of the basic concepts out of the way, let's take a brief look at two of the most common meter readout displays.

The VU Meter

The traditional signal level indicator for analog equipment is the VU meter (Figure 12.21). The scale chosen for this device is calibrated in volume units (hence its name) and is designed to display a signal's average rms level over time. 0 VU is considered to be the standard operating level for most consoles, mixers, and analog tape machines. Although VU meters do the job of indicating rms volume levels, they ignore the short-term peaks that can overload tape (Figure 12.22), which can be 8 to 14 dB higher in level than the indicated rms value. This means that the professional console systems must often be designed so that unacceptable distortion doesn't occur until at least 14 dB above 0 VU. The typical VU meter specification is provided in Table 12.2.

Since recording is an art form, I have to rise to the defense of those who prefer to record certain instruments (particularly drums and percussion) at levels that bounce or even "pin" VU needles at higher levels than 0 VU. When recording to a professional analog machine, this can actually give a track a "gutsy" feel that can add impact to a performance. This is rarely a good idea when recording instruments that contain high-frequency/high-level signals (such

Figure 12.21. A VU meter's upper scale is calibrated in volume units (used for recording), while the lower scale is measured in percentage of modulation (often used in broadcast).

Figure 12.22. A VU meter reads the rms level and ignores instantaneous peaks that don't contribute to loudness.

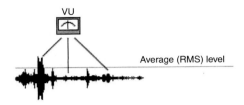

Table 12.2. VU meter specifications.

Characteristic	Specification
Sensitivity	Reads 0 VU when fed a +4-dBm signal (1.228 V into a 600-Ω circuit)
Frequency response	±0.2 dB from 35 Hz to 10 kHz; ±0.5 dB from 25 Hz to 16 kHz
Overload capability	Can withstand 5 times 0-VU level (approximately +18 dBm) continuously and 10 times 0-VU level (+24 dBm) for 0.5 sec

as a snare or cymbals), as the peak transients will probably distort in a way that's hardly pleasing. Always be aware that you can often add distortion to a track at a later time (using any number of ingenious tricks), but you can't remove it from an existing track. As always, it's wise to talk such moves over with the producer and artist beforehand.

The Averaging/Peak Meter

While many analog devices display levels using the traditional VU meter, most digital hardware and software devices display theirs using an LED, LCD, or on-screen display that emulates a VU's ballistics while also showing the program's peak levels. This best-of-both-worlds system makes sense in the digital world as it gives us a traditional readout that visually corresponds to what our ears are hearing, while providing a quick-and-easy display of the peak levels at any point in time. Often, the peak readout is frozen in position for a few seconds before resetting to a new level (making it easier to spot the maximum levels) . . . or it is permanently held at that level until a higher level comes along to bump it up. Of course, should a peak level approach the clipping level, a red peak indicator will light showing that it's time to back off the levels.

The idea of "pinning" the input levels of any digital recording device is a definite no-no. The dreaded "clip" indicator of a digital meter means that you've reached the saturation point, with no headroom to spare above this point. Unfortunately, digital standard operating levels are far more ambiguous than their analog counterparts, and you should consult the particular device manual for level recommendations (if it says anything about the subject). In light of this, I actually encourage you to slightly pin the meters on several digital devices (make sure the monitors aren't turned up too far), just to find out how obnoxious even the smallest amount of clipping can sound . . . and how those levels might vary from device to device. Pretty harsh, huh?

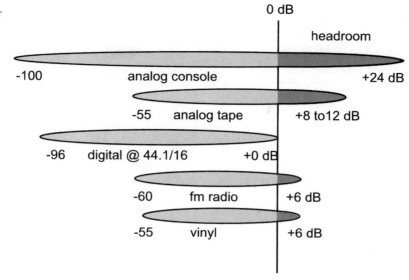

Figure 12.23. *Dynamic ranges of various audio media.*

0 dB

headroom

-100   analog console   +24 dB

-55   analog tape   +8 to12 dB

-96   digital @ 44.1/16   +0 dB

-60   fm radio   +6 dB

-55   vinyl   +6 dB

Dynamic Range Processors

The overall dynamic range of music is potentially on the order of 120 to 140 dB (Figure 12.23), whereas the overall dynamic range of a compact disc is often 80 to 90 dB, and analog magnetic tape is on the order of 60 dB (excluding the use of noise-reduction systems, which can improve this figure by 15 to 30 dB). However, when working with 20- and 24-bit digital word lengths, a system, processor, or channel's overall dynamic range can actually approach or exceed the full range of hearing. Even with such a wide dynamic range, unless the recorded program is played back in a noise-free environment, either the quiet passages would get lost in the ambient noise of the listening area (35–45 dB SPL for the average home and much worse in a car) or the loud passages would simply be too loud to bear. Similarly, if a program of wide dynamic range were to be played through a medium with a limited dynamic range (such as the 20- to 30-dB range of an AM radio or the 40- to 50-dB range of FM), a great deal of information would get lost in the background noise. To prevent these problems, the dynamics of a program can be restricted to a level that's appropriate for the reproduction medium (radio, home system, car, etc.). This gain reduction can be accomplished either by manually riding the fader's gain or through the use of a *dynamic range processor* that can alter the range between the signal's softest and loudest passages.

The concept of automatically changing the gain of an audio signal (through the use of compression, limiting, and/or expansion) is perhaps one of the most misunderstood aspects of audio recording. This can be partially attributed to the fact that a well-done job won't be overly obvious to the listener. Changing the dynamics of a track or overall program will often affect the way in which it will be perceived (either unconsciously or consciously) by making it "seem" louder, by reducing its volume range to better suit a particular medium, or by making it possible for a particular sound to ride at a better level above other tracks within a mix.

Figure 12.24. *Universal Audio 1176LN limiting amplifier: (a) 1176LN limiting amplifier; (b) 1176LN and 1176SE limiting plug-ins for Universal Audio's UAD-1 and Pro Tools. (Courtesy of Universal Audio, www.uaudio.com.)*

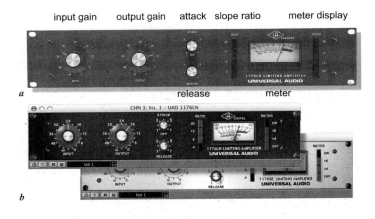

Compression

A *compressor* (Figure 12.24), in effect, can be thought of as an automatic fader. It is used to proportionately reduce the dynamics of a signal that rise above a user-definable level (known as the *threshold*) to a lesser volume range. This process is done so that:

◆ The dynamics can be managed by the electronics and/or amplifiers in the signal chain.

◆ The range is appropriate to the overall dynamics of a playback or broadcast medium.

◆ An instrument better matches the dynamics of other recorded tracks within a song or audio program.

Since the signals of a track, group, or program will be automatically turned down (hence the terms *compressed* or *squashed*) during a loud passage . . . the overall level of the newly reduced signal can now be amplified. In other words, since the dynamics have been reduced downward, the overall level can now be boosted, such that the range between the loud and soft levels is less pronounced (Figure 12.25). We've not only raised the louder signals back to a prominent level but we have also turned up the softer signals. In effect, we've turned up the softer signals that would otherwise be buried in the mix or ambient background noise.

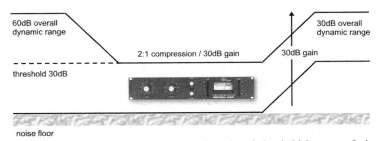

Figure 12.25. *A compressor reduces input levels that exceed a selected threshold by a specified amount. This reduced dynamic range signal can then be boosted in level at the output, thereby allowing the softer signals to be raised above other program or background sounds.*

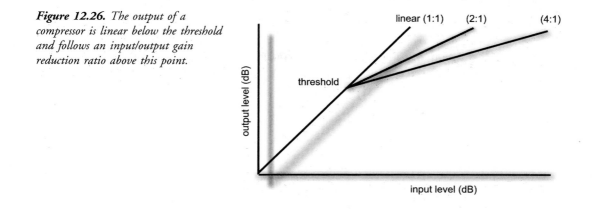

Figure 12.26. *The output of a compressor is linear below the threshold and follows an input/output gain reduction ratio above this point.*

The most common controls on a compressor (and most other dynamic range devices) include input gain, threshold, output gain, slope ratio, attack, release, and meter display:

◆ *Input gain*—This control is used to determine how much signal will be sent to the compressor's input stage.

◆ *Threshold*—This setting determines the level at which the compressor will begin to proportionately reduce the incoming signal. For example, if the threshold is set to −20 dB, all signals that fall below this level will be unaffected, while signals above this level will be proportionately attenuated . . . thereby reducing the overall dynamics. On some devices, varying the input gain will correspondingly control the threshold level. In this situation, raising the input level will lower the threshold point . . . and thus reduce the overall dynamic range. Most quality compressors offer a hard and soft knee threshold option. A *soft knee* widens or broadens the threshold range, making the onset of compression less obtrusive, while the *hard knee* setting causes the effect to kick in quickly above the threshold point.

◆ *Output gain*—This control is used to determine how much signal will be sent to the device's output. It's used to boost the reduced dynamic signal into a range where it can best match the level of a medium or be better heard in a mix.

◆ *Ratio*—This control determines the slope of the input-to-output gain ratio. In simpler terms, it determines the amount of input signal (in decibels) that's needed to cause a 1-dB increase at the compressor's output (Figure 12.26). For example, with a ratio of 4:1 . . . for every 4-dB increase at the input there will be only a 1-dB increase at the output; an 8-dB input increase will raise the output by 2 dB, while a ratio of 2:1 will produce a 1-dB increase in output for every 2-dB increase at its input. Get the idea?

◆ *Attack*—This setting (which is calibrated in milliseconds; 1 msec = 1 thousandth of a second) determines how fast or how slowly the device will turn down signals that exceed the threshold. It is defined as the time it takes for the gain to decrease to a percentage (usually 63%) of its final gain value. In certain situations (as might occur with instruments that have a long sustain, such as the bass guitar), setting a compressor to instantly turn down a signal might be audible (possibly creating a sound that pumps the signal's dynamics). In this situation, it would be best to use a slower attack setting. On the other hand, such a setting might not give the compressor time to react to sharp, transient sounds (such as a hi-hat). In this case, a fast attack time would probably work better. As you might expect, you'll need to experiment to arrive at the fastest attack setting that won't audibly color the signal's sound.

◆ *Release*—Similar to the attack setting, release (which is calibrated in milliseconds) is used to determine how slowly or quickly the device will restore a signal to its original dynamic level once it has fallen below the threshold point (defined as the time required for the gain to return to 63% of its original value). Too fast a setting will cause the compressor to change dynamics too quickly (creating an audible pumping sound), while too slow a setting might affect the dynamics during the transition from a loud to a softer passage. Again, it's best to experiment with this setting to arrive at the slowest possible release that won't color the signal's sound.

◆ *Meter display*—This control changes the compressor's meter display to read the device's output or gain reduction levels. In some designs, there's no need for a display switch, as readouts are used to simultaneously display output and gain reduction levels.

As was previously stated, the use of compression (and most forms of dynamics processing) is often misunderstood, and compression can easily be abused. Generally, the idea behind these processing systems is to reduce the overall dynamic range of a track, music, or sound program or to raise its overall perceived level... without adversely affecting the sound of the track itself. It's a well-known fact that overcompression can actually squeeze the life out of a performance by limiting the dynamics and reducing the transient peaks that can give life to a performance. For this reason, it's important to be aware of the general nuances of the controls we've just discussed.

Do-It-Yourself Tutorial: Compression

1. Go to the "Tutorial" section of www.modrec.com, and download the tutorial sound files that relate to compression (which include instrument/music segments in various dynamic states).

2. Listen to the tracks. If you have access to an editor or DAW, import the files and look at the waveform amplitudes for each example. If you'd like to DIY, then...

3. Record or obtain an uncompressed bass guitar track and monitor it through a compressor or compression plug-in. Increase the threshold level until the compressor begins to kick in. Can you hear a difference? Can you see a difference on the console or mixer meters?

4. Set the levels and threshold to a level you like and then set the attack time to a slow setting. Can you hear the signal pumping? Now, select a faster setting and continue until it sounds natural. Try setting the release to its fastest setting. Does it sound better or worse? Now, select a slower setting and continue until it sounds natural.

5. Repeat the above routine and settings using a snare drum track. Were your findings different?

During a recording or mixdown session, compression can be used in order to balance the dynamics of a track to the overall mix or to keep the signals from overloading preamps, the recording medium, and your ears. Compression should be used with care for any of the following reasons:

◆ Minimize changes in volume that occur whenever the dynamics of an instrument or vocal are too great for the mix. As a tip, a good starting point might be a 0-dB threshold setting at a 4:1 ratio, with the attack and release controls set at their middle positions.

◆ Smooth out momentary changes in source-to-mic distance.

◆ Balance out the volume ranges of a single instrument. For example, the notes of an electric or upright bass often vary in volume from string to string. Compression can be used to "smooth out" the bass line by matching their relative volumes. In addition, some instruments (such as horns) are louder in certain registers because of the amount of effort that's required to produce the notes. Compression is often useful for equalizing these volume changes. As a tip, you might start with a ratio of 5:1 with a medium-threshold setting, medium attack, and slower release time. Overcompression should be avoided to avoid pumping effects.

◆ Reduce other frequency bands by inserting a filter into the compression chain that causes the circuit to compress frequencies in a specific band (frequency-selective compression). A common example of this is a de-esser, which is used to detect high frequencies in a compressor's circuit so as to suppress those "SSSS," "CHHH," and "FFFF" sounds that can distort or stand out in a recording.

◆ Reduce the dynamic range and/or boost the average volume of a mix so that it appears to be significantly louder (as occurs when a television commercial seems louder than the show).

Although it may not always be the most important, this last application often gets a great deal of attention, as many producers strive to cut their recordings as "hot" as possible. That is, they want the recorded levels to be as far above the normal operating level as possible without blatantly distorting. . . . In this competitive business, the underlying logic behind the concept is that louder recordings (when broadcast on top-40 radio, played on a multiple CD changer, or an MP3 player) will stand out from the softer recordings in a playlist. In fact, reducing the dynamic range of a song or program's dynamic range will actually make the overall levels appear to be louder. By using a slight (or not-so-slight) amount of compression and limiting to squeeze an extra 1- or 2-dB gain out of a song, the increased gain will also add to the perceived bass and highs because of our ears' increased sensitivity at louder levels (remember the Fletcher–Munson curve?). To achieve these hot levels without distortion, multiband compressors and limiters often are used during the mastering process to remove peaks and to raise the average level of the program. You'll find more on this subject in Chapter 16 (Mastering).

Compressing a mono mix is done in much the same way as one might compress a single instrument. Adjusting the threshold, attack, release, and ratio controls, however, is more critical in order to prevent the pumping of prominent instruments within the mix. Compressing a stereo mix gives rise to an additional problem: If two independent compressors are used, a peak in one channel will only reduce the gain on that channel and will cause sounds that are centered in a stereo image to shift (or jump) toward the channel that's not being compressed (since it will actually be louder). To avoid this center shifting, most compressors (of the same make and model) can be linked as a stereo pair. This procedure of ganging the two channels together interconnects the signal-level sensing circuits in such a way that a gain reduction in one channel will cause an equal reduction in the other (thereby preventing the center information from shifting in the mix).

Multiband Compression

Multiband compression (Figure 12.27) works by breaking up the audible spectrum into various frequency bandwidths through the use of multiple bandpass filters. This allows each of the bands to be isolated and processed in ways that strictly minimizes the problems or maximizes the benefits in a particular band. Although this process is commonly done in the final mastering stage, multiband techniques can be used on an instrument or grouping. For example:

◆ The dynamic upper range of a slap bass could be lightly compressed, while heavier amounts of compression could be applied to the instrument's lower register.

◆ An instrument's high end can be brightened simply by adding a small amount of compression. This can act as a treble boost while accentuating some of the lower-level high frequencies.

More information on the subject can be found in Chapter 16 (Mastering).

Figure 12.27. Waves Linear Phase Multiband Compressor plug-in. (Courtesy of Waves, Ltd., www.waves.com.)

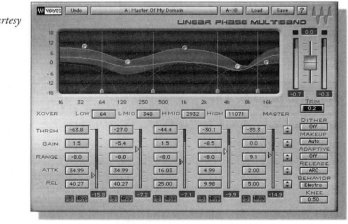

Figure 12.28.
*Waves L1
Ultramaximizer
Limiting/
Quantization plug-
in. (Courtesy of
Waves, Ltd.,
www.waves.com.)*

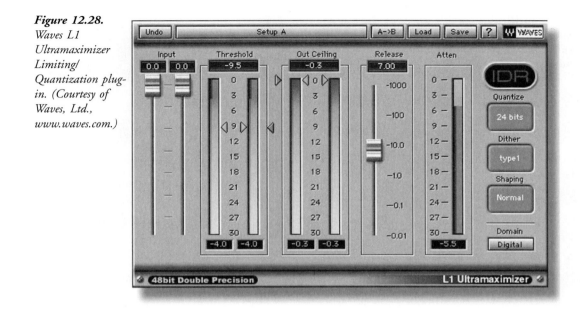

Limiting

If the compression ratio is made large enough, the compressor will actually become a *limiter* (Figure 12.28). A limiter is used to keep signal peaks from exceeding a certain level in order to prevent the overloading of amplifier signals, recorded signals onto tape or disc, broadcast transmission signals, and so on. Most limiters have ratios of 10:1 or 20:1 (Figure 12.29), although some have ratios that can range up to 100:1. Since a large increase above the threshold at the input will result in a very small increase at its output, the likelihood of overloading

Figure 12.29. *The output of a
limiter is linear below the
threshold and follows a high
input/output gain reduction
ratio (10:1, 20:1, or more)
above this point.*

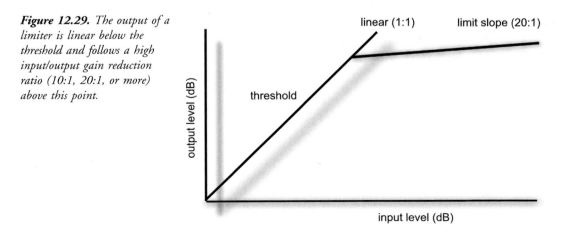

Figure 12.30. *Should only a few peaks exist in a file, a digital audio editor can be used to lower the excessive waveform.*

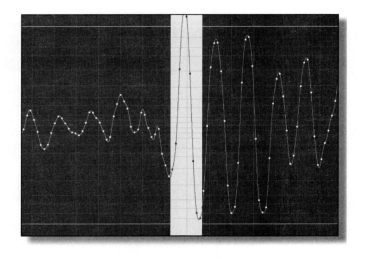

any equipment that follows the limiter will be greatly reduced. Commonly, limiters have two basic functions:

◆ Prevent signal levels from increasing beyond a specified level. Certain types of audio equipment (often those used in broadcast transmission) are often designed to operate at or near their peak output levels. Significantly increasing these levels beyond 100% would severely distort the signal and possibly damage the equipment. In these cases, a limiter can be used to prevent signals from significantly increasing beyond a specified output level.

◆ Prevent short-term peaks from reducing a program's average signal level. Should even a single high-level peak exist at levels above the program's rms average, the average level can be significantly reduced. This is especially true whenever a digital audio file is normalized at any percentage value, as the peak level will become the normalized maximum value and not the average level. Should only a few peaks exist in the file, they can easily be zoomed in on and manually reduced in level (Figure 12.30). If multiple peaks exist, then a limiter should be considered.

◆ Prevent high-level, high-frequency peaks from distorting analog tape. When recording to certain media (such as cassette and videotape), high-energy, transient signals actually don't significantly add to the program's level . . . relative to the distortion that could result from their presence (if they saturated the tape) or from the noise that would be introduced into the program (if the signal was recorded at such a low level that the peaks wouldn't distort).

Unlike the compression process, extremely short attack and release times are often used to quickly limit fast transients and to prevent the signal from being audibly pumped. Limiting a signal during the recording and/or mastering phase should only be used to remove occasional high-level peaks, as excessive use would trigger the process on successive peaks and would be noticeable. If the program contains too many peaks, it's probably a good idea to reduce the level to a point where only occasional extreme peaks are be detected.

Do-It-Yourself Tutorial: Limiting

1. Go to the "Tutorial" section of www.modrec.com, click on "Ch. 12—Limiting," and download the sound files (which include instrument/music segments in various states of limiting).
2. Listen to the tracks. If you have access to an editor or DAW, import the files and look at the waveform amplitudes for each example. If you'd like to DIY, then . . .
3. Feed an isolated track or entire mix through a limiter or limiting plug-in.
4. With the limiter switched out, turn the signal up until the meter begins to peg (you might want to turn the monitors down a bit).
5. Now reduce the level and turn it up again . . . this time with the limiter switched in. Is there a point where the level stops increasing, even though you've increased the input signal? Decrease and increase the threshold level and experiment with the signal's dynamics. What did you find out?

Expansion

Expansion is the process by which the dynamic range of a signal is proportionately increased. Depending on the system's design, an *expander* (Figure 12.31) can operate either by decreasing the gain of a signal (as its level falls below the threshold) or by increasing the gain (as the level rises above it). Most expanders are of the first type, in that as the signal level falls below the expansion threshold the gain is proportionately decreased (according to the slope ratio) . . . thereby increasing the signal's overall dynamic range (Figure 12.32). These devices can also be used as noise reducers. This is done by adjusting the device so that the noise is downwardly expanded during quiet passages, while louder program levels are unaffected or

Figure 12.31. *The Aphex Model 622 Logic-Assisted Expander/Gate. (Courtesy of Aphex Systems, Inc., www.aphex.com.)*

Figure 12.32. *Commonly, the output of an expander is linear above the threshold and follows a low input/output gain expansion ratio below this point.*

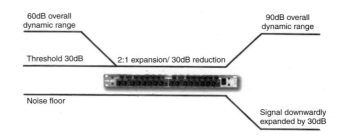

60dB overall dynamic range

90dB overall dynamic range

Threshold 30dB

2:1 expansion/ 30dB reduction

Noise floor

Signal downwardly expanded by 30dB

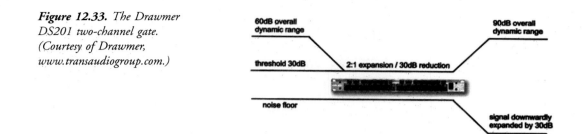

Figure 12.33. *The Drawmer DS201 two-channel gate. (Courtesy of Drawmer, www.transaudiogroup.com.)*

only moderately reduced. As with any dynamics device, the attack and release settings should be carefully set to best match the program material. For example, choosing a fast release time for an instrument that has a long sustain can lead to audible pumping effects. Conversely, slow release times on a fast-paced, transient instrument could cause the dynamics to return to its linear state more slowly than would be natural. As always, the best road toward understanding this and other dynamics processes is through experimentation.

The Noise Gate

One other type of expansion device is the *noise gate* (Figure 12.33). This device allows a signal above a selected threshold to pass through to the output at unity gain and without dynamic processing; however, once the input signal falls below this threshold level, the gate acts as an infinite expander and effectively mutes the signal by fully attenuating it (Figure 12.34). In this way, the desired signal is allowed to pass while background sounds, instrument buzzes, leakage, or other unwanted noises that occur between pauses in the music aren't. A few reasons why a noise gate might be used include the following:

◆ Reduce leakage between instruments. Often, parts of a drum kit fall into this category; for example, a gate can be used on a high-tom track in order to reduce excessive leakage from the snare.

◆ Eliminate noise from an instrument or vocal track during silent passages.

Figure 12.34. *The output of a gate is linear above the threshold and follows an infinite expansion slope (i.e., is turned off) below this point.*

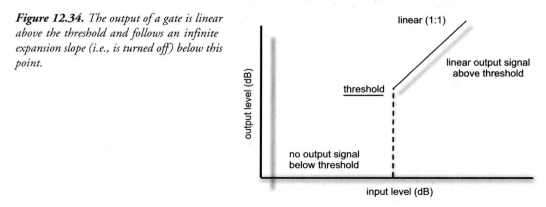

Figure 12.35.
Diagram of a basic keyed-input noise gate. (a) The signal is passed whenever a signal is present at the key input. (b) Without a key signal, the signal is gated (not allowed to pass).

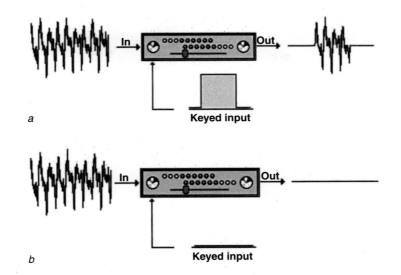

The general rules of attack and release apply to gating as well. Fortunately, these settings are a bit more obvious during the gating process than with any other dynamic tool. Setting the attack and release times either too fast or too slow will often be immediately obvious as the sound will cut in and out at inappropriate times—when listening to the instrument or vocal track either on its own (solo) or within a mix.

Commonly, a *key input* (Figure 12.35) is included as a side-chain path to a noise gate. A key input is an external control that allows an external analog signal source (such as a miked instrument, synthesizer, or oscillator) to trigger the gate's audio output path. For example, a mic or recorded track of a kick drum could be used to key a low-frequency oscillator. Whenever the kick sounds, the oscillator will be passed through the gate. By combining the two, you could have a deep kick sound that'll make the room shake, rattle, and roll.

Time-Based Effects

Another important effects category that can be used to alter and/or augment a signal revolves around delays and regeneration of sound over time. These *time-based effects* often add a perceived depth to a signal or change the way that we perceive the dimensional space of a recorded sound. Although a wide range of time-based effects exist, they are all based on the use of delay (and/or regenerated delay) to achieve such results as:

◆ Delay
◆ Chorus
◆ Flanging
◆ Reverberation

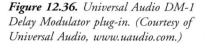

Figure 12.36. Universal Audio DM-1
Delay Modulator plug-in. (Courtesy of
Universal Audio, www.uaudio.com.)

Even though discrete and regenerated echoes can be created in either the acoustic or analog realm (using either tape machines or electronic devices), one of the many wonders of digital technology deals with the ease with which it can produce and manipulate delays according to algorithms that range from being straightforwardly simple to astoundingly rich and complex. For this reason (and for simplicity's sake), this section will generally concentrate of the manipulation of sound in the digital domain.

Delay

One of the most common effects used in audio production today alters the parameter of time by introducing various forms of delay into the signal path. Creating such a delay circuit is a relatively simple task to accomplish digitally. Although dedicated delay devices (often referred to as digital delay lines, or DDLs) are readily available on the market, most multifunction signal processors and time-related plug-ins are capable of creating this straightforward effect (Figure 12.36). In its basic form, digital delay is accomplished by storing sampled audio directly into RAM. After a defined length of time (usually measured in milliseconds), the sampled audio can be read out from memory for further processing or direct output (Figure 12.37). Using this basic concept, a wide range of effects can now be created simply by assembling circuits and/or program algorithms into blocks that can introduce delays and/or regenerated echo loops. Of course, these circuits will vary in complexity as new blocks are introduced.

Short-Term Delays

Probably the best place to start looking at the delay process is at the sample level. By introducing delays downward into the microsecond (1 millionth of a second) range, control over a signal's phase characteristics can be introduced to the point where selective equalization actually begins to occur. This is actually how EQ is carried out in the digital domain!

Figure 12.37. A digital delay device stores sampled audio into RAM, where it can be read out at a later time.

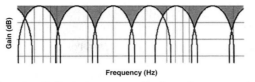

Figure 12.38. *Peaks and dips in a signal's frequency response (as shown in gray areas) result from the combination of several short-term delays that shift over time to create the effects of phasing or flanging.*

Whenever delays that fall below the 15-msec range are slowly varied over time and then are mixed with the original undelayed signal, a *combing* effect is created. Combing is the result of changes that occur when equalized peaks and dips appear in the signal's frequency response (Figure 12.38). By either manually or automatically varying the time of one or more of these short-term delays, a constantly shifting series of effects known as *phasing* or *flanging* can be created. Depending on the application, this effect (which makes a unique "swishing" sound that's often heard on guitars, vocals, etc.) can range from being relatively subtle (phasing) to having moderate to wild shifts in time and pitch (flanging).

By combining two identical (and often slightly delayed) signals that are slightly detuned in pitch from one another, another effect known as *chorusing* can be created. Chorusing is an effects tool that's often used by guitarists, vocalists, and other musicians to add depth, richness, and harmonic structure to their sound. Increasing delay times into the 15- to 35-msec range will create signals that are spaced too closely together to be perceived by the listener as being discrete delays. Instead, these closely spaced delays create a "doubling" effect (Figure 12.39) when mixed with an instrument or group of instruments. In this instance, the delays actually fool the brain into thinking that more instruments are playing than actually are. The effect, at least subjectively, increases the sound's density and richness. This effect (known as *doubling*) can be used on background vocals, horns, string sections, and other grouped instruments to make the ensemble sound as though it has doubled (or even tripled) its actual size. This effect also can be used on foreground tracks, such as vocals or instrument solos, to create a larger, richer, and fuller sound.

Some delay devices introduce slight changes in delay times in order to create a more natural, "humanized" sound. Should time or budget become an issue, it's also possible to create this doubling effect by actually recording a second pass to a new set of tracks. For example, a ten-piece string section could actually sound like a much larger ensemble by recording a second pass to available tracks. This process automatically gives vocals, strings, keyboards, and other legato instruments a more natural effect than by using an electronics effects device. In truth, however, these devices can actually go a long way toward duplicating the effect.

Figure 12.39. *In certain instances, doubling can fool the brain into thinking that more instruments are playing than actually are.*

Figure 12.40. *A DDL introduces one or more discrete repeats of the input signal at user-defined intervals: (a) single delayed echo; (b) numerous, repeated echoes created by signals fed back into memory.*

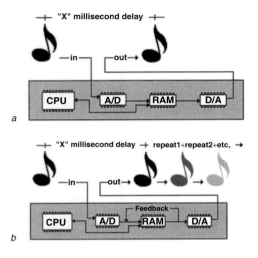

. . . Hey, you could even add a doubled effect to a double-pass recording to fill it out even more or to add space to a surround mix.

When the delay time is increased beyond the 35- to 40-msec point, the listener will begin to perceive the sound as being a discrete echo (Figure 12.40a). When mixed with the original signal, this effect can add depth and richness to an instrument or range of instruments. Care should be used, however, when adding delay to an entire musical program, as the program could easily begin to sound muddy and unintelligible.

By feeding the delayed signal back into the circuit a repeated series of echo . . . , echo . . . , echoes can be made to simulate the delays of yesteryear . . . is Elvis still in the house (Figure 12.40b)?

Do-It-Yourself Tutorial: Delay

◆ Go to the "Tutorial" section of www.modrec.com, click on "Ch. 12—Delay," and download the sound file (which includes segments with varying degrees of delay).

◆ Listen to the track. If you'd like to DIY, then . . .

◆ Insert a digital delay unit or plug-in into a program channel and balance its output mix so that the input signal is set equally with the delayed output signal. *Note:* If there is no mix control, route the delay unit's output to another input strip and combine delayed/undelayed signal at the console.

◆ Playback various instrument and voice sources (soloed ones, if possible) while slowly sweeping the delay settings.

◆ First, vary the settings over the 1- to 10-msec range. Can you hear any rough EQ effects?

◆ Vary the settings over the 10- to 35-msec range. Can you simulate a rough phasing effect?

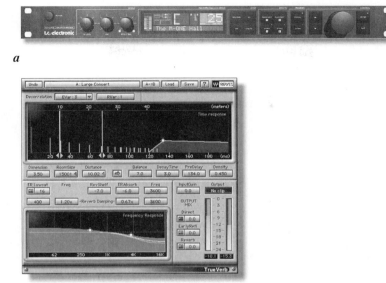

Figure 12.41. *Digital reverb devices: (a) TC Electronic M-One effects processor (courtesy of TC Electronic, www.tcelectronic.com); (b) Waves True Verb reverb plug-in (courtesy of Waves Ltd., www.waves.com).*

a

b

◆ Increase the settings above 35 msec. Can you hear the discrete delays?

◆ If the unit has a phaser setting, turn it on . . . how does it sound different?

◆ Now change the delay settings a little faster to create a wacky flange effect. If the unit has a flange setting, turn it on. Try playing with the time-based settings that affect its sweep rate. . . . Fun, huh?

Reverb

In professional audio production, natural acoustic reverberation is an extremely important tool for the enhancement of music and sound production. A properly designed acoustical environment can add a quality and natural depth to a recorded sound that often affect the performance as well as its overall sonic character. In those situations where there is little, no, or substandard natural ambience, a high-quality reverb device or plug-in (Figure 12.41) can be extremely helpful in filling out and giving the production a sense of dimensional space and perceived warmth. As we learned in Chapter 3 (Studio Acoustics and Design), *reverb* is closely spaced and random multiple echoes that are reflected from one boundary to another within a determined space (Figure 12.42). This effect helps give us perceptible cues as to the size, density, and nature of a space (even though it might have been artificially generated). These cues can be broken down into three subcomponents:

◆ Direct signal

◆ Early reflections

◆ Reverberation

Figure 12.42. *Signal level versus reverb time.*

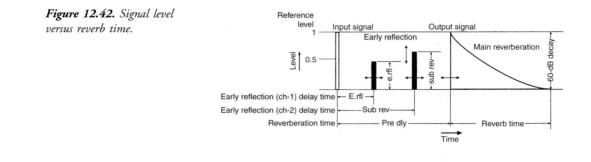

The *direct signal* is heard when the original sound wave travels directly from the source to the listener. *Early reflections* are the first few reflections that bounce back to the listener from large, primary boundaries in a given space. Generally, these reflections are the ones that give us subconscious cues as to the perception of size and space. The last set of reflections makes up the signal's *reverberation* characteristic. These sounds are comprised of zillions of random reflections that travel from boundary to boundary within the confines of a room. These reflections are so closely spaced in time that the brain can't discern them as individual reflections, so they're perceived as a single, densely decaying signal.

Reverb Types

By varying program and setting parameters, a digital reverb device can be used to simulate a wide range of acoustic environments, reverb devices, and special effects. A few popular categories include:

◆ *Hall*—Simulates the acoustics of a concert hall. This often is a diffuse, lush setting with a longer RT^{60} decay time (the time that's required for a sound to decay by 60 dB)

◆ *Chamber*—Simulates the acoustics of an echo chamber. Like a live chamber, these settings often simulate the brighter reflectivity of tile or cement surfaces.

◆ *Room*—As you might expect, these settings simulate the acoustics of a mid- to large-sized room. It's often best suited to intimate solo instruments or a chamber atmosphere.

◆ *Live (stage)*—Simulates a live performance stage. These settings can vary widely but often simulate long early-delay reflections.

◆ *Spring*—Simulates the low-fidelity "boingyness" of yesteryear's spring reverb devices.

◆ *Plate*—Simulates the often-bright diffuse character of yesteryear's metallic plate reverb devices. These settings are often used on vocals and percussion instruments.

◆ *Reverse*—These backward-sounding effects are created by reversing the decay trail's envelope so that the decay increases in level over time and quickly cut off at the tail-end . . . yielding a sudden break effect. This can also be realistically created in a DAW by reversing a track or segment, applying reverb . . . and then reversing it again to yield a true backward reverb trail.

Figure 12.43. *Altiverb sampled acoustics processor plug-in. (Courtesy of Audio Ease, www.audioease.com.)*

◆ *Gate*—Cuts off the decay trail of a reverb signal. These settings are often used for emphasis on drums and percussion instruments.

Do-It-Yourself Tutorial: Reverb Types

1. Go to the "Tutorial" section of www.modrec.com, click on "Ch. 12—Reverb Types," and download the sound file.
2. Listen to the track.

Impulse Response Reverb

Impulse response reverb devices and plug-ins (Figure 12.43) are capable of accurately simulating the sampled acoustics of real halls, cathedrals, bathrooms, missile silos . . . virtually any acoustic environment . . . in a real-time processing environment. This process is carried out through the modeling of an acoustic environment in the digital domain from a recorded *impulse response*. This reverb "footprint" is created by firing a starter pistol or by playing a sine wave sweep from a speaker into an auditorium, church, or other acoustic space. The resulting reverberation decay is then recorded as a digital audio file, and (using a process called *deconvolution*) the dry source signal is then extracted from the audio file. The resulting acoustic fingerprint can then be used to re-create the actual acoustics of that particular environment within a mono, stereo, or surround soundfield.

Since the process of creating an impulse response isn't overly difficult, an ever-growing number of concert halls, reverb devices, microphone simulations, and bathrooms are showing up on the Web as a download. In addition, certain plug-ins will even let you insert a .jpg picture or slideshow of pictures into the preset, so as to show mic positions and other details of the sampled space.

Pitch- and Time-Shift-Related Effects

Another time-based effect that is best carried out in the digital domain is the alteration of pitch and time within audio. A wide range of effects devices, plug-ins, and DAW program applications can be used to carry out these functions, although it should be noted that not all speed/pitch processors are created equal, and (depending on the hardware or software) most systems sound best when the effect is carried out within a limited pitch- or time-shift range. That is, a hardware system or program might be able to correct or alter program material over a limited shift range . . . however, beyond this point, certain amounts of granularity or harmonic distortion could be introduced into the signal (although these figures have definitely improved in recent years). This range often varies from system to system and with different types of program material (such as voice, music, and complex waveforms). In the final analysis, your ears must judge whether the drawbacks of induced distortion (if any) outweigh the benefits of the effect itself. Currently, the options for pitch and time shifting include changing pitch without changing duration, changing duration without changing pitch, and changing both duration and the pitch.

Pitch Shifting

Ever had a perfectly good vocal take that was spoiled by just one or two flat notes? . . . Or had a project come in the door with a guitar track that was out of tune? . . . Or needed to change the key on a 30-second radio spot? It's times like these that pitch shifting can save your day! Pitch shifting can be used to vary the pitch of a signal or sound file (either upward or downward) in order to transpose the relative pitch of an audio program without affecting its duration. This process can take place either in real time or in non-real time. Although it's not important to know all of the details, pitch shifting works by writing sampled audio data to a temporary memory, where it is resampled to either a higher or a lower sample rate (according to the desired final pitch). Once done, the processor either adds to (lowers pitch) or subtracts samples from (raises pitch) the resampled data to return it back to the original output rate (while keeping the altered pitch intact). Figure 12.44 gives two basic examples of how this is often carried out.

A degree of caution should be used when changing the pitch of a program or audio segment. Whenever uneven or minute interval changes are made, the interpolation of samples doesn't always fall perfectly into place. This can lead to digital artifacts that add unacceptable amounts of harmonic distortion. If the track is in the background, there shouldn't be a problem; however, care should be taken with upfront instruments and vocals. It's important to keep in mind that large pitch changes might be more noticeable. As always, your ears are the best judge.

Pitch Shifting to the Rescue

Although pitch shifting can be used to fix any number of problems or create a wide range of effects, here are a few concrete examples of how this process can work for you:

◆ When working with a tape-based system (analog or digital), you can route the problem track through a pitch shifter and re-record its output onto another track. You can then use the shifter's bypass switch to switch the effect on to correct the offending note or passage. It's often wise to keep the original track intact, just in case.

Figure 12.44. *Two pitch shift examples with an initial 1-kHz digital signal and a sample rate of 44.1 kHz. (a) The signal can be halved in pitch (to 500 Hz) by internally downsampling to a new rate of 22.05 k. In order to return the output rate to 44.1 (while retaining the 500-Hz pitch), new sample points must be added into each dropped position. (b) The signal can be doubled in pitch (to 2 kHz) by internally upsampling to a new rate of 88.2 k. In order to return the output rate to 44.1 (while retaining the 2-kHz pitch), every other sample point must be dropped.*

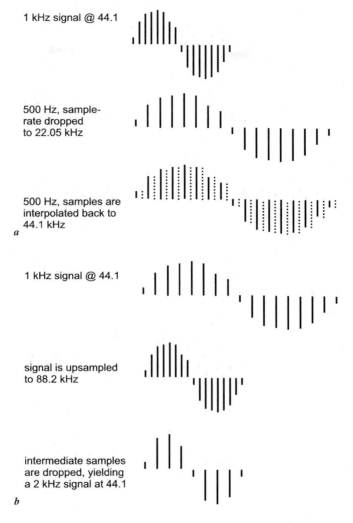

- ◆ Correcting hard-disk pitch problems is even easier, as most DAWs provide for pitch shifting. Simply highlight the segment and apply the proper amount of correction. As noted above, it's probably best to apply non-destructive correction. If this isn't possible, it's wise to make alterations to a copy of the original instrument file.
- ◆ By inserting a 10- to 25-msec digital delay and pitch shifter into a circuit, it's a simple matter to create a vocal or instrument track that can harmonize along with the original by a 3rd, 5th, or other interval.
- ◆ Certain pitch shift devices (Figure 12.45) have been specifically designed for instrumentalists and vocalists in that presets can easily be dialed up according to commonly used intervals of the music scale. These no-brainer tools can be used in the studio or on stage at a moment's notice.

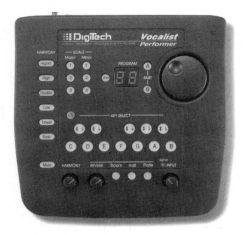

Figure 12.45. *Digitech studio vocalist. (Courtesy of Digitech, a Harmon International Company, www.digitech.com.)*

Time and Pitch Changes

By combining variable sample rates and pitch shifting techniques, it's possible to alter a program's duration (varying the length of a program by raising or lowering its playback sample rate) or to alter its relative pitch (either up or down). In this way, three possible combinations of time and pitch changes can occur:

◆ *Time change*—A program's length can be altered without affecting its pitch.

◆ *Pitch change*—A program's length can remain the same while pitch is shifted either up or down.

◆ *Both*—Both a program's pitch and length can be altered (using simple resampling techniques).

These processing functions have become an important part of the signal processing and music production arsenal for use in the audio-for-video, film, and broadcast industries. These tools help give producers control over the running time of film video and audio soundtracks while maintaining the original, natural pitch of voice, music, and effects. For example, using a DAW, we could add a 5-second trailer onto the end of an existing 30-second public service radio spot simply by time compressing the 30-second spot to 25 seconds (while keeping the pitch intact) . . . and then adding the trailer.

In addition to the basic time/pitch techniques that are commonly used in music production (most often by electronic musicians), this technology has allowed for the huge explosion in loop-based music composition and production. These popular programs and music plug-ins involve the use of recorded sound files that are encoded with headers that include information on their native tempo and length (in both samples and beats). By setting the loop program to a master tempo (or a specific tempo at that point in the score), once a loop segment has been imported, it can go about the process of recalculating its pitch and tempo to match the current session tempo . . . and *voila*! The file's in sync with the song! Should you wish to check it out for yourself, a number of loop-based demo programs can be found on the Web.

Figure 12.46. *204 Aural Exiter with optical Big Bottom. (Courtesy of Aphex Systems, www.aphex.com.)*

Psycho-Acoustic Enhancement

A number of signal processors rely on psycho-acoustic cues in order to fool the brain into perceiving a particular effect. The earliest and most common of these devices are those that enhance the overall presence of a signal or entire recording by synthesizing upper-range frequency harmonics and inserting them into a mix in order to brighten the perceived sound Figure 12.46. Although the additional harmonics won't significantly affect the program's overall volume, the effect is a marked increase in its perceived presence. Other psychoacoustic devices that make use of complex harmonic, phase, delay, and equalization parameters have become standard production tools in the field of mastering in order to shape the final sound into one that's interesting, with a sonic character all its own.

In addition to synthesizing harmonics in order to change or enhance a recording or track, other digital psycho-acoustic processors deal exclusively with the subject of spatialization (the placement of an audio signal within a three-dimensional acoustic field), even though the recording is being played back over stereo speakers. By varying the parameters of a stereo or multiple input source, this processing function creates phase and amplitude paths that can fool the brain into perceiving that the stereo image is actually emanating from a soundfield that's wider than the physical speaker positions. In practice, care should be taken when using these devices, as the effect is often carried off with degrees of success that vary from system to system. In addition, the use of phase relationships to expand the stereo soundfield can actually cause obvious cancellation problems when the program is listened to in mono.

Multiple-Effects Devices

Since most digital signal processors are by nature multifunctional chameleons, it follows that most hardware and certain plug-in processors can be easily programmed to perform various functions. For this reason, many digital systems have been designed to perform as multiple-effects devices (Figures 12.47 and 12.48). Multiple effects, in this case, can have several basic meanings:

- ◆ A single device might offer a range of processing functions but allow only one effect to be called up at a time.
- ◆ A single device might offer a range of processing functions that can be "stacked" to perform a number of simultaneous effects.
- ◆ An effects device might have multiple ins and outs, each of which can perform several processing functions (effectively giving you multiple processors that can be used in a multichannel mixdown environment).

Figure 12.47. Remote for the tc electronics Reverb 6000 multiple-effects processor. (Courtesy of tc electronic, www.tcelectronic.com.)

Figure 12.48. Lexicon 960L multichannel digital effects system. (Courtesy of Lexicon, Inc., www.lexicon.com.)

Each of the above types are invaluable tools that are most often found in a hardware-based professional or project studio environment. (Plug-ins tend to be more specific in their function; however, since any number of them can be plugged into a computer-based project at any point in time . . . this limitation simply doesn't matter.) Table 12.3 lists just a few of the possible effects that can be offered by these versatile devices.

Table 12.3.

Type	Audible Effect
Reverb	Delay, chorus, phasing, and flanging
Equalization	Compression, limiting, expansion, and gating
Pitch shift	Time/pitch change
Sample rate conversion	Spectral and spatial enhancement
Sampling or one-shot sampling	Overdrive distortion
Wah pedal	Rotary speaker and auto-panning
Tremolo and vibrato	Effects morphing

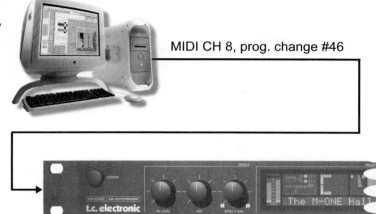

Figure 12.49. *Effects settings can be automated throughout a system with the use of MIDI program-change commands.*

MIDI CH 8, prog. change #46

Dynamic Effects Automation and Editing via MIDI

A common way to automate changes in a hardware effects device during a mix is through the use of MIDI program-change commands. In the same way that a favorite sound patch can be stored into an instrument's memory for later recall, most hardware effects devices will let you store effects types as a "patch" that can be stored into its internal memory. These patch changes can be automated from a MIDI sequencer or on stage during a live performance from a MIDI controller by transmitting a program change command; more information on program change commands can be found in Chapter 7 (MIDI and Electronic Music Technology). Often, making such a scene change is as simple as transmitting the desired program change number (from a sequencer or controller) on a port and MIDI channel number that's being sent to the device (Figure 12.49).

In addition to changing program numbers in real time, any number of effects parameters such as delay, reverb time, depth, or EQ can often be changed in real time through the use of system-exclusive messages (see Chapter 7). Control over these messages is generally carried out through the use of a *patch editor* (Figure 12.50), which can be a dedicated hardware device, generic MIDI controller, or software application. Once the desired effect or multiple-effect has been assembled and fine-tuned, these settings can be saved into as a preset for recall at a later time. Generic hardware MIDI controllers have gained in popularity due to the simple fact that you can actually get your hands on a set of knobs or data sliders. By assigning a controller to an individual function, it's possible to tweak the effects (or electronic instrument) settings in real time . . . often a very satisfying and straightforward experience.

Once a device's preset bank has been filled, it's often possible to transmit these settings as a sys-ex bulk dump or program dump back to a MIDI sequencer, where the data can be stored

Figure 12.50.
Effects parameters can be dynamically controlled in real time via system-exclusive messages.

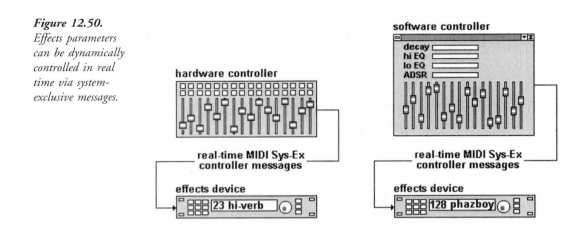

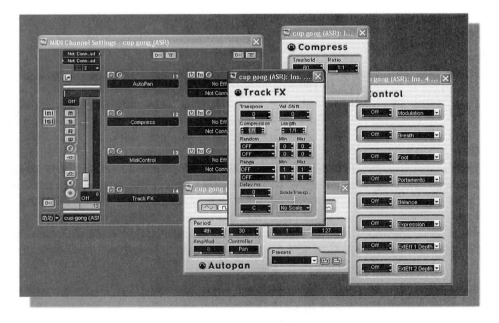

Figure 12.51. *Many DAW and sequencer packages include extensive MIDI effects. (Courtesy of Steinberg Media Technologies GMBH, www.steinberg.net.)*

as a named file. Using this method, multiple preset banks can be stored and recalled back to the device, thereby allowing a much larger number of effects patches to be stored in a computer-based library. As with electronic instrument patches, effects patches can be acquired from a number of different sources. Among these are patch data cards (device-specific ROM cards or cartridges that contain patch data from manufacturers or second-party developers), patch data disks (computer files containing patch data from manufacturers or second-party developers), and, of course, the Web (you'd be surprised at how many device-specific sys-ex patch dumps are available in cyberspace).

Dynamic Effects Automation within a DAW/Sequencer Environment

In closing, an almost unlimited degree of effects control is available to us through most high-level (and many entry-level) digital audio workstations. You may have noticed that the above section on hardware automation didn't talk about plug-ins at all. . . . Fortunately for us, external automation of a plug-in isn't necessary at all (unless you're talking about making a quick-and-easy setting change from a DAW hardware controller). The vast majority of plug-in effects can be directly and dynamically automated within the computer's DAW program—just grab the control and make the change (with the read/write automation turned on). It's that simple. For further information, take a look at your favorite DAW manual and start reading!

On the MIDI front, many advanced sequencers offer a number of MIDI plug-ins that can act as a virtual MIDI effects device (Figure 12.51). Examples include MIDI echo (which simply repeats MIDI notes according to such parameters as session tempo, number of repeats, decay, etc.), dynamics (which can compress, limit, or expand MIDI velocity values within a selected track or range), autopan, chorder (creates chords at determined intervals), etc. Again, for further info, take a look at your favorite DAW or sequencer manual and start reading!

CHAPTER

13

◆

Noise Reduction

An increase in dynamic range and a demand for better quality sound has been brought about by newer microphone technologies, digital audio, the compact disc, and surround-sound home movies. Because of this, it's more important than ever for those in all forms of audio production to pay close attention to the background noise levels that are produced by amplifier self-noise, analog magnetic tape, and the like. The overall dynamic range of human hearing roughly encompasses a full 140 dB. This often can't be fully captured by the recording medium for several reasons:

◆ The medium itself might be incapable of encoding a wide dynamic range.

◆ An acoustic or electronic weak link in the chain could introduce noise and/or limit the program's dynamic range.

Although the first scenario can easily crop up when transmitting a signal through the airwaves or a telephone line (and the solutions are often the same), the first part of this chapter will largely focus on reducing tape noise that's a natural by-product of the analog recording/playback process

The roughly 60-dB signal-to-noise (S/N) limitation that's imposed on conventional analog ATR and VTR audio tracks is dictated by tape hiss (which is heard when the overall signal level is too low), and by the recorded level of the program (which is limited at the high end by tape saturation). Should an optimum level produce an unacceptable amount of noise, the engineer is faced with several options: record at a higher level (with the possibility of increased distortion), change the signal's overall dynamic range by raising low-level signals above the noise, or introduce a system that can change the dynamic range of the recording medium itself.

Not all the blame, however, can be put on our older technology friends. Even though a 16-bit digital recording has a theoretical dynamic range of 96 dB, and a 24-bit system can actually encode 144 dB, noises can (and often will) crop up from such modern-day sources as mic preamps, effects & outboard gear, background noise, analog communication lines, and poorly designed digital audio converters, not to mention from recordings that were made under adverse and/or noisy acoustic conditions. For these reasons, the latter half of the chapter will deal with single-ended and digital noise reduction processors that can help reduce noises that have been introduced into the recording chain from any number of gremlin sources.

Analog Noise Reduction

Analog tape noise might not be a limiting factor when dealing with one or two tracks in an audio production, but the combined noise and other distortions that can occur when combining 8, 16, 24, or 48 tracks ranges from being bothersome to downright unacceptable. The following types of noises are often the major contributors to the problem:

◆ Tape and amplifier noise
◆ Crosstalk between tracks
◆ Print-through
◆ Modulation noise

Modulation noise is a high-frequency component that causes "frizziness" as well as sideband frequencies that can distort the signal (Figure 13.1). This noise is due, in part, to irregularities in the coating of magnetic recording tape and only occurs when a signal is present and increases as signal levels rise (due to its interaction with the applied magnetic field). It's interesting to note that this noise is often higher in level than you might expect, and when combined with *asperity noise* (sideband frequencies that are introduced by the analog record/ playback process), these effects definitely play a role in what could be called the "analog sound."

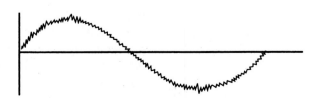

Figure 13.1. *Modulation noise of a recorded analog sine wave.*

Do It Yourself Tutorial: Analog Tape Modulation and Asperity Noise

1. Feed a 0-VU, 1-kHz test tone to a track on a professional analog recorder.
2. Listen to the recorder's source (input) signal through the monitors at a moderate level. Does it sound like 1 kHz?
3. Switch the recorder to monitor the track's playback (tape) signal. Does it sound different?

These analog-based noises can be reduced to acceptable levels by using different combinations of the following actions:

◆ Improving the dynamic range of professional recording tape and/or increasing tape speed in order to record at higher flux levels

◆ Using ATRs with wider tracks (i.e., higher record/playback levels and reduced crosstalk specs

◆ Using a thicker tape base to reduce print-through

By using tape formulations that combine low noise and high output (which have generally increased the S/N ratio by 3 dB or more), noise levels can be reduced even further.

The Compansion Process

When all is said and done, making most or all of the above improvements often proves to be too costly and/or impractical—and even if they're implemented, it's not possible to get around the fact that noise is simply an inherent part of the analog recording process.

In order to reduce the effects noise on analog recording, a process called *compansion* was created. The compansion encode—decode process gets its name from the fact that the incoming signal is first (COMP)ressed before being recorded onto tape. Then on playback, the signal is reciprocally exp(ANDED) downward to its original dynamic range (because the tape noise that's introduced by the machine is also downwardly expanded, it will be likewise reduced). This is a reciprocal process, in that the encoder must be placed between the console (or other desired source) and the tape track on record, and the decoder must be placed between the tape output and the console (or desired destination) upon playback.

To better understand this process, let's take a look at Figure 13.2. In this example, an input signal's overall dynamic range is restricted (COMPressed) before being sent to the recorder's track input. This is done so that the newly compressed signal can be recorded at levels that are much higher than the tape noise. During playback, the signal is then downwardly expANDED back to its original dynamic range. Fortunately, the tape noise will likewise be expanded downward to lower (and ideally inaudible) levels.

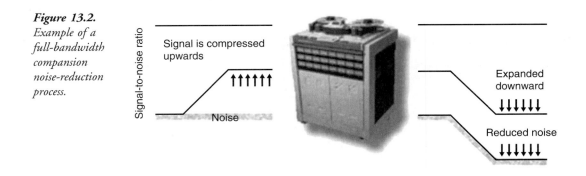

Figure 13.2.
Example of a full-bandwidth compansion noise-reduction process.

The dbx Noise-Reduction System

The dbx system (Figure 13.3) is a full-bandwidth compansion system that's able to provide between 20 and 30 dB of noise reduction. The compression chain uses a 2:1 ratio across the entire frequency bandwidth between the levels of −90 dBm and +25 dBm (with its unity gain point occurring at +4 dBm or 0 VU). Because all signals are compressed and expanded at a 2:1 ratio (regardless of signal level), the system isn't overly sensitive to variations in level (although it can be sensitive to frequency calibrations).

From the example in Figure 13.4, let's assume that we're recording a program that has a 60-dB dynamic range onto a tape machine that also has a 60-dB S/N range. During the softest program passages, the tape noise will be just as loud as the program and therefore will be very audible. With the dbx system, the program passes through the encoder and is compressed (by 2:1) into an overall dynamic range of 30 dB before being recorded onto tape. (It's important to note that the recorded tape noise is now 30 dB below the softest music passage.) On playback, the decoding expander will proportionately turn down softer signals until the dynamics are restored back to their original 60-dB range. Because the noise is 30 dB

Figure 13.3. *The dbx 140X type II noise-reduction system. (Courtesy of dbx Professional Products, www.dbxpro.com.)*

Figure 13.4. *Expansion of dynamic range from 60 to 90 dB when using dbx noise reduction.*

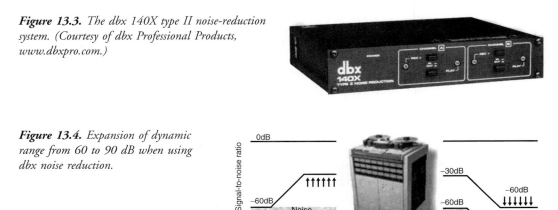

below the softest encoded program signal, it will also be reduced in level by 30 dB to a nearly inaudible level of −90 dB.

A 2:1 ratio was chosen over a higher value because of the effect that tape dropouts have on the expansion process. Because the changes in level are 1:2 expanded on playback, a tape dropout of 2 dB would ultimately result in a level drop of 4 dB. A 1:2 ratio would cause a 6-dB dropout, and while a 1:1.5 expansion would make the dropout problem less noticeable; it would do so at the cost to the noise reduction process (only 20 dB would be possible). Therefore, a 2:1 compansion ratio was considered a compromise between the best noise reduction level and oversensitivity to dropouts.

The Dolby Noise-Reduction System

Although similar in theory to the previous example, the Dolby family of analog noise-reduction systems operates by breaking the audio signal into various frequency bandwidths. This makes it possible for frequency ranges that contain louder passages to be unaffected, while those containing softer passages (especially those in the upper noise-perception range) can be appropriately processed.

Dolby companders are available for audio production in four popular flavors: the professional Dolby SR, Dolby A, and the consumer Dolby B and C. Dolby spectral recording (SR) uses an encode–decode process that readily lends itself to any analog audio recording or transmission application. This system (Figure 13.5) reduces the influence of noise and nonlinearity on reproduced sound by improving tape noise figures by as much as 24 dB.

The Dolby SR signal-shaping processor uses a sidechain that runs parallel to the device's main audio path. The output of this sidechain is either added to or subtracted from the main signal, depending on whether the circuit is enabled as an encoder or decoder. Although most of the Dolby SR electronics are used for spectral analysis, its principal operating system consists of five groups of fixed- and sliding-band filters with gentle slopes that are arranged by level and frequency. Those with fixed bandwidths are electronically controlled to vary gain; those with fixed gain can be adjusted to cover different frequency ranges. These filters are crosslinked by a technique known as action substitution, which allows both types of filters to be selected in the proper proportion.

Figure 13.5. *The Dolby XPSR SR 24tk noise-reduction rack. (Courtesy of Dolby Laboratories, Inc, www.Dolby.com.)*

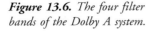
Figure 13.6. *The four filter bands of the Dolby A system.*

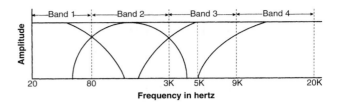

By selecting and combining filters, the SR control circuit can process the signal using an infinite number of filter shapes during the encoding process. Upon decoding, these filter shapes are recreated in a reciprocal fashion, which reduces noise in a way that preserves the signal's original level, phase, and frequency. In short, an SR channel optimizes these gain and EQ settings to dynamically compress the signal during recording and conversely expand the signal on playback.

Each tape track has its own dedicated SR channel that's used for both record and playback processing. Switching between these modes is usually carried out automatically by the recorder's transport switching logic.

Dolby's original type A professional compansion system reduces tape noise approximately 10–15 dB by dividing the audio spectrum into four separate bands (Figure 13.6). Each frequency band has its own dynamic-range processor, thus the presence of a high-level signal in one band won't interfere with the detection circuits of another band.

The four filters and processors are combined in such a manner that low-level signals (below −40 dBm) are boosted by 10 dB from 20 Hz to 5 kHz, with the boost gradually rising between 5 and 14 kHz to yield a maximum noise reduction of 15 dB. These bands aren't sharply defined, so when the noise reduction in Band 2 is disabled by the presence of loud signals between 80 Hz and 3 kHz, some noise reduction (in addition to masking) will be provided by Band 1 and above 1.8 kHz by Band 3. If Band 3 also has its noise reduction turned off by a loud signal between 3 and 9 kHz, Band 4 will continue to add noise reduction above 5 kHz. Noise reduction in Bands 1 and 4 are rarely shut down (except by very loud organ tones or heavy cymbal crashes). Although the actual amount of noise reduction throughout the audio spectrum changes from one moment to the next, the perceived noise level will remain constant.

To ensure that the tape is played back at the same level at which it was recorded, a 400 or 700 Hz, 0-VU level tone is recorded at the beginning of each tape, which allows levels to be properly adjusted during playback. This ensures that processing thresholds will be the same during recording and playback. A 3-dB tolerance is allowable before differences in the playback signal can be noticed. If the Dolby signal is played back too loudly, too little expansion will take place, because the limiter is blocked by the high-level signal and it will sound compressed and overly bright. If the signal is played back at too soft a level, the expander will expand too much and the signal will sound dull and have too great a dynamic range. The recorded signal level doesn't matter as long as the playback level matches.

Dolby B (which is commonly found in consumer cassette decks) is designed solely to reduce tape hiss. The Dolby B system acts only to compand the upper frequency component and has no operating effect on lower frequency noises, such as hum or rumble. The effect of type B noise reduction is 3 dB at 600 Hz, rising to 10 dB at 5 kHz, at which point it levels off in a shelving fashion. Dolby C is a refinement of Dolby B, while Dolby S is the consumer version of SR, which offers up to 20 dB of overall noise reduction when used on cassette tapes.

Single-Ended Noise-Reduction Process

Since a compansion system must process a signal before recording and after playback to achieve noise reduction, noise can't be removed from the original source material: it can only be prevented from being introduced by the recording medium. The *adaptive filter* or *single-ended noise-reduction* process is different. It extracts noise from an audio source by combining a downward dynamic-range expander with a variable low-pass filter. In other words, a single-ended device can be used to dynamically analyze, process, and EQ an existing program to reduce the noise content with little or no audible effects (or the best possible compromise, in extreme cases).

Single-ended noise reduction systems work by breaking up the audio spectrum into a number of frequency bands, such that whenever the signal level within each band falls below a user-defined threshold, the signal will be attenuated. This downward expansion/filtering process accomplishes noise reduction by taking advantage of two basic psycho-acoustical principles:

◆ Music is capable of masking lower level noise that exists within the same bandwidth.
◆ Reducing the bandwidth of an audio signal reduces the perceived noise.

It's a psycho-acoustical fact that our ears are more sensitive to noises that contain a greater number of frequencies than to those containing fewer frequencies. In light of this, whenever the program's high-frequency content is reduced or restricted to a certain bandwidth, the dynamic filter examines the incoming signal and reduces the bandwidth accordingly. Whenever the high-frequency signal returns, the filter opens back up as far as necessary to pass the signal.

Noise Gates

A noise gate can be a very effective noise-reduction device when used to reduce background noise on certain program material. A noise gate effectively passes signals that fall above a user-defined threshold at unity gain, while turning off signals that fall below this threshold. This makes it a useful tool for eliminating noise, leakage, and other gremlins from a track or tracks within a mix. Obviously, it wouldn't be a good idea to gate the main output busses of a mix, as the entire program would simply cut out below any reasonable threshold.

When "gating" a specific track, it's often necessary to take time out to fine-tune the device's attack and release controls. This is done to reduce or eliminate any unwanted "pumping" or "breathing" of the noise floor and/or leakage as the signal falls and rises around the threshold point.

Digital Noise Reduction

Digital signal processing (DSP) is often used to reduce the noise levels within a recorded soundfile. These noises include artifacts such as tape hiss, hum, obtrusive background ambience, needle ticks, pops, and even certain types of distortion that are present in the original recording. Although stand-alone digital noise processors do exist, most systems exist as plug-in software applications for computer-based digital audio editors and workstations. These algorithms can be used to reduce the noise on one or more recorded tracks within a multitrack session, or they can be used during a mastering setting to reduce the overall noise on a master tape. Let's not forget a major application of digital noise and clean-up processing for professional and novice users: cleaning up and restoring older 45s, LPs, and analog tapes for transfer to CD.

Fast Fourier Transform

Most noise reduction plug-ins and programs (Figures 13.7 and 13.8) make use of a mathematically intense algorithm known as *Fast Fourier Transform* (FFT). FFT-based noise reduction. These applications and plug-ins are able to analyze the amplitude/frequency domain of an audio signal in order to reduce hum, tape hiss, and other extraneous noises from your recordings. This digital analysis begins by taking a digital "snapshot" of a short snippet of the offending noise (a brief section that contains only the noise will yield the best results). This noise template can then be digitally subtracted from the original soundfile or region segment in varying amounts and under the control of various program parameters. Under the best of conditions, the noise level will be reduced, while the desired signal will be unaltered.

Figure 13.7. *Noise-reduction plug-in from Waves X-Noise. (Courtesy of Waves Ltd., www.waves.com.)*

Figure 13.8. Sonic NoNoise TDM noise-reduction plug-in. (Courtesy of Digidesign, a division of Avid Technology, Inc., www.digidesign.com.)

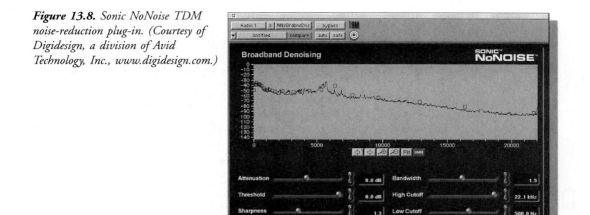

Most FFT-based noise reduction plug-ins are able to process one or more tracks or an entire mix in either real-time during playback, mixdown, or in non-real time (allowing the processed file to be saved and archived).

Before closing this section, it's important that we briefly discuss a few of the unfortunate artifacts that can occur when using FFT-based noise reduction. The most notable of these is *"chirping."* This audible artifact most often occurs whenever too much FFT processing is applied. It literally sounds like a flock of small chirping birds that can either be heard in the background or in a way that sounds like an Alfred Hitchcock movie.

If you find yourself running for cover, it's best to pull back on the FFT settings (and/or increase the processing quality level) until the artifacts are less noticeable. At this point the noise will begin to creep back into the file. From this, it's easy to tell that the process isn't an exact science and is often a balancing act.

Because of these and other band-limiting artifacts, it's generally not a good idea to process and overwrite your original file. Instead, it's best to apply the noise reduction in real-time, or, alternately, save both the original file and the processed file for future changes, decisions, and/or technological advances.

Do It Yourself Tutorial: FFT-Based Noise Reduction

1. Load a track containing an excessive amount of noise into the DAW of your choice.
2. Call up an FFT-based noise reduction plug-in, select a short segment of noise, and follow the application's user directions.
3. Apply varying amounts of noise reduction in real-time and listen to the results. Can you make it chirp? Did you achieve acceptable results?

Should chirping and/or bandwidth limitations become a problem, you might consider using a single-ended noise reduction plug-in to do the job. Because these algorithms use an adaptive filter to intelligently change the program's bandwidth, no FFT-type artifacts will be introduced. The application of digital noise reduction is often more art and experimentation than an exact science.

Finally, it's a misconception that an FFT-based noise reduction application can only be used for reducing noise. Literally, ANY sound can be used as a footprint, and as a result, vocal formants, snare hits, etc., can be pulled from a soundfile to create unique and interesting effects. The sky is literally the limit!

Digital Single-Ended Noise-Reduction Systems

Single-ended noise-reduction systems are currently available in both analog and digital flavors. In addition to the handful of stand-alone hardware devices that are on the market (Figure 13.9), single-ended noise reduction can be found as a pre-set algorithm on a number of hardware multi-effects processors. Several single-ended software plug-ins are also available that can be used with a digital audio editor (Figure 13.10). As seen above in the Digital Noise Reduction section, whenever traditional FFT-based noise-reduction techniques fail (due to the introduction of audible artifacts), a single-ended plug-in can be an effective and important tool for reducing noise from a digital file.

Figure 13.9. *Behringer's Denoiser system Model SNR 2000. (Courtesy of Behringer International GMBH, www.behringer.de.)*

Figure 13.10. *Ray Gun 3.0 noise-reduction plug-in for the Mac/PC. (Courtesy of Arboretum Systems, Inc., www.arboretum.com.)*

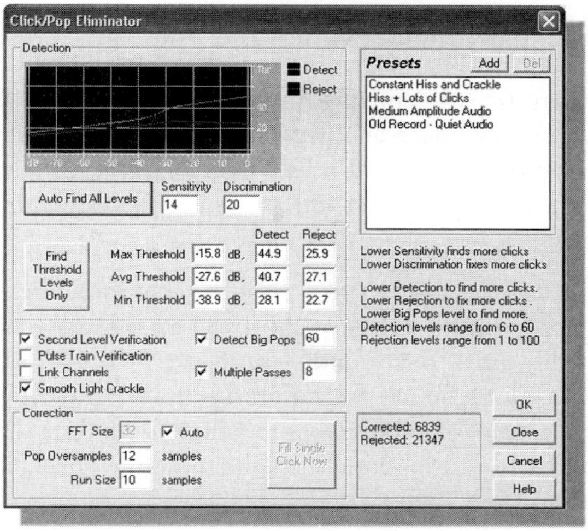

Figure 13.11. *Click/pop eliminator application within Adobe's Audition 1.5. (Courtesy of Adobe Systems, Inc., www.adobe.com.)*

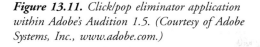

De-clicking and De-popping

In addition to removing noise, programs and DAW plug-ins also exist for removing *clicks* and *pops* from vinyl and older recordings (Figure 13.11). Although FFT analysis is often involved in the process, click removal differs slightly from FFT noise reduction. This two-step process begins by detecting high-level clicks (or those that exceed a user-defined threshold) that exist within a soundfile or defined segment. Once the offending clicks are detected, the program performs a frequency analysis (both before and after the click), which is just enough to make a plausible guess as to what the damaged amplitude/frequency content should sound like. Finally, it pastes this resynthesized "guess" over the nasty offender, ideally rendering it less noticeable or gone, and moves onto the next click.

Although the nature of pops are often quite different, both in their duration and frequency makeup, the reduction process is often similar in form and function. As a result, noise-reduction plug-ins will usually offer sections that are specifically suited to reducing both clicks and pops.

Dither

Although it's not a system or program algorithm for removing high levels of noise from an existing soundfile, *dither* can be used to increase the overall bit resolution (and therefore low-level noise and signal clarity) of a mix. Technically, dither is the addition of small amounts of randomly generated noise to an existing bitstream that allows the S/N and distortion figures to actually fall to levels that approach their theoretical limits. This actually makes it possible for signals to be encoded at levels that are actually less than the data's least significant bit (less than a single quantization step).

Dither can be applied internally to an application or process to reduce the quantization errors, slight increases in noise, and/or fuzziness that could otherwise creep into the bitstream. For example, when mixing tracks together within a DAW, it's not uncommon for digital data to be internally mixed and processed at 32- and 64-bit depths. In situations like this, dither is often used to smooth and round the data values, so that the least amount of resolution will be lost when being reduced to the 24- and 16-bit depths that can be played back by most hardware systems.

Dither can also be manually applied to soundfiles that have been saved at 20- and 24-bit depths (and possibly greater by the time you read this). Applications and DAW plug-ins can be used to apply dither to a soundfile or master mix that's to be downsized to a lower bit depth (to reduce the effects of lost resolution due to the truncation of least significant bits). For example, it's often wise to apply dither to a high-resolution file before saving it as a 16-bit master copy. This reduces noise and increases the soundfile's overall clarity. Further information on dither can be found in Chapter 6: Digital Audio Technology.

CHAPTER 14

Monitoring

Within the recording process, our ability to judge and adjust sound is primarily based on what's heard through the monitor speakers in a project studio or control room environment (Figure 14.1). In fact, within the audio and video industries, the word *monitor* refers to a device that acts as a subjective professional standard or reference by which program material can be judged.

Despite steady advances in design, speakers are still one of the weakest links in the audio chain. This weakness is generally due to potential nonlinearities that can exist in a speaker system's frequency response. In addition, interactions with a room's frequency response often lead to peaks and dips that can affect a speaker's sonic character in ways that are difficult to predict. Add to this the factors of personal "tastes" in the sound, size, and design types, and you'll quickly find that speakers are also one of the most subjective tools in a production studio.

Speaker and Room Considerations

Unless you have several rooms with matching dimensions, materials, and furnishings (an unlikely scenario), you can bet your bottom buck that an identical speaker system will sound different in different room environments. That's to say, it'll interact with each environmental factor to exhibit a unique frequency response curve.

Figure 14.1. Example of a professional monitoring system. (Mega Studios, Paris. Courtesy of Solid State Logic LTD., www.solid-state-logic.com.)

Although variations between production rooms often play a big part in giving a facility its own particular sound, extreme variations in a room's frequency response can lead to difficulties that can be heard in the final product. For this reason, certain basic principles (which are covered in Chapter 3 in the section Control Room Design) have become common knowledge to many who attempt the art of control room design. A few examples of these include:

- ◆ Reducing standing waves to help reduce erratic frequency response characteristics within a room
- ◆ Reducing excessive bass buildup in room corners through the use of bass traps
- ◆ Keeping the room/equipment layout symmetrical throughout a room so the left/right, and front/rear imaging is consistent
- ◆ Using absorptive and reflective surfaces to help "shape" a room's sonic character

Fortunately for us, production and mixdown room designs have greatly improved over the last few decades. This is due to an added awareness of careful room design and the increased availability of acoustical products that can help shape a room's sound. Because of the untold number of acoustic variables that are involved, a project that's been recorded in one facility will often sound quite different when played and/or mixed in another, even when high acoustical construction standards are followed.

Beyond careful acoustic design and construction, a professional facility might choose to further reduce variations in frequency response by *tuning* (equalizing) their speakers to the room's acoustics so that the adjusted frequency response curve will be reasonably flat and, therefore, reasonably compatible with most other control rooms.

Tuning a speaker system to a room can be carried out in either of two ways: by altering settings on the speaker itself or by equalizing the monitor output lines.

Figure 14.2. Rear controls for the Mackie HR824 active monitor speaker. (Courtesy of Loud Technologies, Inc., www.mackie.com.)

Most actively-powered speaker systems offer basic EQ setting controls that can be roughly tuned by the user to match the speakers to their application or layout placement (Figure 14.2). Often these settings can be used to compensate for bass buildup (whenever speakers are placed in or near a corner or other large boundary).

Larger, passive monitors (often a farfield pair) can be tuned by placing a $\frac{1}{3}$-octave bandwidth graphic equalizer (Figure 14.3) between each of the console's control-room monitor outputs and the power amplifier.

By feeding pink noise (which contains a frequency curve that's flat throughout the audio range) into each speaker, the acoustic output can be measured and adjusted (in $\frac{1}{3}$-octave increments) using an instrument known as a *spectrum analyzer*. This instrument is used to visually display the speaker's frequency response as measured through a specially calibrated omnidirectional condenser microphone (Figure 14.4). Initial readings are taken by placing the mic at the console's central listening position, and then by placing the mic at spots throughout the listening area (as the response curve of room will almost always vary from one spot in a room to another). It should be noted that pink noise, rather than individual sine waves, is used for testing as it is random in nature and doesn't stimulate standing waves within a room (which would otherwise introduce false readings whenever the mic is moved).

The frequency response and delay/reflection response of a speaker/room combination can be more accurately measured by using a TEF Time Delay Spectrometer (TDS).

Figure 14.3. Rane GE 130 single-channel, 1/3-octave graphic equalizer. (Courtesy of Rane Corporation, www.rane.com.)

Figure 14.4. *A real-time spectrum analyzer can be used to help adjust a graphic equalizer's response curve in order to determine a speaker's optimum frequency response at the listening position.*

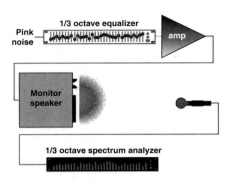

This computer-based system is capable of accurately reading frequency- and time-related response curves within a control room, studio, performance hall, or auditorium, and then assist with or automatically tune a room using the proper response variables (Figure 14.5).

Speaker Design

Just as the sound of a speaker system will vary when heard in different acoustic environments, speakers of different design and operating type will usually sound very different from one another, even when heard in the same room. Enclosure size, number of components and driver size, crossover frequencies, and design philosophy contribute greatly to these differences in sound quality.

Figure 14.5.
Frequency- and time-related measurements can be made with great precision using modern DSP hardware and a personal computer. (Courtesy of Gold Line, www.gold-line.com.)

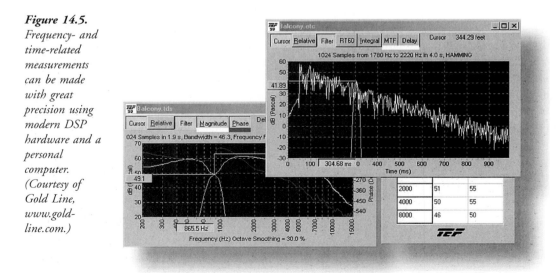

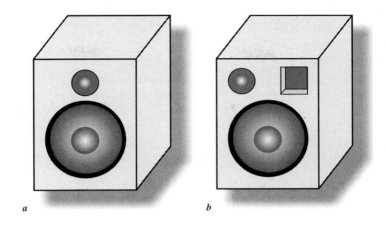

Figure 14.6. Speaker enclosure designs: (a) air suspension and (b) bass reflex.

a   b

Usually professional speaker enclosures are one of two design types: air suspension and bass reflex. An air-suspension speaker enclosure (Figure 14.6a) is an airtight system that seals the air in its interior from the outside environment. This system type (which is often used in "bookshelf" designs) generally provides a strong, "tight" bass response, while often being rolled off at the extreme low end. The bass-reflex or vented-box design (Figure 14.6b), makes use of a tuned bass porthole that is designed into the front or rear of the speaker enclosure. This allows the air mass inside the enclosure to mix freely with the outside air in such a way as to act as a tuned resonator (which serves to acoustically boost the speaker's output at the extreme lower octaves).

Crossover Networks

Because individual speaker elements (drivers) are more efficient in some frequency ranges than in others, different driver sizes and types are often used in combination to give the desired frequency response and level output. For example, large-diameter drivers (such as 15″ and 30″ units) produce low-frequency information more efficiently than at high frequencies; medium-sized speakers (such as 4″ and 5″ units) operate best in the midrange frequencies; and small speakers (such as ½″ to 1½″ diaphragm sizes) reproduce highs more effectively.

These speakers are often connected by passive crossover networks, which prevent signals outside a certain frequency range from being applied to its assigned speaker. Passive networks make use of frequency-selective inductors and capacitors to split the frequency range into several frequency bands. This design provides a smooth transition from speaker to speaker by routing input signals above the crossover frequency to the mid- and/or high-frequency driver, while routing signals below the crossover frequency to the bass driver or drivers (Figure 14.7).

If a speaker system has only one crossover frequency, it is called a two-way system, because the signals are divided into two bands. Likewise, if the signal has two crossover frequencies, it's called a three-way system. The Westlake Audio BBSM-10 monitor speaker (Figure 14.8), for example, is a ported three-way system that includes two 10″ woofers for the bass, a 6.5″

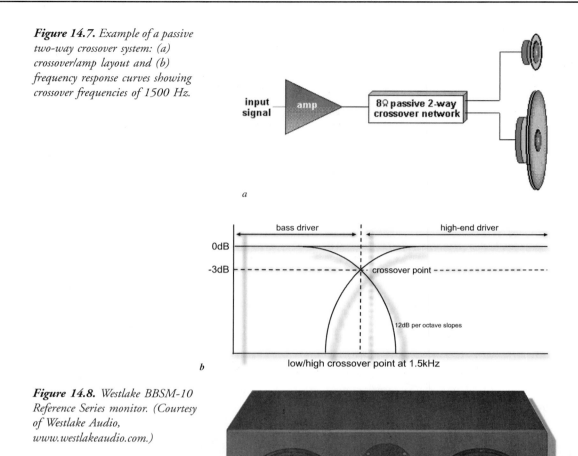

Figure 14.7. *Example of a passive two-way crossover system: (a) crossover/amp layout and (b) frequency response curves showing crossover frequencies of 1500 Hz.*

Figure 14.8. *Westlake BBSM-10 Reference Series monitor. (Courtesy of Westlake Audio, www.westlakeaudio.com.)*

driver for the mid-frequencies, and a 1.5″ diaphragm dome tweeter for the highs with the crossover frequencies being tuned at 600 Hz and 4 kHz, respectively.

Certain designs incorporate crossover level controls that determine how much energy is to be sent to the middle- and high-frequency drivers. This lets you compensate for various

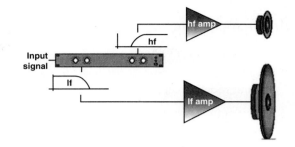

Figure 14.9. *Example of an active two-way crossover system.*

room environments and/or deficiencies (for example, an absorptive room might require more high-frequency energy than a live room).

Electronic crossover networks (Figure 14.9), called active crossovers, differ from conventional passive crossover systems in that the line level audio signal is split into various frequency bands. Each equalized signal is then fed to its own power amp, which in turn is used to drive the respective bass, mid-, and/or high-driver elements. Such a system is generally referred to as being bi-amplified or tri-amplified, depending on the number of crossovers and power amps that are required per channel. Such a system has several advantages:

◆ The crossover signals are low in level, meaning that inductors (which can introduce audible ringing and inter-modulation distortion) can be eliminated from the design.

◆ Power losses (due to the inductive resistance within the passive crossover network) can be eliminated.

◆ Each frequency range has its own power amp, so the full power of each amplifier in the respective speaker efficiency range will be available (meaning that the drawing of excessive current in one range won't affect the sound in another frequency range).

For example, let's assume that we're feeding a 100-W power amp through a passive crossover network to a low- and high-frequency speaker system. If the low frequencies are pulling 100 watts of power from the amp and a high-frequency signal comes along that requires an additional 25 W of power, the amplifier won't be able to supply it. Both the lows and the highs will become distorted. These monitoring requirements, however, could be met without incurring distortion by using an active crossover network to feed a 100-W amp for the low-end speaker and a separate 25-W amp for the highs.

The crossover points for either a passive or an active crossover network are generally 3 dB down from the flat portion of the response curve. Frequency ranges outside the filter's passband will be attenuated according to a slope (usually 6, 12, 18, or 24 dB per octave with a 12-dB/octave slope being the most common). The system's crossover points will be determined by the driver design and type; however, the more commonly selected frequencies are 500, 800, 1200, 5000, and 7000 Hz.

Powered vs. Passive Speaker Design

It goes without saying that many of the more popular monitor types that are in use today incorporate an active-powered amplifier into its design. These cost-effective systems have become widely accepted by the professionals and project communities due to their:

◆ Compact design

◆ High-quality sound (often these systems are bi- or tri-amplified)

◆ Expandability (additional speakers can be cost-effectively added for surround-sound monitoring)

◆ Lack of a need for an external power amplifier

For these reasons, these systems are often ideal for project- and DAW-based facilities and are steadily increasing in popularity.

Passive systems might be bi- or tri-amplified. However, this is most often not the case due to the cost constraints of designing separate crossover and multiple amp systems into the monitor design. Although many opt to use a passive speaker amp system for desktop and nearfield monitoring, these types are almost exclusively found in large farfield monitor systems as a high degree of power is required to drive these large speakers to full power. In the end, the choice is strictly up to you, your preferences, and your pocketbook.

With so many variables to consider in a speaker and room combination, it quickly becomes clear that there's no such thing as the "ideal" monitor system. The choice is often more of a matter of personal taste and current marketing trends than one of subjective measurements. Monitors that are widely favored over a long period of time tend to become regarded as the industry standard; however this can easily change as preferences vary. Again, the best judge of what works best for you should be your own ears and personal sense of style.

Speaker Polarity

A common oversight that can drastically effect the sound of a multi-speaker system is to wire them out-of-phase with respect to each other. *Speaker polarity* is said to be electrically in-phase (Figure 14.10a) whenever one signal that's equally applied to both speakers will cause their cones to move in the same direction (either positively or negatively). When the speakers are wired out-of-phase (Figure 14.10b), one speaker cone will move in one direction while the other moves in the opposite direction.

Speaker polarities can be easily tested by applying a mono signal to both or all of the speakers at the same level. If the signal's image appears to originate from a point directly between the speakers, they have been properly wired (in-phase). If the image is hard to locate and appears to originate beyond the outer boundaries of a stereo speaker pair, or shifts as the listener moves his or her head, it's a good bet that the speakers have been improperly wired (out-of-phase).

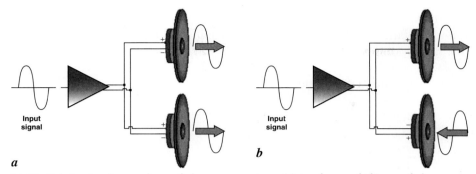

a
b

Figure 14.10. *Relative in-phase and out-of-phase cone motions: (a) in-phase and (b) out-of-phase.*

An out-of-phase speaker condition can be easily corrected by checking the speaker wire polarities. The "hot" lead (+ or red post) leading from each amp channel should be secured to the same lead on its respective speaker (Figure 14.11). Likewise, the negative lead (− or black post) should be connected to its respective lead for each speaker in the system.

Speaker cable is generally color-coded, with white or red being positive (+) and black being negative (−). If no color-coding is present, heavy-duty power cable or other cabling types that are suitable for speakers can be used. These cables will often have a notched ridge (or set of ridges) or have a printed white band that's generally connected to the negative lead post.

Speaker wire gauges should always be as heavy duty as is possible or practical. The #18 wire is considered to be the minimum for lengths of less than 25′–50′, while #14 is considered the minimal length that should be used for 50′ and 100′ runs. (Note: The smaller the gauge number, the thicker the wire; therefore #14 is thicker than #18.) Two reasons for increasing the thickness of the conductor as cable length increases are:

◆ All cable has resistance, which will increase with length. Thinner cables generally have greater resistance values, meaning that more power will be dissipated in the cable and will therefore be unavailable to drive the speaker.

Figure 14.11. *Banana plug binding posts are often used to properly connect wires to the speaker leads.*

◆ The higher the cable resistance, the lower the amplifier's effective damping factor. Damping factor is related to how well the amplifier is able to control the motion of the speaker cone. A lowered damping will often result in a loss of tightness, definition, and clarity in the low end. Again, thicker conductors will have less resistance and thus help to minimize damping problems.

Monitoring

When mixing, it's important that the engineer be seated as closely as possible to the center of the soundfield (making allowances for the producer, musicians, and others who are also doing their best to be in the "sweet spot") and that all the speaker volumes are adjusted equally. For example, if the engineer is closer to one speaker than another, that speaker will sound louder and the engineer may be tempted either to pan the instruments toward the far speaker or boost that entire side of the mix to equalize the volumes. The resulting mix would sound properly centered when played in that room, but in another environment, the mix might be off-center. As a quick check against this, the engineer should always make sure that an audible volume difference between speakers is accompanied by a corresponding visual difference on the main output VU or display meters. Another guard against off-center levels is to monitor pink noise (or a test tone signal) from each speaker in the soundfield to check that they're equally loud (either by doing a quick audible check or by placing a mic in the center listening position). In the latter case, the output level from each speaker can be read and matched using an SPL meter or VU meter on a spare console input.

Here are a few additional pointers that can help you get the best sound from your control room monitors:

◆ Make sure that the room's reverb time is both low and smooth over the audible range (absorptive and diffusive materials can help).

◆ Keep large reflections to a minimum within the room (again, absorptive and diffusive materials can help reduce reflections to a level that's at least 20 dB down from the direct signal).

◆ Keep all room boundaries and reflections as symmetrical as possible along the L/R and front/back axis of the mixing soundfield.

◆ If diffusers are used, place them at the rear part of the room.

◆ If the speakers are mounted in soffits, make sure the front wall has a hard, smooth surface and that the speakers are acoustically isolated from the surface (as much as possible, see Chapter 3 for more information).

◆ Angle the monitors symmetrically toward the listening position in both the horizontal and vertical planes.

◆ If nearfield monitors are used, place them on medium-density foam blocks to reduce console- and desk-borne vibrations.

Monitor Volume

Before continuing, I'd like to revisit another important factor — volume. During the record and mixdown stage, it's important to keep in mind that the Fletcher-Munson curves will always have a direct effect on the frequency balance of a mix. Because our ears perceive recorded sound differently at various monitoring levels, when monitoring at loud levels our ears will easily perceive the extreme high and low frequencies in the mix (sounds good doesn't it?). However, when the mix is played back at lower levels (such as over the radio, TV, or computer), our ears will be much less sensitive to these frequencies and the bass and extreme highs will probably be deficient (leaving the mix sounding distant and lifeless).

Unlike during the 1970s, when excruciatingly high SPLs tended to be the rule in most studios, recent decades have seen the reduction of monitor levels to a more moderate 75–90 dB SPL. (A good rule of thumb is that if you have to shout to communicate in a room, you're monitoring too loud.) These moderate levels offer a good compromise for mixing, as they more accurately represent listening levels that are likely to be encountered in the average home (i.e., the Fletcher-Munson curves will be more closely matched). Ear fatigue and potential ear damage due to prolonged exposure to high SPLs by industry professionals can also be avoided at these levels. For more information on safe monitor levels and hearing conservation, contact the House Ear Institute at www.hei.org.

Monitoring Configurations

In addition to getting the best overall sound out of a mix, another monitoring concern that has gained importance is the need to tailor the mix to the intended room/speaker configuration (i.e., mono-stereo, mono-surround, and stereo-surround).

It's important to remember that a large percentage of your potential customers may first hear your mix over a computer or AM/FM radio in mono. Therefore, if a recording sounds good in stereo but poor in mono, it might not sell as well because it failed to take these media into account. The same might go for a surround-sound mix of a music video or feature release film in which proper attention wasn't paid to phase cancellation problems in mono and/or stereo (or vice versa). The moral of this story is simply this: To prevent potential problems, a mix should be carefully checked in all its release formats in order to ensure that it sounds good and that no out-of-phase components are included that would cancel out instruments and potentially degrade the balance.

The most commonly accepted speaker configurations are mono, stereo, and surround sound.

Mono

Even in this day and age, much of the buying public will first experience a mix in *monaural* (mono) sound (Figure 14.12). That is to say, they'll hear your song over the radio, on TV, in an elevator, on the computer, etc. For this reason record companies, producers, and everyone

508

Figure 14.12. *Example of a monaural monitoring configuration.*

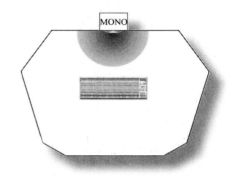

else involved in the process will often place a great deal of importance on mono compatibility and the overall sound of a mono mix. In fact, it's not uncommon for a separate mono mix to be made to ensure that it'll sound as good as it can for its intended medium.

Stereo

Ever since the practical development of the 45°/45° record cutting process, *stereophonic* (stereo) sound (Figure 14.13) has ruled the turntable. Of course, over the years, stereo has also grown to rule FM radio, the CD player, and TV. For these reasons, the creation of a quality stereo mix is extremely important with relation to L/R balance, overall frequency balance, dynamics, depth, and effect.

It has been stated throughout this book that the mixing environment should be acoustically and physically symmetrical (within reason) in order to ensure that the L/R balance and overall imagery is accurate within the stereo soundfield. Beyond this, it's always a good idea to check for mono compatibility. Phase cancellations can cause instruments or frequencies in the spectrum to simply disappear whenever a mix is summed to mono. The best tools for reducing phase errors are good mic technique, a phase meter or X/Y scope display and, of course, your ears.

Figure 14.13. *Example of a stereo monitoring configuration.*

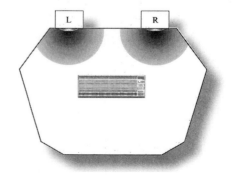

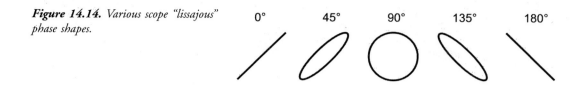

Figure 14.14. *Various scope "lissajous" phase shapes.*

It should be noted that many larger consoles and DAW programs and/or plug-ins will often include a phase scope display. This readout can be used to show phase by sending the left channel to the vertical (Y) trace and right to the horizontal (X) trace inputs (Figure 14.14). The resulting display results in a "lissajous" trace that produces such shapes as:

◆ A slope that rises from left to right at a 45° angle shows a L/R signal, showing that the combined signal is in-phase.

◆ A slope that rises from right to left at a 45° angle shows a L/R signal, showing that the combined signal is out-of-phase.

Of course, audio signals in the real world of stereo are rarely purely in or out of phase and will result in a complicated, elongated (and fun to look at) trace that follows these general shapes.

Surround Sound

With the advent of 5.1 surround playback in home and audio "theaters," surround sound (Figure 14.15) has grown into a major professional and consumer entertainment market. The 5.1 name refers to the five, full-range channels (left, center, right, surround left, and surround right), plus a sixth sub-bass channel (containing a narrow frequency response of 5–125 Hz).

DVD videos and audio discs commonly use 5.1 encoding in the form of Dolby Digital (a scheme that encodes the discrete 5.1 information into a single bitstream, known as AC-3). Most players are able to decode or to route this serial bitstream to an external decoder in various digital surround formats.

Figure 14.15. *Example of a surround-sound monitoring configuration.*

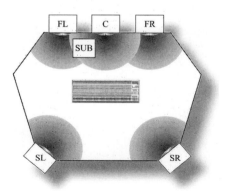

The debate over how to mix for surround has raised the temperature of many a panel forum and Web chat discussion. Those who dive into these discussions tend to fall into three camps:

◆ Those who advocate that music should be focused in the front L/C/R mix, while placing the ambient and special effects into the rear surround field

◆ Those who prefer to leave such traditional concepts behind and look at the soundfield as a full 360° environment, where instruments and mix elements can fall anywhere in the surround soundfield

◆ Those who tend to leave the decision up to the artist, producer, engineer, and record company

When dealing with discrete 5.1, compatibility issues aren't usually a major problem, as it's understood that media will be played back on a discrete playback system. In such a situation, it's common for a separate stereo/mono-compatible mix to be built up from the soundtrack or production mix. However, should the mix be encoded in Dolby ProLogic (a scheme that encodes the surround information into a L/R stereo track using complex phase relationships), care should be taken to ensure surround, stereo, and mono playback compatibility. Additional and detailed information can be found in Chapter 15.

Monitor Level Control

As monitor systems grow to accommodate various production and playback formats, controlling the monitor level and switching can become problematic. Many multi-bus consoles and high-end DAW systems are capable of handling the monitoring requirements of the various formats (including surround sound); however, even these can fall short when multiple sources, level trims, and other straightforward access are taken into consideration. For this reason, many have tuned to using high-quality preamp systems or dedicated studio monitor management systems (Figures 14.16 and 14.17) in order to handle the monitoring needs of a professional or project studio. As an example, the dbx DriveRack Studio Monitor management system is a single rackspace stereo system that features dual XLR inputs, six XLR outputs, dual-channel graphic EQ, dbx compression, multiple crossovers, parametric EQ, stereo output linniters, built-in real-time audio analyzer, and output switching capabilities. With the addition of a reference mic, it can automatically compensate for the frequency response curve of a studio. In addition, the processor comes with 24 presets that are specifically written and designed to work in tandem with popular studio reference monitors.

Figure 14.16. Big Knob Studio command system. (Courtesy of Loud Technologies Inc., www.mackie.com.)

Figure 14.17.
Grace Design
M906 surround
monitor control
system. (Courtesy
of Grace Design,
www.gracedesign.
com.)

Spectral Reference

Even if your monitor speakers have perfect time/frequency response readings, few people who buy recordings will have a perfectly accurate reproduction system (if there is such a thing). As a result, they won't hear the exact mix that was heard in the control room. The buying public will often hear different frequency balances, due to response variances between the almost limitless types of speaker/listening room combinations. Faced with this fact, the best we can do as professionals is to rely on our judgment, our experience, and our ears to create a balanced mix that'll do the best possible justice to a project under a wide range of listening conditions.

In addition to our best set of tools — our experience, judgment, and ears — a visual tool (known as a *spectral analyzer*) is often used to give visual cues as to the overall balance of frequencies that's contained in an audio program at any point in time. These devices (which can be either stand alone or found in a DAW program or suite of plug-ins) give a visual bar readout of a signal's level at various frequencies throughout the audible band (Figure 14.18). Obviously, such a tool can help an engineer or producer to zero in on an offending or deficient frequency and/or bandwidth simply by looking at the display over time.

As was hinted above, your DAW, audio editor or suite of plug-ins might have a phase meter and/or spectral analyzer built into its software (Figure 14.18). During both the record and mix phases, these tools can help point out and avoid potential phase and spectral problems. Those having a dual-monitor DAW display might choose to have these visual tools neatly tucked away in a monitor corner as a visual guide. Those who don't have a spectral analyzer can check out a graphic application that roughly resembles such a readout within Windows

Figure 14.18. *Software spectral display. (Courtesy of Steinberg Media Technologies GmbH, www.steinberg.de.)*

Figure 14.19. *Windows Media Player quick-n-dirty spectral display.*

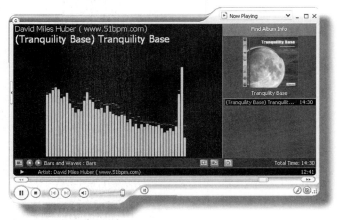

XP by opening a file with Windows Media Player and choosing View/Visualizations/Bars & Waves/Bars. Now play a file (Figure 14.19).

Monitor Speaker Types

In order to obtain the best possible compromise in sound balance, several alternative monitor speaker options are often available as a reference during a session and/or mix (Figure 14.20). Quite often, a console or monitor control system will let you select between speaker/monitor

Figure 14.20. *Alternate control room monitoring combinations: (a) farfield monitoring, (b) nearfield monitoring, (c) small speaker monitoring, and (d) headphones.*

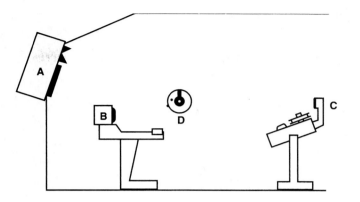

types, with each set commonly having its own associated amplifier for power and level matching flexibility. These types include: farfield, nearfield, small-speaker, and headphone monitoring.

Farfield Monitoring

Farfield monitors often involve large, multi-driver loudspeakers that are capable of delivering relatively accurate sound at moderate to high volume levels. Because of their large size and basic design, the enclosures are generally soffit mounted (built into the control room wall to reduce reflections around and behind the enclosure and to increase overall speaker efficiency). Further reading on soffit design and construction can be found in Chapter 3.

These large-driver systems (Figures 14.21 and 14.22) are often used during the recording phase because of their ability to safely handle high sound levels (which can come in handy should a microphone drop or a vocalist decide to be cute and scream into a mic). They're

Figure 14.21. *Westlake BBSM-15 Reference Series monitor. (Courtesy of Westlake Audio, www.westlakeaudio.com.)*

Figure 14.22. *Tannoy System 215 DMTII main reference monitor. (Courtesy of Tannoy/TGI North America, Inc., www.Tannoy.com.)*

also great for listening to a mix at loud levels in order to hear the impact that it'll have on the dance floor or in a souped-up car system. In fact, certain types of music rely on bass levels that can only be supplied by such a system at moderate-to-high SPLs. (Note: It's important to be aware of the danger of long-term exposure to such sound levels.)

Nearfield Monitoring

Although farfield monitors are generally the best reference at high listening levels, few systems are equipped with speakers that can deliver "clean" sound at such high SPLs. For this reason, most professional and project studios use *nearfield monitors* that more realistically represent the type of listening environment that John and Jill Q. Public will most likely have.

The term nearfield refers to the placement of small to medium-sized bookshelf speakers on each side of a desktop working environment or on (or slightly behind) the metering bridge of a production console. These speakers (Figures 14.23 through 14.27) are generally placed at

Figure 14.23. *Yamaha NS-10M studio monitor speakers. (Courtesy of Yamaha Corporation of America, www.yamaha.com.)*

Figure 14.24. *Genelec 1031A bi-amplified monitor system. (Courtesy of Genelec OY, www.genelec.com, art designer, Herbie Kastemaa.)*

Figure 14.25. *Mackie HR824 active monitor speaker. (Courtesy of Loud Technologies, Inc., www.mackie.com.)*

closer working distances, allowing us to hear more of the direct sound and less of the room's overall acoustics.

In recent times, nearfields have become an accepted standard for monitoring in almost all areas that relate to audio production for the following reasons:

◆ Quality nearfield monitors more accurately represent the sound that would be reproduced by the average home speaker system.

Figure 14.26. M-Audio Studiophile BX-8 active monitor speaker. *(Courtesy of M-audio, Inc., www.m-audio.com.)*

Figure 14.27. Altec Lansing ASC-22C multimedia speakers. *(Courtesy of Altec Lansing Technologies, Inc., www.alteclansing.com.)*

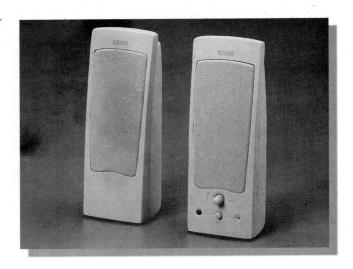

◆ The placement of these speakers at a position closer to the listening position reduces unwanted room reflections and resonances. In the case of an un-tuned room, this helps to create a more accurate monitoring environment.

◆ These moderate-sized speaker systems cost significantly less than their larger studio reference counterparts (not to mention the reduced amplifier cost because less wattage is needed).

As with any type of speaker system, nearfields vary widely in both construction and fundamental design philosophy. It almost goes without saying that care should be taken when choosing the speaker system that best fits your production needs and personal tastes.

Small Speakers

As radio, television, and Web airplay are major forces in audio production and help in the distribution and sales of recordings, it's often a good idea to monitor your final mix through a small, inexpensive speaker set that mimics the nonlinearities, distortion, and poor bass response of those media. Such speakers can either be bought or easily made. Occasionally, these small speakers are incorporated into console and two-track ATR designs for easy monitoring.

Before listening to a mix over such small speakers (Figure 14.27), it's often a good idea to take a break in order to allow your ears and your brain to recover from the prolonged exposure of listening to higher sound levels over larger speakers.

Headphones

Headphones (Figure 14.28) are also an important monitoring tool, as they remove you from the room's acoustic environment. Headphones also offer excellent spatial positioning in that they let the artist, engineer, or producer place a sound source at critical positions within the stereo field without reflections or other environmental interference from the room. Because they're portable, you can take your favorite headphones with you to quickly and easily check out a mix in an unfamiliar environment.

It should be noted that while headphones eliminate the acoustics of a room from the monitoring situation, they don't always give a true representation of how sounds will behave through loudspeakers (especially with regard to imaging). Monitoring through headphones will also often emphasize low-level sounds like reverb and other effects more than loudspeakers in a room. As a result, listening to a mix over both monitor types will often be beneficial.

Figure 14.28. Sony MDR-7506 professional dynamic stereo headphones. (Courtesy of Sony Electronics, Inc., www.sony.com/proaudio.)

Your Car

Last, but not least, your (or any other) car can be a big help in determining how a mix will sound in one of modern society's most popular listening environments. You might try it out on a basic car system, as well as a souped-up, window-shakin' bass bomb. Hey; you might even want to take your mix out for a spin.

Monitoring in the Studio

In addition to the need for an accurate monitoring environment in the control or production studio, musicians will often have special needs for playing back or monitoring the sound that's being recorded during a session.

Headphones in the Studio

Monitoring over headphones in the studio is by far the most common way to monitor sound during a recording session.

When recording, it's generally best to use sealed headphones to prevent or minimize the monitor feed from leaking back into the newly recorded track. The number of headphones will vary with the particular session requirements. An orchestral film overdub, for example, could easily use upward to 60 pair on one session and need only one pair on the next solo project.

Such variations also place demands on the distribution of headphone power and the number of required feeds. As you might imagine, the power that would be needed to run 60 headphones is quite considerable, requiring that a power amplifier be used to drive any number of headphone distribution boxes throughout the room. On the other hand, the power that would be required to drive one or two headphones in a project studio might be so small that they could be driven from the console/mixer's internal headphone or a basic headphone distribution amp (Figure 14.29).

Likewise, the need for separate monitor mix feeds will vary from studio to studio, as well as with varying session requirements. For example, a straightforward session involving

Figure 14.29. *Powerplay Pro-8 8-channel headphone distribution amp. (Courtesy of BEHRINGER Spezielle Studiotechnik GmbH, www.behringer.com.)*

Figure 14.30. *HRM-16 16 channel personal headphone mixing station. (Courtesy of Furman Sound, Inc., www.furmansound.com.)*

only a few musicians will most likely require a single headphone monitor mix. During a tracking or a more complicated session, two or more separate mixes could be made available to the musicians by sending separate "cue" feeds from the mix outputs of multiple auxiliary sends.

For those who wish to allow the musicians to create their own mix, headphone systems (Figure 14.30) exist that can draw separate instrument or submix feeds from various bus or cue feed outputs. This gives the musician complete control over volume, pan, and instrument mix within his or her headphones.

Figure 14.31. *Out-of-phase, mono cue setup for monitoring/overdubbing without a headphone.*

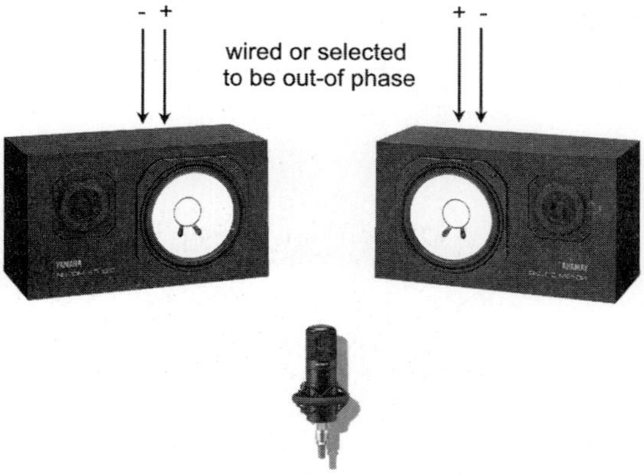

wired or selected
to be out-of phase

Speakers in the Studio

Often, there's not enough time for the musicians to leave their instruments and walk into the studio to hear a playback. For this reason, studio monitors might be mounted in the studio for immediate playback. These are usually larger monitors, as they will need to be driven to levels that can adequately fill the studio and will often need to withstand unintentional abuse (from feedback, too loud a level, etc.)

Using Speakers During a Session

In situations where a musician might be uncomfor wearing a pair of headphones, a pair of speakers may be used as a cue monitor (Figure 14.31). In such a case, two speakers will need to be fed by a single mono source and be wired out-of-phase from each other. By placing them in a studio or vocal booth at exact distances from a vocal mic, the out-of-phase signals will cancel at the mic position. This allows the vocalist/musician to hear the monitor feed, while minimizing any leakage that might feed into the pickup.

CHAPTER 15

Surround Sound

Whether you are an advocate or an adversary of the concept of surround sound, one thing is for sure, it exists in the here and now of music and audio-for-visual media, and it is certain to play an important role in the media technologies of tomorrow.

I think that at this point I have to break from my role as a neutral author and state flat out that I am a HUGE surround proponent. FOR ME, the ability to compose and mix in surround has been an uplifting experience. I clearly remember as a kid, placing two album covers behind my ears and listening as the music came to life around my head (go ahead, try it)! The ability to augment music and visual media by placing sounds within a 360° circle has literally opened up new dimensions in mixing and effects-placement technologies.

Most of the people that I know who are ideologically closed to the idea of surround haven't worked with the medium. I urge that you keep an open mind to the process, watch movies, and listen to music in surround (either in ProLogic or any discrete 5.1 format) and, if at all possible, take the time to familiarize yourself with the process of producing sound in surround.

Before I put my neutral writers cap back on, I'd like to present one last argument for becoming familiar with the production techniques of surround—job opportunities. I have friends living in the technological heart of the San Francisco Bay area who are completely unfamiliar with any and all forms of surround. It never occurred to them that they could increase their client base and perceived prestige in the fields of mixing music, soundtracks for movies and gaming, by installing a surround monitoring system and learning the basic tools and techniques of mixing and mastering media for surround. If for no other reason, the ability to understand and work in new and upcoming technologies can help give your career a marketing edge.

Surround Sound: Past to the Present

Of course, it all started with the movies. Once talkies eclipsed the use of a musical score in the late 1920s (which was played by an orchestra, organ, or piano, depending on the theater's budget), all of the soundtracks were played back in mono.

On November 13, 1940, Walt Disney's *Fantasia* opened up the soundfield to stereo, when it premiered at New York's Broadway Theater. Although it wasn't the first film to be recorded using the "multiple channel recording" process (that distinction went to the Deanna Durbin film *One Hundred Men and a Girl*, which was ultimately released by Universal in mono) . . . Fantasia was the first to make use of multichannel sound. (Unlike the two-channel format that was adopted for home playback, film "stereo" sound started out with, and continues to use, a minimum of four, discrete channels.)

The final mix of Fantasia was printed onto four master optical tracks for playback using a special RCA system called "Fantasound." This multi-speaker setup placed three horns behind the screen and 65 smaller speakers around the walls of the theater. Due to the outlandish setup costs (estimated at about $85,000 for each theater), RCA stopped making this fantastic system after setting another one up at the Carthay Circle Theater in Los Angeles.

In the early 1950s, the first commercially successful multichannel sound formats came onto the scene with the development of CinemaScope (four-track 35 mm) and Todd-AO (six-track 70 mm). Both of these formats made use of magnetic tracks that existed alongside the release print picture, and required that the projector be fitted with special playback heads, amps, and speakers.

In the early 1970's, the home consumer stereo market was gaining in popularity and audio quality. With the development of higher quality amps, speakers, and record turntables came new experimentations in systems design that eventually led to the development of Quadraphonic Sound (Quad). This playback system made use of four speakers (Figure 15.1) that were placed in the four corners of a room, which enveloped the home listener in a FL-FR-RL-RR listening experience.

Although analog reel-to-reel and cassette tape machines were used in homes, they were still relatively expensive. By elimination, this meant that playback would have to be carried out by the most popular medium of the day—the LP record.

Reproducing four channels from a record wasn't easy, given the technology of the day and the various encode/decode systems that were adopted. Often, the task of encoding four channels onto the two walls of a vinyl record was done with relative phase or by using a complex, high-frequency carrier tone that was used to modulate the sum and difference channels.

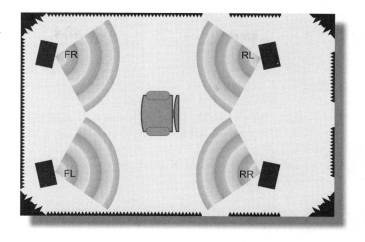

Figure 15.1. Example of a
Quadraphonic (Quad) speaker setup.

After several years of competition between incompatible encode/decode formats, that were both expensive and prone to deterioration over a short period of time (the high-frequency signals on modulated records would literally wear away), the quad revolution died away.

Stereo Comes to Television

Since its inception, surround sound had been used in motion picture soundtrack production with great success. With the introduction of Dolby noise reduction and multichannel audio in the theater, good sound was not only appreciated, it was expected! Up until the early 80s, television sound was strictly a lo-fi, mono experience. Sound was strictly an afterthought to the visual image. However, with the adoption of the video cassette recorder (VCR) and later HI-FI stereo sound from a VCR, discriminating audiences began to appreciate the higher-quality sound that accompanied the almighty image. With the dawning of the music video (I Want my MTV!), stereo broadcast television, and the stereo VCR...TV was finally thrust into the era of the higher-quality, multichannel, visual experience.

Theaters Hit Home

By 1982, Dolby Labs introduced "Dolby Surround," an extension of their professional Dolby Film Sound Project. However, by 1987, millions of homes were fitted with consumer receivers and high-end audio systems that were integrated with video to usher in the era of the surround home theater.

With the introduction of Dolby Pro Logic, it was possible to use a straightforward system of extracting phase information from the two tracks of a stereo program in order to reproduce the L-C-R-S channels of a surround soundfield. In 1992, with the introduction of Dolby's AC3 surround encode/decode system (Dolby Digital), it became possible for discrete 5.1 surround sound to be encoded directly with the new visual entertainment medium of the early 21st century—the DVD!

Surround in the Not-Too-Distant Future

With the recent introduction of the Acura TL car with an Elliott Sheiner (ELS) surround-sound DVD-A system, many in the audio world await the open acceptance of surround sound into the domain of the open highways. In fact, many feel that the automobile, when joined with the home theater system, will greatly help to propel surround into the open market mainstream. Meanwhile, video games and audio codecs (most notably WMA9 and MP3) allow for surround-sound music to be encoded and distributed over the Web for personal use with the millions of surround systems that are now installed on home personal computers.

Monitoring in 5.1 Surround

With the recent proliferation of surround-sound speaker options for both the home and studio, options exist at all levels of quality, functionality, and cost-effectiveness for installing a surround monitoring environment. As with most new technologies, it's important that your existing facility be taken into account, so as to maximize control over monitor levels and monitor format choices (discrete surround, Pro Logic, stereo, and mono), as well as its integration with your current console and/or DAW system. Before choosing a 5.1 speaker system/setup, it would be wise to consider the following:

◆ What are the commercial advantages to producing audio in surround?—Are there any new clients or business ventures that can make use of this technology?

◆ What is the budget for such a system?

◆ Can your existing speakers be integrated into the surround system?—Many powered monitor systems can be upgraded to surround by adding matching or similar speakers from the same product line.

◆ Can your console produce audio in surround sound? If your console has 6 or more output busses (8 bus +), your system can output surround in some manner; however, surround panning and surround monitor control are often essential to true surround production.

◆ Can your DAW produce audio in surround sound?—As above, certain DAWs are capable of routing audio to multiple output busses (6 bus +); however, surround panning and surround monitor control are often essential to true surround production.

◆ How do I plan to monitor in surround?—If the console and/or DAW offers true surround monitor capabilities, you're in luck. If not, a surround monitor control system (Figure 15.2), software monitor mixer (Figure 15.3), or a surround preamp will be necessary.

◆ In what types of surround mastering tools should I invest?—Creating a surround sound mix is only part of the battle. Often the real challenge is to master the 6 tracks into a final format that can be played on a commercial playback system.

Figure 15.2. *The 2380 SPL Surround Monitor Controller (SMC). (Courtesy of SPL electronics GmbH, www.soundperformancelab.com.)*

Figure 15.3. *Surround software monitor mixer in the M-Audio Sonica Theater USB audio interface. (Courtesy of M-Audio, www.m-audio.com.)*

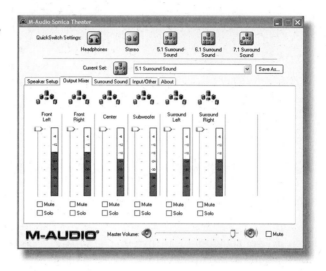

Once these and other considerations have been taken into account, the task of choosing and installing a 5.1 surround system into the production control room can get underway. This can be a daunting task, requiring technical expertise and acoustical knowledge, or it can be a straightforward undertaking that requires only basic placement and system setup.

5.1 Speaker Placement

As defined by the International Telecommunications Union (ITU), the "official" 5.1 speaker setup is made up of five full-range monitors that are positioned in a circular arc, with the speakers being placed at equal distances to the listener (at the center position). Three of the speakers are placed to the front with the center speaker being placed dead center (0°) and the left/right speakers being placed at ?30° arcs to the center point. The surround speakers are then placed behind the listener at ?110° arcs to the center point (Figure 15.4).

The full-range monitors are augmented in the bass-end through the use of a low-pass low-frequency effects (LFE) channel, which is then sent to the system's sub-woofer. In film

Figure 15.4. The ITU "official" 5.1 speaker setup.

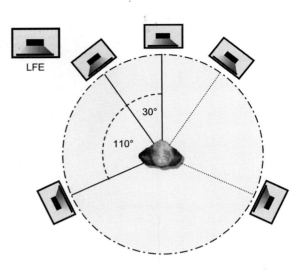

mixing, the low-pass crossover value has been defined to be 120 Hz (although lower values, such as 80 Hz are often chosen).

The LFE's placement within the room isn't usually as critical, although placing it in a corner or in a position that's effected by a harmonic node in the room could greatly affect its response. If possible, the sub should be placed on the floor, near dead center (0°), at a reasonable distance from the front wall (to prevent excessive bass buildup). Finally, most active subs will offer full control over gain and crossover frequency in order to best match the response to the room and main speaker set.

From a level standpoint, all five of the surround speakers should be gain-adjusted to provide the same sound output level. This level matching can be done in many ways. For example, most surround preamps and certain surround monitor systems are able to output white or pink noise to the speakers at equal levels. These signals are sequentially switched from speaker to speaker throughout the room. This allows the levels to be individually matched (either by ear or by using a level meter).

Practical Placement

From a practical standpoint, I have often found surround to be far more forgiving in placement than the above "spec" suggests. For example, it's often not practical to place the three front speakers in an equidistant arc on most consoles or DAW desks as there simply isn't room. Usually, this means placing the speakers in a straight line (while angling the speakers for the best overall soundfield coverage).

Placing three matched speakers on the front bridge of a console, on floor/ceiling mounts, or flush in a soffit generally isn't difficult. However, I found that matching the center speaker on a DAW desk is sometimes a challenge. This relates to the fact that the computer monitor (or monitors) is commonly placed at the center position. If matched speakers are used, where

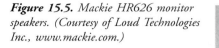

Figure 15.5. *Mackie HR626 monitor speakers. (Courtesy of Loud Technologies Inc., www.mackie.com.)*

can the center speaker be placed? You might be able to place the center speaker on its side. In addition, certain manufacturers make a dedicated, low-profile center speaker that allows the computer monitor to be placed on top (or under a shelf that can hold two monitors, as shown in Figure 15.5).

In cases where space, budget, and time are a consideration, it might not be possible to exactly match the rear speakers to those in the front. In such cases, your best judgment should be used to match the general characteristics of the speaker sets. For example, if a company makes a speaker series that comes in various sizes, you might try putting a pair of their smaller monitors in the rear. If this isn't an option, intuition and ingenuity are your next best friends.

Active/Passive Monitors in Surround

As you might recall from Chapter 14, active monitors include a powered amplifier(s) within their design, whereas passive monitors require that an external power amplifier be used to power their drivers.

One of the results of the rise in popularity of powered speakers is the ability to upgrade a stereo system to a surround environment by simply adding three monitors to a suitably equipped production system. Often, adding extra passive speakers to a system adds the extra expense of having to install 2 extra stereo amps (for C, LS, RS), although consumer and professional 5-channel surround amplifiers can be readily found on the market.

For project studios on a budget that would also like to gain the added benefit of having extensive control over surround monitoring, a logical choice might be to invest in a high-quality surround receiver system (which integrates preamp, monitoring, and surround amp features into a single unit). Such a unit would offer monitoring features that are actually difficult to attain in most professional studios (i.e., mono, stereo, surround, Pro Logic, multiple input switching, record/play, dub switching, etc.).

Figure 15.6. *Revolution 7.1 Surround Sound PCI card.*
(Courtesy of M-audio, Inc., www.m-audio.com.)

With regard to the LFE channel, almost all surround sub speakers are actively self-powered. Although many of these systems provide crossover outputs for diverting the low-frequency energy to the sub speakers, while sending the highs out to the 5 surround (or more likely, the front stereo channels), care should be taken when considering the use of this crossover network. The following should only be taken as a recommendation when setting up your system. There are a number of powered/passive variables, and it's best that you fully research your options for obtaining the best sound:

◆ If the system is a "satellite" system (meaning that the surround speakers are not full-range, but only contain mid-high drivers), the crossover should be installed, as per the manufacturer's instructions.

◆ If the surround speakers are full-range, you might consider not routing the signals through the active sub's crossover, as this would effect the natural low-end of your monitors and could adversely effect the overall sound.

Surround Interfacing

Because surround requires that you have at least 6 output channels, it logically follows that your audio interface should be either:

◆ A dedicated surround sound interface (often with 2 inputs and 6 outs)

◆ A multichannel audio interface (for example, an 8 in × 8 out)

In recent years, a number of dedicated surround cards (and USB/FireWire interfaces) have come onto the market (Figure 15.6). These devices allow audio to be easily and cost-effectively routed from your host DAW or application to an external amp or active monitor system. In certain cases, these audio cards might include a driver (Figure 15.7) that has extensive monitor controls for varying setup levels, switching between modes (mono, stereo, surround), inserting surround processing, and generating setup tones—often well worth the investment.

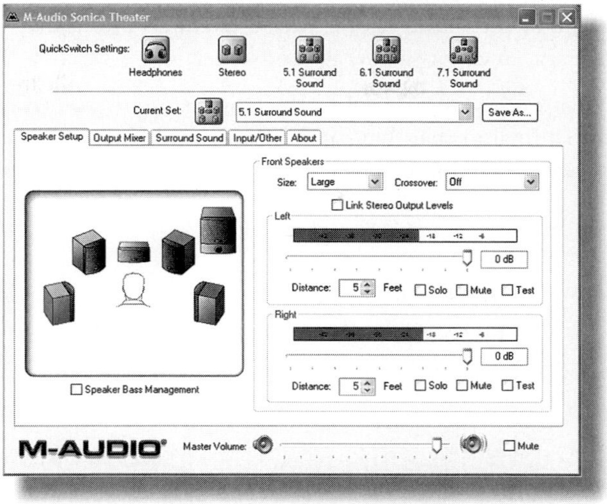

Figure 15.7. *M-Audio Surround card control panel. (Courtesy of M-audio, Inc., www.m-audio.com.)*

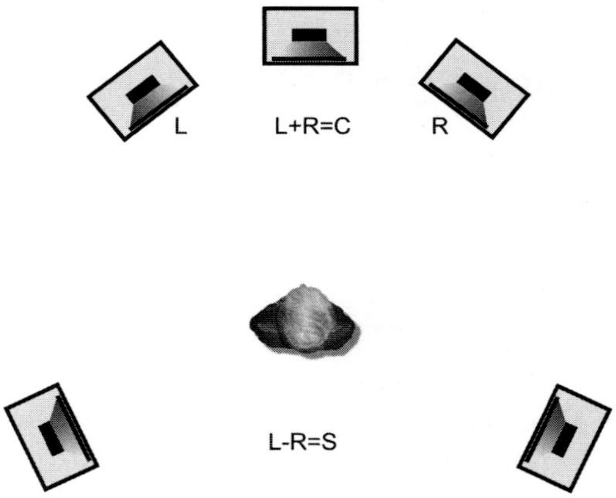

Figure 15.8. *In its most basic form, Dolby Pro Logic is able to derive surround by using the simple phase-related mathematics of L = left, R = right, L + R = center, and L − R = surround.*

The various I/O ports of a multichannel audio interface system can be accessed from within most surround capable DAW systems. Usually the output busses of a DAW are designed to route to the first 6 outputs in a L-R-C-LFE-LS-RS configuration (although this scheme might change from one DAW platform to another, and can often be changed to fit the program and mastering requirements).

Surround Formats

One of the underlying fallacies of surround-sound production is that you can take a console with at least 6 output busses, connect 5.1 speakers to their outputs . . . and then believe that you're ready to fully produce and distribute audio in surround. In fact, there are many hardware and software considerations that must be taken into account, including considerations that relate to the mastering of these 6 channels into a final format that can be played by both the average, and not-so-average consumer.

This process often involves the conversion of a set of master recorded tracks or DAW session into a final format that can be distributed and reproduced using existing surround technology. Several of the production formats that are in current use include:

◆ Dolby Pro Logic
◆ SRS
◆ Dolby Digital (AC3)
◆ DVD-Audio
◆ DTS www.dtsonline.com
◆ SACD
◆ WMA9
◆ MP3
◆ 3D Sound

Dolby Pro Logic

One of the more common delivery formats for getting surround sound to the masses is through the use of *Dolby Pro Logic*. In its simplest form, Pro Logic (Figure 15.8) is able to derive a surround soundscape from a standard pair of stereo channels by using a phase matrix system that operates on the basic principal that:

◆ Left channel = Left channel information only
◆ Right channel = Right channel information only
◆ Center channel = Left + Right channels (summed information with a 3-dB level drop to maintain a constant mix level).
◆ Surround channels = Left − Right channels (difference information is derived via 90° phase shifts, which are applied to create a 180° differential between the L & R signals. In addition, the signal is bandlimited from 100 Hz to 7 kHz and is then encoded with a modified form of Dolby B-type noise reduction.

Figure 15.9. A Pro Logic encoder/decoder should be placed into the record monitor path to ensure that "what you hear is what you get."

This simple system for deriving surround information from a standard stereo signal is what allows us to:

- Listen to a movie from a regular stereo VCR tape in surround
- Listen to our favorite TV show in surround
- Listen to the radio in surround or pseudo-surround
- Listen to your favorite stereo CD or record in surround or pseudo-surround

Yes, you read that last one right! When Pro Logic is switched in, it's possible to listen to a stereo program (of any type) in surround. There is one catch:

- If the stereo information is totally or largely in-phase, most of the sound will be summed (L + R) into the center speaker. If this happens, simply take Pro Logic out of circuit and listen in stereo.
- If the stereo information is partially out-of-phase (L + R, and you'd be surprised how much of music is), the sound will open up into the surround field in a way that lets you hear old standards in a whole new way.

On the production side, a surround project can be encoded into Type I or II (a newer process that offers stereo, full-bandwidth surrounds, and superior sound quality compared to Pro Logic decoding), using either a hardware encode/decode process or a software encode plug-in (when used with a suitable DAW system).

Encoding a surround mix into Pro Logic in the hardware domain can be a tricky venture. It usually requires that the hardware encoder and decoder be installed directly into the signal path with the encoder being placed just before the 2-channel recorder, and the recorder's outputs be routed to the decoder and then into to the surround monitor path (Figure 15.9). This is often required as the results from a Pro Logic encode/decode process can be fraught with unintentional side effects. As a result, any attempts to create a discrete 4.0 or 5.1 surround

Figure 15.10. Steinberg's Matrix Encoder is capable of encoding a discrete surround mix into a phase-encoded surround form. (Courtesy of Steinberg Media Technologies GMBH, www.steinberg.net.)

Figure 15.11. *Steinberg's Matrix Encoder is capable of encoding a discrete surround mix into a phase-encoded surround form. (Courtesy of Steinberg Media Technologies GMBH, www.steinberg.net.)*

mix at one facility and then encode the surround mix into a hardware system can easily create unintended and unacceptable results.

On the other hand, certain DAW systems are capable of encoding a 5.1 mix into a form that's compatible with commercial phase-related surround decoders in the software domain (Figure 15.10). Often these software plug-ins can export an accurate Pro Logic compatible file without the need for an expensive hardware encoder and decoder, and often with more accurate results.

Pro Logic's Phase Compatibility

Because the surround or rear components of Pro Logic are an intentional (and often unintentional) by-product of "?90° phase shifts, which are applied to create a 180° differential between the L & R signals," this surround format may be prone to compatibility issues when summed to mono.

For this reason, it's often wise to make periodic checks as to Pro Logic stereo and mono compatibilities during the course of a mix. Although the problems are often less than one might think, adjustments in balance and placement might need to be made within a music and/or visual soundtrack mix. At times this requires that a compromise be made between the surround effect and mono compatibility.

It's interesting to note that the more you listen to unencoded stereo mixes in surround, the more you'll come to realize that phase is a definite component of recorded music and sound. One of the best examples of this is the effect of reverb. By definition, reverb is made up of random reflections. This means that 50% of these reflections will be in phase and 50% will be out-of-phase. As a result, when listening to a Pro Logic mix, much of the reverb component will be steered to the surround speakers, often in a very natural way. Synthesized sound patches will often be awash in phase components that can create spacious soundscapes that are amazing to experience.

3D Sound

Although it's not strictly a surround format, *3D Sound* has been adopted by the gaming industry (and by some in music) as a pseudo-surround sound effect. This process makes use of complex phase relationships between the L & R channels of a stereo speaker pair to fool the brain into perceiving that sounds are emanating from the speakers at widths that are actually wider than their actual placement. In addition, 3D Sound is capable of fooling the brain into perceiving that certain sounds are actually arriving from behind the listener, when only 2 speakers are being used!

The ability to gain a limited (but definitely perceptible) surround-sound effect from two speakers is often attributed to the outer folds in our ear and the ways in which they interact with complex phase relationships, which allows us to perceive direction. More information on the ear and the perception of sound can be found in Chapter 2 (Sound and Hearing).

So, why am I bringing this obscure psycho-acoustic effect to your attention? Interestingly enough, this 3D effect is often a by-product of Pro Logic surround, when listening to the stereo fold down mix over two speakers. Often, if a sound has been placed in the rear speakers, they will be perceived (in an ethereal, phasey, and intriguing way) as coming from slightly behind the head.

You'll notice that I keep saying speakers and not headphones. Unlike binaural, where the surround effect is perceived when using headphones, the 3D effect is only heard when speakers are used.

SRS

The *Circle Surround* format from SRS technologies is a 6.1 channel matrix surround sound encode/decode system (Figure 15.11) that makes use of signal processing and phase steering technology (the foundation of the technology used in Dolby Pro Logic).

Figure 15.12. Dolby Digital Cinema System setup. (Courtesy of Dolby Laboratories, inc., www.dolby.com.)

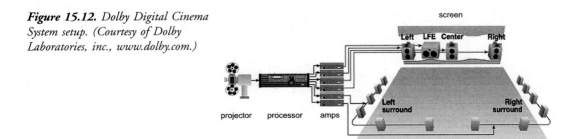

When listening to Circle Surround encoded material, a number of "spatial" enhancements can be taken advantage of from a stereo or surround setup:

◆ The SRS TruSurround XT process makes use of signal processing and phase steering technology in a way that allows us to perceive surround effects from a stereo speaker pair. This technology (which has its basic foundation in 3D) lets us hear sounds that emanate from points beyond the physical width of the stereo speakers, as well as from behind the listener's head.

◆ SRS Circle Surround II allows SRS or Dolby Pro Logic encoded DVDs to be reproduced from a surround-sound speaker system. It also allows stereo sources (such as CDs and MP3s) to be spatially enhanced, allowing stereo program material to be heard in a pseudo-surround.

Dolby Digital (AC3)

Dolby Digital (a technical spec that's also known as AC3), is a popular codec that is used to encode digital audio into a multichannel (mono through 5.1) bitstream through the use of perceptual coding techniques. See Chapter 8 (Multimedia) for more information.

By eliminating unused audio data that is masked or imperceptible, the amount of data that's encoded within the bitstream is greatly reduced (often by a factor of 10), when compared to its uncompressed counterpart.

Dolby Digital is used in the encoding of surround audio with the ever-present DVD-video disc, and has also been adopted for use in high definition television (HDTV) production, digital cable, and satellite transmissions.

The traditional flavor of Dolby Digital found on almost all DVDs provides for 5, full-bandwidth channels in the L-R-C-Sub-LS-RS format. An LFE channel provides a low-pass channel (usually cutting off frequencies above 120 Hz, although lower values, such as 80 Hz are often chosen) for special effects and action sequences in movies (Figure 15.12).

Figure 15.13. *Nuendo Dolby Digital Encoder plug-in. (Courtesy of Steinberg Media Technologies GMBH, www.steinberg.net.)*

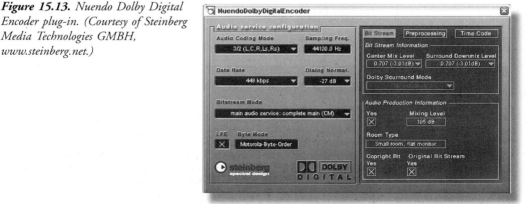

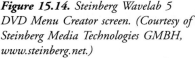

Figure 15.14. *Steinberg Wavelab 5 DVD Menu Creator screen. (Courtesy of Steinberg Media Technologies GMBH, www.steinberg.net.)*

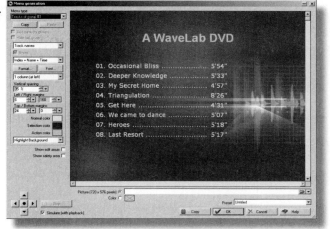

Not all members of the home audience are able to listen to Dolby Digital in discrete 5.1, so a downmixing feature was implemented into the codec to ensure compatibility with any mono or stereo playback device (meaning that additional tracks in various channel formats aren't necessary). Of course, by now, most of you are aware that the compressed nature of this format allows for options of another sort, i.e., soundtracks in various languages and director/actor/background tracks can easily be included as a bonus.

The encoding a 6-channel audio master into a Dolby Digital encoded bitstream can take place through the use of a Dolby licensed hardware encoder or through the use of an encoding software program or plug-in (Figure 15.13). Unlike Pro Logic, since Dolby Digital is a discrete encoding process, the 6-channel master can be converted into an AC3 bitstream without the need for monitoring through the encode/decode process during the mixdown stage.

DVD-Audio

By using Dolby Digital compression, full 5.1 audio is able to accompany a full-feature length video on a DVD-Video disc. However, numerous audiophile and record company concerns have pushed for the commercial adoption of DVD-Audio (Figures 15.14 and 15.15), which allows for uncompressed stereo or multichannel surround-sound audio to be encoded onto a DVD disc in a wide range of bit and samplerate specifications (up to 24 bits/96 kHz, or even 192 kHz).

In addition to audio, a DVD-Audio disc is able to include a limited amount of video, photo stills (for displaying album art, behind the scenes pics) and text (for displaying lyrics and notes). Up to 16 graphic stills can be associated with each track and on-screen displays can be used to display art and navigate around the disc. As an option, the DVD-Audio spec allows for audio tracks that have been encoded in the DVD-Video formats (Dolby Digital and DTS) to ensure compatibility with existing DVD-Video players.

Figure 15.15. Diskwelder Bronze
DVD-Audio Tool. (Courtesy of
Minnetonka Audio Software, inc.
www.discwelder.com.)

Figure 15.16. Dolby Digital release print. (Courtesy of Dolby
Laboratories, inc., www.dolby.com.)

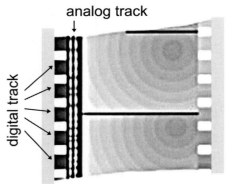

It almost goes without saying that standard CD players will not recognize and play a
DVD-Audio disc. In fact, most standard DVD-Video players aren't compatible with
the uncompressed DVD-Audio data. Newer units that are compatible will display both the
DVD-Video and DVD-Audio logos.

DTS

Digital Theater Sound (DTS) is an audio encoding scheme that supports up to 6.1 channels
of discrete audio from a single datastream for use in cinema sound, DVD/home theater,
and multimedia. The DTS codec exists in several format flavors (depending upon its intended
application) and, as of this writing, these include:

◆ DTS Digital Surround—Provides 5.1 channels of discrete digital audio for commercial
  movie soundtracks, consumer electronics products, and software content.

- ◆ DTS-ES—A digital audio format for delivering 6.1 channels of discrete audio in the consumer electronics market. DTS-ES is also fully backward-compatible with DTS decoders that aren't "Extended Surround" equipped.

- ◆ DTS Neo:6—Provides up to six channels of matrix decoding from stereo matrix material.

- ◆ DTS 96/24—Offers 24 bit/96 kHz samplerate encoding for multichannel sound on DVD-Video.

- ◆ DTS Interactive—Delivers real-time, discrete multichannel interactive playback for PlayStation2 games when connected to a DTS-equipped device (such as an AV receiver).

- ◆ DTS Virtual—Down-converts DTS 5.1- or 6.1-channel soundtracks to stereo while processing a realistic simulation of surround sound for two-channel equipment.

DTS is used to encode audio data at rates of 1.5 Mbit/sec or 754 kbit/sec (rates that are higher than the 448 kbit/sec or 384 kbit/sec rates of Dolby Digital). Although far fewer titles have been released in DTS over its Dolby Digital counterpart, the new 754 kbit/sec rate has allowed a number of movie production studios to offer both soundtrack formats (Dolby Digital 5.1 and DTS 5.1) on their major release DVDs.

Unlike Dolby Digital (which optically encodes the soundtrack as data blocks between the sprocket hoes of the film), DTS stores the encoded data onto separate CD-ROM discs, which are then synchronized to the film via a SMPTE time code film track (Figure 15.16).

WMA9

Through the use of plug-in encoders from various DAW manufacturers, it's possible to encode 2-channel, as well as discrete 5.1 and 7.1 surround-sound audio into a bitstream that

Figure 15.17. *Adobe Audition WMA9 surround encoder. (Courtesy of Adobe Systems Inc., www.adobe.com.)*

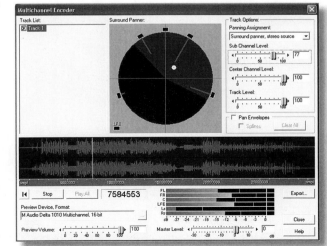

can be played back using Microsoft's Windows Media 9 player. This codec, which was designed for digital download and multimedia formats, has been optimized for streaming or download-and-play delivery at bit rates that range from 128 kilobits per second (Kbps) to 768 Kbps, at depths of 16 or 24 bits using samplerates of 44.1, 48, 88.2, or 96 kHz (Figure 15.17).

When set up properly, true surround sound can be played from a standard Windows XP computer, using an off-the-shelf surround sound card/speaker system or from a multichannel audio interface (Figure 15.18).

Using WMA9 (or a higher version), it's possible to playback surround-encoded files from a stereo computer or media player through the use of a down mix codec that proportionately mixes the center and rear channels into the L-R channels.

Figure 15.18. *Windows 5.1 speaker setup dialog box.*

Figure 15.19. *Surround sound panners within the Nuendo 2.0 mixer. (Courtesy of Steinberg Media Technologies GMBH.www.steinberg.net.)*

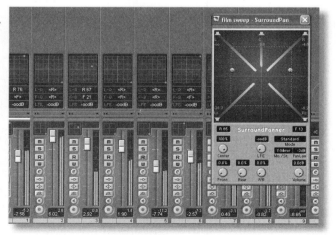

MP3 Surround

MP3 surround has now been developed by Fraunhofer IIS in collaboration with Agere Systems (www.agere.com), which allows for the reproduction of multichannel sound at bit rates that are comparable to those currently used to encode stereo MP3 material.

Using a codec that's based on the MP3 standard—MP3 Surround technology encodes multichannel sound by transmitting a stereo audio signal that carries a compatible stereo down mix of the multichannel material. Along with the traditional audio tracks (which are touted as being fully compatible with traditional MP3s), a small amount of side information is transmitted that is used to "steer" the center/surround information to its appropriate speakers.

Currently, for example, a number of commercial music download services offer their music in MP3 stereo. Such services can be seamlessly extended to provide multichannel MP3, and yet remain fully compatible with existing stereo players. The same applies to Internet radios that will be able to stream multichannel content with nearly the same effort as for stereo broadcasting.

Mixing in Surround

Any number of multiple bus consoles (typically having 8 or more busses) can be used to create a surround sound mix, but the important question of the day is: How easily can signals be routed, panned, and effected in a surround environment to create a 5.1 mix without having to pull teeth.

Whether you're working in the analog hardware, digital hardware, or digital software mixing environment, the ability to pan mono or stereo sources into a surround soundscape, place effects in the 5.1 scape, and monitor multiple output formats without difficulty can make the difference between a difficult and compromised mix and one that lifts your spirits.

Virtual Surround Mixers

As you might imagine, designing an analog console for use with full-scale surround features is almost always a massive and costly undertaking. Digital consoles, on the other hand, are better suited to controlling the spatial and effects panning functions of numerous inputs through the use of a central control panel.

Virtual surround-sound software mixers that are often designed into high-level production DAWs are often able to effectively and cost-effectively route, pan, effect, and automate signals in the software domain (Figure 15.19). The ability to mix, automate, effect, and then export a full-surround mix into a final format is often a thing of beauty that's extremely hard to match in the hardware domain, regardless of the price.

Mixing Philosophies

After having mixed many projects in surround and in talking with many engineers and producers, I've come to realize that the first rule of recording—"There are no rules, only guidelines"—definitely applies to surround sound mixing.

The ways in which sound can be placed in the soundscape are as varied as those who place them. For example:

◆ There are those who believe that the sound should be primarily placed in the front (L-C-R), while the reverb, natural ambient, and/or audience should be placed in the rear.

◆ Others feel that there literally are no rules, and that sounds can be placed anywhere that sound good, interesting, or just plain "right."

◆ Others will place monophonic sounds at pinpoint positions throughout the field, while others will place multiple stereo fields throughout the soundscape (i.e., the L-R fields of an XY mic or stereo outs of a synth, etc.).

◆ Many will not use the LFE (sub) channel, believing that it adds an unnaturalness to the low bass that's better handled by the full-range speakers.

◆ Others will not use the center speaker, believing that it messes with the center image in a (L-R-LS-RS) way that's reminiscent of the quadraphonic era.

It's my strong belief that how you choose to convey your music through the art of surround-sound mixing is up to you and the muses. It's a fun medium that should be played with.

Re-issuing Back Catalog Material

On a final note, one of the unintended by-products of older standards that are being re-issued into surround by the record companies is the resurrection and the rescue of many of the old analog (and digital) masters that are so old they require monumental and heroic restoration efforts.

In an effort to transfer these tracks to 24/96 digital archive soundfiles, it's not uncommon to hear of stories of 16- and 24-track masters that have to be reconditioned and carefully baked, then played onto an analog machine, and then watched during the one-pass transfer as the iron oxide sheds and separates from its backing as it plays onto the floor.

Rescue stories like these are varied and awesome (in the truest form of the word). Sometimes the masters have already deteriorated past the point of playability and the safety backup must be used. Suffice it to say that surround reissues are definitely doing their part toward helping to keep the original master tracks alive and kickin' into the 21st century, not to mention the fact that they are breathing new life into the tacks in the form of a killer, new surround-sound reissue.

Listening to a major surround project over killer speakers in the studio (or a home theater system) can rank way up there with chocolate, motorcycle ridin', and sex—absolutely the best!

CHAPTER

Mastering

The *mastering* process is an art form that uses specialized, high-quality audio gear in conjunction with one or more sets of critical ears to help the artist, producer, and/or record label attain a particular "sound" and "feel" for the finished manufactured product. Working with such tools (Figures 16.1 and 16.2), a mastering engineer or experienced user can go about the task of ordering and shaping the various cuts of a project into a final form that can be replicated into a salable product.

In past decades, when vinyl records ruled the airwaves and spun on everyone's sound system, the art of transferring high-quality sound from a master tape to a record was as much of a carefully-guarded art form as it was technology—and it still is. Because this field of expertise was (and still is) well beyond the abilities of almost every recording engineer and producer, the field of vinyl mastering was left to a very select few.

The art of transferring sound to CD is also still very much an art form that's often best left to those who are familiar with the tools and trade of getting the best sound out of a project. However, recent advances in computer and effects processing technology have made it much easier for producers, engineers, and musicians to own high-quality hard- and software tools that are fully capable of creating a professional-sounding final product in the studio or on a desk/laptop computer.

Figure 16.1. EngineRoom Audio
Mastering facility, NYC. (Courtesy
of EngineRoom Audio,
www.engineroomaudio.com.)

Figure 16.2. Colossal Mastering,
Chicago, IL. (Courtesy of Colossal
Mastering, www.colossalmastering.com.)

The Mastering Process

In addition to the concept of capturing the pure artistry of a production onto tape, hard disk, or other medium, one of the many goals during the course of recording a project is the overriding concept that the final product will have a certain "sound." This sound might be "clean," "punchy," "gutsy," or any other sonic adjective that you might be striving for. Of course, it's everyone's goal that once all of the cuts have been mixed, you'll be able to sit back and say, "Yeah, that's it!" If this isn't the case, having an experienced mastering engineer "shape" the sonic character of your project through the careful use of level balancing, dynamics, and EQ could save your sonic Technicolor day.

Another factor that can affect a project's sound is the reality that recordings may have been recorded and/or mixed in several studios, living rooms, bedrooms, and/or basements over the course of several months or years. This could mean that the cuts would actually sound different from each other. In situations like this, where a unified, smooth sound might be hard to attain, it's even more important that someone who's experienced at the art of mastering be sought out.

In essence, the process of successfully mastering can help a project:

- ◆ Sound "right"—This is often accomplished through the use of careful EQ matching and dynamics processing. As was previously mentioned, this process takes not only the right set of processing gear, but requires experienced ears that intuitively know how the project will most likely sound under a wide range of playing conditions.

- ◆ Be in the right order—Choosing a project's song order is best done by the artist and/or producer to convey the overall "feel" of a project. In addition to order, the process of setting the gap times between songs can also make the difference between awkward pauses and a project that "flows" smoothly from one cut to the next.

- ◆ Playback at optimum levels—Traditionally, the industry as a whole tends to set the average level of a project at the highest possible value. This is often due to the fact that record companies will always want their music to "stand out" above the rest when played on TV, the radio, or the Web. This is usually accomplished by applying compression to the track or overall project. Again, this is an artistic technique that often requires experience. Over-compression can lead to audible artifacts that can affect the "sound" that you worked so hard for. In fact, light compression or even no compression at all is an alternative. Classical music lovers, for example, often spend big bucks to hear a project's full dynamic range. To tamper with this might affect sales. Although the choice is up to you, the producer, and/or the record company, care must be taken when dealing with the subject of dynamics.

- ◆ Match levels throughout the project—In addition to getting the best overall level; it's often important that levels match when transitioning from one song to the next. If you keep songs from sticking out like a sore thumb it will help to improve the flow and professionalism of a project.

The equipment that deals with the art and technology of creating a finished "master" are available in many different guises. Often, top-level mastering engineers will use specially designed EQ, dynamics, and level matching gear that often won't be found in the recording studio environment. Having said this, with the advent of CD and DVD burning software, dedicated hard- and software processing systems are currently on the market that give musicians, producers, and engineers a greater degree of control over the final mix and/or finished master than ever before.

To Master or Not to Master—Was That the Question?

As was mentioned, the process of mastering often requires specialized technical skills, audiophile equipment, a carefully tuned listening environment, and talented ears in order to pull a rabbit out of a problematic hat or even one that could simply use some dressing up.

When approaching the question of whether to have a project professionally mastered or not, it's important that you objectively consider the following questions:

◆ Is the final mix adequate (or more than adequate) for its intended purpose?

◆ If not, would the project benefit most from an outside set of professional ears or could it be done in house?

◆ Does the budget allow for the project to be professionally mastered?

If the decisions favor that the services of a professional mastering house and engineer be sought, the next question to ask is "Who'll do the mastering?" In this instance, it's important that you take a long hard look at the experience level of the one who will be doing the job and again objectively consider the following:

◆ What is their mastering track record?

◆ Are you familiar with examples of their work? If not, ask for a client list, so you can listen to their work.

◆ Are they familiar with your music genre as well as the type of "sound" that you're hoping to achieve?

◆ What are their hourly or project rates? Does this fit your budget?

Bottom line: Beware of the inexperienced mastering engineer, even if that person is you. Once the choice has been made, it's often wise for you/your band and the producer to personally sit in on the mastering session. Make sure that there are several ears around to listen to the project and listen over several types of systems. And above all, be patient, be critical of the project's sound, and listen to the opinions of others. Sometimes you get lucky and the mastering process can be quick and painless; at other times it takes the right gear, keen ears, and lots of careful attention to detail.

Mastering the Details of a Project

From the mastering engineer's standpoint (as well the artists and producer who are overseeing the project), the artistic craftwork that's required in creating a master recording lay in the numerous details of how the project is carefully finessed into its final, approved form. The decision-making steps that are involved in this process often include:

◆ Choosing the proper song order

◆ The application of EQ to improve the sound and overall "flow" of the project

◆ The judicial use of dynamics to balance out the sound and increase the project's overall level

Important note: Should it be decided that an outside, professional mastering engineer will be used on a particular project? It's ALWAYS wise to consult with that person about the general specifications of the final product before beginning the mixdown (or even the recording) process. For example, that person might prefer that the files be recorded and/or mixed at a certain bit rate and bitdepth, as well as in a specific format.

In most instances, a mastering engineer would prefer that the final master be changed or processed as little as possible with regard to normalization, fade changes, overall dynamic changes (compression), and applied dither. These processing functions are best done in the final mastering process by a qualified engineer.

Sequencing: The Natural Order of Things

Whether the master is to be assembled using analog tape, or in a digital audio workstation/ editor, the running order that the songs of a project will be played often affects the overall flow and tone of a project. The considerations for which song follows what is infinitely varied, and can only be garnered from experience and an artistic "feel" of how their order and interactions will affect the listening experience. A number of variables that'll directly affect the sequenced order of a project include:

- Total length—How many songs will be included on the disc or album? If you've recorded extra songs, will you include Bonus Tracks on the disc? Is it worth adding a weaker song, just to fill up the CD?
- Running order—Which song should start? Which should close? What order feels best and supports the overall mood and intention of the project?
- Transitions—Altering the transition times between songs can actually make the difference between an awkward silence that jostles the mood and a transition that keeps with the pace and feel of the project. The Red Book CD standard calls for 2 seconds of silence as a default setting between tracks. Although this is necessary before the beginning of the first track, it isn't at all the law for spacings that fall between any of the songs. Most editors will allow you to alter the index spacings between tracks from 00 seconds (butt splice) to longer spaces that help maintain the transitional mood.
- Crossfades—In certain situations, the transition from one song to the next is best served by crossfading from one track directly to the next. Such a fade could seamlessly intertwine the two pieces, providing the glue that can help convey any number of emotional ties.

A good friend of mine, Craig Anderton, actually maps out the up and down flow of a musical project over its course by using a spreadsheet (although a document file would work equally as well). Methods such as these give you an overview indication of weak areas that could use a bit of special attention . . . as well as notation that gives crossfade times, index times, gap lengths, crossfade in/outs, and other information that would be extremely helpful to the mastering engineer.

Relative Volumes

In addition to addressing the overall volume levels of a project, one of the tasks of the mastering process is to smooth out the relative volume differences between songs over the course of the disc or album. These differences could occur from a number of sources, including general variations in mixdown and program content levels and levels between projects that have been mixed at different studios.

Cues as to smoothing out the relative rms and peak differences can be obtained by:

◆ Looking at the general attributes of a soundfile from within a digital audio editor
◆ By carefully watching the master output meters on a recorder or editor
◆ By watching the graphic levels of the songs as they line up in a digital audio editor
◆ By using your ears to fine-tune the volume levels from song to song

Contrary to popular belief, the use of the standard normalization tool can't smooth out these level differences, as this process detects the peak level within a soundfile and raises the overall level to a determined value. The average (rms) and peak levels will often vary widely between the songs of a project. In certain instances, however, certain CD burning software and digital audio editors will include a special normalizing tool that can be used to smooth out the various levels between songs.

EQ

As is true in the mixdown process, *equalization* is often an extremely important tool for boosting, cutting, or tightening up the low end; adding presence to the midrange; and tailoring the high end of a song or overall project. Again, EQ could be used as a tool to smooth out differences between cuts or for making changes that effect the overall character of the entire project. Of course, wide ranges of hardware and software plug-in EQ systems are available for applying the final touches both within a studio and project-based setting (Figures 16.3 and 16.4).

Figure 16.3. Neve 1073 outboard EQ rack. (Courtesy of AMS Neve plc, www.ams-neve.com.)

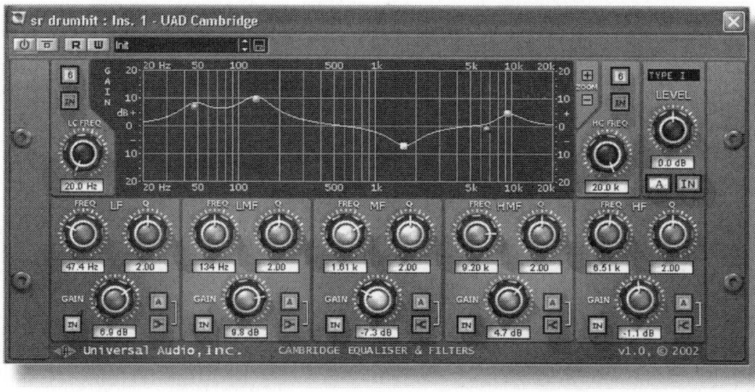

Figure 16.4. *Cambridge EQ plug-in for the UAD-1 card. (Courtesy of Universal Audio, www.uaudio.com.)*

Dynamics

One of the most commonly used (and overused) tools within the mastering process relates to *dynamics* processing, or specifically, compression.

Although the general name-of-the-game is to achieve the highest overall average level within a song or complete project, care must be used not to apply so much compression that the life gets dynamically sucked right out of the sound. As with the 1st rule in recording—"There are no rules"—the amount of dynamics processing is entirely up to those that are creatively involved in the final mastering process. However, it's important to keep in mind the following guidelines:

◆ Depending upon the program content and genre, the general dynamic trend is toward raising the overall levels to as high a point as possible.

◆ When pushed to an extreme, this will often have the (intended or unintended) side effect of creating a sound that has been "squashed," giving a "wall-of-sound" character that's thick (that's good) and potentially lifeless (that's bad).

◆ When compression is not applied (or little is used), the sound levels will often be lower, thinner (that might be bad), and full of dynamic life (that's good).

From all of this, you'd be correct if this process is entirely subjective and relative! The use of compression can help add a strong presence to a recording, while it's overuse can actually pull the dynamic life out of it, so use it wisely!

Multiband Dynamic Processing

The modern-day mastering process often makes use of multiband dynamic processing (Figures 16.5 and 16.6) in order to break the frequencies of the audio spectrum into bands that can be individually processed. For example, such a hard- or software system could be used to strongly compress the low frequencies of a song using a specific set of parameters, while a small degree of compression is applied to the sibilance at its upper end. Depending upon the system, up to 5 distinct bands might be available for processing the final signal.

Figure 16.5. *TC Finalizer 96K Studio Mastering Processor. (Courtesy of t.c.electronic, www.tcelectronic.com.)*

Figure 16.6. *Waves Linear Phase Multiband compression, EQ, and limiting plug-in. (Courtesy of Waves Ltd., www.waves.com.)*

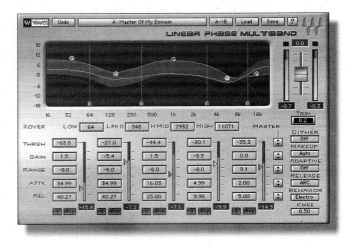

Noise Reduction

Digital *noise reduction* (NR) is often a tool that's used in the mastering process in order to reduce background hum, tape hiss, preamp noise, and other gremlins. Two types of digital NR can be used:

◆ FFT
◆ Single-ended

FFT NR

Most digital NR plug-ins, programs, and hardware systems (Figure 16.7) make use of a mathematically intense algorithm known as *Fast Fourier Transform* (FFT). FFT-based NR applications are able to analyze the amplitude/frequency domain of an audio signal in order to reduce hum, tape hiss, and other extraneous noises from your recordings.

This digital analysis begins by taking a digital "snapshot" of a short snippet of the offending noise (a brief section that contains only the noise will yield the best results). This noise template can then be digitally subtracted from the original soundfile or region segment in varying amounts and under the control of various program parameters. Under the best of conditions, the noise level will be reduced, while the desired signal will be unaltered.

Important note: When sending a soundfile to a mastering engineer that's in need of being noise reduced, it's important that at least a short sample of the noise be preserved at the

Figure 16.7. *Bias Sound Soap Pro Audio restoration/ cleaning plug-in. (Courtesy of Bias, Inc., www.bias-inc.com.)*

beginning (or end) of the file. Engineers will often make the mistake of editing out the noise at the beginning and end of a song. In order for the mastering engineer to noise reduce the file (using the FFT method), a short sample of just the offending noise must be available for sampling by the NR processor.

Single-Ended NR

Single-ended NR systems work by breaking the audio spectrum up into a number of frequency bands, such that whenever the signal level within each band falls below a user-defined threshold, the signal will be attenuated. This downward expansion/filtering process accomplishes NR by taking advantage of two basic psycho-acoustical principles:

◆ Music is capable of masking lower level noise that exists within the same bandwidth.

◆ Reducing the bandwidth of an audio signal reduces the perceived noise.

It's a psycho-acoustical fact that our ears are more sensitive to noises that contain a greater number of frequencies than to those containing fewer frequencies. In light of this, whenever the program's high-frequency content is reduced or restricted to a certain bandwidth, the dynamic filter examines the incoming signal and reduces the bandwidth accordingly. Conversely, whenever the high-frequency signal returns, the filter opens back up as far as necessary to pass the signal.

File Resolution

The samplerate and bit rate resolution that a project has been recorded in is a personal matter (as with many things in sound recording). Given that high-quality converters and a DAW with a high-quality bit-processing structure are used, many believe that recordings made with 16 bit/44.1 kHz file resolutions are sufficient to capture the nuances of sound (for both the tracks and for the final mixdown.) Others believe that rates upward to 24 bits/96 kHz (although 24 bit/192 kHz is possible) are necessary to capture the extended high frequencies and increased resolution of music.

Regardless of which rates are chosen, it's generally best to deliver the final master recording to the mastering lab at the bit- and samplerate resolution that the session files were originally recorded (i.e., if the session was recorded at 24/96, the final mixdown resolution should be at 24/96). The reasoning behind this is that, in almost every circumstance, the mastering engineer will be able to process the mix at a higher bit-resolution and will have better tools by which to dither the soundfiles down to the necessary 16/44.1 rate for final transfer to CD.

The Digital Audio Editor in the Mastering Process

By far, the most commonly used system for modern-day mastering is the digital audio editor (Figures 16.8 through 16.10). These 2-channel and multichannel systems make use of the personal computer's existing processing, disk storage, and data I/O hardware to perform a wide range of audio editing, processing, and mastering production tasks. By offering a graphic, on-screen environment, these software programs allow individual songs to be imported, processed, and exported in a file format that can be burned to disc—either directly from within the program or through the use of an independent CD burning application.

In addition to using a 2-channel editor for the purpose of mastering a project to disc, it's also possible to load the various songs within a project into a multichannel DAW in order to put the finishing touches on a project. Using this approach, each of the various songs (or processed sections) can be loaded into their own track (Figure 16.11), which allows each track to be independently processed according to it's needs, while also allowing overall effects to be added to the master output section (allowing EQ, dynamics, dither, etc., to be applied to the entire project mix).

Figure 16.8. Bias Peak 2-channel editor for the Mac. (Courtesy of Bias, Inc., www.bias-inc.com.)

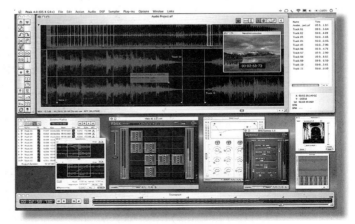

Figure 16.9. *Adobe's Audition for the PC. (Courtesy of Adobe Systems Inc., www.adobe.com.)*

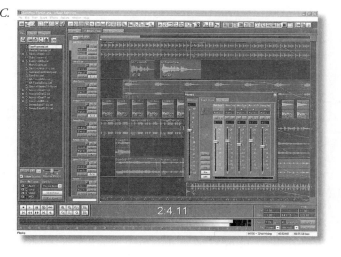

Figure 16.10. *SADIE DSD8 8-channel mastering and authoring System. Courtesy of Studio Audio & Video Ltd., www.sadie.com.)*

Figure 16.11. *Steinberg Wavelab 5. (Courtesy of Steinberg Media Technologies GMBH, www.steinberg.net.)*

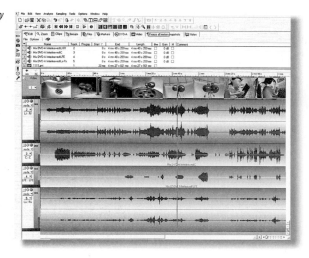

Once the mix has been finessed into its final form, it can be saved as a session file for recall at a later time should changes need to be made at a later time. This final version can also be saved to disk as a continuous, processed file (or as separate songs that can be butt-spliced together in a single chain) for import into a CD burning program.

CD/DVD—A Mastering and Burning Software

Once the files are ready, there are a number of ways that a project (and/or it's various soundfiles) can be imported into a program and burned to a final master CD. As one can imagine, burning software systems come in various flavors of features and complexity, ranging from simple burning programs that come bundled with most home computer systems to high-end production systems for preparing and burning discs within a professional facility.

A number of CD authoring/burning programs (Figure 16.12) allow a single soundfile (or a number of soundfiles) to be imported into its on-screen environment in a way that lets you to place index markers at any point within the soundfile, process various segments (or the entire soundfile) and then burn the final master to disc.

On a Final Note

Many of the topics that have been covered within this section have also been covered (often in greater detail) within Chapters 6 (Digital Audio Technology), 12 (Signal Processing), 13 (Noise Reduction), and 17 (Product Manufacture).

In closing, it's important to remember, whether you have decided to have your latest project mastered at a professional facility or in house . . . always ask for a proof copy(s) of the final master recording. If at all possible, take a week and listen to it in your car, on your boom box, home theater system, in another studio—virtually everywhere! As a musician, producer, or record label, it will be your calling card for quite some time to come. Once you're satisfied with the finished product, then the real work begins (sales, marketing, touring, and general all-round schmoozing). All the best to you and yours!

CHAPTER

Product Manufacture

One of greatest misconceptions surrounding the music, visual, and other media-related industries is the idea that once you walk out the door of a studio with your final master in hand, the creative process of producing a project is finally over. All that you have left to do is hand the file, CD, or other medium over to a duplication facility and—Ta-Dah!—the buying public will be clamoring for your product, Web site, and merchandise. Obviously, this scenario is almost always far from the truth. Now that you have the program content in hand, you have to think through and implement your master plan, if your product is to make it into the hands of the consumer.

Early in this book, I told you about the 1st rule of recording...that there are no rules, only guidelines. This actually isn't true. There is one rule: If you don't pre-plan and follow through with these plans once the project is recorded, you can be fairly sure that your project will sit on a shelf, or worse, you'll have a 1000 CDs sitting in your basement that'll never be heard—a huge shame given the blood, sweat n' tears that went into making it.

Once the recording and mixing phases of a project have been completed (assuming that you've done your homework as to your audience, distribution methods, live and Web marketing presence, production budgeting, etc.), the next step toward getting the product out to the people is to transform the completed song or project into a form that can be mass-produced, distributed, marketed, and "SOLD." Given the various technologies that are available today, this could take the form of a compact disc, DVD, CD-ROM, vinyl record, or encoded file. Each of these media types has its own set of manufacturing and distribution needs that require a great deal of careful attention throughout each step of the manufacturing and/or creation process.

Choosing the Right Facility and Manufacturer

Just as recording studios have their own unique personalities and particular "sound," the right mastering and duplication facilities may also have a profound effect on the outcome of a project. If a project is being underwritten and distributed by an independent or major record label, they will generally be fully aware of their production needs and will certainly have an established production and manufacturing network in place. If, however, you are distributing the project yourself, the duty of choosing the best facility or manufacturing organization that'll fit your budget and quality needs is all yours.

A number of resources exist that can help you find such manufacturers. For starters, Billboard Online (www.billboard.com) provides numerous services for searching out media mastering and manufacturing facilities. They offer resource magazines (such as the *Billboard Tape/ Disc Directory*), as well as a free online search database for *Billboard* magazine subscribers. *The Mix Master Directory* (Intertec Publishing, 6400 Hollis St., Suite 12, Emeryville, CA 94608; 1-510-653-3307, www.mixonline.com) publishes an annual directory of industry-related products and services. This directory (which is sent out as a supplement to the January Mix issue) provides a comprehensive listing of manufacturers, recording studios, producers, engineers, music business services, etc., and is cross-referenced by product and service categories. The *Recording Industry SourceBook* (artistpro.com, 447 Georgia Street, Vallejo, CA 94590; 1-707-554-1935, http://isourcebook.com or www.artistpro.com) also includes a full listing for these companies. Another simple but effective resource is to look at the back-page ads in most music- and audio-related magazines.

Manufacturing facilities come in two types: those that perform and offer all of their services "in-house" (on the premises), and those that "out-source" (contract with other business or individuals to perform various services). Neither of these types is good or bad. On one hand, in-house facilities are able to handle all of the phases of producing a finished product, from beginning to end; these facilities are often large and expensive to equip (meaning that one may not be located nearby). On the other hand, manufacturers and duplicators that farm out projects may not have total control over their production timeline, but are often able to offer personalized, one-on-one service.

As with any part of the production process, it's always wise to do a full background check on a production facility and even compare prices and services from at least 3 manufacturing houses. A good way to check out a place is to ask for a promotional pack (which includes product and art samples, facility, service options, and a price sheet). You might want to ask about former customers and their contact information (so you can e-mail them about their experiences). Most important, once you've chosen a mastering and/or manufacturing facility, it's extremely important that you be given art proofs and test pressings BEFORE the final products are mass duplicated.

Making a test pressing is well worth the time and money; the alternative is to receive a few thousand copies at your doorstep, only to find that they're not what you wanted—a far more expensive, frustrating, and time-consuming option. It's never a good idea to assume that a manufacturing or duplication process is perfect and doesn't make mistakes. Remember, Murphy's Law can pop up at any time!

As the mastering process was previously covered, for the remainder of this chapter we will largely concentrate on the manufacturing process for the various mass-market media. At the end of this chapter, we'll briefly discuss highlights of what is likely the most important part of the production process—basic marketing and distribution strategies.

CD Manufacturing

Although a project can wind up in any number of final medium forms, as of this writing, the compact disc (CD) is still the easiest and most widely recognized medium for distributing music. These $4\frac{3}{4}$ " silvery discs (Figure 17.1) contain digitally encoded information (in the form of microscopic pits) that's capable of yielding playing times of up to 74 minutes, at a standard sampling rate of 44.1 kHz.

A CD pit is approximately half a micrometer wide, and a standard manufactured disc can hold about 2 billion pits. These pits are encoded onto the disc's surface in a spiraling fashion, similar to that of a record, except that 60 CD spirals can fit in the groove of a single long-playing record. The CD spirals also differ from a record in that they travel outward from the center of the disc and are impressed into the plastic substrate, which is covered with a thin coating of aluminum (or occasionally gold) so that the laser light can be reflected back to a receiver. When the disc is placed in a CD player, a low-level infrared laser is alternately reflected and not reflected back to a photosensitive pickup. In this way, the reflected data is modulated so that each pit edge represents a binary 1, and the absence of a pit edge represents

Figure 17.1. *The compact disc. (Courtesy of 51bpm.com, www.51bpm.com.)*

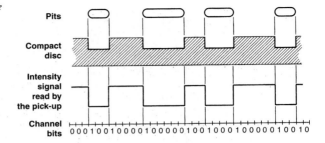

Figure 17.2. *Transitions between a pit edge (binary 1) and the absence of a pit edge (binary 0).*

a binary 0 (Figure 17.2). Upon playback, the data is then demodulated and converted back into an analog form.

Songs or other types of audio material can be grouped on a CD as indexed "tracks." This is done via a subcode channel lookup table, which makes it possible for the player to identify and quickly locate tracks with frame accuracy.

Subcodes are event pointers that tell the player how many selections are on the disc and where their beginning address points are located. At present, eight subcode channels are available on the CD format, although only two (the P and Q subcodes) are used.

Functionally, the CD encoding system splits the 16 bits of information into two 8-bit words and error correction is applied in order to correct for lost or erroneous signals. (In fact, without error-correction, the CD playback process would be so fragile and prone to dropouts that it's doubtful it would've become the dominant medium that it has.) The system then translates these 8-bit words into a 14-bit word format (a process known as eight-to-fourteen modulation or EFM) is then constructed into a methodical code, known as a data frame. Each data frame contains a frame-synchronization pattern (27 bits) that tells the laser pickup beam where it is on the disc. This is followed by a 17-bit subcode word, 12 words of audio data (17 bits each), 8 parity words (17-bits each), 12 more words of audio, and a final 8 words of parity data.

The Process

In order to translate the raw PCM of a music or audio project into a format that can be understood by a CD player, a compact disc mastering system must be used. Modern mastering systems often come in two flavors: those that are used by professional mastering facilities, and CD-R/CD-RW hardware/software systems (Figure 17.3) that allow a personal computer to easily and cost-effectively burn CDs (see Chapter 6 for more information on CD burning technology).

Both system types allow audio to be entered into the system, assemble tracks into the proper order, and enter proper gap times between tracks (in the form of index timings). Depending on the system, cuts might also be processed using crossfades, volume, EQ, and other parameters. Once assembled, the project can be "finalized" into a media form that can be directly

Figure 17.3. *Tascam CD-R624 CD Burning System. (Courtesy of Tascam Corporation, www.tascam.com.)*

accepted by a CD manufacturing facility. In the case of a professional system, this media could take the form of either an Exabyte type data tape, $\frac{3}{4}$ " U-matic videotape (using a Sony PCM-1630 type digital processor), or CD-R. Although digitally encoded tapes are considered to be the most reliable, the equipment that's required to prepare a master is extremely expensive. This is the reason why the majority of final CD masters being received by CD pressing plants are user-created CD-Recordable (CD-R) discs.

Once the manufacturing plant has received the recorded media, the next stage in the process is to cut the original CD master disc. The heart of such a CD cutting system is an optical transport assembly that contains all the optics necessary to write the digital data onto a reusable glass master disc that has been prepared with a photosensitive material.

After the glass master has been exposed using a special recording laser, it's placed in a developing machine that etches away the exposed areas to create a finished master. An alternative process, known as non-photoresist, etches directly into the photosensitive substrate of the glass master without the need for a development process.

After the glass or CD master disc has been cut, the compact disc manufacturing process can begin (Figure 17.4). Under extreme clean-room conditions, the glass disc is electroplated with a thin layer of electro-conductive metal. From this, the negative metal master is used to create a "metal mother," which is used to replicate a number of metal "stampers" (metal plates which contain a negative image of the CD's data surface). The resulting stampers make it possible for machines to replicate clear plastic discs that contain the positive encoded pits, which are then coated with a thin layer of foil (for increased reflectivity) and encased in clear resin for stability and protection. Once this is done, all that remains is the screen-printing process and final packaging. The rest is in the hands of the record company, the distributors, marketing, and you.

Figure 17.4. *Various phases of the CD manufacturing process: (a) the lab, where the CD mastering process begins; (b) once the graphics are approved, the project's packaging can move onto the printing phase; (c) while the packaging is being printed, the approved master can be burned onto a glass master disc; (d) next, the master stamper (or stampers) is placed onto the production line for CD pressing; (e) the freshly stamped discs are cooled and checked for data integrity; (f) labels are then silk screened printed onto the CDs; and (g) finally, the printed CDs are checked before being inserted into their finished packaging. (Courtesy of Disc Makers, inc., www.discmakers.com.)*

a

b

c

Figure 17.4. Continued

Figure 17.4. *Continued*

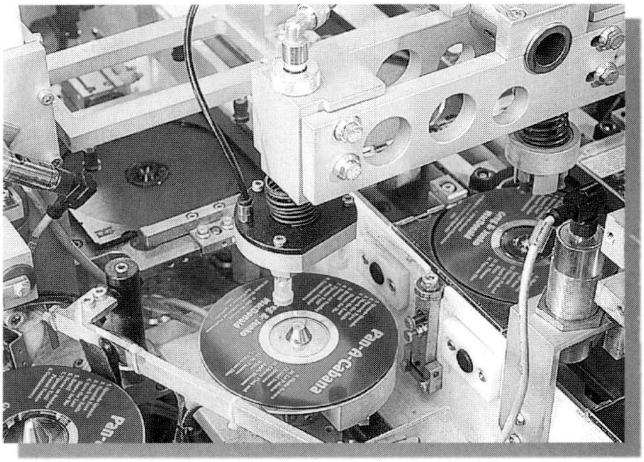

g

CD Burning

Before the proliferation of CD-recording hardware and software (Figures 17.5 through 17.7), the only way to hear how your final CD would sound was to press a "one-off " disc. This meant that the CD manufacturer had to go through the aforementioned process of creating a glass master and "cut" a single or limited run of CDs for the producer, artist, and/or record company as a reference disc. As you could guess, this made it a time-consuming and expensive process that was only available to companies and individuals with big bucks.

Nowadays, the process of burning a CD on a Mac or PC has become so widespread and straightforward that producers, engineers, artists, and the general public use it as the preferred medium for creating and distributing music.

Although most manufacturing plants receive master CDs that have been burned onto a CD-R, it's interesting to note that many of these discs don't pass the basic requirements that have been set forth for creating an acceptable Red Book-Audio CD (the standard industry specification). Some of the problems associated with CD-Rs that have been burned on a desktop system include the following:

◆ Excessive data errors—This can lead to mass-produced CDs that have problems when being played on older or less reliable CD players. These errors could crop up because of such factors as hardware/software reliability problems or media integrity.

Figure 17.5. *CDR-830 "Burnit" Compact Disc Recorder. (Courtesy of HHB Communications Ltd., www.hhb.co.uk.)*

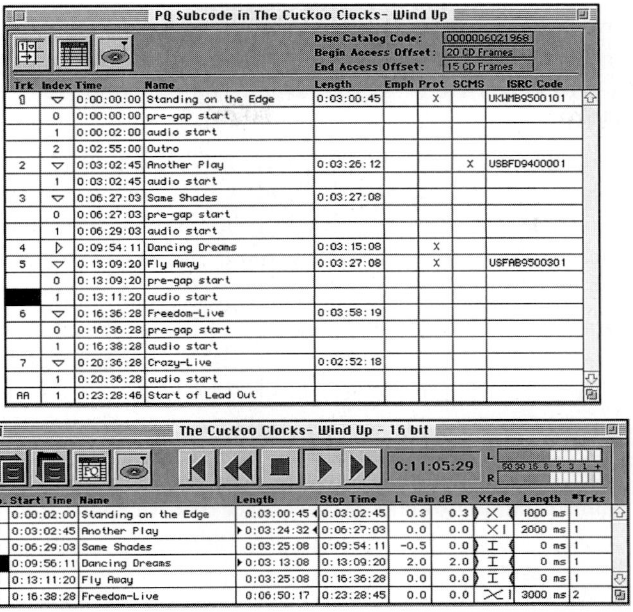

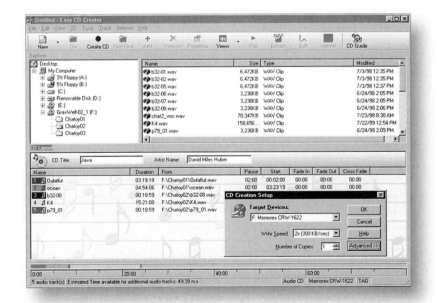

Figure 17.6. *MasterList CD burning Software. (Courtesy of Digidesign, a division of Avid Technology, Inc., www.digidesign.com.)*

Figure 17.7. *Easy CD Creator CD burning software (Courtesy of Roxio, www.roxio.com.)*

- Discs that haven't been "closed"—It's very important that the master disc be closed (a coding process that ensures that no other sessions or data can be added to the disc). Most CD mastering software packages will give you the option of closing or "finalizing" the disc upon burning.
- Multi-session discs—Final master discs should never contain multiple sessions (in which music cuts or program material is added at a later time to an existing CD-R). The disc should be recorded and finalized in the "disk-at-once" mode (meaning the disc was burned from beginning to end, without any interruptions in the laser burning process).
- Inaccurate index marker points—Index markers tell the CD player where the tracks begin and end on a disc. If the markers are wrong, the program could begin early or cut off parts of a song. Once a disc has been cut, always listen to a disc to check for accurate index markers.

In fact, once you've checked the beginning and end marker points, it's always wise to critically listen to the disc from beginning to end. Never forget that Murphy's Law lurks around every corner! Once you've agreed that the CD sounds great, it's always a good idea to burn an extra master that can be set aside for safekeeping, in case something happens to the original production master.

It's interesting to note that many CD burning programs will allow information (such as title, artist name/copyright, and track name field code info) to be written directly into the CD's subcode area. This is often a good idea, as important artist, copyright, and track identifiers can be directly embedded within the CD itself. As a result, illegal copies will still contain the proper copyright and artist info, and discs that are loaded into a computer or media player will often display these fields.

Currently, two types of CD-recording media are commonly found: the CD-R and CD-RW (rewritable) disc. These media use a dye whose reflectivity can be altered, so that data can be burned to disc using a number of available writing options:

- *Disc-at-once*—This mode continuously writes the data onto CD without any interruptions. All of the information is transferred from hard disk to the CD in a single pass, with the lead-in, program, and lead-out areas being written to disc as an uninterrupted event.
- *Track-at-once*—This allows a session to be written as a number of discrete events (called tracks). With the help of special software, the disc can be read before the final session is fixated (a process that "closes" the disc into a final form that can be read by any CD or CD-ROM drive).
- *Multisession*—Discs written in this mode allow several sessions to be recorded onto a disc (each containing its own lead-in, program data, and lead-out areas), thereby allowing data to be recorded onto the free space of a previously recorded CD. It should be noted that older drives might not be able to read this mode and will only read the first available session.

While the altering of the data pits on a CD-R is permanent, CD-RWs can be erased and rewritten any number of times (often figured in the thousands). When using a specially designed CD-RW drive (which can also burn standard CD-Rs), this medium type is excellent for creating data backups and media archiving. In addition, many of the newer CD and MP3 disc players are capable of reading rewritable CD-RW media.

Rolling Your Own

With the rise of Internet audio distribution and the slow but steady breakdown of the traditional record company distribution system, bands and individual artists have begun to produce, market, and sell their own music on an ever-increasing scale. This age-old concept of the "grower" selling directly to the consumer is as old as the town square produce market. However, by using the global Internet economy, independent distribution, fanzines, live concert sales, etc., savvy independent artists are taking matters into their own hands (or are smart enough to combine with the talents of others) by learning the inner-workings of the music business. In short, artists are taking business matters more seriously in order to reap the fruits of their labor and craft . . . something that has never been and never will be an easy task.

Beyond the huge tasks of marketing, gigging, and general business practices, many musicians are also taking on the task of burning, printing, packaging, and distributing their own CDs from the home or business workplace. This homespun strategy allows for small runs to be made in an "on-demand" basis, without tying up financial resources and storage space in CD inventories.

Creating a system for burning CD-Rs for distribution can range from being a simple home computer setup that creates discs on an individual basis to sophisticated replication systems that can print and burn stacks of CD-Rs or DVD-Rs under robot control at the simple touch of a button (Figure 17.8).

Figure 17.8. Elite Pro CD burning and printing system. (Courtesy of Disc Makers, inc., www.discmakers.com.)

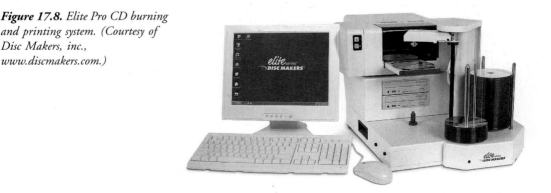

Burning Speeds

Whenever you see the specs on a CD-R or CD-RW burner that look like $32 \times 10 \times 40$ (three numbers that are separated by the letter "x"), the numbers indicate the various read and writing speeds of the CD drive. The "x" stands for the device's data transfer speed as multiples of 150 KB per second. The first number (32 in the above example) indicates the speed that the drive is capable of writing data onto a CD-R disc. In the above example, the drive can write at transfer speeds of up to 32×150 KB/sec = 4800 KB/sec. The second number represents the speed that the drive can rewrite data onto a CD-RW disc (i.e., 10×150 KB/sec = 1500 KB/sec). The final number indicates the top speed that the drive can read at (i.e., 40×150 KB/sec = 6000 KB/sec).

All of the data that's written to a CD uses Cross-Interleaved Reed-Solomon Code (CIRC) for error correction. CIRC is capable of applying two levels of correction, C1 and C2. C1 applies to bit errors that can be easily corrected by the system during the read process, as the data is interleaved and spread over the disc surface in a large arc (this is why CDs should always be cleaned in a straight line from the center out as a circular scratch could cause multiple errors across several data frames that can't be corrected). The larger C2 errors apply to bytes in a frame that can result in uncorrectable errors or ones that can be smoothed over (as in the case of a CD Audio disc) or corrected using additional coding (unless the media is degraded further).

Finding the optimum CD-R burning speed for your computer or replicator is a topic that's best left as a debate between buddies over a pint of lager. There are those that passionately feel that burning at lower speeds will improve the burning process due to improved disc stability and optimum laser performance, while others will argue that newer media dyes, improved laser assemblies, and numerous amounts of published data will prove them wrong—that performance actually improves at higher (though not always the maximum writing speeds). I will bow out of this debate by challenging you to research the data, the articles, and the many message postings that have been dedicated to this subject. A good site to begin your research can be found at www.cdrfaq.org.

CD Labeling

Once you've burned your own CD-R/RW or DVD-R/RW, there are a number of options for printing labels onto the newly burned discs (burning the disc first will often reduce data errors that can be introduced by dust, fingerprints, or scratches due to handling):

◆ Using a felt-tip pen—This is the easiest and fastest way to label a disc. However, water-based ink pens should be used, because permanent markers use a solvent that

Figure 17.9. Neato CD Labeler kit. *(Courtesy of Neato LLC, www.neato.com.)*

can permeate the disc surface and cause damage to either the reflective or dye layer. When properly done, this is an excellent option for archived discs.

◆ Label printing kits—"Stick-on" labels (Figure 17.9) that have been printed using specially designed software and an inkjet or laser printer are one of the least expensive options. Although their design has improved over the years, you should be aware that some adhesives could peel off, leak over time, or contain solvents that might adversely affect the disc. This professional-looking approach is often excellent for use on non-archival products.

◆ CD printers—Specially designed inkjet or laser printers are able to print high-quality, full-color layouts onto the face of a printable (white or silver-faced) disc. This is a cost-effective option for those who burn discs in small batch runs and want a professional look and feel.

Although stand-alone programs are available, most of the above-mentioned printing kits and CD printers include a label printing program for creating and printing professional-looking CDs, CD books, and trays (as well as labels for DATs, cassettes, and almost any other medium you can think of). These programs let you import graphics and position text to create and print out personalized, professional-looking labels (Figure 17.10). In addition to these programs, word processing templates are often available (most often for MS Word) that let you import graphics and position text as a document that can be opened and printed directly from the word processor, without the need for a special program.

Figure 17.10. Mediaface label printing program. (Courtesy of Neato LLC, www.neato.com.)

CD and DVD Handling and Care

Here are a few basic handling tips for CDs and DVDs (including the recordable versions) from the National institute of Standards and Technology:

DO:

- Handle the disc by the outer edge or center hole (your fingerprints may be acidic enough to damage the disc).
- Use a felt-tip permanent marker to mark the label side of the disc. The marker should be water- or alcohol-based. In general, these will be labeled "non-toxic." Stronger solvents may eat though the thin protective layer to the data.
- Keep discs clean. Wipe with a cotton fabric in a straight line from the center of the disc toward the outer edge. If you wipe in a circle, any scratches may follow the disc tracks, rendering them unreadable. Use a CD/DVD-cleaning detergent or isopropyl alcohol to remove stubborn dirt.
- Return discs to their cases immediately after use.
- Store discs upright (book-style) in their cases.
- Open a recordable disc package only when you are ready to record.
- Check the disc surface before recording.

DON'T:

- Touch the surface of a disc.
- Bend the disc (as this may cause the layers to separate).

- Use adhesive labels (as they can unbalance or warp the disc).
- Expose discs to extreme heat or high humidity, for example, don't leave them in sun-warmed cars.
- Expose discs to extreme rapid temperature or humidity changes.
- Expose recordable disc to prolonged sunlight or other sources of ultraviolet light.

ESPECIALLY DON'T:

- Scratch the label side of the disc (it's often more sensitive than the transparent side).
- Use a pen, pencil, or fine-tipped marker to write on the disc.
- Try to peel off or reposition a label (it could destroy the reflective layer or unbalance the disc).

DVD Burning

Of course, on a basic level, DVD burning technology has matured enough to be available and affordable to the general Mac and PC public.

From a technical standpoint, these CD-drive compatible discs differ from the standard CD format in several ways. The most basic of these are:

- An increased data density due to a reduction in pit size (Figure 17.11)
- Double-layer capabilities (due to the laser's ability to focus on two layers of a single side)
- Double-side capabilities (which again doubles the available data size)

In addition to the obvious benefits that can be gained from increasing the data density of a standard CD from 650 Mbyte to a maximum of 17 Gbytes, DVD discs allow for much higher data transfer rates, making DVD the ideal medium for the following applications:

- The simultaneous decoding of digital video and surround-sound audio
- Multichannel surround sound
- Data- and access-intensive video games
- High-density data storage

Figure 17.11. *Detailed relief showing standard CD and DVD pit densities.*

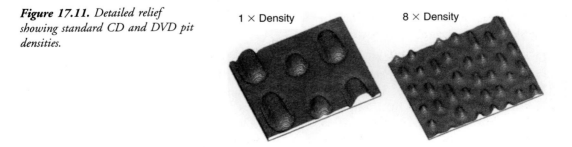

1 × Density   8 × Density

Using extensive, high-quality compression techniques, this technology has breathed new life into the home entertainment industries, allowing computer fans to have access to increased game and multimedia storage capacity and home viewers to enjoy master-quality audio and video programs in a digital surround-sound environment.

As DVD-ROM and writable drives have become commonplace, affordable data backup and mastering software has come onto the market that brought the art of DVD mastering to the masses. Even high-level DVD production is now possible in a desktop environment, although creating a finished product for the mass-markets is often an art that's often best left to professionals who are familiar with the finer points of this complex technology. More information on the finer points of codec data compression and media technologies relating to both CD and DVD technologies can be found in Chapter 8 (Multimedia and the Web).

Cassette Duplication

In many world markets, the pre-recorded music cassette is still a strong, cost-effective medium for getting commercial music out to the masses. Contrary to public misconception, a great deal of artistry and quality control often goes into the duplication of voice and music cassettes. Currently, there are three basic methods of cassette duplication: real-time duplication, bin-loop high-speed duplication, and high-speed, in-cassette duplication.

Real-Time Duplication

In real-time cassette duplication, slave machines are used in a system to record a program at their normal speed of $1\frac{7}{8}$ ips. Thus, the single side of a program lasting 30 minutes will take 30 minutes to duplicate, with the number of copies being dependent on the number of slave decks.

Most industry insiders agree that this format yields the highest reproduction quality, as the slave machines are operating at the optimum speed for the medium and the audio signal is kept within the audio bandwidth and not shifted into a higher one.

Using this method, the recorded tape (called the duplication master or dupe master) is played back from a copy of the original program master onto any number of duplication tape drives (Figure 17.12). The final dupe master can exist in any format (including hard disk, DAT, reel-to-reel tape, or cassette). Dual-cassette tape decks or multi-transport setups are often the easiest to use, as you simply place a master cassette into the playback drive and place the slave(s) into record.

Often, professional facilities will format the duplication (dupe) master in a 4-track stereo format onto hard disk or open-reel, 4-track tape format. Using this mode means that side A's stereo program material can be played on the forward direction, while side B will be played back in reverse. By fitting the slave recorders with 4-track record heads, both directions of the tape can be recorded at once, thereby cutting the duplication time in half (Figure 17.13).

Figure 17.12. *Example of a hard disc-base real-time cassette duplication system.*

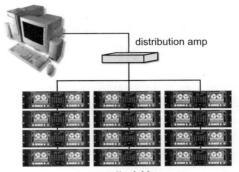

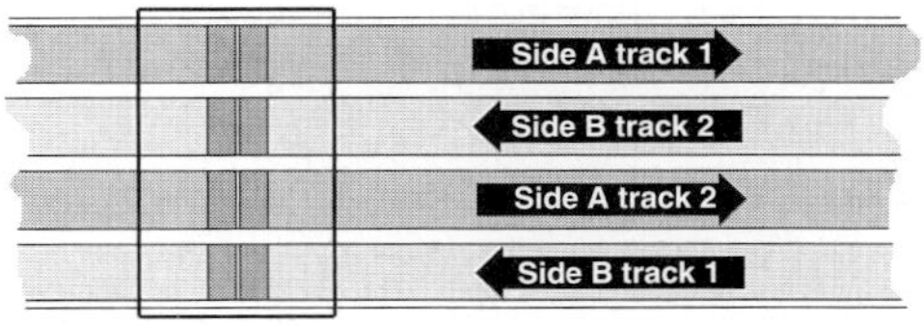

distribution amp

cassette dubbers

Figure 17.13. *By recording side A of a dupe master in the forward direction and side B backwards, it's possible to duplicate both sides of a cassette tape in a single pass.*

Side A track 1
Side B track 2
Side A track 2
Side B track 1

In-Cassette High-Speed Duplication

In-cassette high-speed duplication makes use of high-speed ratios (2×, 8×, and 16×) by copying a dupe master (often a physical cassette) to a set of cassette slave recorders (Figure 17.14).

These duplication units often are self-contained, with the master and several slaves being located in the same unit. Extra slaves can often be added, letting you cost-effectively expand the system. The biggest drawback to this system is the resonances, distortions, and wow and flutter that are introduced as the tape moves at high speeds within the tape housing.

Figure 17.14. *Telex XGEN in-cassette duplicator. (Courtesy of Telex Communications, Inc., www.telex.com.)*

High-Speed Duplication

High-speed duplication is probably the most common mass-production system and takes place before the tape is physically loaded into the cassette shells. Using this method, the duplicated tape is recorded on reel-to-reel machines (Figure 17.15) that handle tape in a higher-quality fashion than most in-cassette units.

High-speed dupe masters are recorded to hard disk or a 4-track master tape that includes a recorded tone signal that corresponds to a 5–15 Hz tone at $1\frac{7}{8}$ ips. The program is then repeatedly recorded onto any number of open-deck reel-to-reel recorders (which are designed to accept $\frac{1}{8}$ " cassette-grade tape on bulk $10\frac{1}{2}$" pancake reels) at extremely high-speed ratios of up to $160\times$.

Since the duplication process occurs at ratios that are many times the normal speed, the frequency spectrum is shifted upward into a high-frequency range that's well beyond the audio spectrum. The record heads, frequency response, and bias currents must also be specially tailored for this demanding application.

The next stage in this duplication process is to load the prerecorded programs (which are now repeatedly recorded onto bulk tape) into their cassette housings. This is accomplished using a machine known as a self-feeding cassette loader (Figure 17.16). The duplicated bulk tape is loaded into the device and a cassette-feed magazine is filled with C-0 cassettes (a cassette that's only loaded with leader tape). Next, the C-0s are dropped into the loading section and the recorded tape is automatically spliced onto the cassette leader at the beginning point where the sensing tone appears. The loader then fast-forwards the tape, loading it into the cassette until the next tone is sensed. Next, the loader splices the program's end onto the cassette's tail leader and ejects the cassette, at which point, the process repeats.

After the tapes have been loaded into their shells, they are finally labeled and packaged for sales and distribution. With a large-scale bin-loop production system, it's possible to produce tens of thousands of cassettes each day.

Quality control is of great importance during this process. The major emphasis often rests on the quality of the duplication master and the master–slave alignment. Distortion and saturation can often be dealt with through the moderate use of peak limiting and compression

Figure 17.15. *1000 Series tape duplicating system. (Courtesy of Versadyne Internationsl, www.versadyne.com.)*

Figure 17.16. Tapematic 2002 self-feeding cassette tape loader. (Courtesy of Tapematic USA, Inc., www.tapematic.com.)

during the mastering process. If dynamic or EQ changes are required, the artist, producer, and/or label should be consulted.

The problem of noise might be reduced through careful mastering, by using a digital master, and/or by encoding the dupe master with Dolby noise reduction (thereby creating a Dolby B- or C-encoded cassette). Whenever possible (as with all manufactured products), always insist on listening to a duplicated "proof " copy before duplicating large quantities, as the results might not be what you expected or hoped for.

Vinyl Disc Manufacture

Although the popularity of vinyl has waned in recent years (as a result, of course, of the increased marketing, distribution, and public acceptance of the CD), the vinyl record is far from dead. In fact, for consumers that range from Dance DJ hip-hipsters to die-hard classical buffs, the record is still a viable medium. However, the truth remains that many record pressing facilities have gone out of business over the years, and there are far fewer mastering labs that are capable of cutting "master lacquers." It may take a bit longer to find a facility that fits your needs, budget, and quality standards, but it's definitely not a futile venture.

Disc Cutting

The first stage of production is the disc-cutting process. As the master is played from a digital source or on a specially designed tape playback machine, its signal output is fed through a disc-mastering console to a disc-cutting lathe. Here the electrical signals are converted into the mechanical motions of a stylus and are cut into the surface of a lacquer-coated recording disc.

Figure 17.17. The 45/45 cutting system encodes stereo waveform signals into the grooves of a vinyl record.

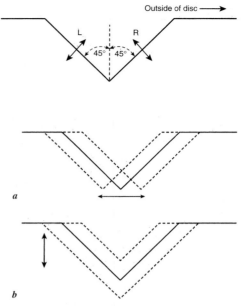

Figure 17.18. Groove motion in stereo recording. (The solid line is the groove with no modulation.) (a) In-phase, (b) out-of-phase.

Unlike the compact disc, a record rotates at a constant angular velocity, such as 33⅓ or 45 revolutions per minute (rpm), and has a continuous spiral that gradually moves from the disc's outer edge to its center. The recorded time relationship can be reconstructed by playing the disc on a turntable that has the same constant angular velocity as the original disc cutter.

The system that's used for recording a stereo disc is the 45/45 system. The recording stylus cuts a groove into the disc surface at a 90° angle, so that each wall of the groove forms a 45° angle with respect to the vertical axis. Left-channel signals are cut into the inner wall of the groove and right-channel signals are cut into the outer wall, as shown in Figure 17.17. The stylus motion is phased so that L/R channels that are in-phase (a mono signal or a signal that's centered between the two channels) will produce a lateral groove motion (Figure 17.18a), while out-of-phase signals (containing channel difference information) will produce a vertical motion that changes the groove's depth (Figure 17.18b). Because mono information relies only on lateral groove modulation, an older disc that has been recorded in mono can be accurately reproduced with a stereo playback cartridge.

Disc-Cutting Lathe

The main components of a vinyl disc-cutting lathe are the turntable, lathe bed and sled, pitch/depth control computer, and cutting head. Basically, the lathe (Figure 17.19) consists of a heavy, shock-mounted steel base (A). A weighted turntable (B) is isolated from the base by an oil-filled coupling (C), which reduces wow and flutter to extremely low levels. The lathe bed (D) allows the cutter suspension (E) and the cutter head (F) to be driven by a screw feed

Figure 17.19. *A disc-cutting lathe with automatic pitch and depth control.*

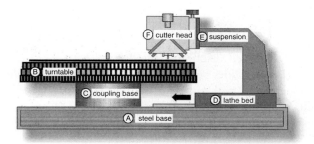

that slowly moves the record mechanism along a sled in a motion that's perpendicular to the turntable

Cutting Head

The cutting head translates the electrical signals that are applied to it into mechanical motion at the recording stylus. The stylus gradually moves in a straight line toward the disc's center hole as the turntable rotates, creating a spiral groove on the record's surface. This spiral motion is achieved by attaching the cutting head to a sled that runs on a spiral gear (known as the lead screw), which drives the sled in a straight track.

The stereo cutting head (Figure 17.20) consists of a stylus that's mechanically connected to two drive coils and two feedback coils (which are mounted in a permanent magnetic field) and a stylus heating coil that's wrapped around the tip of the stylus. When a signal is applied to the drive coils, an alternating current flows through them creating a changing magnetic field that alternately attracts and repels the permanent magnet. Because the permanent magnet is fixed, the coils move in proportion to a field strength that causes the stylus to move in a plane that's 45° to the left or right of vertical (depending on which coil is being driven).

Figure 17.20. *Simplified drawing of a stereo cutting head.*

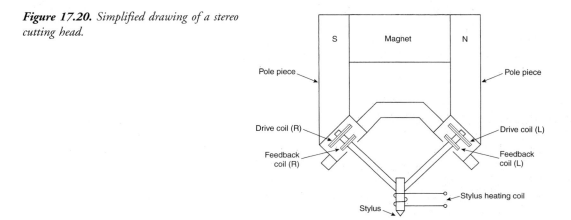

Pitch Control

The head speed determines the "pitch" of the recording and is measured by the number of grooves, or lines per inch (lpi), that are cut into the disc. As the head speed increases, the number of lpi will decrease, resulting in a corresponding decrease in playing time. Groove pitch can be changed by:

◆ Replacing the lead screw with one that has a finer or coarser spiral

◆ Changing the gears that turn the lead screw, so as to alter the lead screw's rotation speed

◆ Vary the lead screw's rotation by changing the motor's speed (a common way to vary the program's pitch in real-time)

The space between grooves is called the land. Modulated grooves produce a lateral motion that's proportional to the in-phase signals between the stereo channels. If the cutting pitch is too high (causing too many lines per inch, which closely spaces the grooves) and high-level signals are cut, it's possible for the groove to break through the wall into an adjacent groove (causing over cut) or for the grooves to overlap (twinning). The former is likely to cause the record to skip when played; while the latter causes either distortion or a signal echo from the adjacent groove (due to wall deformations). Groove echo can occur even if the walls don't touch and is directly related to groove width, pitch, and level.

These cutting problems can be eliminated either by reducing the cutting level or by reducing the lines per inch. A conflict can arise here as a louder record will have a reduced playing time, but will also sound brighter, punchier, and more present (due to the Fletcher-Munson curve effect). Because record companies and producers are always concerned about the competitive levels of their discs relative to those that are cut by others, they're reluctant to reduce the overall cutting level.

The solution to these level problems is to continuously vary the pitch so as to cut more lines per inch during soft passages and fewer lines per inch during loud passages. This is done by splitting the program material into two paths: undelayed and delayed. The undelayed signal is routed to the lathe's pitch/depth-control computer (which determines the pitch needed for each program portion and varies the lathe's screw motor speed). The delayed signal (which is usually achieved by using a high-quality digital delay line) is fed to the cutter head, thereby giving the pitch/depth control computer enough time to change the lpi to the appropriate pitch.

Pitch is divided into two categories: coarse (which refers to spacing between 96 and 150 lpi) and microgroove (which is between 200 and 300 lpi, or more). Microgroove records have less surface noise, wider frequency range, less distortion, and greater dynamic range than do coarse-pitched recordings. They can also be tracked with lower stylus pressure, resulting in a longer life; however the stylus is more likely to skate across the record if the turntable isn't level. The playback stylus for a stereo microgroove record must have a tip radius of 0.7 mil or less (compared to 2.5 mils $\pm$ 0.1 for coarse-groove records). Older 78-rpm and early $33\frac{1}{3}$ -rpm records were recorded with a coarse pitch; however, virtually all current records are microgroove (having an average pitch of 265 lpi). At maximum pitch, the playing time of one side of a 12-inch disc, with no modulation in the grooves, is about 23 to 26 minutes, while the duration of a variable-pitch 12-inch disc cut at average levels is about 45 minutes per side.

Recording Discs

The recording medium used on the lathe is a flat aluminum disc that's coated with a film of lacquer, which is dried under controlled temperatures, coated with a second film, and then dried again. The quality of these discs (called lacquers) is determined by the flatness and smoothness of the aluminum base. Any irregularities in this surface (such as holes or bumps) will cause similar defects in the lacquer coating. Lacquers are always larger in diameter than the final record, which makes it easy to handle them without damaging the grooves. For example, a 12-inch album is cut on a 15-inch lacquer and a 7-inch single is cut on a 10- or 12-inch lacquer. As always, it's wise to cut a reference test lacquer in order to hear how the recording will sound after being transferred to disc.

The Mastering Process

Once the mastering engineer sets a basic pitch on the lathe, a lacquer is placed on the turntable and compressed air is used to blow any accumulated dust off the lacquer surface. A chip suction vacuum is started and a test cut is made on the outside of the disc to check for groove depth and stylus heat. Once the start button is pressed, the lathe moves into the starting diameter, lowers the cutting head onto the disc, starts the spiral and lead-in cuts, and begins playing the master production tape. As the side is cut, the engineer can fine-tune any changes to the previously determined console settings. Whenever an analog tape machine is used, a photocell mounted on the deck senses white leader tape between the selections on the master tape and signals the lathe to automatically expand the grooves to produce track bands. After the last selection on the side, the lathe cuts the lead-out groove and lifts the cutter head off the lacquer.

This master lacquer is never played, as the pressure of the playback stylus would damage the recorded soundtrack (in the form of high-frequency losses and increased noise). Reference lacquers (also called reference acetates or simply acetates) are cut to hear how the master lacquer will sound.

After the reference is approved, the record company assigns each side of the disc a master (or matrix) number that the cutting room engineer scribes between the grooves of the lacquer's ending spiral. This number identifies the lacquer in order to eliminate any need to play the record, and often carries the mastering engineer's personal identity mark. If a disc is remastered for any reason, some record companies retain the same master numbers; others add a suffix to the new master to differentiate it from the previous "cut."

When the final master arrives at the plating plant, it is washed to remove any dust particles and is then electroplated with nickel. Once the electroplating is complete, the nickel plate is pulled away from the lacquer. If something goes wrong at this point, the master will be damaged, and the master lacquer must be re-cut.

Vinyl Disc Plating and Pressing

The nickel plate that's pulled off the master (called the matrix) is a negative image of the master lacquer (Figure 17.21). This negative image is then electroplated to produce a nickel positive image called a mother. Because the nickel is stronger than the lacquer disc, several mothers can be made from a single matrix. Since the mother is a positive image, it can be played as a test for noise, skips, and other defects. If it's accepted, the mother can be electroplated several times, producing stampers that are a negative images of the disc (a final plating stage that's used to press the record).

The stampers for the two sides of the record are mounted on the top and bottom plates of a hydraulic press. A lump of vinylite compound (called a biscuit) is placed in the press between the labels for the two sides. The press is then closed and heated by steam to make the vinylite flow around the raised grooves of the stampers. The resulting pressed record is too soft to handle when hot, so cold water is circulated through the press to cool it before the pressure is released. When the press opens, the operator pulls the record off the mold and the excess (called flash) is trimmed off after the disc is removed from the press. Once done, the disc's edge is buffed smooth and the product is ready for packaging, distribution, and sales.

Producing for the Web

In this day of surfing and streaming media off the Web, it almost goes without saying that the WWW has become an important marketing tool for cost-effectively getting downloadable

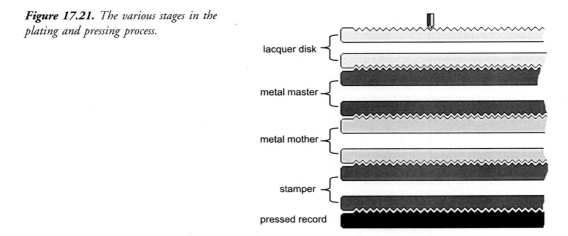

Figure 17.21. *The various stages in the plating and pressing process.*

lacquer disk

metal master

metal mother

stamper

pressed record

Table 17.1. DVD video/audio formats.

Format	Sample Rate	Bit Rate	Bit/s	Ch	Common Format	Compression
PCM	48, 96 kHz	16, 20, 24	Up to 6.144 mbps	1 to 8	48 kHz, 16 bit	None
AC3	48 kHz	16, 20, 24	64 to 448 kbps	1 to 6.1	192 kb/s, stereo	AC3 and 384 kbps, 448 kbps
DTS	48, 96 kHz	16, 20, 24	64 to 1536 kbps	1 to 7.1	377 or 754 kbps for stereo and 754.5 or 1509.25 kbps for 5.1	DTS coherent acoustics
MPEG-2	48 kHz	16, 20	32 to 912 kbps	1 to 7.1	Seldom used	MPEG
MPEG-1	48 kHz	16, 20	384 kbps	2	seldom used	MPEG
SDDS	48 kHz	16	Up to 1289 kbps	5.1, 7.1	seldom used	ATRAC

songs, promotional materials, touring info, and liner notes out to mass audiences. As with other media, mastering for the Internet can either be complicated, requiring professional knowledge and experience, or it can be a straightforward process that can be carried out from a desktop computer. It's a matter of meeting the level of professionalism and development that's required by the site.

In this iPod world of MP3s, Windows Media, desktop video, Internet radio stations, and who knows what other types of streaming media, the rule that all cyber-producers live by is bandwidth. Basically, the bandwidth of a media and delivery/receiving system refers to the ability to squeeze as much data (often compressed data) through a wire, wireless, or optical pipeline in as short a time as is possible. Transmitting the highest audio and/or video feed over a limited bandwidth will often require specialized (and often accessible) production tools. Beyond this, an even more important tool is mastery of the medium, mass marketing, and good eyes and ears for design layout and media management. More info on streaming 'media and producing for cyber space can be found in Chapter 8 (Multimedia and the Web).

CHAPTER 18

Studio Session Procedures

One of the most important concepts to be gained from this book (beyond an understanding of the technology and tools of the trade) is the fact that there are no rules for the process of recording. This rule holds true insofar as inventiveness and freshness tend to play a major role in keeping the creative process of making music (and music productions) alive and exciting. There are, however, guidelines and procedures that can help you have a smoother flowing, professional-sounding recording session, or at the very least, help you solve potential problems when used in conjunction with three of the best tools for guiding you toward a successful project:

- Preparation
- Creative insight
- Common sense

Preparation

Probably the most important step that one can make to help ensure that a recording project will be successful and that it has a chance at being marketable to its intended audience is careful *preparation and planning*. By far the biggest mistake that a musician or group can make is to go into the studio, spend a lot of money and time, press a few thousand CDs, make a template Web site . . . and then sit back and expect an adoring audience to spring out of thin air! It ain't gonna happen! Beyond a good dose of business reality and added experiences, the artist(s) will have the dubious distinction of joining the throngs that have thousands of CDs sittings in their closet or basement.

In order to help avert many of the more common mistakes, the first half of this chapter is devoted to laying out many of the tools that can be gathered to help ensure that a project will go smoothly, be on budget, sound good, and, of course, sell? The latter half will be devoted to the actual process of recording in a professional or project studio. To those that choose to dive into the deep end of music and sound production (on either side of the glass)—all the best!

What's a Producer and When Do you Need One?

One of the first steps that can help ensure the success of a project is to seek the advice and expertise of those who have experience in their chosen fields. This might include seeking legal council (for help and advice with legal matters, business contacts, or both). Another important "advisor" can come in the form of that all-important title, *producer*. Basically, the producer of a project can fill one of two roles:

◆ The first type can be likened to a film director, in that his/her role is to be an artistic, psychological, and technical guide that will help the band or artist reach their intended goals of obtaining the best possible song, album, remix, film score, etc. It is the producer's job to stand back and objectively look at the big picture, and to offer up suggestions as to how to shape and guide the performance and to direct the artist or group in directions that will result in the best possible final product.

◆ The second type also encompasses the directorial role, but also has the added responsibilities of being an executive producer. He or she will also be charged with many of the business responsibilities of overall session budgeting, choosing and arranging for all studio and session activities, contracting (should outside musicians and arrangers be needed on the project), etc. This type of producer may even be charged with initiating and/or smoothing contact relations with potential record companies or distributors.

As you can see, this role can be either limited or broad in scope and should be carefully discussed and agreed upon long before any record button is pressed. The importance of finding a producer that can work best with your particular personalities, musical style, and business/marketing needs can't be stressed enough. Finding the right producer for you can be a time-consuming and rewarding experience. Here are a few tips to prepare you for the hunt:

◆ Check out the liner notes of groups or musicians that you love and admire. You never know, their producer just might be interested in taking you on!

◆ Find a local up-and-coming producer that might be right for your music. This could help fast-track your reputation.

◆ Talk with other groups or musicians. They might be able to recommend someone.

A few of the questions to ask when searching out a producer are:

- Does he/she openly discuss ideas and alternate paths that contribute to growth and better artistic expression?
- Is he/she a team player or are the rules laid out in a dictator-like fashion?
- Does the producer know the difference between a creative endeavor and one that wastes time in the studio?
- Does he/she say "Why not?" a lot more often than "Why?"

Although many engineers have spent most of their lives with their ears wide open and have gained a great deal of musical, production, and in-studio experience, it's generally not a good idea to assume that the engineer can fill the role of a producer. For starters, he/she will probably be unfamiliar with the group and its sound, or may not even like your sound! For these and other reasons, it's always best to seek out a producer that is familiar with you, your goals, and your style (or is contacted early enough the he/she has time to become familiar).

Long Before Going into the Studio

Most in the industry realize that music in the modern world is a business. Once you get to the phase of getting your band or your client's band out to the buying public, you'll quickly realize just how true this is. Building and maintaining an audience with an appetite for your product can easily be a full-time business; one where you'll encounter well-intentioned people, and also others who would think nothing of taking advantage of you or your client.

Whether you're selling your products on the street, at gigs, or over the Internet, or whether you're shopping for (or have) a record label, it's often a wise decision to retain the counsel of a trusted music lawyer. The music industry is fraught with its own special legal and financial language, and having someone on your side who has insight into the language, quirks, and inner workings of this unique business can be an extremely valuable asset.

By far, one of the most important steps to be taken when approaching a project that involves a number of creative and business stages, decisions, and risks is preparation. Without a doubt, the best way to avoid pitfalls and to help get you, your client, or your band's project off the ground is to discuss and outline the many factors and decisions that will affect the creation and outcome of that all-important "final product." Just for starters, a number of basic questions need to be asked long before anyone presses the "REC" button:

- How are we going to recoup the production costs?
- How is it to be distributed to the public? Self-distribution? Indy? Record company?
- Will other musicians be involved?
- Do we need a producer or will we self-produce?
- How much practice will we need? Where and when?
- Should we record it in the drummer's project studio or at a commercial studio?
- If we use the project studio and it works out, should we mix it at the commercial studio?

◆ Who's going to keep track of the time and budget? Is that the producer's job? Or will he/she be strictly in charge of creative and contact decisions?

◆ Are we going to need a music lawyer with contacts and contracts? Do we know someone who can handle the job?

◆ Carefully discuss the artist or group's artistic and financial goals and put them down on paper. Discuss budget requirements and possible rewards as early as possible in the game! This might be the time to discuss matters with the music lawyer we talked about.

These are but a few of questions that should be asked before tackling a project. Of course, they'll change from project to project and will depend on the final project's scope and purpose. However, in the final analysis, asking the right questions (or finding someone who can help you ask the right questions) can help keep you from having to store 10,000 unsold CDs in your basement.

Now that you've answered the questions, here's a list of tasks that are often wise to tackle well before going into the studio:

◆ Practice, practice, and more practice. Need I say more!

◆ Create a "Mission Statement" for you/your group and the project. This can help clue your audience into what you are trying to communicate through your art and music and can greatly benefit your marketing goals. For example, you might want to answer such questions as: Who are you? What are your musical goals? How should the finished project sound? What emotions should it evoke? What is the budget for this project? How will it be sold? What are the marketing strategies?

◆ Start working on the projects artwork, packaging, and Web site ASAP.

◆ Copyright your songs. Form PA is used for the registration of music and/or lyrics (as well as other works of the performing arts), even if your song is on a cassette. This form can be found at www.copyright.gov/forms/formpai.pdf or by searching the Library of Congress at www.loc.gov.

◆ Take the time to check out several studios and available engineers. Which one best fits your style, budget, and level of professionalism?

Before Going Into the Studio

Before beginning a recording session (possibly a week or more before), it's always a good idea to mentally prepare yourself for what lies ahead. It's always a good idea to have a basic checklist that can help answer what type of equipment will be needed, the number and type of musicians/instruments, their particular miking technique (if any), and where they'll be placed. The best way to do this is for you, your group, and the producer (if there is one) to sit down with the engineer and discuss instrumentation, studio layout (Figure 18.1), musical styles, and production techniques. This meeting lets everyone know what to expect during the session and lets everyone become familiar with the engineer, studio, and staff. This is time well spent, as it will invariably come in handy during the studio setup and will help get the session off to a good start. The following tips can also be immensely valuable:

Figure 18.1. *Floor plan layout for Capitol Studios A and B (which can be opened up to create a shared space). (Courtesy of Capitol Studios, Hollywood, CA, www.capitolstudios.com.)*

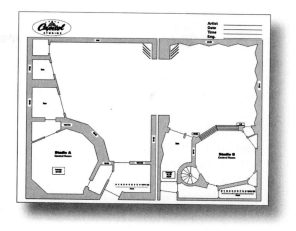

◆ If there is no producer on the project, it's often wise to pick (or at least, consider picking) a spokesman for the group who has the best production "chops." He/she can then work closely with the engineer to create the best possible recordings.

◆ Record your songs during live gigs or rehearsals—It doesn't matter if you record them professionally, with a 4-track, or with a cassette recorder/boom box.

◆ Audition the session's song list before a live audience.

◆ Work out all of the musical and vocal parts before going into the studio—Unrehearsed music can leave the music standing on shaky ground; however, leave yourself open to exploring new avenues and surprises that can be the lifeblood of a magical session.

◆ Rehearse more songs than you plan to record—You never know which songs will end up sounding best or will have the strongest impact.

◆ Meet with the engineer beforehand—Take time for the producer and/or group to get to know him/her, so you'll both know what to expect on the day of the session. You might ask to hear examples of what the engineer has recorded.

◆ Prepare and edit any sequenced, sampled, or pre-recorded material beforehand. In short, be as prepared as possible.

◆ Try working to a metronome (click track) if timing is an issue.

◆ Take care of your body—Try to relax and get enough sleep before and during the session. Eat the foods that are best for you (you might have some health foods or fruits around to keep your energy up). Don't fatigue your ears before a session, keep them rested and clear.

In addition to the above, it's always a wise idea to plan out your tracking arrangement. If you're using tape-based equipment, decide on your tracking order and make sure that there are enough tracks for all of the instruments (with a few to spare). Confer to find out how many instruments are to be used on the song, including overdubs. This helps to determine how many tracks will need to be left open. The number of tracks often influences the number and way that mics are to be assigned to the available tracks. This holds especially true for drums.

If a large number of instruments are to be recorded and overdubbed to a recorder with a limited number of tracks, you might want to consider the following:

◆ Group several instruments together and record the composite mix to a limited number of tracks.

◆ Record the instruments onto separate tracks and then mix the composite mix (or mixes) onto a small group of tracks on a separate master tape leaving tracks open for additional overdubs. Keeping the original tracks intact on a separate tape will at least give you more options should the group mixes need to be changed at a later date (see Bouncing later in the Overdubbing section).

◆ Decide to use (or rent) a recorder that has more tracks or integrate one or two additional modular digital multitracks into the recording setup. In this day and age, there are often several options for adding on tracks.

◆ Use a hard-disk-based DAW that has fewer track limitations.

Setting Up

Once the musicians have shown up at the studio, it's extremely important that all of the technical, musical, and emotional preparation be put into practice in order to get the session off to a good start. Here are a few tips that can help:

◆ Show up at the studio on time or early—At some studios, the billing clock starts on time (whether you're there or not). Ask about their setup policy (Is there another session before yours? Is there adequate setup time to get prepared? Are there any charges for setup? What is the studio's cancellation policy?).

◆ Use new strings, chords, drumsticks, and heads, and bring spares—It's also a good idea to know the location and hours of a local music store, just in case.

◆ Tune up before the session starts, and tune up regularly thereafter.

◆ Don't use new or unfamiliar equipment—Taking the time to troubleshoot or become familiar with new equipment can cost you time and money. The frustration could even result in a lost vibe! If you must use a new toy or tool, it's best to learn it beforehand.

◆ Have the producer and/or group go over the song list with the engineer—This might help smooth out setup problems and help the engineer to better understand the goals of the session.

◆ Take the time to make the studio a comfortable place in which to work—You might want to adjust the lighting to match your mood ring, lay down a favorite rug and/or bean bag, turn on your lava love light, or put your favorite stuffed toys on the furniture. Within reason, the place is yours to have fun with!

From a technical standpoint, the microphones for each instrument are selected either by experience or by experimentation and are then connected to the desired console inputs. The input used and mic type should be noted on a track sheet or piece of paper for easy input and track assignment in the studio and/or for choosing the same mic during a subsequent overdub.

Some engineers find it convenient to standardize on a system that uses the same mic input and tape track for an instrument type at every session. For example, an engineer might consistently plug the kick drum mic into input #1 and record it onto track #1, the snare mic onto #2, and so on. That way, the engineer instinctively knows which track belongs to a particular instrument without having to think much about it.

Once the instruments, baffle, and mics have been roughly placed and headphones that are equipped with enough extra cord to allow free movement have been distributed to each player, the engineer can now get down to business by using his/her setup sheet to label each input strip with the name of the corresponding instrument. Label strips (which are often provided just below each channel input fader) can be marked with an erasable felt marker, or the age-old tactic of rolling out and marking on a strip of paper masking tape could be used (ideally, a kind that doesn't leave a tacky residue on the console surface).

At this point, the mic/line channels can be assigned to their respective tracks, making sure to fully document the assignments and other session info on the song or project's track sheet (Figure 18.2). If a DAW is to be used, make sure to name each input track in the session for easy reference and track identification.

After all the assignments and labeling have been completed, the engineer then can begin the process of setting levels for each instrument and mic input by asking each musician to play

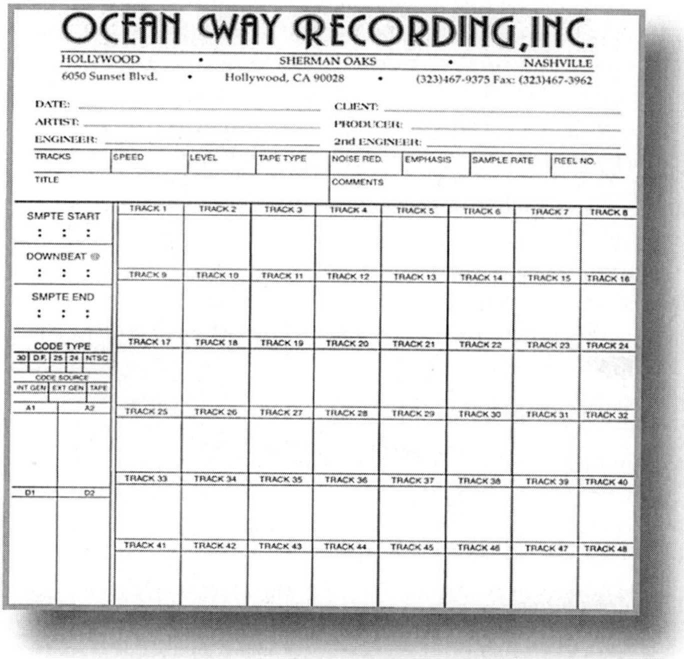
OCEAN WAY RECORDING, INC.

HOLLYWOOD • SHERMAN OAKS • NASHVILLE

6050 Sunset Blvd. • Hollywood, CA 90028 • (323)467-9375 Fax: (323)467-3962

DATE: CLIENT:

ARTIST: PRODUCER:

ENGINEER: 2nd ENGINEER:

TRACKS	SPEED	LEVEL	TAPE TYPE	NOISE RED.	EMPHASIS	SAMPLE RATE	REEL NO.

TITLE COMMENTS

SMPTE START : : :

DOWNBEAT @ : : :

SMPTE END : : :

CODE TYPE 30 DF 25 24 NTSC

CODE SOURCE INT GEN EXT GEN TAPE

A1 A2

D1 D2

TRACK 1	TRACK 2	TRACK 3	TRACK 4	TRACK 5	TRACK 6	TRACK 7	TRACK 8
TRACK 9	TRACK 10	TRACK 11	TRACK 12	TRACK 13	TRACK 14	TRACK 15	TRACK 16
TRACK 17	TRACK 18	TRACK 19	TRACK 20	TRACK 21	TRACK 22	TRACK 23	TRACK 24
TRACK 25	TRACK 26	TRACK 27	TRACK 28	TRACK 29	TRACK 30	TRACK 31	TRACK 32
TRACK 33	TRACK 34	TRACK 35	TRACK 36	TRACK 37	TRACK 38	TRACK 39	TRACK 40
TRACK 41	TRACK 42	TRACK 43	TRACK 44	TRACK 45	TRACK 46	TRACK 47	TRACK 48

Figure 18.2. *Example of a studio track log used for instrument/track assignments. (Courtesy of Ocean Way Recording, www.oceanwayrecording.com.)*

solo or by asking for a complete run-through of the song. By placing each of the channel and master output faders to their unity (0 dB) setting and starting with the EQ settings at the flat position, the engineer can then check each of the track meter readings and adjust the mic preamp gains to their optimum gain while listening for preamp overload. If it's necessary, a gain pad can be inserted into the path in order to eliminate distortion.

After these levels have been set, a rough headphone mix can be made so that the musicians can hear themselves. Mic choice and/or placements can be changed or EQ settings can be adjusted (if necessary) to obtain the sound the producer wants on each instrument, and dynamic limiting or compression can be carefully inserted and adjusted for those channels that require dynamic attention. It's important to keep in mind that it's easier to change the dynamics of a track later during mixdown (particularly if the session is being recorded digitally) than to undo any changes that have been made during the recording phase.

Once this is done, the engineer and producer can listen for extraneous sounds (such as buzzes or hum from guitar amplifiers or squeaks from drum pedals) and eliminate them. Soloing the individual tracks can ease the process of selectively listening for such unwanted sounds, as well as for listening to the natural pickup of an instrument. If several mics are to be grouped into one or more tracks, the balance between them should be carefully set at this time.

After this procedure has been followed for all the instruments, the musicians should do a couple of practice "rundown" songs so that the engineer and producer can listen to how the instruments sound together before being recorded (if tape or disk space isn't a major concern, you might consider recording these tracks, as they might turn out to be your best takes— you just never know). During the rundown, you might consider soloing the various instruments and instrument combinations as a final check, and finally, monitor to all the instruments together. Careful changes in EQ can be made at this time (make sure to note these changes in the track sheet for future reference). These changes should be made sparingly, as final compensations are better made during the final mixdown phase.

While the song is being run down, the engineer can make final adjustments to the recording levels and the headphone monitor mix. He/she can then check the headphone mix either by putting on a pair of headphones connected to the cue system or by routing the mix to the monitor loudspeakers. If the musicians can't hear themselves properly, the mix should be changed to satisfy their monitoring needs, regardless of their recorded levels. If several cue systems are available, multiple headphone mixes can be built up to satisfy those with different balance needs. During a loud session, the musicians might ask you to turn up their level (or the overall headphone level), so they can hear the mix above the ambient room leakage. It's important to note that high sound-pressure levels can cause the pitch of instruments to sound flat, so musicians might have trouble tuning or even singing with their headphones on. To avoid these problems, tuning shouldn't be done while listening through phones. The musicians should play their instruments at levels that they're accustomed to and adjust their headphone levels accordingly (many actually put only one cup over an ear, leaving the other ear free to hear the natural room sound).

The importance of proper headphone levels and a good cue balance can't be stressed enough, as they can either help or hinder a musician's overall performance. The same situation exists in the control room with respect to high monitor-speaker levels: Some instruments might sound out of tune, even when they aren't, and ear fatigue can easily impair your ability to properly judge sounds and relative balance.

During the practice rundown, it's also a good idea to ask the musician(s) to play through the entire song so you'll know where the breaks, bridges, and any point that's of particular importance might be. Making notes and even writing down or entering the timing numbers (possibly into a recorder's transport autolocator or into a DAW as session markers) can help speed up the process of finding a section during a take or overdub. You can also pinpoint the loud sections at this time, so as to avoid any overloads. If compression or limiting is used, you might keep an ear open to ensure that the instruments don't trigger an undue amount of gain reduction (if the tracks are recorded digitally, gain reduction might be applied during mixdown). Even though an engineer might ask each musician to play as loudly as possible, they'll often play even louder when performing together. This fact may require further changes in the mic preamp gain, record level, and compression/limiting threshold. Soloing each mic and listening for leakage can help to check for separation and leakage between the instruments. If necessary, the relative positions of mics, instruments, and baffles can be changed at this time.

Electric and Electronic Instruments

Electric instruments (such as guitars) generally have mid-level, unbalanced, high-impedance outputs that can be directly recorded in the studio without using their amplifiers, via a direct box (DI). As we've already learned, these devices convert line-level, high-impedance output signals into a low-impedance, balanced signal that can be fed directly into a console's mic preamp or external preamp. Instruments often are recorded "direct" in order to avoid instrument leakage problems in the studio, to reduce noise and distortion that can occur when miking an amp, or simply to get a "clean and tight" direct sound.

Electronic instruments (such as synths, samplers, and effects boxes) more closely match the impedance and levels of studio equipment and can be plugged directly into a console or interface line-level input. Within the studio itself, DI boxes are generally still considered to be the best way to insert a signal into the console.

Any of these instrument types can be played in the control room while listening over the main studio monitor speakers without fear of leakage. If you prefer, the signal of a DI can be split into two paths: one that can be directly inserted into the console/interface and another that can be fed to an instrument speaker in the studio (Figure 18.3). This technique makes it possible for the both direct and miked pickup to be combined and blended, giving you the benefits of a clean, direct sound with the added rougher, gutsier sound of a miked amp.

On most guitars, the best tone and lowest hum pickup for a direct connection occurs when the instrument volume control is fully turned up. Because guitar tone controls often use a variable

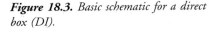

Figure 18.3. *Basic schematic for a direct box (DI).*

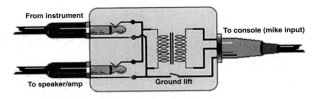

treble rolloff, leaving the tone controls at the treble setting and using a combination of console EQ and different guitar pickups to vary the tone will often yield the maximum amount of control over the sound. If the treble is rolled off at the guitar, boosting the highs with EQ will often increase the pickup noise.

Drums

During the past several decades, drums have undergone a substantial change with regard to playing technique, miking technique, and choice of acoustic recording environment. In the 1960s and 1970s the drum set was placed in a small isolation room called a drum booth. This booth acoustically isolated the instrument from the rest of the studio and had the effect of tightening the drum sound because of the limited space (and often dead acoustics). The drum booth also physically isolated the musician from the studio, which often caused the musician to feel removed and less involved in the action. Today, many engineers and producers have moved the drum set out of smaller iso-rooms and back into larger open studio areas where the sound can fully develop and combine with the studio's own acoustics. In many cases, this effect can be exaggerated by placing a distant mic pair in the room (a technique that often produces a fuller, larger-than-life sound, especially in surround).

Before a session begins, the drummer should tune each drum while the mics and baffles for the other instruments are being set up. Each drumhead should be adjusted for the desired pitch and for constant tension around the rim by hitting the head at various points around its edge and adjusting the lugs for the same pitch all around the head. Once the drums are tuned, the engineer should listen to each drum individually to make sure that there are no buzzes, rattles, or resonant after-rings. Drums that sound great in live performance may not sound nearly as good when being close miked. In a live performance, the rattles and rings are covered up by the other instruments and are lost before the sound reaches the listener. Close miking, on the other hand, picks up the noises as well as the desired sound.

If tuning the drums doesn't bring the extraneous noises or rings under control, duct or masking tape can be used to dampen them. Pieces of cloth, dampening rings, paper towels, or a wallet can also be taped to a head in various locations (which is determined by experimentation) to eliminate rings and buzzes. Although head damping has been used extensively in the past, present methods use this damping technique more discreetly and will often combine dampening with proper design and tuning styles (all of which are the artist's personal call).

During a session, it's best to remove the damping mechanisms that are built into most drum sets, as they apply tension to only one spot on the head and unbalance its tension. These built-in dampers often vibrate when the head is hit and are a chief source of rattles. Removing the front head and placing a blanket or other damping material inside the drum (so that it's pressing against the head) can often dampen the kick drum. Adjusting the amount of material can vary the sound from being a resonant boom to a thick, dull thud. Kick drums are usually (but not always) recorded with their front heads removed, while other drums are recorded with their bottom heads either on or off. Tuning the drums is more difficult if two heads are used because the head tensions often interact; however, they will often produce a more resonant tone. After the drums have been tuned, the mikes can be put into position. It's important to keep the mics out of the drummer's way, or they might be hit by a stick or moved out of position during the performance.

The Final Check

The placement of musicians and instruments will often vary from one studio and/or session to the next because of the number of instruments, isolation (or lack thereof) among instruments, and the degree of visual contact that's needed. If additional isolation (beyond careful microphone placement) is needed, flats and baffles can be placed between instruments in order to prevent loud sound sources from spilling over into other open mikes. Alternatively, the instrument or instruments could be placed into separate isolation (iso) rooms and/or booths, or they could be overdubbed at a later time.

During a session that involves several musicians, the setup should allow them to see and interact with each other as much as possible. It's extremely important that they be able to give and receive visual cues and otherwise "feel the vibe." The instrument/mic placement, baffle arrangement, and possibly room acoustics (which can often be modified by placing absorbers in the room) will depend on personal preferences, as well as on the type of sound the producer wants. For example:

- If the mics are close to the instrument and the baffles are packed in close, a tight sound with good separation will often be achieved.
- A looser, more "live" sound (along with an increase in leakage) will occur when the mics and baffles are placed farther away.
- An especially loud instrument can be isolated by putting it in an unused iso-room or vocal or instrument booth. Electronic amps that are played at high volumes can also be recorded in such a room. Alternatively, the amps and mics can be surrounded on several sides by sound baffles and (if needed) a top can be put on the "box."
- An amp and the mic can be covered with a blanket or other flexible sound-absorbing material (Figure 18.4), so that there's a clear path between the amplifier and the mic.
- Separation can be achieved by placing the softer instruments in an iso-room or by plugging otherwise loud electronic instruments directly into the console via a D.I. (direct injection) box, thereby bypassing the miked amp.

Figure 18.4. *Isolating an instrument amplifier by covering it with a sound-absorbing blanket.*

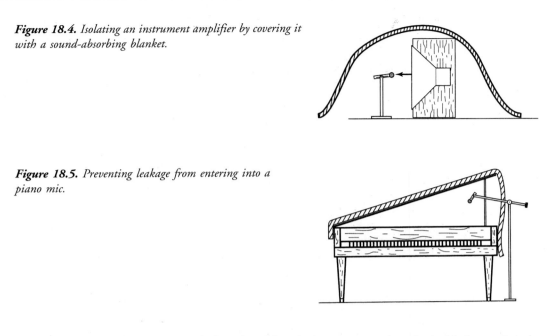

Figure 18.5. *Preventing leakage from entering into a piano mic.*

◆ Piano leakage can be similarly reduced by placing one or more mics inside it, putting the lid on its short support stick and covering it with blankets (Figure 18.5), or by placing it in an iso-room.

Obviously, these examples can only suggest the number of possibilities that can occur during a session. For example, you might collectively choose not to isolate the instruments, and instead, place the instruments in an acoustically "live" room. This approach will require that you carefully place the mics in order to control leakage; however, the result will often yield a live and present sound. As artists, the choices belong to you, the producer, the group, and (possibly) the production/distribution company.

From a technical/engineering standpoint, if analog tape machines are to be used, they should be cleaned, demagnetized, and (if necessary) aligned for the specific tape formulation that's to be used for the session (it's preferable to use the actual session tape itself). Generally it's a good idea to record a 100-Hz, 1-kHz, and 10-kHz tone at 0 VU on all tracks at the beginning of the tape to indicate the proper operating level. If it's necessary to overdub or mix at another studio, the tones will make it easier for the engineer to calibrate the unknown tape machine to your reference tones. SMPTE time code should also be stripped before the session starts (making sure to know the proper TC rate for the session).

If tape-based digital recorders are to be used, make sure that the tapes have been properly formatted well before the session starts. Although it's true that certain recorders can record onto unformatted tapes in a single pass or in a special update mode, it's never as easy as having tapes that are preformatted.

If a hard-disk DAW is to be used, make sure that the proper sample and bit rates are chosen for the session (this should be discussed beforehand with the group and/or producer). If an external drive (or drives) has been brought in for the session, make sure that it has been formatted and is ready to roll. Remember, the idea is to work out the kinks and to simplify technology as much as possible (Murphy's Law is and always will be alive and well in any production facility).

Recording

It's obviously a foregone conclusion that no two recording sessions will ever be exactly alike. In fact, in keeping with the "no rules" rule, they're often radically different from each other. During the *recording session*, the engineer watches the level indicators and (only if necessary) controls the faders to keep from overloading the media. It's also an engineer's job to act as another set of production ears by listening both for performance and quality factors. If the producer doesn't notice a mistake in the performance, the engineer just might catch it and point it out. The engineer should try to be helpful and remember that the producer and/or the band will have the final say, and that their final judgment of the quality of a performance or recording must be accepted.

Once recording is underway, at the beginning of each performance, the name of the song and a take number are recorded to tape for easy identification (a process that's often referred to as *slating* the tape). A take sheet should be carefully kept to note the position of the take on a tape. Comments are also written onto this sheet to describe the producer's opinion of the performance, as well as other information of importance. In fact, the concept of documentation can't be overstressed. The diligent documentation of a session might not be one of the most fun things to do, but should changes be needed in the future, knowing which effect was used on a vocal, how an instrument was placed in a surround field, what mic or pickup combo was used to capture that killer guitar riff, etc., would come in mighty handy.

From the musician's standpoint, here are a few additional pointers that can help the session go more smoothly:

- ◆ It's always best to get the right sound and vibe onto tape or disk during the session. If you need to do another take, do it! If you need to change a mic, change it. "Fixing in the mix" will usually apply a band-aid to an ailing track.
- ◆ Know when to quit! If you're pushing too hard or are tired, it'll often show.
- ◆ Technology doesn't always make a good track; feeling, emotion, and musicality does.
- ◆ Beware of adding new parts or tracks onto a piece that's "just right." Remember, too much is too much!
- ◆ Leave plenty of time for the vocal track(s). It's not uncommon for a group to spend most of their time and budget on getting the perfect drum or guitar sound. It takes time and a clear focus to get the vocals right.
- ◆ If you mess up on a part, keep going, you might be able to fix the bad part by punching-in. If it's really that bad, the engineer or producer will stop you.

In his *EQ Magazine* article "The Performance Curve: How do you know which take is the one?", my buddy Craig Anderton laid out his experiences on how different musicians will deal with the process of delivering a performance over time. Being in front of a mic isn't always easy and we all deal with it differently. Here's a basic outline of his findings:

◆ *Curves Up Ahead*—With this type of performer, the first couple of takes are pretty good, then start to go downhill before ramping back up again, until they hit their peak before going down hill really fast.

◆ *The quick starter*—This type starts out strong and doesn't improve over time in later performances. Live performers often fall into this category, as they're conditioned to get it right the first time.

◆ *The long ramp-up*—These musicians often take a while to warm up to a performance. After they hit their stride, you might get a killer take or a few great ones that can be composited together into the perfect take.

◆ *Anything goes*—This category can vary widely within a performance. Often snippets can be taken from several takes into a single composite. You want to record everything with this type of performer, cause you just never know what gem (or bad take) you'll end up with.

◆ *Rock steady*—This one represents the consummate pro that's fully practiced and delivers a performance doesn't waver from one take to the next; however, you might record several takes to see which one has the most feeling.

From the above examples, we can quickly draw the obvious conclusion that there are all types of performers, and that it takes a qualified and experienced producer and/or engineer to intuit just which type is in front of the mic and to draw the best possible performance from him or her.

Overdubbing

Overdubbing (Figure 18.6) is used to add more instruments to a performance after the basic tracks have been recorded. These additional tracks are added by monitoring the previously recorded tape tracks (usually over headphones) while simultaneously recording new, doubled, or augmented instruments and/or vocals onto one or more available tracks.

In an overdub (OD) session, the same procedure is followed for mic selection, placement, EQ, and level as they would occur during a recording session. If only one instrument is to be overdubbed, the problem of having other instrument tracks leak into the new track won't exist. However, leakage can occur if the musician's headphones are too loud or aren't seated properly on his or her head.

The natural ambience of the session should be taken into account during an overdub. If the original tracks are made from a natural, roomy ensemble, it could be distracting to hear an added track that was obviously laid down in a different (usually deader) room environment.

Figure 18.6.
Overdubbing allows instruments to be added to existing tracks on a multitrack recording medium.

kick drum
snare
toms L
toms R
overhead L
overhead R
lead guitar
Synth
piano L
piano R
lead vocal (overdub)
background vocals
claps
shakers
big bass
SMPTE

headphone mix of
recorded tracks

If the recorder to be used is analog, it should be placed in the master sync mode (thereby reproducing the previously recorded tracks from the record head in sync). The master sync mode is set either at the recorder or using its autolocator/remote control. Usually, the tape machine can automatically switch between monitoring the source (signals being fed to the recorder or console) and tape/sync (signals coming from the playback or record/sync heads). The control room monitor mix should prominently feature the instrument(s) that's being recorded, so mistakes can be easily heard. During the initial rundown, the headphone cue mix can be adjusted to fit the musician's personal taste.

Punching-In

Should a mistake or bad take be recorded onto the new track, it's a simple matter to start over and re-record over the unwanted take. If only a small part of the take was bad, it's easy to *punch-in* (Figure 18.7), a process that can:

◆ Silently enter the record mode at a pre-determined point
◆ Record over the unwanted portion of the take
◆ Silently fall back out of record at a pre-determined point

A punch can be manually performed on most recording systems; however, newer tape machines and DAW systems can be programmed to automatically go into and fall out of record at a pre-determined time.

When punching-in, any number of variables can come into play. If the instrument is an overdub, it's often easy to punch the track without fear of any consequences. If an offending instrument is part of a group or ensemble, leakage from the original instrument could find its way into adjacent tracks, making a punch difficult or impossible. In such a situation, it's usually best to re-record the piece, pick up at a point just before the bad section

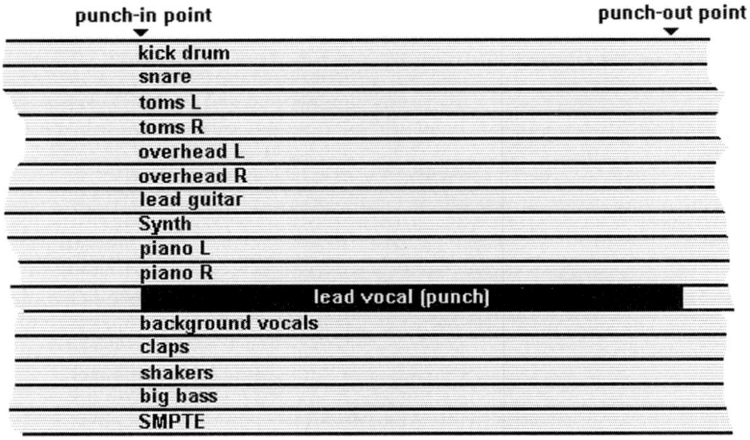

Figure 18.7. *Punch-ins let you selectively replace material and correct mistakes.*

and splice or insert it into the original recording, or attempt to punch the section using the entire ensemble.

It's often best to punch-in on a section immediately after the take has been recorded, as changes in mic choice, mic position, or the general mood of the session can lead to a bad punch that will be hard to correct. If this isn't possible, make sure that you carefully document the mic choice, placement, preamp type, etc., before moving onto the next section: You'll be glad you did.

Performing a "punch" on a tape-based recorder should always be done with care. Allowing the tape to record too long a passage could possibly cut off a section of the following, acceptable track and require that the following section be likewise overdubbed. Stopping it short could cut off the natural reverb trail of the final note. Performing a punch using a DAW is often far easier, as the leading and trailing edge of the punch can often be manually adjusted to expose or hide sections after the punch has been performed. Most modern recording media can also be programmed to perform the punch under automated control. This built-in repeatability can go a long way toward reducing operator error and its associated tension.

An additional technique that can be used in conjunction with (or as an alternative to) punching-in over a recorded track is to overdub the instrument onto another available track or set of tracks. The advantage of recording onto another track (if they're available) is that a good or marginal take can be saved and the musician can try to improve the performance, rather than having to erase the previous take in order to improve it. When several tracks of the overdub have been saved, different sections of each take might be better than others. These can be combined into a single, composite (comp) take by playing the tracks back in the sync mode, mixing and muting them together at the console (often with the help of automation), and recording them on another tape track (Figure 18.8). It should also be noted that the job of recording multiple takes and then combining then into a single composite is also well suited to a DAW, which can often record unlimited takes (without using up physical tape tracks)

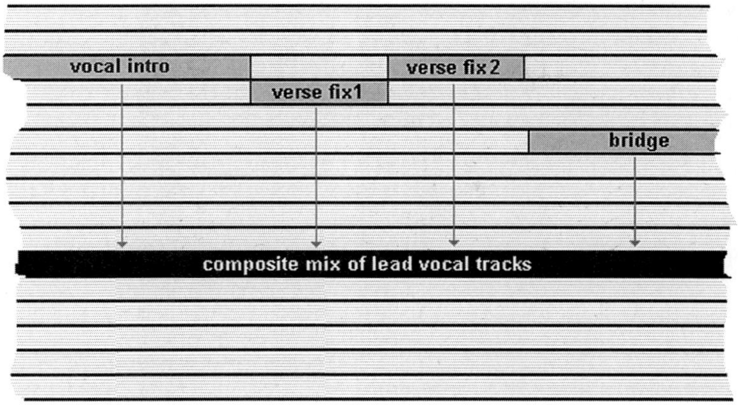

Figure 18.8. *A single composite track can be created from several partially acceptable takes.*

and then combine them into a composite track with relative ease. (Note: it's often a good idea to export the comp takes into a single, composite track that can be easily archived and imported from one DAW platform to another.)

Bouncing Tracks

Another procedure that's an extension of the composite process (when using a tape-based system) is the concept of track *bouncing* or *ping-ponging*. The process of bouncing tracks is often used to mix a group or an entire set of basic tracks down to one track, a stereo track pair, or grouping of several tracks usually, when the number of available tape tracks are limited.

Bouncing can be performed either to make the final mixdown easier (by grouping instruments together onto one or more tracks, as seen in Figure 18.9a) or to open up needed tape tracks by bouncing similar instrument groups down to one or more tracks. This frees up the originally recorded tape tracks for overdubs.

For example, suppose that we have two MDM digital recorders with which to do a demo recording (giving us a total of 16 tracks). During the basic track session, let's say that we decided to record the drums onto all 8 tracks of MDM #1 and then record a stereo piano, bass, and lead guitar (4 tracks) onto separate tracks of MDM #2. After these 12 tracks have been recorded, we could then go back and mix the drums down to a stereo submix pair of the 4 available tracks on MDM #2. Once this is done, the original drum tape can be set aside (whenever possible, it's always a good idea to keep the original recordings intact, just in case you need them in the future or in case of an accidental punch error—Murphy's Law, you know!). Once we've put a new tape into MDM #1, we can go about the task of recording onto 10 tracks (8 newly opened tracks of MDM #1 and 2 tracks on #2, as shown in Figure 18.9b).

Figure 18.9. *Track bouncing is a common production technique that's used in tape-based multitrack recording. (a) Instruments can be grouped together onto one or more tracks or (b) bouncing can be used to expand the number of available tracks and thus allow for additional overdubs.*

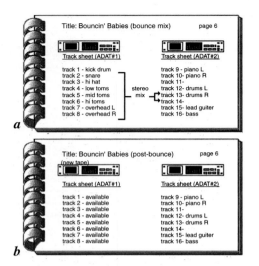

When using a DAW, bouncing isn't usually necessary as there are large numbers of virtual tracks that can be used. In this situation, any number of tracks can be corralled together into groups that can be easily controlled in level from a single group fader.

Mixdown

After all the tracks for a song have been recorded, the DAW, multitrack tape or other medium can be mixed down to mono, stereo, or surround for subsequent duplication and distribution to the consumer.

When beginning a mixdown using an analog multitrack and/or mastering recorder, it's customary to demagnetize, clean, and align the machine. Once this is done, the engineer should record 0-VU level tones at the head of the mixdown reel at 1 kHz, 10 kHz, and 100 Hz. This makes it possible for the mastering engineer to align the tape playback machine at these reference frequencies, resulting in a proper playback EQ and levels.

Once you're ready, the console can be placed into the mixdown mode (or each input module can be switched to the line or tape position) and the fader label strips can be labeled with their respective instrument names. Channel and group faders should be set to unity gain (0 dB) and the master output faders should likewise be set with the monitor section being switched to feed the mixdown signal to the appropriate speakers. If a DAW is being used to create the final mix, a basic array of effects can be programmed into the session mixer and, if available, a controller surface can be used to facilitate the mix by giving you hands-on control.

The engineer can then set up a rough mix of the song by adjusting the levels and the spatial pan positions. The producer then listens to this mix and might ask the engineer to make specific changes. The instruments are often soloed one by one or in groups, allowing any necessary EQ changes to be made. The engineer, producer, and possibly the group can then begin the cooperative process of "building" the mix into its final form. Compression and limiting can be used on individual instruments as required (either to make them sound fuller, more consistent in level, or to prevent them from overloading the mix when raised to the desired level). At this point, the console's automation features can be used (if available). Once the mix begins to take shape, reverb and other effects types can be added to shape and add ambience in order to give close-miked sounds a more "live," spacious feeling, as well as to help blend the instruments.

If the mix isn't automation-assisted, the fader settings will have to be changed during the mix in real-time. This means that the engineer will have to memorize the various fader moves (often noting the transport counter to keep track of transition times). If more changes are needed than the engineer can handle alone, the assistant, producer, or artist (who probably knows the transition times better than anyone) can help by controlling certain faders or letting you know when a transition is coming up.

It's usually best, however, if the producer is given as few tasks as possible, so that he/she can concentrate fully on the music rather than the physical mechanics of the mix. The engineer then listens to the mix from a technical standpoint to detect any sounds or noises that shouldn't be present. If noises are recorded on tracks that aren't used during a section of a song, these tracks can be muted until needed. After the engineer practices the song enough to determine and learn all the changes, the mix can be recorded and faded at the end. The engineer might not want to fade the song during mixdown, as it will usually be performed after being transferred to a DAW (which can perform a fade much more smoothly than even the smoothest hand). Of course, if automation is available or if the mix is performed using a DAW, all of these moves can be performed much more easily and with full repeatability.

It's usually important that levels should be as consistent as possible between the various takes and songs, and it's often wise to monitor at consistent, moderate listening levels. This is due to the variations in our ear's frequency response at different sound-pressure levels, which could result in inconsistencies between song balances. Ideally, the control room monitor level should be the same as might be heard at home, over the radio, or in the car (between 70 and 90 dB SPL), although certain music styles will "want" to be listened to at higher levels. Once the final mix or completed project master is made, you might want to listen to it over different speaker systems (ranging from the smallest to the biggest/baddest you can find). It's usually a good idea to run off a few copies for the producer and band members to listen to at home and in their cars. In addition, the mix should be tested for mono–stereo/surround compatibility (when using any mix format) to see what changes in instrumental balances might have occurred. If there are any changes in frequency balances or if phase becomes a problem when the mix is played in mono, the original mix might have to be modified.

Sequence Editing

After all the final mixes for a recording have been completed, a final edited master can be assembled. The producer and artists begin the process by listening to the songs and deciding on a final sequence, based on their tempos, musical keys, how they flow into one another, and which songs will best attract the listener's attention. Once this is done, the process of assembling the final master can begin.

Mastering

Although the subject of mastering has been covered in-depth within Chapter 16, the fact is that the producer and/or artist will be faced with the question of whether to hire the services of a qualified or well-known mastering engineer to put the finishing touches on a master recording, or to make use of the talents at-hand (i.e., producer, engineer, and artist) to finesse the various songs into a final product statement. These questions should be thoroughly discussed in the pre-planning phase, allowing for an on-the-spot change of plans. Who knows, you just might try mastering the project in-house. If it doesn't work out, you can always work with a trusted mastering engineer to work your masterpiece into its best possible final form.

Digital Sequence Editing

With the advent of the DAW, the relatively cumbersome process of sequencing music tracks in the analog domain using magnetic tape has given way to the faster, easier, and more flexible process of editing the final masters from hard disk. Using this system, all the songs can be loaded from their final media form to disk or, if the project was recorded entirely to disk, each song in the session can be exported as individual mixdown files that can then be loaded into a workstation, digital audio editor, or CD burning program for assembly into its final, edited form (Figure 18.10).

Figure 18.10. *A digital audio program or burner can be used to edit, sequence, and process the final takes into a finished project. (Courtesy of Adobe Systems Inc., www.adobe.com.)*

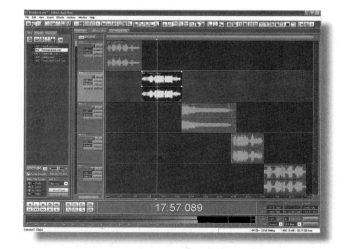

Whenever the various soundfiles of a project are loaded into a DAW, audio editor, or certain CD burning programs, each song or program segment can be tailored in a number of ways:

- ◆ The start and end points can be located and defined for each song.
- ◆ Each song can then be individually processed using EQ, overall level, dynamics, etc.
- ◆ The proper fades and silence gaps can be entered into the edit session.
- ◆ The completed master can then be transferred to a CD-R or final mastering media.

Whenever a basic CD burning program is used to create a final master (or the data file for mass-producing any number CD-R discs), each soundfile will need to be loaded into the playlist in its final processed and edited form. Most of these programs will let you enter gaps of silence (index marker gaps) between each song in seconds and possibly subdivisions of a second, while leaving a 2-second gap before the start of the program (which is necessary as it is part of the Redbook CD-Audio standard).

When sequencing a project, the length of time between the end of the song and the beginning of the next can be constant, or the timings can vary according to the musical relationship between the songs. Decreasing the time between them can make a song seem to blend into the next (if they're similar in mood), or could create a sharp contrast with the preceding song (if the moods are dissimilar). Longer times between songs help the listeners get out of the mood of the previous song and prepare them for hearing something that might be quite different.

It's always a good idea to make at least one master copy of the final product from the original files as a backup. It's always a good idea to make a backup of the final mix soundfile and session data, just in case the record company, producer or artist wants to make changes at a later date. This simple precaution could save you a lot of time and frustration. When it comes to backing up your session and mixdown data, follow the instructions on the tee-shirt and "Just do it!"

Analog Sequence Editing

Although the process of assembling a final master entirely in the analog domain occurs less frequently than in the digital realm, it is still done. During this process, the engineer edits the original mixes out from their respective reels and begins the process of splicing them together into a final sequence on a master reel set. At this time, the level test tones (which were laid down at the beginning of the mixdown session) should be placed at the beginning of side one. Once this is done, the mix master in/out edits should be tightened (to eliminate any noise and silence gaps), by listening to the intro and outro at high volume levels, while the heads are in contact with the tape (this might require that you place the transport into the edit mode). The tape can then be moved back and forth (a process known as "jogging" or "rocking" the tape) to the exact point where the music begins (intro) and after it ends (outro). Once the in (or out) point is positioned over the playback head, the exact position is marked with a grease pencil. If there's no noise directly in front of this spot, it's a good practice to cut the tape half an inch before the grease pencil mark as a safety precaution against editing out part of the first sound. If there is noise ahead of the first sound, the tape should be cut at the mark and the leader should be inserted at that point. Paper (rather than plastic) leader tape is used because plastic often causes static electricity pops.

The tail of the song might need to be monitored at even higher volume levels because it's usually a fade-out or an overhang from the last note, and is therefore much softer than the beginning of the song. The tape is marked and cut just after the last sound dies out to eliminate any low-level noises and tape hiss.

As was stated in the digital section, the length of time between the end of the song and the beginning of the next can be constant in a sequenced project, or the timings can vary according to the musical relationship between the songs. Decreasing the time between them can make a song seem to blend into the next (if they're similar in mood), or could create a sharp contrast with the preceding song (if the moods are dissimilar). Longer times between songs help the listeners get out of the mood of the previous song and prepare them for hearing something that might be quite different.

When the sequencing is complete, one or two analog or DAT backup copies should be made of the final, sequenced master before it leaves the studio. These copies serve as a backup in case the original mixes are lost or damaged. Several CD-R, DAT, and/or cassette copies should also be made for the producer, artist, and record company executives for final approval.

Once all of the recording, mixing, and mastering is done and you've left the studio, then the real work begins—getting the final product into the hands of the buying public—a process that's truly not for the faint of heart! (Note: it should be mentioned that if the artist or group is to retain the copyrights to the final product, the Form SR should be filed with the Library of Congress at this time. This form copyrights the "performance and production of a particular recording" and can be found at www.copyright.gov/forms/formsri.pdf or by searching the Library of Congress at www.loc.gov.)

Marketing and Sales

Although this section is mentioned last, it's by far one of the most important areas to be dealt with when contemplating the time, talent, and financial effort that are involved in creating a recorded product. For starters, the following questions (and more) should all be answered long before the record button is pressed and the first downbeat is played:

- ◆ Who is my audience?
- ◆ Will this be distributed by a record company or will I try and sell it myself?
- ◆ What should the final product look and sound like?
- ◆ How much is this going to cost me?

In this short section, I won't even attempt to cover this extremely important and complex topic. It has been fully discussed in such well-crafted and highly recommended books as:

◆ *How to Make and Sell Your Own Recording*, 5th Edition, Diane Sward Rapaport, Jerome Headlands Press, 1999 www.dianerapaport.com
◆ *The Guerrilla Music Marketing Handbook*, Bob Baker, 2004, www.bob-baker.com
◆ *This Business of Music Marketing and Promotion*, Tad Lathrop, Watson-Guptill Publications; 2nd edition (2003), www.watsonguptill.com
◆ "The Sound of Money," in *Audio Recording for Profit*, Chris Stone, Focal Press, 2000, www.focalpress.com

These and many other books on this subject discuss in detail the three primary methods by which a finished recording can be distributed and sold:

◆ Through a major-label record company
◆ Through an independent record label
◆ By selling the product yourself.

Each of these marketing and sales tactics represents varying degrees of financial and time outlay, as well as artistic and distribution control. No matter which avenue you choose, it's important that you fully and carefully investigate every aspect of a deal before any binding commitments are made.

A Final Word on Professionalism

Before we close this chapter, there's one more subject that I'd like to touch upon—perhaps the most important one of all—*professional demeanor*. Without a doubt, the life and job of a typical engineer, producer, or musician isn't always an easy one. It often involves long hours and extended concentration with people who, more often than not, are new acquaintances. In short, it can be a high-pressure job. On the flip side, it's one that's often full of new experiences, with demands changing on almost a daily basis, and often involves you with exciting people who feel passionately about their art and chosen profession.

It's been my observation (and that of many I've known) that the best qualities that can be exhibited by anyone in "The Biz" are:

◆ Having an innate willingness to experiment
◆ Being open to new ideas (flexibility)
◆ Having a sense of humor
◆ Having an even temperament (this often translates as patience)
◆ Willing to communicate with others
◆ Being able to convey and understand the basic nuances of people from all walks of life and with many different temperaments

The best advice I can possibly give is to *be open, be patient, and above all, be yourself.* Also, be extra patient with yourself. If you don't know something, ask. If you made a mistake (trust me, you will, we all do), admit it and don't be hard on yourself. It's all part of the process of learning and gaining experience.

This last piece of advice might not be as popular as the others, but it may come in handy someday: It's important to be open to the fact that there are many, many aspects to the music and sound production, and you may find that your calling might be better served in another branch of the biz. That's totally OK! Change is an important part of any creative process as it and taxes are the only constants you can count on!

CHAPTER 19

Yesterday, Today, and Tomorrow

I'm sure you've heard the phrase, "Those were the good old days." I've usually found it to be a catch-all term that refers to a time in one's life that had a sense of great meaning, relevance, and all-around fun. Personally, I've never met a group of people who seem to bring that sense of relevance and fun with them into the present more than music and audio professionals, enthusiasts, and students. The fact that many of us refer to the tools of our profession as "toys" says a lot about the way we view our work. Fortunately, I was born into that clan and have reaped the benefits all my life.

Music and audio industry professionals, by necessity, tend to keep their noses to the workaday grindstone. But market forces and personal visions often cause them to keep one eye focused on future technologies; whether these are new developments (such as advances in digital audio technologies), rediscovering retro trends that are decades old (such as the reemergence of tube technology and the reconditioning of older devices that sound far too good to put out to pasture), or future vaporware technologies that excite the imagination. Such is the time paradox of a music and audio professional, which leads me to the book's final task: addressing the people and technologies in the business of sound recording in the past, present, and future.

Yesterday

I've always looked at the history of music and sound technology and applied techniques with a sense of awe and wonder, although I really can't explain why. Like so many in this industry, I tend to get shivers when I see a wonderful old tape machine or an original tube compressor. For no reason whatsoever, I get all giggly and woozy when I read about the Redd 37 console that was used to record many of the early Beatles albums at Abbey Road (including Sergeant Pepper) or see an original Ampex 200 (the first commercially available professional tape machine). I experience the same sense of awe when I read about my personal historical heroes such as Alan Dower Blumlein (Figure 19.1), who was instrumental in developing stereo mic techniques, the 45°?/45°? stereo disc-cutting process, the TV camera, and radar. To many, his list of accomplishments is second only to those of Edison. Mary C. Bell (Figure 19.2) who was probably the first woman sound engineer also comes to mind, along with another unsung hero, the late John (Jack) T. Mullin (Figure 5.1), who stumbled across a couple of German Magnetophones at the end of WWII and was smart enough to send them back to San Francisco. With the help of Alexander M. Poniatoff (founder of "AMP"ex) and Bing Crosby, Jack and his machines played a crucial role in bringing the magnetic tape recorder into commercial existence (Figure 19.3).

Every once in a techno-blue moon, major milestones come along that affect almost every facet of information and entertainment technology. Such milestones have ushered us from the Edison and Berliner era of acoustic recordings, into the era of broadcasting, electrical recording, and tape, to the environment of the multitrack recording studio (Figures 19.4 and 19.5), and finally into the age of the computer, digital media, and the Web.

When you get right down to it, the foundation of the modern information and digital age was laid with the invention of the integrated circuit (IC). The IC has likewise drastically changed the technology and techniques of present-day recording by allowing circuitry to be easily

Figure 19.1. The life of Alan Dower Blumlein (truly, one of my long-time heroes) has been published by Focal Press. (Courtesy of Focal Press, www.focalpress.com.)

Figure 19.2. Mary C. Bell in NBC's *dubbing room #1 (April 1948) inspecting broadcast lacquer discs for on-air programs. (Courtesy of Mary C. Bell.)*

Figure 19.3. Early Ampex tape machines. *(Courtesy of Mary C. Bell.)*

designed and mass produced at a fraction of the size and cost of equipment made with tubes or discrete transistors.

Advances in digital hard-and software have brought about new developments in equipment and production styles that have affected the ways in which music is created. Integrating cost-effective yet powerful production computers with digital mixing systems, modular digital multitracks, MIDI synths/samplers, plug-in effects and instruments, digital signal processors, etc., gives us the recipe for having a powerful production studio in your home, bedroom, or on the bus. Such laptop and desktop music studios have made it possible for more and more people to create and distribute their own music with an unprecedented degree of ease, quality, and cost-effectiveness.

Figure 19.4. Gilfoy Sound Studios, Inc., circa 1972. Notice that the room is set up for quad! (Courtesy of Jack Gilfoy, www.jackgilfoy.com.)

Figure 19.5. Little Richard at the legendary LA Record Plant's Studio A (circa 1985) recording "It's a Matter of Time" for the Disney film Down and Out in Beverly Hills. (Courtesy of the Record Plant Recording Studios, photo by Neil Ricklen.)

Peter Gotcher (Digidesign co-founder) was one of the first to envision the creation of a cost-effective "studio-in-a-box" (Figure 19.6). This conceptual spark, which started a present-day goliath, helped to create a system that would offer the power of professional hard-disk-based audio at a price that most music, audio, and media producers could afford (you have to realize that previous to this, systems started at over $100,000!). His goal (and that of countless others since) has been to create an integrated system that would link together the many facets that go into audio and audio-for-visual production, via a personal computer. Years later, this dream has transformed the very way in which music and audio is produced.

One of the cooler by-products of this digital age is an upsurge in "retro-future" trends in music and technology that blend together older and newer technologies. With this, has come an increased interest and awareness in museums and Web sites that help us to learn about our

Figure 19.6. *Digidesign's Control console/ Pro Tools-based recording system. (Courtesy of Digidesign, a division of Avid Technology, Inc., www.digidesign.com.)*

musical and technological roots throughout music production history. A few of the physical- and cyber-sites that can help give you a better understanding of our roots are:

Museums:

◆ Computer History Museum
   1401 N Shoreline Blvd.
   Mountain View, CA 94043
   1-650-810-1010
   info@computerhistory.org

◆ Experience Music Project
   325 5th Ave. N. (on the Seattle Center Campus)
   Seattle, WA 98109
   1.877.EMPLIVE (1-877-367-5483)
   www.emplive.com

◆ Museum of Sound Recording
   580 8th avenue, 21st floor
   New York, NY 10018
   1 212 997 9279
   www.lovesphere.org/mosr/

◆ National Music Museum
   The University of South Dakota
   414 East Clark Street
   Vermillion, SD 57069
   www.usd.edu/smm

◆ Rock and Roll Hall of Fame and Museum
   One Key Plaza
   Cleveland, Ohio 44114
   (216) 781-ROCK
   www.rockhall.com

Sites:

- ◆ Audio Engineering Society Historical Committee (www.aes.org/aeshc)
- ◆ Museum of Sound (www.museumofsound.org)
- ◆ Synthmuseum.com (www.synthmuseum.com)
- ◆ Tinfoil.com (www.tinfoil.com)

When it comes to understanding the tools, toys, and techniques of our trade, I've always felt that there are a lot of benefits to be gained from looking back into the past as well as trying to gaze into the future. A wealth of experience in design and application has been laid out for us. It's simply there for the taking; all we have to do is search it out and put it to good use.

Today

"Today" is a really difficult subject to talk about, since new equipment literally comes out on a monthly basis. However, it safely goes without saying that the two buzz terms that best sum up the dawn of the 21st century are digital audio and the Web.

1s and 0s

The grouping of digital "1s" and "0s" into words that represent alphanumeric values, sampled voltage levels over time (digital audio), or pixilated color and brightness values on a screen grid (digital graphics) have changed communications and creative production forever. Literally, even the most diehard analog fanatic isn't able to escape its far-reaching grasp on human communication.

With the advent of the personal computer, DAW, cost-effective peripherals, application and effects plug-ins, and the CD/DVD (in all its content and data format forms), music creation isn't only more cost-effective than ever before it also offers a degree of power, flexibility, and portability that literally boggles the mind!

Fortunately, this newfound digital technology has spawned a primal urge, almost a frenzied lust to marry cutting-edge hardware/software with vintage gear or newly re-issued toys that are based on decades-old technologies, particularly tube devices (such as tube condenser mics, preamps, and signal processing gear). So why embrace this older technology? Well, for starters, tube electronics inherently have a sound that's very different than their IC- or transistor-based counterparts. When a tube is overdriven to the point of clipping, the square edge of the distorted signal is actually rounded off. This tends to yield a smoother, fuller, more "rounded" distortion, when compared to a sharply distorted, harsh edge that's usually exhibited by transistor and IC circuit designs. In addition, tubes generate even-order harmonics, which are far more musical sounding in nature than the harsher odd-order distortions that are generated by its modern counterparts.

The WWW

Beyond the overall concept of digital audio, another obvious mover and shaker of everyday life is the Web. Cyberspace made the creation of this book much easier for me as a writer (search engines and company sites make research is a relative breeze, and photos can be quickly e-mailed to me.). Although many of the music share sites that contain ripped downloads of major releases have been shut down, major legitimate pay-per-download sites have sprung into mega-buck action. Personal and indy music sites have allowed upstart and established artists to directly sell their music, inform their fans about upcoming tours, and publish fanzine info that keeps their public begging for more.

As I write this, one of the major issues still facing the music industry relates to the distribution of copyrighted commercial music over the Web. Even the cloudiest crystal ball can foresee that cyberspace is the new frontier for distributing media (of both the "for fee" and "for free" types). The question is simply a matter of how, using what medium, and how fees and royalties would be collected. With the advent of the Secure Digital Media Initiative (SDMI), methods for collecting and administering fees and royalties for copyrighted material are beginning to fall into place. Unwilling to give up their stronghold on the traditional brick-n-mortar distribution and sales of CDs, major labels have consistently lagged behind in the search for building a uniform Web distribution model. This has paved the way for new commercial Web ventures to boldly go where most record companies dread to go making it possible for major-, indy-, and self-produced artists to get the word out and get paid at the same time. In short, although we're still at the dawning of the Web distribution age, cyberspace has already proven itself to be a source of increased visibility, viability, and hope for the budding and serious artist, as well as for the established corporate world of the music biz.

Tomorrow

Usually, I tend to have a decent handle on the forces that might or might not help shape the sounds and toys of tomorrow...but it's getting increasingly harder to make specific predictions in this fast-paced world. Today, there are simply far more choices in information and entertainment than reading a book or watching Lucy and Desi on the tube. Now, we can interact with others in a high-speed, networked environment (Figure 19.7) that lets us be more than just spectators; it lets us participate and share our thoughts with others, which leads to creative discourse (one could only hope), a faster pace of communication, and growth.

This idea of intercommunication through the WWW has already begun the drive of almost every developing technology toward an e-based commerce that's based on the distribution and sharing of goods, information, and media-on-demand. With the advent of high-speed, broadband technology one of the greatest concerns facing intellectual and media properties is the need for copy protection: the concept that even though it only took 5 seconds to copy a file, the intellectual copyrights of a song, video, book, or program has a value that must be compensated for. Without that concept, media providers are in for a harder ride than they're

Figure 19.7. *Reaching out into cyberspace.*

already having. The flip side to this, of course, is that downloading a song, program, demo, or whatever will become familiar with the product and will pay for it. One thing's for sure, copy protection issues are still up for grabs, for the most part, and it'll be interesting to see how they become resolved.

As I write this, the wireless (Wi-Fi) networking revolution is beginning to catch on. Technological crystal balls are already foreseeing a near future where Wi-Fi transmitters can be installed in a stationary blimp or robot plane high above a metro area that can blanket a large region with a network "mushroom." This high-speed network could be used to communicate WWW, radio, music, and/or video on demand, e-mail, and Internet-phone communications (cell technology would be made obsolete). In such a case, information could be communicated on a pay-per-use basis that could be directly billed to you Internet Provider (IP) address to a laptop, PDA, or Blackberry type wireless device.

On the music production front, digital audio has and will continue to become smaller, more virtual, and more powerful. Even now, I have a killer laptop-a-go-go production system that fits snuggly into my M-Audio backpack. Although my main surround sound studio is still centered around keyboards, synths, and music controllers of various types, I've definitely welcomed the continued march toward quality virtual instruments and useful plug-ins that seamlessly integrate into the DAW. I think I'll always marvel at a computer's ability to be a chameleon: one moment it's a music production system, next it's a video editor, then a word processor, then a graphic workstation, then a partridge in a pear tree. This facet literally frees us to be creative in an amazing number of ways that's a thing of joy.

One of my greatest hopes for the not-so-distant future is that the audio production community would embrace surround sound more fully. The leap from stereo to surround in a studio environment can range from being simple and straightforward (with certain DAW systems) to being a financial and logistical challenge; however, the rewards and the gained knowledge that comes with learning surround production, mastering, and media delivery are well worth it. I remember as a kid, holding double album jackets behind my head to fill in the rear reflections (try placing reflective surfaces behind your two ears and angle them toward the stereo speakers. Can you hear the difference?). Similarly, those of you who have a home theater system might try playing your favorite stereo record or CD, while in the Dolby Pro-Logic surround mode. The sound will either get sucked into the center speaker (in which case, it's best to switch back to stereo) or it'll widen out into the surround field in a way that might cause your mouth to drop.

Happy Trails

Before we wrap up the 6th edition, I'd like to take a moment to honor one of the greatest forces driving humanity today (besides sex): information and the dissemination thereof. Through the existence of quality books, trade magazines, university programs, workshops, and the Web, a huge base of information on almost any imaginable subject is now being distributed to and understood by a greater number of aspiring artists and technicians than ever before. These resources often provide a strong foundation for those who are attending accredited schools, as well as those who are attending the school of hard knocks. No matter what your goals are in life (or in the business of music), I urge you to jump in and read, surf, skip through pages, and keep your eyes and ears open for new sounds, ideas, technologies, and experiences. The increased knowledge and gained skills will always be well worth the expended time and effort.

On a final note, I'd like to paraphrase Max Ehrmann's "Desiderata," when he urges us to keep interested in our own career, as it's an important possession in the changing fortunes of time. Through my work as a musician, writer and educator, I've been fortunate enough to know many fascinating, talented, and fun people. For some strange reason, I was born with a strong drive to have music and production technology in my life. By "keeping interested in my own career" and working my butt off, while having several brushes with extreme luck, I've been able to turn this passion into a successful career.

To me, all of this comes from following your bliss (as some might call it), listening to reason (both your own and that of others you trust), and doing the best work that you can (whatever it might be). As you know, thousands of able and aspiring bodies are waiting in line to make it as an engineer, a successful musician, a producer, etc.

So how does that one person make it? By following the same directions that it takes to get to Carnegie Hall — practice! Or as the tee-shirt says, "Just do it!" Through perseverance, a good attitude, and sheer luck, you'll be led through paths and adventures that you never thought were possible.

APPENDIX

The following sections, Real-Time and Non-Real-Time DSP and DSP Basics are excerpts from the book *Hard Disk Recording for Musicians* by David Miles Huber, 1995, Amsco Publications, NYC, and have been reprinted with permission.

DSP Basics

The scope and capabilities of digital signal processing are limited only by speed, number-crunching power, and human imagination. Yet the process itself is made up of only three basic building blocks (Figure A-1):

- ◆ Addition
- ◆ Multiplication
- ◆ Delay

One of the best ways to understand the building blocks of digital logic is through application. Just for fun, let's try building a simple 4-in × 1-out digital mixing "app" that could be used to combine the output channels of a device.

Addition

As you might expect, a digital adder sums together the various bits at the input of the circuit in order to create a single combined result. With this straightforward building block, the word value of each input is mathematically

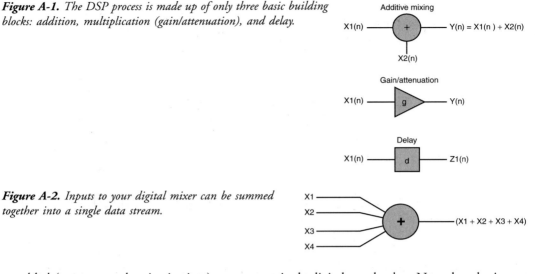

Figure A-1. *The DSP process is made up of only three basic building blocks: addition, multiplication (gain/attenuation), and delay.*

Figure A-2. *Inputs to your digital mixer can be summed together into a single data stream.*

added (at one samplepoint in time) to create a single digital word-value. Now that the inputs can be easily combined, we have the first building-block of a simple mixer that combines the four input signals into one output channel (Figure A-2).

Multiplication

The multiplication of sample values by a numeric coefficient allows the gain (level) of digitized audio to be changed either up or down. Whenever a sample is multiplied by a factor of 1, the result is unity gain or no change in level. Multiplication by a factor of less than 1 yields a reduction in gain (attenuation), while the multiplication by a number greater than 1 will result in an increase in gain.

Now that we have this, we can add gain controls to our mixer. This will finally give us some real control, as digital faders can be used to modify the gain of each of our mixer's channel (Figure A-3).

Before moving on, let's take a look at how multiplication can be applied to the everyday production world by calculating how the gain of a recorded soundfile can be changed over time. Examples include: fade-ins, fade-outs, crossfades, and (of course) gain changes.

Figure A-3. *Variable gain can be added by using multiplication to determine the mix ratio of each of the mixer's channels.*

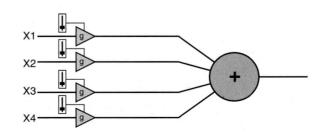

Figure A-4. (a) Original soundfile, (b) defined area to
be faded is calculated and written to disk as a separate
file, and (c) faded file is tagged to the original file and
is reproduced with no audible break or adverse effect.

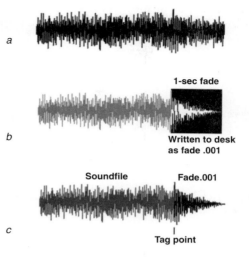

In order to better understand this process, let's look at how a non-real-time fade might be
created using a hard-disk-based system. Suppose that we have a song that wasn't faded during a
mixdown session but now definitely needs to be faded out over the final chorus section. The
first task is to define the part of the song that needs to be faded (Figure A-4)a, call up the fade
function, and then perform it.

While the fade is being performed, it's the processor's job to continually multiply the affected
samples by a diminishing coefficient. The result will be a file that's reduced in gain over the
duration of the defined region. After the region has been recalculated, the results are
then automatically written to disk as a separate file (Figure A-4b). Upon playback, the fade is
then digitally spliced onto the original soundfile at the appropriate point or points
(Figure A-4c).

Delay

The final DSP building block deals with time, that is, the use of delay over time in order to
perform a specific function or effect. In the world of DSP, delay is used in a wide variety of
applications. This discussion, however, focuses on two types:

◆ Effects-related delay
◆ Delay at the sample level

Most modern musicians and those associated with audio production are familiar with the way
different delay ranges can accomplish a wide range of effects. They're also often familiar with
the use of digital delay for creating such sonic effects as doubling and echo. These effects
(discussed later in this chapter) are created from discrete delays that are 35 milliseconds or
more in length (1 millisecond, or ms, equals 1 thousandth of a second).

Figure A-5. *A digital delay device stores sampled audio into RAM memory, where it can be read out at a later time.*

Figure A-6. *The effect of mixing a short-term delay (that varies in delay over time) with the original undelayed signal creates a comb (multiple notch) filter response that goes by the generic effect name of flanging.*

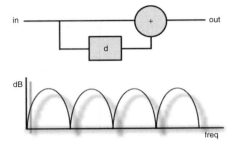

In its basic form, digital delay is accomplished by storing sampled audio directly into RAM. After a defined length of time (in milliseconds or seconds), this sampled audio can be read out from memory for output or further processing (Figure A-5).

As this delay time is reduced below the 10-ms range, however, a new effect begins to take hold. The effect of mixing a variable short-term delay with the original undelayed signal creates a unique series of peaks and dips in the signal's frequency response. This effect (known as *flanging*) is the result of a selective equalization (Figure A-6) that occurs when the delay time is varied. If you have a digital delay hanging around, you can check out this homemade flange effect by combining the delayed and undelayed signal and listening for yourself.

By further reducing the delay times downward into the microsecond range (1 microsecond or ms equals 1 millionth of a second), you can begin to introduce delays that affect the digitized signal at the sample level. In doing so, control over the phase characteristics can be introduced to the point that selective equalization is accomplished. Figure A-7 shows examples of two basic EQ circuits that provide low- and high-frequency shelving characteristics in the digital domain.

Figure A-7. *Simple EQ circuits and possible response curves.*

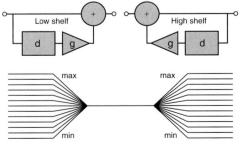

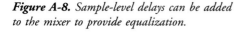

Figure A-8. Sample-level delays can be added to the mixer to provide equalization.

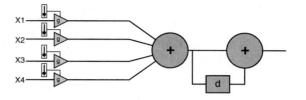

The amount of equalization to be applied (either boost or cut) depends on the multipliers that control the amount of gain that's to be fed from the delay modules. By adding more delay and multiplication stages to this basic concept, it's possible to create digital equalizers that are more complex and parametric in nature.

It should be pointed out that delays of such short durations aren't created using RAM delay-type circuits (which are used in creating longer delays). Rather, sample-level delay circuits (known as shift registers) are used, which are better suited to the task, because they're simpler and more cost-effective in design.

Getting back to our digital mixer example, we can now finish the project by adding some form of equalization to the final stage. For example, Figure A-8 shows the project as having a very simple high-pass filter. This filter could be used to cut out any low-end rumble that could get into the system courtesy of a loud air conditioner or a local transit authority bus.

Do It Yourself Tutorial: Digital Delay

1. Get out a digital delay unit and balance its output mix (so that the input signal is set equally with the delayed output signal). Note: If there is no mix control, mix the delayed/undelayed signal at the console.
2. Playback various instrument sources (i.e., solo instruments and vocal tracks) while slowly sweeping the delay settings...
3. First in the 35-ms and greater range. Can you hear the discrete delays?
4. Now slowly vary the settings over the 10- to 35-ms range. Can you simulate a phasing effect? If the unit has a phase setting, turn it on. Does it sound different?
5. Now change the delay settings a little faster to create a wacky flange effect. If the unit has a flange setting, turn it on. Does that sound different? Try playing with the time-based settings that affect its sweep rate. Fun, huh?

Echo and Reverberation

Now that we've looked at single delay effects, let's add to our bag of effects tricks by successively repeating delays to create echoes. Repeated echoes are created by feeding a portion of a delayed signal's output back into itself (Figure A-9). By adding a multiplier stage into this

Figure A-9. *By adding a simple feedback loop to a delay circuit,*
it's possible to create an echo, echo, echo effect.

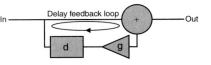

loop, it's possible to vary the amount of feedback gain and thus control both the level and the number of repeated echoes.

Although reverb could be placed in a separate category, it's actually nothing more than a series of closely spaced echoes. In nature, acoustic reverb can be broken down into three subcomponents:

- Direct signal
- Early reflections
- Reverberation

The *direct signal* is heard when the original sound wave travels directly from the source to the listener. *Early reflections* are the first few reflections that are reflected back to the listener from large, primary boundaries in a given space. Generally, these reflections are the ones that give us subconscious cues as to the perception of size and space.

The last set of reflections makes up the signal's *reverberation* characteristic. These sounds are broken down into zillions of random reflections that travel from boundary to boundary within the confines of a room. They're so closely spaced in time that the brain can't discern them as individual reflections, so they are perceived as a dense, single decaying signal.

By designing a system that uses a number of delay lines that are carefully controlled in both time and amplitude, it's possible to create an almost infinite number of reverb types, parameters, and characteristics. To illustrate this point, let's return to our basic building block approach to DSP and build a crude reverb processor (Figure A-10).

As you now know, the direct signal is the first signal to arrive at the listener. Thus, it can be represented as a single data line that flows from the input to the output. Included with this first DSP stage are a number of individually tunable delay lines. By tuning these various modules to different times between a range of 10 ms to over a second, early reflections can be simulated (being perceived as ranging from bathroom size to the Grand Canyon). Following this section, one or more delay lines with echo feedback loops (which are designed to repeat echoes at a very fast rate and decay over a pre-determined time) can be placed into the system.

Figure A-10. *A crude reverb processor design.*

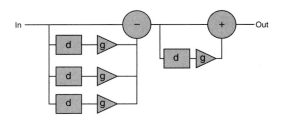

Such a reverb unit might sound crude since its simplicity would severely limit the sound type and quality. However, adding more stages and placing the gain controls under microprocessor and user-defined algorithms command could create a large library of high-quality room sounds.

The Real World of DSP Design

As we've seen from the preceding discussion, the basics of DSP are fairly straightforward. In application, however, these building blocks can be programmed into a processing block that uses rather complex combinations and elaborate algorithms to arrive at a final result.

In addition to complexities that can exist, there are restrictions and overhead requirements that must be used to safeguard the process from cranking out erroneous, degraded, or even disastrous results. For example, whenever a number of digital samples are mixed together using an adding circuit, the results could easily equal a large number that's beyond the system's maximum signal limit. Without proper design safeguards, a condition known as "bit wrap-around" could occur, causing the signal to output a loud "pop!" Fortunately, modern design has removed this and other obvious gremlins. However, the deeper you dig into DSP, the more you'll find that the challenge in designing a quality system isn't in getting it to work but in eliminating the pesky glitches that often stem from performing complex sonic functions.

In addition to debugging a system, word accuracy is often of concern. For example, whenever signal processors are called on to add samples together or to multiply numeric values by lengthy coefficients, it's possible for errors to accumulate or for the final results to be greater than 16 bits in word length. To reduce these errors to acceptable levels or to prevent the "chopping off" of potentially important least-significant-bit values, most high-quality processors are capable of calculating word length values with a resolution of up to 24 or 32 bits.

APPENDIX

◆

Tax Tips for Musicians

by Jeffrey P. Fisher

Everybody complains about taxes. But few do anything about it. Well, there are several actions you can take to improve your tax situation right now. If you're making even the tiniest amount of music-related scratch, there's no reason to pay more taxes than you have to. To reap the most tax benefits, start running your music career as a legal small business. The IRS loves small business. According to the Small Business Administration, there are 25 million small businesses in the US today. And a large percentage of them are sole proprietorships—one person shops. As a sole proprietor, you report your music business income as part of your personal income using Schedule C (and a few other forms).

A Business Of Your Own

It makes real financial sense to run your project studio as legal business. Follow these basic steps:

◆ *Set up your business by choosing its legal structure (sole-proprietorship, partnership, corporation, etc.)*—Consult with a tax adviser for details about the financial aspects of each form. Contact a legal adviser for answers to liability issues.

◆ *Get legal*—Make sure you meet any regulations that pertain to running a business in your town. For example, you may have to get a business license from your local clerk's office.

◆ *File a doing-business-as (dba) with your local government if you call your business something other than your legal name*—You may need a separate tax ID for your business and some states require a sales tax ID number.

◆ *Open a business checking account*—Deposit your project studio income into it and pay your business expenses using checks drawn on it. Also, use a credit card only for business purchases and pay it off on time from your business checking account.

◆ *Use bookkeeping software to track all your business income and expenses*—This makes tax preparation and monitoring your financial situation easier. Understand the various tax consequences of your business, too. You'll probably need to make quarterly tax payments in addition to yearly tax preparation.

◆ *Protect yourself through health and property insurance*—Also, consider life, disability, and liability insurance if it makes sense for your situation.

It all comes down to income and expenses—the money you make and the money you spend. The more you make, the more you pay in taxes. Simple, right? Even the most convoluted of IRS instructions make that point painfully clear. That means the inverse is also true. Since the IRS only taxes your business profits, cut back on the profit and pay less taxes.

"But, Dude, I Gotta Eat."

Whoa there. I'm not saying that you should earn less. Instead, look for all the possible ways to convert some of your everyday expenses into legitimate business deductions. Even some personal expenses may be deductible against the business. The more expenses you have, the more you reduce your taxable income. And since you were going to spend the money anyway, you might as well realize some tax benefits, too.

Write-Offs

Basically, all the expenses you incur to run your little music business are deductible. To be fully deductible, however, these business expenses must be "ordinary and necessary" according to the IRS. That's just fuzzy enough to be dangerous. Ordinary means the expenses must be typical for the business. Buying a new guitar could apply; buying a dishwasher wouldn't. Necessary just means the expense is vital to the success of your business. Office equipment, postage, phone charges, graphic design charges, recording studio fees, duplication, dues, magazine subscriptions, and other such related items are definitely necessary for the success of the typical music business.

Here is a basic list of business deductions for musicians:

◆ Advertising and promotion costs

◆ Car and truck expenses

◆ Commissions and fees you pay to other people/businesses

◆ Depreciation and section 179 deduction (see article)

- Insurance (except health insurance—see article)
- Interest (on business loans)
- Legal and professional fees (hire a lawyer; deduct the charge)
- Office expenses
- Rent/lease payments
- Repairs and maintenance
- Supplies
- Taxes and licenses such as a business license
- Travel costs
- Meals and Entertainment (see article)
- Utilities (see home-office deduction in article)
- Wages, salaries paid

For every $100 you earn, you pay approximately $45.30 in taxes (if you're in the 27% tax bracket, pay the 15.3% self-employment tax, and send an additional 3% to your state). Of course, that also implies that for every legitimate $100 business expense you incur, you also save the $45.30 you would otherwise pay in taxes. Hey, that's like getting everything you buy at a discount.

Why does the IRS let you deduct all these expenses? They want you to succeed. So, they let you invest money (spend it) in your business as incentive for you to earn more. And the more money you make, the more you'll pay in taxes. You see they have an ulterior motive; they ain't jus' bein' neighborly.

But there's a caveat (isn't there always?). You need to be gainfully engaged in making a buck. You must turn a profit in your business three out of every five years or your business will be classified as a hobby, and you forfeit the expense deductions. Bottom line: very bad news and very high tax bill. (And one dollar in profit those three years ain't gonna make you popular down at the Treasury Department neither).

Since the burden of proof falls solely to you, it's vital that you record all your music business income and expenses diligently. A shoebox full of receipts does not a bookkeeping system make. Get help setting up your books or look for a software solution to help document your business financial transactions.

Another important gotcha: If you're just launching your music business, startup expenses can't be deducted all at once. You must amortize them over five years by taking 20% portions of the total expenses and deducting them over five consecutive years.

Gear Lust = Tax Savings

Did you know that the gear you buy for making your music magic could be a sweet tax deduction? Under section 179 of the tax code you can deduct or "expense" up to $100,000 of tangible property and write it all off when you prepare your taxes. For tangible property think expensive, long-lasting items, such as a new computer. This amount can be above and beyond many other normal business expenses you might incur.

If you've had a particularly strong earnings year, you might want to offset some of that gain by deducting all the cost of large purchases in one year (up to the limit). Alternately, you can choose to depreciate what you buy and deduct a portion of those costs over the next several years.

Home Sweet Home

If you do the majority of your music work in your home office, you can deduct a portion of the same expenses that now do little or nothing to lessen your tax burden. You can write off rent or mortgage interest, property taxes, utilities (gas, electricity, water/sewer), insurance, repairs, and depreciation. First, dedicate a portion of your home entirely for your music business. Keep it free of personal items and make it your primary business location. Beware that if you do most of your work elsewhere (gigging, for instance), and only use this home office occasionally, your deduction may be limited or entirely verboten.

Here's how to figure your deductions. Total up the square footage of this exclusive and principal place and compare it to the total square footage of your crib. Say your math works out to 10% (100 square feet of a 1000 square foot home—you need a bigger place!). You can then deduct 10% of the aforementioned expenses using Form 8829: Expenses for Business Use of Your Home. The total deduction then flows through to your Schedule C reducing your income, and therefore your taxes, considerably.

There is a recapture clause (which doesn't apply to renters). If you sell your home and make a profit, those profit dollars become taxable business income at the same percentage rate as your deduction. Score a $50,000 gain (good for you!), and, following the above example, $5,000 of it belongs to the business (subject to self-employment tax and regular income tax, of course). It's important to note that the personal income you make from a house sale is generally not taxed, though! Stop taking the home-office deduction for two tax years prior to the home sale and this recapture clause doesn't apply.

Self-Employment Tax

Yes, we self-employed have a special tax just for us. Actually every worker pays the same tax—funding for Social Security and Medicare—it's just a little different when you're on your own. You must contribute both the employee and employer contributions which total up to a whopping 15.3%. Yep, just over fifteen pennies on every buck you earn goes right into the Social Security kitty. This is, of course, before you start paying any regular income taxes. Ouch! And you have to pay the self-employment taxes (along with income taxes) quarterly. You need to predict what you are going to earn this year, and the taxes that would be due on that dollar amount. Then, you send in 25% of that money on April 15, June 15, September 15, and January 15 of the next year. These estimated tax payments are important, because if you don't pay enough, there's a penalty due the next April 15.

Health Insurance

You can deduct the premiums paid for health insurance, too. This doesn't come off the Schedule C, but is actually a front page deduction on your personal 1040. Self-employed individuals can deduct 100% of the premiums they pay (unless Congress changes its collective mind). Other typical medical costs are deductible on Schedule A (if you qualify).

Eat, Drink, And Be Merry

When you entertain your clients, the money you spend is another write-off. However, these meals and entertainment are subject to a 50% limitation. Spend a $100 on a pizza party, and take $50 off on Schedule C. Give clients gifts, up to $25 per client, and you can take that as a full deduction, though.

When you travel as part of your music business, those expenses are deductible including airfare, lodging, and meals. You must support your travel and lodging deductions with receipts. However, instead of keeping track of your meals, you can take the government's standard per diem allowance of $30 for Meals and Incidentals. Other cities may have higher amounts so check the official Web site (www.policyworks.gov/perdiem). Meals on the road are, of course, still subject to the 50% limit. The IRS figures you gotta eat anyway, so they limit the expense. Bummer.

Vehicular Deductions

Yes, that old beater is worth money! Keep track of actual vehicle expenses (gas, repairs, etc.) or take the standard mileage rate (which changes every year; check with the IRS). In either case, you must document the miles you drive for business, date, and purpose of trips, along with expenses incurred. A dedicated notebook/diary earns a gold star from the IRS.

Even if you use your ride for business and personal use, the business portion of your expenses is still deductible. Determine your business percentage by dividing your business miles by the total miles driven (2500 business/10,000 total = 25%). If you just use the standard mileage rate, multiply your business miles driven by that rate (2500 × $/mile = deduction) to arrive at your deductible expense amount. You can also deduct the full cost of tolls and parking fees incurred while on business. And the loan interest on the car is deductible (subject to the business use percentage, if it applies).

Feed The Nest Egg And Save, Too

You also save money by contributing to a qualified retirement plan. The IRS makes it easy to sock away some cash for a rainy day and rewards you with a nice, fat deduction each year. This is another 1040 deduction, not Schedule C. IRAs are the first method that pop up.

However, they're limited to $2250 (which changes regularly). With a SEP (Simplified Employee Pension) you can deduct as much as 15% of your business income topping out at $30,000 total per year. The more you put away, the more you save. And since you're really helping yourself down the road, it's a smart way to manage your taxes and your retirement. For some of us, a Roth IRA may be more prudent. Roths give you no up front deduction, but the earnings are tax-free. Talk to a financial planner.

EOY Tax Tips

At the end of each year, you have another opportunity to reduce your tax burden: accelerate expenses and decelerate income. First, spend some cash on business expenses. Don't just blow the wad. Make sensible purchases this year that will reduce your taxable income. Ideal last-minute purchases include postage, equipment, general office supplies, and promotions. You can also pay your mortgage and health insurance premium before the year-end to realize some other tax savings on the personal side. Second, put off collecting money this December to January by billing your clients a little later. Though you'll have to pay taxes on the money eventually, you defer that payment for a whole year.

Get Help From The IRS

Surf on over to the always exciting IRS Web site (www.irs.gov) and download the free guides that explain the specific tax benefits for small business owners.

- ◆ #334, Tax guide for small business
- ◆ #463, Travel, entertainment, gift, and car expenses
- ◆ #533, Self-employment tax
- ◆ #535, Business expenses
- ◆ #583, Starting a business and keeping records
- ◆ #587, Business use of your home.

Final Word

While we all have to pay taxes, we are only required to pay our fair share. Make sure you are not throwing money out the window. Take advantage of these and all the other tax breaks available to you. And put more music money in your pocket. . . where it belongs!

Author Bio

Jeffrey P. Fisher provides audio, video, music, writing, training, and media production services. He also writes about music, sound, and video for print and the Web. He's published six books including: *Instant Sound Forge* (CMP, 2004), *Moneymaking Music* (Artistpro, 2003), and *Profiting From Your Music and Sound Project Studio* (Allworth Press, 2001). He teaches at the College of DuPage in Glen Ellyn, IL. Also, Jeffrey co-hosts the Acid, Sound Forge, and Vegas forums on Digital Media Net (www.dmnforums.com). For more information visit his Web site at www.jeffreypfisher.com or contact him at jpf@jeffreypfisher.com.

Index

S